Methods in Molecular Biology

Series Editor
John M. Walker
School of Life and Medical Sciences
University of Hertfordshire
Hatfield, Hertfordshire, UK

For further volumes:
http://www.springer.com/series/7651

For over 35 years, biological scientists have come to rely on the research protocols and methodologies in the critically acclaimed *Methods in Molecular Biology series.* The series was the first to introduce the step-by-step protocols approach that has become the standard in all biomedical protocol publishing. Each protocol is provided in readily-reproducible step-by step fashion, opening with an introductory overview, a list of the materials and reagents needed to complete the experiment, and followed by a detailed procedure that is supported with a helpful notes section offering tips and tricks of the trade as well as troubleshooting advice. These hallmark features were introduced by series editor Dr. John Walker and constitute the key ingredient in each and every volume of the Methods in *Molecular Biology series.* Tested and trusted, comprehensive and reliable, all protocols from the series are indexed in PubMed.

Polyploidy

Methods and Protocols

Edited by

Yves Van de Peer

Department of Plant Biotechnology and Bioinformatics, Ghent University, Ghent, Belgium

Humana Press

Editor
Yves Van de Peer
Department of Plant
Biotechnology and Bioinformatics
Ghent University
Ghent, Belgium

ISSN 1064-3745 ISSN 1940-6029 (electronic)
Methods in Molecular Biology
ISBN 978-1-0716-2560-6 ISBN 978-1-0716-2561-3 (eBook)
https://doi.org/10.1007/978-1-0716-2561-3

This Humana imprint is published by the registered company Springer Science+Business Media, LLC part of Springer Nature.
The registered company address is: 1 New York Plaza, New York, NY 10004, U.S.A.

Preface

Polyploidy, resulting from the duplication of the entire genome of an organism or cell, affects genes and genomes, cells and tissues, organisms, populations, and even entire ecosystems. The fact that many cells, tissues, or whole organisms often possess, or have possessed, multiple copies of their genome has intrigued scientists for many years. More recently, the advent of genomics and genome sequencing has reignited interest in polyploidy and whole-genome duplication (WGD). Consequently, recent years have seen a surge in the development of novel approaches and technologies to sequence, analyze, and study polyploid organisms or polyploid tissues. Here, we have tried to provide a breadth of methods and tools to unveil, identify, and analyze the causes and consequences of both ancient and recent polyploidization.

The book is organized into six parts. In Part I, we focus on finding evidence for (remnants of) ancient WGDs using comparative genomics approaches, as well as on the computational analysis of genome evolution after such ancient WGD events. Phylogenetic and phylogenomic methods also play an important role in studying both ancient and more recent polyploidy. Not only does polyploidy (and the often-associated hybridization events) present several considerable challenges to systematists that need to be addressed but phylogenetic methods also provide insights into the timing of ancient WGDs and the evolution of gene duplicates after WGD. In addition, phylogenetic methods based on models of chromosome number evolution have proved indispensable for the study of macroevolutionary patterns of polyploidy. Phylogenetic and phylogenomic methods of this sort are the subject of Part II of this volume. Having additional copies of genomes, and therefore genes, has consequences for gene expression and the study thereof using high-throughput sequencing. Therefore, methods are needed that are specifically geared towards polyploids to enable accurate quantification and analysis of gene expression and regulation, forming the topic of Part III. Part IV focuses on population genomics approaches to study polyploidy and methods for analyzing genetic data from polyploid populations. Polyploidy has since long been a topic of interest to population geneticists, and the revolutions in sequencing technologies have had a huge impact on population genomics and polyploidy as well. While population genomic approaches allow a detailed view of genetic variation across polyploid and mixed-ploidy populations, polyploidy also presents considerable challenges to standard methodology, which is almost exclusively based on population genetics theory developed for diploid organisms which does not readily generalize to polyploids. Also discussed in Part IV are methods to infer polyploid origins and inheritance modes. As different evolutionary histories predict different patterns of genetic variation in extant populations, large-scale population genomic data enables powerful methods to unveil the plausible evolutionary history of extant polyploids. While the focus of most chapters up to Part IV has been on bioinformatics and computational approaches to study polyploidy, another way to study polyploidy and its consequences is through experimental evolution, where the "duplication tape of life" is replayed. In Part V, two different organisms, a unicellular alga (Chlamydomonas) and a fast-growing duckweed (Spirodela), are proposed as model organisms for studying genome duplications under controlled conditions and environments. As stated, polyploidy not only always affects the entire organism but also affects individual cells and tissues, and Part V also presents several approaches to identify, visualize, and measure such cellular polyploidy. Dealing with multiple copies of a genome, different subgenomes, and multiple copies of genes because of polyploidy also has consequences for functional genomics. For instance, CRISPR/Cas has been widely used for genome editing in plants, but genome editing of polyploid genomes poses specific challenges and demands adaptation of established methods for diploid organisms, as discussed in the final Part VI of this book.

Ghent, Belgium *Yves Van de Peer*

Acknowledgments

We are very thankful to all authors who dedicated some of their precious time to contribute to this collection. Additional thanks go to Dr. Annick Bleys for editorial help.

Contents

Part III Analysis of Gene Expression and Regulation in Polyploids

Part IV Population Genomics Approaches to Study Polyploidy

Part V Experimental Approaches to Study Polyploidy

Part VI Sequencing, Assembling, Editing and Engineering Polyploid Genomes

Contributors

VICTOR A. ALBERT • *Department of Biological Sciences, University at Buffalo, Buffalo, NY, USA*

KSENIA ARZUMANOVA • *Center for Theoretical Evolutionary Genomics, Department of Integrative Biology, University of California, Berkeley, Berkeley, CA, USA*

EYLEM AYDOGDU • *Department of Plant Biotechnology and Bioinformatics, Ghent University, Ghent, Belgium; VIB Center for Plant Systems Biology, VIB, Ghent, Belgium*

QUINTEN BAFORT • *Department of Plant Biotechnology and Bioinformatics, Ghent University, Ghent, Belgium; VIB Center for Plant Systems Biology, VIB, Ghent, Belgium; Department of Biology, Ghent University, Ghent, Belgium*

MICHAEL S. BARKER • *Department of Ecology & Evolutionary Biology, University of Arizona, Tucson, AZ, USA*

RENATA BASTO • *Institut Curie, PSL Research University, CNRS, UMR144, Biology of Centrosomes and Genetic Instability Laboratory, Paris, France*

SARA VANESSA BERNHARD • *Department of Molecular Genetics, Paul Ehrlich Strasse 24, Kaiserslautern, Germany*

CAMILLE BERTHELOT • *Institut de Biologie de l'Ecole Normale Supérieure (IBENS), Ecole Normale Supérieure, CNRS, INSERM, Université PSL, Paris, France*

J. LUCAS BOATWRIGHT • *Advanced Plant Technology, Clemson University, Clemson, SC, USA; Department of Plant and Environmental Sciences, Clemson University, Clemson, SC, USA*

MAGDALENA BOHUTÍNSKÁ • *Department of Botany, Faculty of Science, Charles University, Prague, Czech Republic; Institute of Botany of the Czech Academy of Sciences, Průhonice, Czech Republic*

LAURA BUTTITTA • *Molecular, Cellular and Developmental Biology, University of Michigan, Ann Arbor, MI, USA*

GRIET CASTELEYN • *Department of Plant Biotechnology and Bioinformatics, Ghent University, Ghent, Belgium; Department of Biology, Ghent University, Ghent, Belgium; VIB Center for Plant Systems Biology, VIB, Ghent, Belgium*

HENGCHI CHEN • *Department of Plant Biotechnology and Bioinformatics, Ghent University, Ghent, Belgium; VIB Center for Plant Systems Biology, VIB, Ghent, Belgium*

JAMES W. CLARK • *Bristol Palaeobiology Group, School of Biological Sciences, University of Bristol, Bristol, UK*

DELISA E. CLAY • *Department of Pharmacology and Cancer Biology, Duke University Medical Center, Durham, NC, USA*

JEREMY E. COATE • *Department of Biology, Reed College, Portland, OR, USA*

GAVIN C. CONANT • *Department of Biological Sciences, North Carolina State University, Raleigh, NC, USA; Bioinformatics Research Center, North Carolina State University, Raleigh, NC, USA; Program in Genetics, North Carolina State University, Raleigh, NC, USA*

OLIVIER DE CLERCK • *Department of Biology, Ghent University, Ghent, Belgium*

PHILIP C. J. DONOGHUE • *Bristol Palaeobiology Group, School of Earth Sciences, University of Bristol, Bristol, UK*

DONALD T. FOX • *Department of Pharmacology and Cancer Biology, Duke University Medical Center, Durham, NC, USA*
WILLIAM A. FREYMAN • *23andMe Inc., Sunnyvale, CA, USA*
RUI-HAI GAN • *State Key Laboratory of Freshwater Ecology and Biotechnology, Institute of Hydrobiology, The Innovation Academy of Seed Design, Chinese Academy of Sciences, Wuhan, China; University of Chinese Academy of Sciences, Beijing, China*
SIMON GEMBLE • *Institut Curie, PSL Research University, CNRS, UMR144, Biology of Centrosomes and Genetic Instability Laboratory, Paris, France*
LARS GRØNVOLD • *Department of Animal and Aquacultural Sciences, Faculty of Biosciences, Norwegian University of Life Sciences, Ås, Norway*
JIAN-FANG GUI • *State Key Laboratory of Freshwater Ecology and Biotechnology, Institute of Hydrobiology, The Innovation Academy of Seed Design, Chinese Academy of Sciences, Wuhan, China; University of Chinese Academy of Sciences, Beijing, China*
BERNARD A. HAUSER • *Department of Biology, University of Florida, Gainesville, FL, USA*
TORGEIR R. HVIDSTEN • *Faculty of Chemistry, Biotechnology and Food Science, Norwegian University of Life Sciences, Ås, Norway*
MENGWEI JIANG • *Center for Genomics and Biotechnology, Fujian Provincial Key Laboratory of Haixia Applied Plant Systems Biology, Key Laboratory of Genetics, Breeding and Multiple Utilization of Crops, Ministry of Education, College of Life Sciences, Fujian Agriculture and Forestry University, Fuzhou, China*
FILIP KOLÁŘ • *Department of Botany, Faculty of Science, Charles University, Prague, Czech Republic; Institute of Botany of the Czech Academy of Sciences, Průhonice, Czech Republic*
TREVOR J. KRABBENHOFT • *Department of Biological Sciences, University at Buffalo, Buffalo, NY, USA*
WENLONG LEI • *Center for Genomics and Biotechnology, Fujian Provincial Key Laboratory of Haixia Applied Plant Systems Biology, Key Laboratory of Genetics, Breeding and Multiple Utilization of Crops, Ministry of Education, College of Life Sciences, Fujian Agriculture and Forestry University, Fuzhou, China*
FAY-WEI LI • *Boyce Thompson Institute, Ithaca, NY, USA; Plant Biology Section, Cornell University, Ithaca, NY, USA*
JIA LI • *Department of Plant Biotechnology and Bioinformatics, Ghent University, Ghent, Belgium; VIB Center for Plant Systems Biology, VIB, Ghent, Belgium*
ZHEN LI • *Department of Plant Biotechnology and Bioinformatics, Ghent University, Ghent, Belgium*
CARMEN MARTÍN-PIZARRO • *Departamento de Mejora Genética y Biotecnología, Instituto de Hortofruticultura Subtropical y Mediterránea (IHSM), Universidad de Málaga – Consejo Superior de Investigaciones Científicas, Departamento de Biología Molecular y Bioquímica, Facultad de Ciencias, UMA, Málaga, Spain*
ITAY MAYROSE • *School of Plant Sciences and Food Security, Tel Aviv University, Tel Aviv, Israel*
MICHAEL T. W. MCKIBBEN • *Department of Ecology & Evolutionary Biology, University of Arizona, Tucson, AZ, USA*
PATRICK G. MEIRMANS • *Institute for Biodiversity and Ecosystem Dynamics (IBED), University of Amsterdam, Amsterdam, The Netherlands*
PATRICK MONNAHAN • *Department of Pediatrics, University of Minnesota, Minneapolis, MN, USA*
SHYAMA NANDAKUMAR • *Molecular, Cellular and Developmental Biology, University of Michigan, Ann Arbor, MI, USA*

Annelore Natran • *Department of Plant Biotechnology and Bioinformatics, Ghent University, Ghent, Belgium; VIB Center for Plant Systems Biology, VIB, Ghent, Belgium*
Polina Yu Novikova • *Department of Chromosome Biology, Max Planck Institute for Plant Breeding Research, Cologne, Germany*
John Pannell • *Department of Ecology and Evolution, Biophore Building, University of Lausanne, Lausanne, Switzerland*
Elise Parey • *Institut de Biologie de l'Ecole Normale Supérieure (IBENS), Ecole Normale Supérieure, CNRS, INSERM, Université PSL, Paris, France; INRAE, LPGP, Rennes, France*
David Posé • *Departamento de Mejora Genética y Biotecnología, Instituto de Hortofruticultura Subtropical y Mediterránea (IHSM), Universidad de Málaga – Consejo Superior de Investigaciones Científicas, Departamento de Biología Molecular y Bioquímica, Facultad de Ciencias, UMA, Málaga, Spain*
Lucas Prost • *Department of Plant Biotechnology and Bioinformatics, Ghent University, Ghent, Belgium; VIB Center for Plant Systems Biology, VIB, Ghent, Belgium; Department of Biology, Ghent University, Ghent, Belgium*
Anna Rice • *School of Plant Sciences and Food Security, Tel Aviv University, Tel Aviv, Israel*
Hugues Roest Crollius • *Institut de Biologie de l'Ecole Normale Supérieure (IBENS), Ecole Normale Supérieure, CNRS, INSERM, Université PSL, Paris, France*
Rori V. Rohlfs • *Department of Biology, San Francisco State University, San Francisco, CA, USA*
Carl J. Rothfels • *University Herbarium and Department of Integrative Biology, University of California, Berkeley, Berkeley, CA, USA*
Camille Roux • *Univ. Lille, CNRS, UMR 8198 - Evo-Eco-Paleo, Lille, France*
Carlos Sánchez-Gómez • *Departamento de Mejora Genética y Biotecnología, Instituto de Hortofruticultura Subtropical y Mediterránea (IHSM), Universidad de Málaga – Consejo Superior de Investigaciones Científicas, Departamento de Biología Molecular y Bioquímica, Facultad de Ciencias, UMA, Málaga, Spain*
Simen R. Sandve • *Department of Animal and Aquacultural Sciences, Faculty of Biosciences, Norwegian University of Life Sciences, Ås, Norway*
Peter Schafran • *Boyce Thompson Institute, Ithaca, NY, USA*
Alison Dawn Scott • *Department of Chromosome Biology, Max Planck Institute for Plant Breeding Research, Cologne, Germany*
Shengchen Shan • *Florida Museum of Natural History, University of Florida, Gainesville, FL, USA*
Douglas E. Soltis • *Florida Museum of Natural History, University of Florida, Gainesville, FL, USA; Department of Biology, University of Florida, Gainesville, FL, USA; Biodiversity Institute, University of Florida, Gainesville, FL, USA*
Pamela S. Soltis • *Florida Museum of Natural History, University of Florida, Gainesville, FL, USA; Biodiversity Institute, University of Florida, Gainesville, FL, USA*
Zuzana Storchova • *Department of Molecular Genetics, Paul Ehrlich Strasse 24, Kaiserslautern, Germany*
Benjamin M. Stormo • *Department of Pharmacology and Cancer Biology, Duke University Medical Center, Durham, NC, USA*
Marius A. Strand • *Department of Animal and Aquacultural Sciences, Faculty of Biosciences, Norwegian University of Life Sciences, Ås, Norway*
Haibao Tang • *Center for Genomics and Biotechnology, Fujian Provincial Key Laboratory of Haixia Applied Plant Systems Biology, Key Laboratory of Genetics, Breeding and Multiple*

Utilization of Crops, Ministry of Education, College of Life Sciences, Fujian Agriculture and Forestry University, Fuzhou, China

YVES VAN DE PEER • *Department of Plant Biotechnology and Bioinformatics, Ghent University, Ghent, Belgium*

JOZEFIEN D. VAN DE VELDE • *Department of Chromosome Biology, Max Planck Institute for Plant Breeding Research, Cologne, Germany*

ANTOINE VAN DE VLOET • *Department of Plant Biotechnology and Bioinformatics, Ghent University, Ghent, Belgium; Department of Biology, Ghent University, Ghent, Belgium; VIB Center for Plant Systems Biology, VIB, Ghent, Belgium*

XAVIER VEKEMANS • *Univ. Lille, CNRS, UMR 8198 - Evo-Eco-Paleo, Lille, France*

JAKUB VLČEK • *Department of Botany, Faculty of Science, Charles University, Prague, Czech Republic*

YIBIN WANG • *Center for Genomics and Biotechnology, Fujian Provincial Key Laboratory of Haixia Applied Plant Systems Biology, Key Laboratory of Genetics, Breeding and Multiple Utilization of Crops, Ministry of Education, College of Life Sciences, Fujian Agriculture and Forestry University, Fuzhou, China*

TIAN WU • *Department of Plant Biotechnology and Bioinformatics, Ghent University, Ghent, Belgium; VIB Center for Plant Systems Biology, VIB, Ghent, Belgium*

BING YANG • *Division of Plant Sciences, University of Missouri, Columbia, MO, USA; Donald Danforth Plant Science Center, St. Louis, MO, USA*

JIAXIN YU • *Center for Genomics and Biotechnology, Fujian Provincial Key Laboratory of Haixia Applied Plant Systems Biology, Key Laboratory of Genetics, Breeding and Multiple Utilization of Crops, Ministry of Education, College of Life Sciences, Fujian Agriculture and Forestry University, Fuzhou, China*

XINGTAN ZHANG • *Shenzhen Branch, Guangdong Laboratory for Lingnan Modern Agriculture, Genome Analysis Laboratory of the Ministry of Agriculture, Agricultural Genomics Institute at Shenzhen, Chinese Academy of Agricultural Sciences, Shenzhen, China*

LI ZHOU • *State Key Laboratory of Freshwater Ecology and Biotechnology, Institute of Hydrobiology, The Innovation Academy of Seed Design, Chinese Academy of Sciences, Wuhan, China; University of Chinese Academy of Sciences, Beijing, China*

ARTHUR ZWAENEPOEL • *Department of Plant Biotechnology and Bioinformatics, Ghent University, Ghent, Belgium*

Part I

Comparative Genomics to Study (Paleo-)Polyploidy

Chapter 1

Inference of Ancient Polyploidy from Genomic Data

Hengchi Chen and Arthur Zwaenepoel

Abstract

Whole-genome sequence data have revealed that numerous eukaryotic organisms derive from distant polyploid ancestors, even when these same organisms are genetically and karyotypically diploid. Such ancient whole-genome duplications (WGDs) have been important for long-term genome evolution and are often speculatively associated with important evolutionary events such as key innovations, adaptive radiations, or survival after mass extinctions. Clearly, reliable methods for unveiling ancient WGDs are key toward furthering understanding of the long-term evolutionary significance of polyploidy. In this chapter, we describe a set of basic established comparative genomics approaches for the inference of ancient WGDs from genomic data based on empirical age distributions and collinearity analyses, explain the principles on which they are based, and illustrate a basic workflow using the software "wgd," geared toward a typical exploratory analysis of a newly obtained genome sequence.

Key words Ancient whole-genome duplication, Paranome age distribution, Synteny, Collinearity, Comparative genomics

1 Introduction

With the increasing availability of whole-genome sequence data, many eukaryotes have been found to retain signatures of ancient whole-genome multiplications (WGMs, often colloquially referred to as whole-genome duplications or WGDs, as in the remainder of this chapter) [1], reinvigorating questions concerning the long- and short-term evolutionary significance of polyploidy. Inference of ancient WGD events remains however a challenging task. In this chapter, we will discuss some of the common approaches for the inference of WGDs based on comparative genomic analyses and exemplify a basic workflow using the software package wgd [2] for a typical initial analysis of a freshly obtained genome sequence.

The most commonly adopted methods for detecting signatures of ancient WGDs can be roughly divided in three categories, making use of (1) age distributions of gene duplicates, (2) patterns in the conservation of genome structure (synteny and collinearity),

Yves Van de Peer (ed.), *Polyploidy: Methods and Protocols*,
Methods in Molecular Biology, vol. 2545, https://doi.org/10.1007/978-1-0716-2561-3_1,

and (3) phylogenomic methods based on gene counts [3, 4], chromosome numbers [5] (*see* also Rice & Mayrose in this volume), or gene trees [3, 6, 7]. For the remainder of this chapter, we will discuss mainly categories (1) and (2), which are most relevant in the situation where a newly obtained genome sequence is to be probed for remnants of ancient WGD events, and only briefly consider phylogenomic methods.

2 Whole-Paranome Age Distributions

Since the early days of evolutionary genomics, the *paranome*, defined as the collection of all gene families in a single genome, has been recognized as an important source of information for the inference of genomic evolutionary processes [8–10]. In this context, a *gene family* is understood as a group of genes consisting of all paralogs derived from a single common ancestral gene in some ancestral genome at time T in the past. Such a gene family is associated with a dated intragenomic phylogeny, or *gene tree*, and the *age distribution* associated with a paranome is then the collection of all divergence times in this collection of gene trees. Clearly, the age distribution thus conceived is not something one can readily observe from genome sequence data. By an *empirical age distribution*, we then refer to some estimate of the true age distribution, acquired using phylogenetic methods. Different modes and rates of key genome evolutionary processes affect the size and shape of paranome age distributions differently, and hence (empirical) age distributions can be used to learn about these processes. WGD events in particular leave distinctive traces in age distributions [11, 12], making them particularly useful for the inference of ancient polyploidy events [13–16].

In order to use empirical age distributions for the purpose of WGD inference, it is important to understand how these are expected to behave with and without the occurrence of WGD. In other words, we need some plausible model of "background" genome evolution by small-scale gene duplication and loss (SSDL) (assumed to be the only major source of gene duplicates in the absence of WGD) and some idea of what the age distribution for such a model would look like. As indicated above, the age distribution is essentially a function of all the dated intragenomic gene trees with the same stem age that makes up an extant genome. Assuming different gene families evolve independently, what we are essentially after is a *model of gene family evolution* relevant for the paranome. Early studies proposing such models include Huynen and Van Nimwegen (1998) [8], Lynch and Conery (2003) [17], and Maere et al. (2005) [12]. Here, we consider a stochastic variant of the model of Lynch and Conery (2003) [17] which explains many of the features of empirical age distributions. We assume birth–death-like evolution of duplicate genes within a gene family, such that, at any point in time, the probability that a gene duplicates in a short interval is approximately $\lambda\Delta t$ per gene. Furthermore, every *duplicated gene* has a probability of getting lost in a similar short time interval. Denoting by the number of *duplicate* genes per family at time, i.e., so that the total family size is $X(t) + 1$, these transition probabilities can be written as

$$P\{X(t+\Delta t) = j | X(t) = i\} = \begin{cases} (i+1)\lambda\Delta t + o(\Delta t) & j = i+1 \\ i\mu\Delta t + o(\Delta t) & j = i-1 \\ 1-(i(\lambda+\mu)+\lambda)\Delta t + o(\Delta t) & i = j \\ o(\Delta t) & |i-j| > 1 \end{cases}$$

Here, λ and μ are of course the per-gene duplication and loss rate, respectively. We remark that this model is essentially similar to a simple linear birth–death process conditioned on nonextinction, as well as a simple linear birth–death immigration model with immigration rate equal to λ. We note that the same birth–death process is used for modeling the insertion and deletion dynamics in the famous TKF91 model used for evolutionary sequence alignment [18]. Elementary Markov chain theory shows that when $\lambda < \mu$, this model has a geometric stationary distribution with mean $E[X] = 1 - \lambda/\mu$. In other words, on average a family observed in the paranome will consist of $\mu/(\mu - \lambda)$ genes. Notably, for certain forms of rate variation of λ and μ across families, this model recovers the famous observation that gene family sizes are distributed according to a power law [8, 19, 20].

One can further show that for this simple model, the distribution function for the age distribution of duplicated genes retained in the extant genome is

$$F(t) = \frac{\mu\left(e^{\lambda t} - e^{\mu t}\right)}{\lambda e^{\lambda t} - \mu e^{\mu t}}$$

and the probability density function is

$$f(t) = \frac{\mu(\lambda - \mu)^2 e^{(\lambda+\mu)t}}{(\lambda e^{\lambda t} - \mu e^{\mu t})^2}$$

when $\lambda \ll \mu$, the density simplifies to an exponential distribution $f(t) \approx \mu e^{-\mu t}$, which is the approximation used by Lynch and Conery (2003) [17] in their discrete-time deterministic variant of the present model. This model is illustrated using simulated data in Fig. 1. As expected, for all practical purposes, the age distribution has an exponential shape, with most duplicates retained in the genome being of fairly recent origin. Clearly, this model has one major short-coming in that it ignores the processes of sub- and neofunctionalization, which cause duplicated genes to become stably established in the genome, whereas in the present model, all duplicated genes remain prone to loss at the same rate μ. Under the

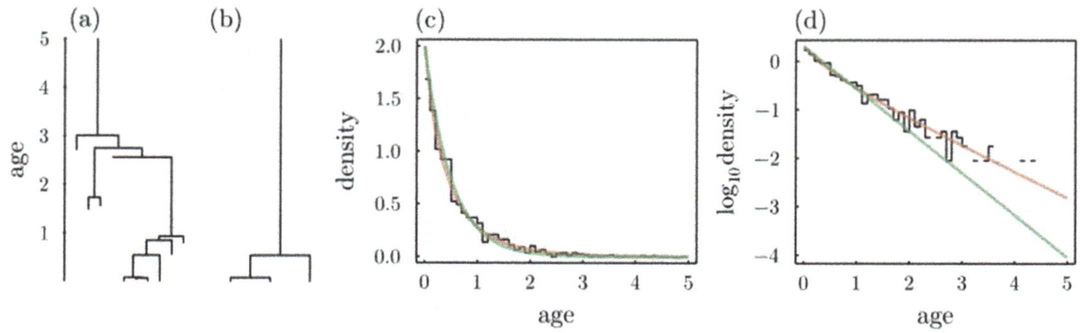

Fig. 1 A simple birth–death model for a whole-paranome age distribution. (**a**) A single random gene tree simulated from the birth–death model with per-gene duplication rate and loss rate $\mu = 2.0$ per duplicate gene. (**b**) The reconstructed tree for the gene tree in (**a**). (**c**) The whole-paranome age distribution based on a simulation of 5000 independently evolving gene families (according the same model as (**a**) and (**b**)). The orange line shows the probability density function for the model, whereas the green line shows the exponential approximation $\mu e^{-\mu t}$ thereof. (**d**) as in (**c**) but on a $\log_{10}$ scale

assumption of stationarity however, and assuming that duplicated genes become established by sub- or neofunctionalization processes with constant rate, one can show that the resulting age distribution is a simple mixture of the same quasiexponential distribution and a uniform component (*see* also below).

Having established what we may reasonably expect an age distribution to look like when paralogous families evolve by simple gene duplication and loss, we now consider the effect of a whole-genome duplication occurring at time t_{wgd} before the present. Naively, we may model this simply by the simultaneous duplication of all lineages at this particular time point, leading to a peak of gene duplicates in the age distribution around t_{wgd}. However, the processes underlying WGD are more subtle. Many WGD-derived duplicates are retained over much longer time spans than expected under the simple birth–death model of gene family evolution with simultaneous duplication of extant genes at t_{wgd}, indicating that duplicates derived from WGD are, in general, not subject to the same loss dynamics as small-scale duplicates. These differences in duplicate retention patterns have been the subject of many studies [11, 12, 21, 22], shedding light on the many biological factors determining whether genes are more likely to be retained after WGD or after small-scale duplication events. For the purpose of modeling the whole-paranome age distribution however, it suffices to assume that a certain proportion q of WGD-derived duplicates are retained, so that they effectively establish new "subfamilies."

A more important caveat for the inference of ancient WGD from empirical age distributions is the mode of polyploidization. In the case of *autopolyploidy* (Fig. 2a), WGD occurs within a population of a single species. The two subgenomes (assuming tetraploidization) that come together will not have diverged substantially,

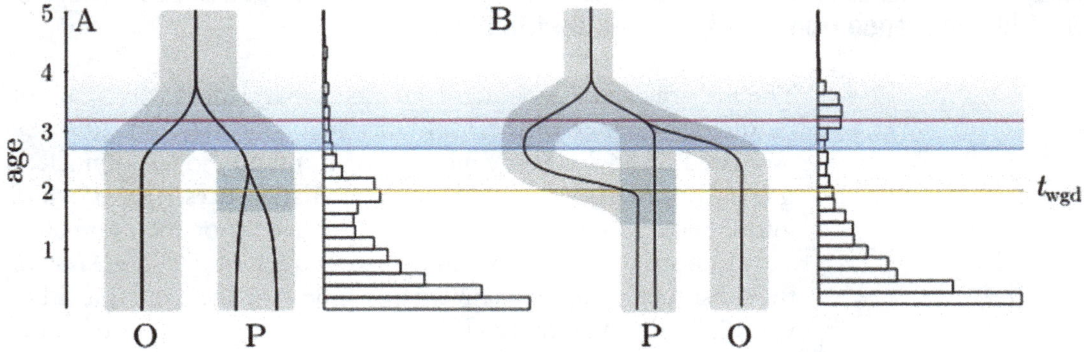

Fig. 2 WGD signatures for auto- and allopolyploidy events in empirical age distributions. (**a**) Autopolyploidy at time $t_{wgd} = 2$ in the lineage leading to *P* after divergence from outgroup *O*. The histogram shows a simulated empirical distribution for species *P* under this scenario ($\lambda = 1$, $\mu = 2$, $q = 0.1$). (**b**) Allopolyploid hybridization at time $t_{wgd} = 2$, bringing together two subgenomes which diverged at different time points from outgroup *O*. Again, a simulated empirical age distribution is shown for species *P* under this scenario

such that all four copies of a chromosome potentially undergo recombination during meiosis (*see* for instance Scott et al. in this volume). Over time, stable diploid inheritance is re-established, a process referred to as rediploidization [23]. The actual processes responsible for the establishment of the tetraploid lineage (dependent on, e.g., the number of founding tetraploid individuals, the frequency of interploidy crosses, the time to rediploidization, the size of the tetraploid population, etc.) determine at what time point WGD-derived gene duplicates coalesce around the actual WGD event (*see* also Roux et al. in this volume). Nevertheless, on usual phylogenetic timescales, most autopolyploidy-derived gene duplicates are expected to diverge around the actual time of WGD, giving a straightforward interpretation to empirical age distributions. In Fig. 2a, for instance, a peak of divergence times around an age of 2 is observed, postdating the divergence from the outgroup *O*. This would result in correctly inferring a lineage-specific polyploidization for *P*. In the case of *allopolyploidy* (Fig. 2b), WGD is the result of hybridization and duplication of two distinct lineages, such that WGD-derived duplicates are expected to coalesce in the ancestral population of the two hybridizing lineages. There is therefore the possibility of a considerable mismatch between the peak observed in the age distribution and the actual timing of the allopolyploid hybridization event, rendering the interpretation of empirical age distributions potentially challenging, as exemplified in Fig. 2b, where the divergence of duplicates derived from a WGD specific to the lineage leading to *P* precedes the divergence from the outgroup *O* (marked by the blue gradient), while the latter does not share the WGD.

3 WGD Inference from Empirical Age Distributions

The age of a pair of duplicates is an unknown quantity which cannot be directly obtained from molecular sequence data, but must be estimated by phylogenetic means. For common models of nucleotide substitution, a pair of homologous sequences only provides information about the evolutionary distance (*d*, or the number of substitutions per site) between the two sequences, i.e., a *product* of the substitution rate *r* and divergence time *t* (or age), and not either variable independently [24]. Without an external source of information about either quantity (e.g., in the field of divergence time estimation, one typically uses palaeontological information), inference of the age of duplication may seem impossible. However, when the substitution rate is approximately constant, the evolutionary distance should be proportional to the divergence time. It is well known, that for neutral sites, the nucleotide substitution rate is equal to the mutation rate [25], so that neutral substitution rates will remain approximately constant over time as long as the neutral mutation rate does so too. Substitutions at synonymous sites, i.e., nucleotide substitutions in protein-coding genes which do not incur an amino acid change, are such putatively neutral sites, so that the synonymous distance K_S (the number of synonymous substitutions per synonymous site) may be used as a proxy for the divergence time of a duplicate gene pair.

Ideally, the observation of additional peaks superimposed on the aforementioned SSDL "background" distribution reflects abrupt increases in the number of retained gene duplicates from a particular episode in the evolution of the genome under consideration, potentially stemming from large-scale gene duplications such as segmental duplications, aneuploidies, and WGDs. However, the intrinsic caveats of using a statistical estimate of K_S as a proxy for the divergence time obscure and sometimes even mislead the inference of putative WGDs from empirical age distributions [26]. First, the synonymous distance is of course a random variable, and common molecular clock models (which are Poisson processes) entail that the variance in the evolutionary distance increases with time. Estimates of the synonymous distance, even when perfect, will therefore reflect the associated divergence time less and less precisely for larger timescales. As a consequence, the signature bursts of duplicates associated with older WGD events are expected to be progressively flattened and eventually blend into the background K_S distribution of SSDL origin. Second, the gradual accumulation of multiple substitutions at the same site over time, concealing intermediate substitutions, renders the estimation of K_S also subject to saturation effects, where the true synonymous substitution would be systematically underestimated at the point of saturation. Although the Markov chain models of nucleotide substitution correct for multiple substitutions and are able to recover the number of cryptic nucleotide changes to some extent, the effect of a correction for multiple substitutions is highly dependent on the degree of sequence divergence and sometimes erroneous when considering the heterogeneous substitution rate of different genes [24]. Consequently, K_S estimates for older duplicates could converge on lower K_S values with an artificial "saturation peak" in the tail as a result [26]. Taken together, these intrinsic issues with K_S estimation could blur the signature of a true older WGD peak and could lead to spurious peaks due to saturation of the synonymous distance, limiting the usefulness of the method based on empirical age distributions to *relatively* recent WGD events—depending of course on the rate of molecular evolution. Although previously K_S estimates higher than 1 were considered unreliable [27], Vanneste and colleagues suggest that K_S saturation and stochastic effects remain acceptable until at least a synonymous age of 2 and likely even higher [26].

4 Inference of K_S-Based Age Distributions Using "wgd"

Here, as an example, we use the empirical K_S-based age distribution to infer the well-documented ancient hexaploidization event (a whole-genome triplication or WGT) in *Vitis vinifera* [28]. In addition, to show what a typical empirical age distribution probably looks like for species which have not undergone WGD in their relatively recent evolutionary past, we conducted the same analysis for *Amborella trichopoda*, a sister group to all the other angiosperms, which has experienced no WGD after the divergence of angiosperms [29]. The CDS and GFF files for protein-coding genes in *V. vinifera* and *A. trichopoda* were downloaded from PLAZA [30]. We use the wgd dmd and wgd ksd programs in wgd [2] to infer paralogous gene families and estimate and plot the K_S distribution, respectively. Note that throughout, we assume to be working in a Linux-based environment. We refer to the documentation (available at https://github.com/arzwa/wgd) for installation instructions and detailed documentation of the used commands and options.

The first step is to obtain the paranome, i.e., infer paralogous gene families. To infer the *V. vinifera* paranome using the wgd package, we use the following command:

```
$ wgd dmd ./Vvi.cds -I 3.0 -o ./Vvi_wgd_dmd
```

and similarly for *A. trichopoda*:

```
$ wgd dmd ./Atr.cds -I 3.0 -o ./Atr_wgd_dmd
```

Using wgd dmd, we infer paralogous gene families by conducting an all-versus-all amino acid level similarity search for the set of protein-coding sequences of the species of interest using the diamond program [31], and by clustering the resulting sequence similarity graph using the Markov clustering algorithm mcl [32]. The clustering algorithm is governed by a single parameter, the so-called "inflation factor," which determines the coarseness of the resulting clusters, with larger values resulting in smaller clusters on average. Here, we use an empirical inflation factor of $I = 3$, using the -I option in wgd dmd. We note that in principle, any gene family inference method (e.g., OrthoFinder [33], OrthoMCL [34], or InParanoid [35]) can be used at this stage, but that for the inference of gene families from a single genome, these should in general give very similar results.

After having obtained the paranome, an empirical K_S-based age distribution can be obtained using wgd ksd:

```
$ wgd ksd -n 4 --pairwise ./Vvi_wgd_dmd/Vvi.cds.tsv ./Vvi.cds
-o Vvi_wgd_ksd
```

and again, similarly for *Amborella*

```
$ wgd ksd -n 4 --pairwise ./Atr_wgd_dmd/Atr.cds.tsv ./Atr.cds
-o Atr_wgd_ksd
```

The pipeline implemented in wgd ksd has been described before in detail in, e.g., [2, 26], but its outline is worth reiterating in brief here. The following three main steps are conducted by wgd ksd for each paralogous gene family:

1. A codon-level multiple sequence alignment (MSA) is obtained by inferring an amino acid MSA using the mafft program [36] which is then back-translated to a codon-level nucleotide MSA.
2. For each pair of genes in the family, a maximum likelihood estimate for the pairwise synonymous distance (K_S) is obtained using the codeml program from the PAML package [37].
3. An estimate of the phylogenetic tree topology of the paralogous gene family is obtained and rooted using midpoint rooting. The nodes of the tree are labeled, and for each gene pair in the family, the label of the most recent common ancestor (MRCA) node in the gene tree is recorded.

Details of default parameter settings and the various program options can be found in the online documentation and source code. Based on the information acquired using this pipeline, an empirical age distribution, as defined above, is then constructed by either summarizing the pairwise K_S estimates by gene tree node (for instance using the mean) or by associating with each pairwise K_S

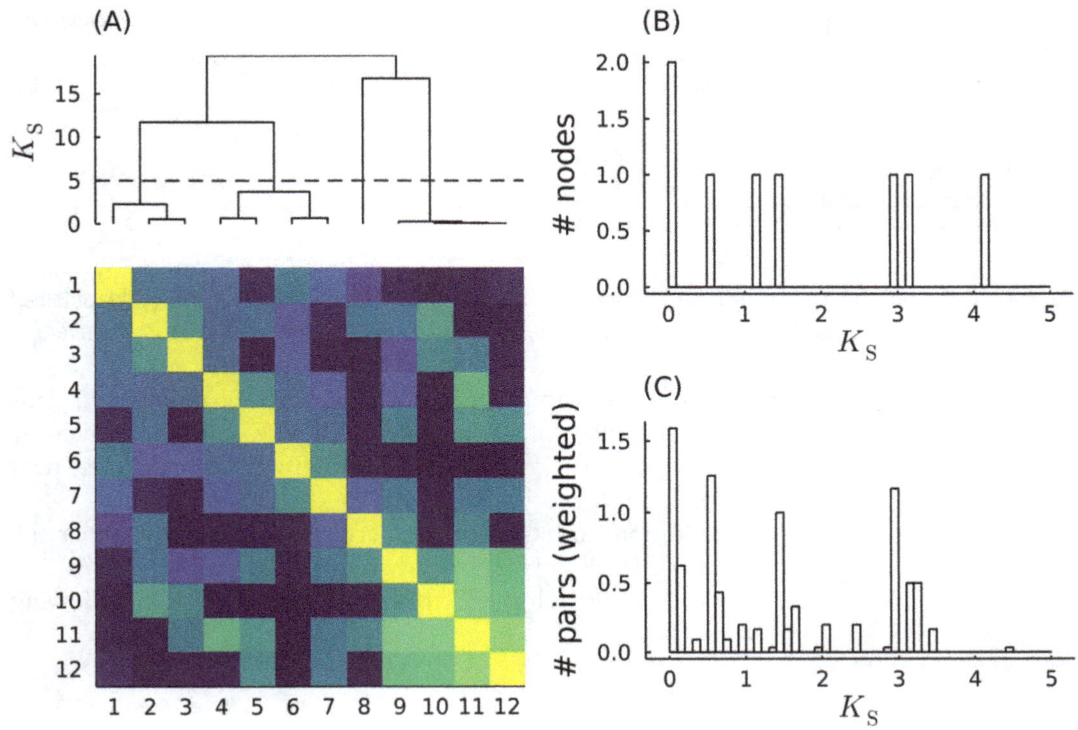

Fig. 3 Illustration of the contribution of a single gene family in *V. vinifera* to the whole-paranome empirical age distribution. (**a**) K_S-based gene tree and distance matrix, where yellow and dark blue indicate the low and high ends of the relevant K_S range, respectively. The gene tree was inferred using average linkage clustering (the default setting in "wgd"). The dotted line marks $K_S = 5$. (**b**) Node-averaged histogram for this particular family. Note that only eight distinct data points end up in the histogram, associated with the eight nodes with age lower than 5 in (**a**). (**c**) Node-weighted histogram for the same family. Note that each pairwise K_S estimate below $K_S = 5$ is displayed but that the total mass in the histogram is the same as in (**b**)

estimate a weight derived from the node depth of its associated gene tree node, and adding each estimate to the whole-paranome distribution with its associated weight. Histograms of these distributions are referred to as *node-averaged* and *node-weighted* histograms, respectively. The difference between these two approaches for aggregating the information from pairwise K_S estimates into a proper empirical age distribution is illustrated in Fig. 3.

Figure 4 shows the resulting node-weighted whole-paranome K_S distributions for our example data sets. *A. trichopoda* did not undergo any recent WGD events (only a putative very ancient WGD event shared by all angiosperms, seed plants, or both [6, 7, 38]; thus, its K_S distribution is expected to follow the uniform/quasiexponential mixture as discussed above. We can readily observe the expected pattern in Fig. 4 . In contrast, *V. vinifera* shows a considerable enrichment of gene duplicates within the K_S range marked by the shaded area in Fig. 4, reflecting a possible

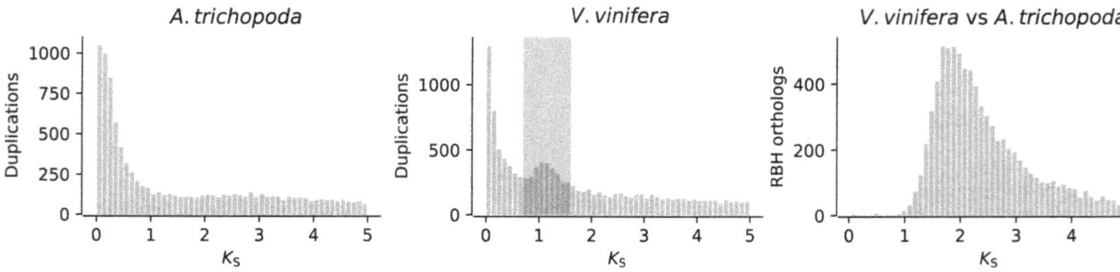

Fig. 4 K_S-based paranome and one-to-one ortholog age distributions for *V. vinifera* and *A. trichopoda* obtained using the wgd ksd pipeline. The shaded area marks the K_S range corresponding to the WGT in *V. vinifera*

WGD event (or any other multiplication level of course). This serves as an initial indication of WGD in *V. vinifera,* which we shall now seek to corroborate by additional means (*see* next section).

In addition, to have a preliminary knowledge on whether the putative WGD of *V. vinifera* is shared with *A. trichopoda*, we can construct an ortholog K_s distribution using the following command:

```
$ wgd dmd ./Vvi.cds ./Atr.cds -I 3.0 -o ./Vvi_Atr_wgd_dmd
$ wgd ksd -n 4 ./Vvi_Atr_wgd_dmd/Vvi.cds_Atr.cds.rbh.tsv
./Vvi.cds ./Atr.cds -o ./Vvi_Atr_wgd_ksd
```

Note that when providing wgd dmd with two CDS files, it will infer one-to-one orthologs between the two species using reciprocal best hits (RBH). The obtained RBH orthologs can then be used with wgd ksd to construct an ortholog K_s-based age distribution. The RBH ortholog K_s distribution shown in Fig. 4 has a mode near $K_s \approx 1.8$, exceeding the WGD signature peak K_s value in the paranome age distribution of *V. vinifera*. This suggests that the speciation event predates the putative WGD event, compatible with the lack of a WGD signature in *Amborella*. However, this interpretation hinges on the assumption that substitution rates in *Vitis* and *Amborella* have been, on average, similar since their divergence. If this assumption is violated, inferring the phylogenetic placement of WGDs from comparisons of K_s distributions can be misleading. One approach to mitigate such issues is to perform a relative rate correction based on multiple interspecific comparisons. This has recently been explored in depth in [16] in the context of phylogenetic placement of WGDs, and we do not consider this further here. Their approach is implemented in the software "ksrates," which builds on the "wgd" package discussed in the present chapter.

5 Synteny, Collinearity, and Anchor Pairs

So far, we used only whole-genome sequence information to infer putative WGD events using empirical age distributions, but this approach can be used for transcriptome data as well. However, when high-quality genome assemblies are available, we can also use positional information to infer potential WGD events. Indeed, duplication of the entire genome is expected to instantaneously generate an additional copy of each chromosome, retaining both the *gene order* and *gene content* of the original copy. Despite chromosomal rearrangements and rampant gene loss following WGD [39], gene order and gene content are expected to be more or less retained on at least some chromosomes for reasonable time frames. Historically, the property of two or more sets of genes from different genomic regions being colocated on the same chromosome was referred to as *synteny*, while the property of two or more sets of genes showing conserved gene order was referred as *collinearity* [40]. Today, the concepts of synteny and collinearity are often confused. Moreover, synteny is often used to refer to *ancestral* colocation, i.e., one designates a set of regions in an extant genome or multiple extant genomes as syntenic when they show conserved gene content, which results from these regions being derived from the same ancestral chromosomal region. We use "syntenic" in this latter sense. A corollary of this is that synteny is a "weaker" concept than collinearity, the latter implying the former but not vice versa. Collinear or syntenic regions *within* a genome are assumed to have originated from the duplication of a common ancestral genomic region, and are as such deemed strong evidence for WGD.

A straightforward way to profile synteny and collinearity within a genome is to draw a whole-genome dot plot, where both the *x*-axis and *y*-axis represent the same genome, and each square represents a single chromosome-to-chromosome comparison. Figure 5a shows such an intragenomic dotplot for *V. vinifera,* obtained using wgd syn in wgd, which uses i-ADHoRe 3.0 [41] to infer collinear blocks. The following command can be used to perform intragenomic collinearity searches and generate whole-genome dot plots:

```
$ wgd syn -f mRNA -a ID ./Vvi_wgd_dmd/Vvi.cds.tsv ./Vvi.gff3
-o ./Vvi_wgd_syn -ks ./Vvi_wgd_ksd/Vvi.cds.tsv.ks.tsv
```

Where the -f and -a options serve to indicate the relevant features in the gff file. Here, homologs are shown as dots, while anchor pairs, defined as homologous pairs in collinear blocks, are marked in a distinct color. When multiple anchors are located adjacently, a red diagonal line, reflecting a collinear block, can be

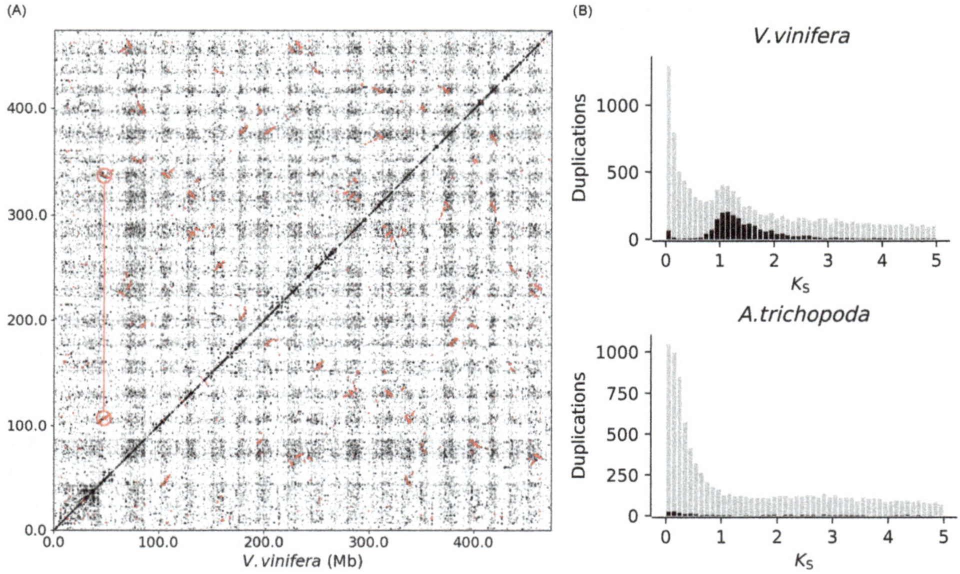

Fig. 5 (**a**) Whole-genome dot plots and intragenomic collinearity for *V. vinifera*. The gray dots represent homologs, while red dots represent anchor pairs (homologs in collinear blocks). The red circles highlight a typical triplicated block in the *Vitis* genome. (**b**) Whole-paranome K_S distributions (gray) and anchor-pair K_S distributions (black) for *V. vinifera* and *A. trichopoda*

observed. We can find quite some collinear blocks in the intragenomic dot plot of *V. vinifera* in Fig. 5a, and a close examination reveals that for many chromosomal regions, we find two copies showing conserved gene order on other chromosomes, while for *A. trichopoda*, anchor pairs are negligible. Figure 5b shows the K_S distributions for anchor pairs against the whole-paranome background, highlighting the correspondence of the duplicated collinear blocks with the whole-paranome WGD signature. Together, the empirical age distribution and collinearity results already allow us to claim with reasonable confidence that *V. vinifera* derives from an ancestral hexaploid, where the whole-genome triplication occurred around a time associated with $K_S \approx 1.1$.

We can further conduct an intergenomic comparison with *A. trichopoda* to validate the hypothesis of hexaploidy again using wgd syn in wgd. Here, we use as input gene families orthogroups inferred using OrthoFinder [33]. With the fasta files containing the protein-coding genes for both species in the current working directory, the following command can be used to infer orthogroups with OrthoFinder using default settings:

```
$ orthofinder -f $PWD -og
```

For more details on using OrthoFinder, we refer to the excellent documentation of this software (https://github.com/

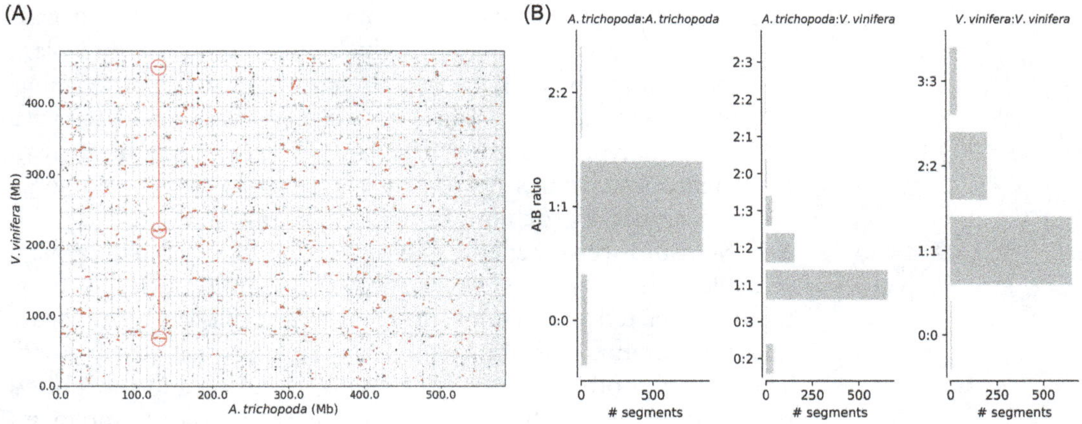

Fig. 6 (**a**) Interspecific dotplot between *V. vinifera* and *A. trichopoda*. The red circles highlight a typical 3:1 collinear block between *V. vinifera* and *A. trichopoda*. (**b**) Intraspecific and interspecific syntenic depths between *V. vinifera* and *A. trichopoda*

davidemms/OrthoFinder). The resulting gene families can then be used to conduct intergenomic collinearity analyses using wgd syn as follows:

```
$ wgd syn -f mRNA \
./OrthoFinder/Results*/Orthogroups/Orthogroups.tsv \
./Atr.gff3 data/Vvi.gff3 -o ./Vvi_Atr_wgd_syn
```

As can be seen in Fig. 6a, b, by comparing *V. vinifera* and *A. trichopoda*, except for an obvious 1:1 ratio, mainly a ratio of 2: 1 and 3:1 is among the syntenic regions between *V. vinifera* and *A. trichopoda*, indicating that up to three homologous blocks are found in *V. vinifera* for a given segment in *A. trichopoda*. Combined with the results of the intragenomic comparisons, we can conclude with confidence that *V. vinifera* experienced a paleo-hexaploidization event after its divergence of *A. trichopoda*.

6 Other Approaches and Conclusion

In the above, we have illustrated a basic workflow to unveil ancient WGDs using comparative genomics approaches available in the "wgd" package based on age distributions and analyses of synteny and collinearity. In many cases, these approaches suffice for uncovering WGD events, although their proper phylogenetic placement may remain a challenge using these methods alone (but *see* [16]). Evidence for ancient WGDs can also be obtained through the use of phylogenomic methods, in particular through gene tree–species tree reconciliation approaches (e.g., [6, 7, 42–44]). These approaches may not only allow the inference of WGD events directly in a phylogenetic context, but may also enable the characterization of very ancient events, such as the putative seed plant and/or angiosperm WGD events. In addition, such methods may be helpful to reveal the relevant hybridization event associated with ancient allopolyploidization events [42]. The reliability of many of these methods remains however challenging to assess, and some results obtained by such means have been contested [6, 7, 38, 45, 46]. Statistical phylogenetic approaches as in [3, 7] may prove a valuable way forward, as these enable the quantification of uncertainty associated with certain inferences and allow to investigate under which model assumptions particular inferences are obtained. The lack of good models of gene family evolution and the limited power to unveil highly eroded ancient events without employing genome structure information present however significant challenges to these approaches. In general, for lack of a fail-safe method, and while awaiting the development of better statistical methods for the analysis of genome evolution, we advocate an integrative approach for the inference of ancient WGD events, combining the fairly straightforward comparative genomic methods described in this chapter with phylogenetic tools where needed.

Acknowledgments

We wish to thank Yves Van de Peer and Zhen Li for their support and helpful feedback. Hengchi Chen and Arthur Zwaenepoel acknowledge the PhD Fellowship of the Research Foundation—Flanders (FWO).

References

1. Van de Peer Y, Mizrachi E, Marchal K (2017) The evolutionary significance of polyploidy. Nat Rev Genet 18(7):411–424
2. Zwaenepoel A, Van de Peer Y (2018) wgd—simple command line tools for the analysis of ancient whole-genome duplications. Bioinformatics 35(12):2153–2155
3. Rabier C-E, Ta T, Ané C (2013) Detecting and locating whole genome duplications on a phylogeny: a probabilistic approach. Mol Biol Evol 31(3):750–762
4. Zwaenepoel A, Van de Peer Y (2020) Model-based detection of whole-genome duplications in a phylogeny. Mol Biol Evol 37(9): 2734–2746
5. Glick L, Mayrose I (2014) ChromEvol: assessing the pattern of chromosome number evolution and the inference of polyploidy along a phylogeny. Mol Biol Evol 31(7):1914–1922
6. Jiao Y, Wickett NJ, Ayyampalayam S, Chanderbali AS, Landherr L, Ralph PE, Tomsho LP, Hu Y, Liang H, Soltis PS, Soltis DE, Clifton SW, Schlarbaum SE, Schuster SC, Ma H, Leebens-Mack J, dePamphilis CW (2011) Ancestral polyploidy in seed plants and angiosperms. Nature 473(7345):97–100
7. Zwaenepoel A, Van de Peer Y (2019) Inference of ancient whole-genome duplications and the evolution of gene duplication and loss rates. Mol Biol Evol 36(7):1384–1404
8. Huynen MA, van Nimwegen E (1998) The frequency distribution of gene family sizes in complete genomes. Mol Biol Evol 15(5): 583–589
9. Lynch M, Conery JS (2000) The evolutionary fate and consequences of duplicate genes. Science 290(5494):1151–1155

10. Karev GP, Wolf YI, Rzhetsky AY, Berezovskaya FS, Koonin EV (2002) Birth and death of protein domains: a simple model of evolution explains power law behavior. BMC Evol Biol 2(1):18
11. Blanc G, Wolfe KH (2004) Widespread paleopolyploidy in model plant species inferred from age distributions of duplicate genes[W]. Plant Cell 16(7):1667–1678
12. Maere S, De Bodt S, Raes J, Casneuf T, Van Montagu M, Kuiper M, Van de Peer Y (2005) Modeling gene and genome duplications in eukaryotes. Proc Natl Acad Sci U S A 102(15):5454
13. Van de Peer Y (2004) Computational approaches to unveiling ancient genome duplications. Nat Rev Genet 5(10):752–763
14. Vanneste K, Van de Peer Y, Maere S (2012) Inference of genome duplications from age distributions revisited. Mol Biol Evol 30(1): 177–190
15. Tiley GP, Barker MS, Burleigh JG (2018) Assessing the performance of Ks plots for detecting ancient whole genome duplications. Genome Biol Evol 10(11):2882–2898
16. Sensalari C, Maere S, Lohaus R (2021) ksrates: positioning whole-genome duplications relative to speciation events in KS distributions. Bioinformatics 38(2):530–532
17. Lynch M, Conery JS (2003) The evolutionary demography of duplicate genes. J Struct Funct Genom 3(1):35–44
18. Thorne JL, Kishino H, Felsenstein J (1991) An evolutionary model for maximum likelihood alignment of DNA sequences. J Mol Evol 33(2):114–124
19. Hughes T, Liberles DA (2008) The power-law distribution of gene family size is driven by the pseudogenisation rate's heterogeneity between gene families. Gene 414(1):85–94
20. Zwaenepoel A, Van de Peer Y (2021) A two-type branching process model of gene family evolution. bioRxiv:2021.03.18.435925
21. Li Z, Defoort J, Tasdighian S, Maere S, Van de Peer Y, De Smet R (2016) Gene duplicability of core genes is highly consistent across all angiosperms. Plant Cell 28(2):326–344
22. Tasdighian S, Van Bel M, Li Z, Van de Peer Y, Carretero-Paulet L, Maere S (2017) Reciprocally retained genes in the angiosperm lineage show the hallmarks of dosage balance sensitivity. Plant Cell 29(11):2766–2785
23. Wolfe KH (2001) Yesterday's polyploids and the mystery of diploidization. Nat Rev Genet 2(5):333–341
24. Yang Z (2014) Molecular evolution: a statistical approach. Oxford University Press, Oxford
25. Kimura M (1983) The neutral theory of molecular evolution. Cambridge University Press, Cambridge
26. Vanneste K, Van de Peer Y, Maere S (2013) Inference of genome duplications from age distributions revisited. Mol Biol Evol 30(1): 177–190
27. Li W (1997) Molecular evolution. Sinauer Associates Incorporated, Sunderland
28. Jaillon O, Aury J-M, Noel B, Policriti A, Clepet C, Casagrande A, Choisne N, Aubourg S, Vitulo N, Jubin C, Vezzi A, Legeai F, Hugueney P, Dasilva C, Horner D, Mica E, Jublot D, Poulain J, Bruyère C, Billault A, Segurens B, Gouyvenoux M, Ugarte E, Cattonaro F, Anthouard V, Vico V, Del Fabbro C, Alaux M, Di Gaspero G, Dumas V, Felice N, Paillard S, Juman I, Moroldo M, Scalabrin S, Canaguier A, Le Clainche I, Malacrida G, Durand E, Pesole G, Laucou V, Chatelet P, Merdinoglu D, Delledonne M, Pezzotti M, Lecharny A, Scarpelli C, Artiguenave F, Pè ME, Valle G, Morgante M, Caboche M, Adam-Blondon A-F, Weissenbach J, Quétier F, Wincker P, French-Italian Public Consortium for Grapevine Genome Characterization (2007) The grapevine genome sequence suggests ancestral hexaploidization in major angiosperm phyla. Nature 449(7161):463–467
29. Amborella Genome Project, Albert VA, Barbazuk WB, dePamphilis CW, Der JP, Leebens-Mack J, Ma H, Palmer JD, Rounsley S, Sankoff D, Schuster SC (2013) The *Amborella* genome and the evolution of flowering plants. Science 342(6165):1241089
30. Van Bel M, Diels T, Vancaester E, Kreft L, Botzki A, Van de Peer Y, Coppens F, Vandepoele K (2017) PLAZA 4.0: an integrative resource for functional, evolutionary and comparative plant genomics. Nucleic Acids Res 46 (D1):D1190–D1196
31. Buchfink B, Reuter K, Drost H-G (2021) Sensitive protein alignments at tree-of-life scale using DIAMOND. Nat Methods 18(4): 366–368
32. Van Dongen SM (2000) Graph clustering by flow simulation
33. Emms DM, Kelly S (2019) OrthoFinder: phylogenetic orthology inference for comparative genomics. Genome Biol 20(1):238
34. Li L, Stoeckert CJ Jr, Roos DS (2003) OrthoMCL: identification of ortholog groups for eukaryotic genomes. Genome Res 13(9): 2178–2189
35. Östlund G, Schmitt T, Forslund K, Köstler T, Messina DN, Roopra S, Frings O,

Sonnhammer ELL (2009) InParanoid 7: new algorithms and tools for eukaryotic orthology analysis. Nucleic Acids Res 38(suppl_1): D196–D203

36. Katoh K, Standley DM (2013) MAFFT multiple sequence alignment software version 7: improvements in performance and usability. Mol Biol Evol 30(4):772–780
37. Yang Z (2007) PAML 4: phylogenetic analysis by maximum likelihood. Mol Biol Evol 24(8): 1586–1591
38. Ruprecht C, Lohaus R, Vanneste K, Mutwil M, Nikoloski Z, Van de Peer Y, Persson S (2017) Revisiting ancestral polyploidy in plants. Sci Adv 3(7):e1603195
39. Adams KL, Wendel JF (2005) Polyploidy and genome evolution in plants. Curr Opin Plant Biol 8(2):135–141
40. Tang H, Bowers JE, Wang X, Ming R, Alam M, Paterson AH (2008) Synteny and collinearity in plant genomes. Science 320(5875): 486–488
41. Proost S, Fostier J, De Witte D, Dhoedt B, Demeester P, Van de Peer Y, Vandepoele K (2011) i-ADHoRe 3.0—fast and sensitive detection of genomic homology in extremely large data sets. Nucleic Acids Res 40(2): e11–e11
42. Thomas GWC, Ather SH, Hahn MW (2017) Gene-tree reconciliation with MUL-trees to resolve polyploidy events. Syst Biol 66(6): 1007–1018
43. Yang Y, Moore MJ, Brockington SF, Mikenas J, Olivieri J, Walker JF, Smith SA (2018) Improved transcriptome sampling pinpoints 26 ancient and more recent polyploidy events in Caryophyllales, including two allopolyploidy events. New Phytol 217(2): 855–870
44. Initiative OTPT (2019) One thousand plant transcriptomes and the phylogenomics of green plants. Nature 574(7780):679–685
45. Roelofs D, Zwaenepoel A, Sistermans T, Nap J, Kampfraath AA, Van de Peer Y, Ellers J, Kraaijeveld K (2020) Multi-faceted analysis provides little evidence for recurrent whole-genome duplications during hexapod evolution. BMC Biol 18(1):57
46. Li Z, Tiley GP, Galuska SR, Reardon CR, Kidder TI, Rundell RJ, Barker MS (2018) Multiple large-scale gene and genome duplications during the evolution of hexapods. Proc Natl Acad Sci 115(18):4713

Chapter 2

Navigating the CoGe Online Software Suite for Polyploidy Research

Victor A. Albert and Trevor J. Krabbenhoft

Abstract

The CoGe software suite at genomevolution.org hosts a number of tools that facilitate genomic research on plant and animal whole-genome multiplication—polyploidy. SynMap permits analysis and visualization of two-way syntenic dotplot alignments of genomes, includes many options and data/graphics download possibilities, and even permits three-genome synteny maps and interactive views. FractBias is a tool that operates within SynMap that permits calculation and graphic display of genome fragments (such as chromosomes) of one species mapped to another, displaying both blockwise homology depths and the extent of syntenic gene (syntelog) loss following polyploidy events. SynMap macrosynteny results can segue into the microsynteny tool GEvo, which provides genome-browser-like views of homologous genome blocks. CoGe FeatView allows call-up of given gene features already stored in the CoGe resource, and CoGeBlast permits searches for additional features that can be analyzed or downloaded further. Links from these tools can be fed into SynFind, which can find syntenic blocks surrounding a feature across multiple specified genomes while also simultaneously providing overall genome-wide syntenic depth calculations that can be interpreted to reflect polyploidy levels. Here, we describe basic use of these tools on the CoGe software suite.

Key words Polyploidy, Synteny, Syntelog, CoGe, genomevolution.org, SynMap, FractBias, GEvo, SynFind

1 Introduction

After sequencing, assembling, and annotating the genome of an organism known or suspected to have undergone a whole-genome duplication (WGD) during its evolutionary history, it is useful to have a single, integrated set of tools that can discover, analyze, and illustrate the multiplicated nature of this genome both against itself and in comparison with those of other species. The online CoGe software suite provides such extensive and user-friendly resources to upload and analyze either private or public genomes for their polyploid features [1–3]. Genomic synteny—internally or exter-

Yves Van de Peer (ed.), *Polyploidy: Methods and Protocols*,
Methods in Molecular Biology, vol. 2545, https://doi.org/10.1007/978-1-0716-2561-3_2,

nally comparable gene order homology—provides the principal data for polyploid detection and analysis based on assembled genome data. The basic tool for this is a self:self or self:other genomic alignment into a homologous ordering of either unannotated or feature-annotated genomic fragments. The basic informatic operation that permits assembly of a syntenic plot is all-by-all sequence similarity search (such as a BLAST variant) across the genomic fragments being compared. Such a syntenic plot, if well organized with graphics, data download, and further analytical possibilities, provides the basis for a complete comparative genomic assessment of polyploidy status. Here, we describe the basic use of several features in the CoGe suite that facilitate such comparative discoveries across multiple organisms.

2 Materials

2.1 Genome Assemblies

The basic input to the CoGe suite is an assembled genome, with or without annotated gene features. Public or private upload to genomevolution.org is simple and described in the online CoGepedia resource (https://genomevolution.org/wiki/index.php/Main_Page; see also [3]). It is highly recommended to use existing or new genomic assemblies that are gene-annotated, since due to server constraints and sequence similarity search times, SynMap will not generate syntenic plots for unannotated vs. unannotated data. However, unannotated genomes can be analyzed against annotated ones using SynMap and GEvo, which is very useful at intermediate steps in a genome study, before annotations are generated. General feature format (GFF) files are the basic information for an uploaded genome, along with its FASTA nucleotide sequence file.

2.2 Online Tools

2.2.1 SynMap

SynMap automates the generation of a syntenic dotplot [4]. It incorporates several steps, including data format processing, database generation, filtration of gene features that can confound synteny calculations (local, tandem gene duplicates that are often species- or block-specific), the informatic chaining of syntenic features, and production of flat file and visual outputs. SynMap can also calculate, output, and display synonymous (Ks) and nonsynonymous (Kn) substitutional differences between homologous features. It can generate syntenic-path pseudoassemblies of one genome against another, using the most contiguous assembly as the guide ([5] see an example for the complex *Utricularia gibba* genome in [6]). Many further options exist to manipulate syntenic plot parameters and displays. Pairwise SynMap is available in both "legacy" and SynMap2 interactive views, and three-way analyses are permitted in SynMap3D [4]. SynMap plots permit direct, visual detection of polyploid block interrelationships within and between genome assemblies (Fig. 1), but following counts across these plots

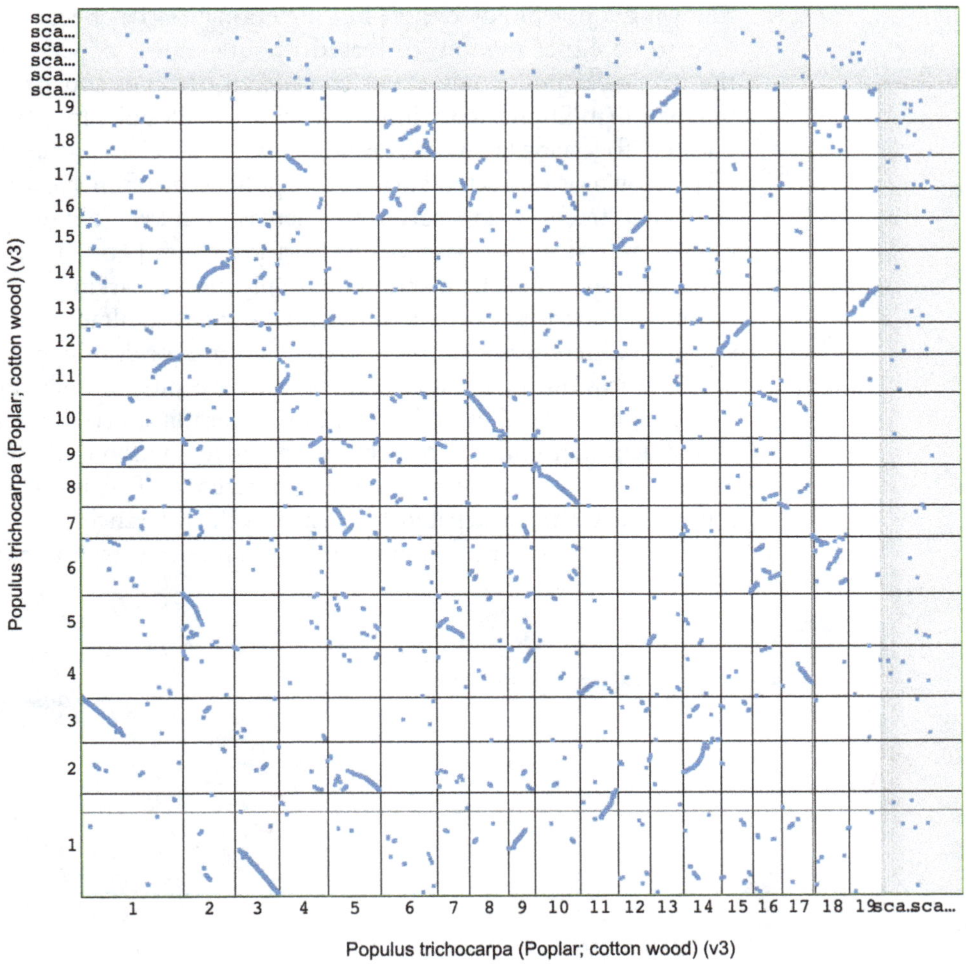

Fig. 1 Basic SynMap2 syntenic dotplot output comparing the gene-annotated *Populus trichocarpa* (cottonwood) genome [7] against itself. Dots represent pairs of genes with shared synteny (syntelogs). The view is in mirror image along the diagonal, which itself represents the exact self:self alignment. The off-diagonal syntenic "lines" are chains composed of individual syntelog pair sequence similarity hits. The longer and less interrupted these syntenic chains of genes are, the less fractionated and less rearranged and therefore the more recent the polyploidy event. Two known paleopolyploidy events are visible in this analysis: the comparatively recent Salicaceae (willow family)-specific event (long syntenic chains) [7, 8] and the considerably more ancient *gamma* hexaploidy event (shorter and chromosomally rearranged chains) that characterizes all core eudicot angiosperms [9–11]

can be challenging, especially when polyploidy events are relatively old and degraded or when substantial genome rearrangements have occurred since duplication events or species splits.

2.2.2 FractBias

FractBias is a CoGe tool connected to SynMap that was originally developed to analyze fractionation differences between paleopolyploid subgenomes, fractionation being the stepwise gene deletion and diploidization process between subgenomes that usually occurs following a polyploidy event [12]. Fractionation has been discovered to be either random or biased by subgenome depending on the particular polyploidy event considered, its age, and its autopolyploid (multiplicated within a species) or allopolyploid (multiplicated following a hybridization between species) nature [13–15]. Known or suspected ancient autopolyploids often show unbiased gene fractionation patterns, whereas even very old allopolyploids often display biased patterns of gene loss [16–18]. Biased fractionation may reflect differential epigenetic masking of transcriptional intensities between syntelogs and the resultant purifying selection differences that might favor retention of dominant suites of duplicates on the most strongly expressing subgenomes. FractBias generates flat files and graphics that permit ascertainment of gene fractionation differences between a given genome of interest analyzed against a target genome as reference. The investigator must specify the anticipated syntenic depth difference (e.g., 2:1; Fig. 2), but this can be manipulated across multiple experiments to

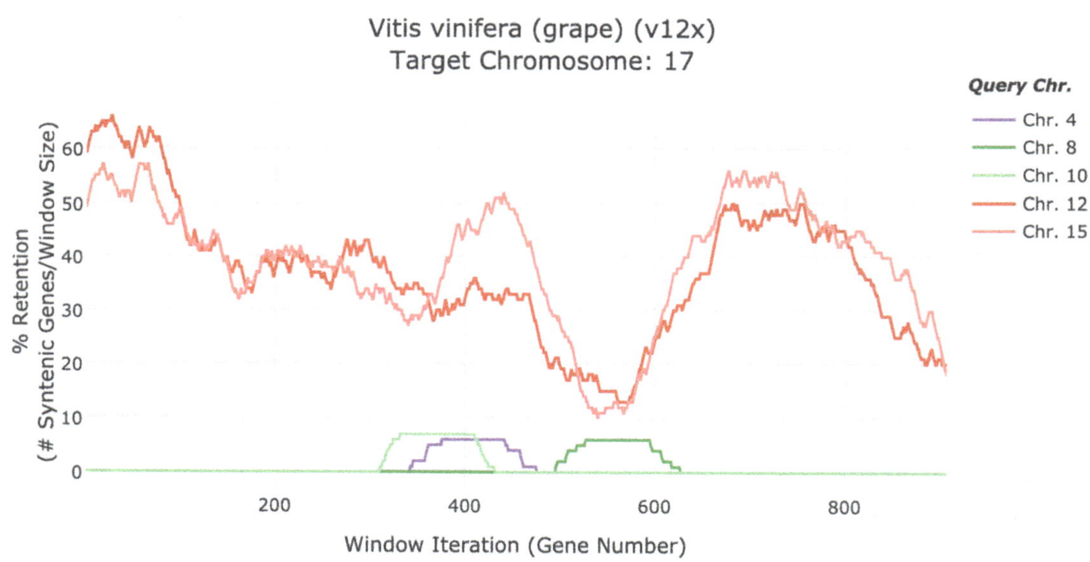

Fig. 2 FractBias output demonstrating the duplicated nature of the *Populus trichocarpa* genome relative to grapevine, *Vitis vinifera*, which has not experienced further polyploidy events past the *gamma* paleohexaploidy common to core eudicots. A 2:1 polyploid level relationship between *Populus* and *Vitis* was prespecified for SynMap via Quota Align settings. *Populus* chromosomes 12 and 15 jointly show syntenic homology as subgenomes following a lineage-specific tetraploidy event with *Vitis* (here, chromosome 17), representing the target genome lacking that event. Fractionation differences between *Populus* chromosomes 12 and 15 (in terms of percent retention of syntenic genes in search windows; see Methods) display little bias following the Salicaceae-specific WGD event, a possible autopolyploidization. The highly fractionated pieces of *Populus* chromosomes 4, 8 and 10 represent additional blocks of that genome that are also syntenically homologous to *Vitis* chromosome 17 via the ancient *gamma* event, with its resultant triplicate genome structure. *Populus* chromosomes 8 and 10 themselves have large, duplicated regions that are syntenic to *Vitis* chromosomes 1 and 5

uncover what appear to be the best-supported intergenomic syntenic levels. FractBias has the additional utility of rather simply and intuitively displaying blockwise relationships between genomes compared, regardless of their fractionation differences, making simple detection of ploidy level differences easier than tracing block:block homologs across a pairwise SynMap syntenic dotplot. Furthermore, due to the species-specific nature of the fractionation process, detection of polyploid block relationships solely through self-genomic comparisons is very limited; use of nonself comparator genomes will almost always uncover more evidence for syntenic homology than internal-only analyses possibly could.

2.2.3 GEvo

The GEvo microsynteny viewer links to SynMap and other CoGe tools [1, 19]. Microsynteny represents local syntenic gene order, whereas macrosynteny refers to larger blockwise relationships within and among genomes. Clicking interactively on SynMap syntelog pairs will call up a GEvo window that can be examined for local synteny patterns, in a genome-browser-like format. Further homologous syntenic regions can be brought into such views (e.g., via CoGeBlast searches) to obtain complete pictures of comparative genome evolution in particular regions. Connectors between sequence similarity hits can be called up to better visualize local syntenic patterns (Fig. 3), including inversions that distinguish two genomic blocks.

2.2.4 FeatView and CoGeBlast

FeatView allows the calling up of features by their ID and species and genome version in order to be used for GEvo comparisons or to start up SynFind runs. CoGeBlast performs various sequence similarity searches on multiple genomes of choice based on either externally supplied sequences or sequences obtained from FeatView.

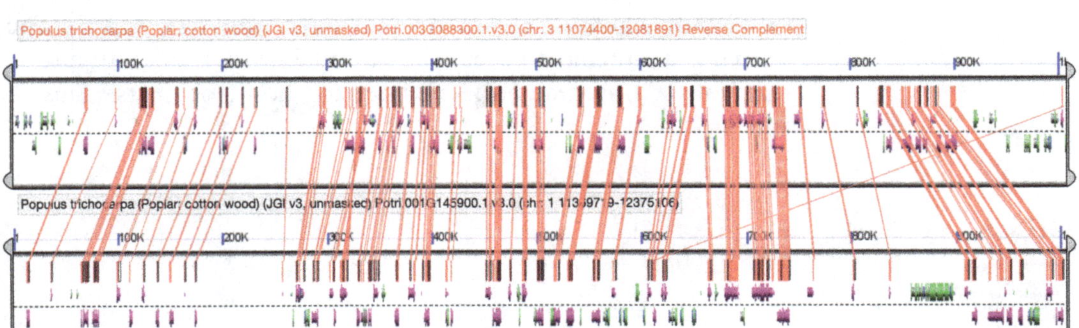

Fig. 3 GEvo microsyntenic view of a portion of *Populus trichocarpa* chromosomes 1 and 3 (see the syntenic chain at lower left of Fig. 1, of which this view is a part). While completely linear synteny is observed in these paralogous regions, each block, representing one of the two subgenomes from the Salicaceae-specific paleotetraploidy, has fractionated to a different extent. Syntenic gene model pairs are shown in purple, whereas gene models distinct to one paralogous block versus the other are shown in green

2.2.5 SynFind

SynFind uses a feature ID, obtained, e.g., from FeatView or CoGe-Blast, and its host genome, to search for homologous syntenic blocks in the same or multiple other genomes using sequence similarity searches in user-specified window parameters [20]. Output includes homologous syntelogs discovered or syntenic regions from which former syntelogs have fractioned (been deleted) out. SynFind also provides a genome-wide syntenic depth calculation, given a reference genome block and its anchoring feature ID, based on numbers of syntenic genes discovered at various depth levels from 0 (none syntenic), 1 (orthologous hits between genomes), to 2 or more (signifying paralogous synteny based on polyploid blocks). Since SynFind syntenic depth calculations operate across entire genome-wide comparisons, suggestions of polyploidy states are more reliable than would be obtained from analyzing copy number status for given (or even a suite of) gene families extracted from homologous coding sequence (CDS) collections obtained, e.g., using clustering methods such as Orthofinder [21].

3 Methods

3.1 SynMap: Basic Default Operation with Ks Calculation

The first step is to select genomes for comparison under the Select Organisms tab (Fig. 4). Genomes can be uploaded from external sources if not publicly available in this list; they can even be set within CoGe to be held privately among given investigators. It is recommended to run annotated genomes against each other (specifying "CDS" if available), but one annotated genome can be run against an unannotated genome sequence (specifying "genomic"; note that in this mode, no Ks calculations can be obtained). To execute Ks calculations with default settings on genomes with coding sequence annotations, the only parameter that needs to be set is "Calculate syntenic CDS pairs and color dots: Synonymous (Ks)" in the CodeML box in the Analysis Options box (Fig. 5). The output is a syntenic dotplot (Figs. 6 and 7) with a coloration scale matching a histogram of Ks values for homolog pairs (Figs. 8 and 9).

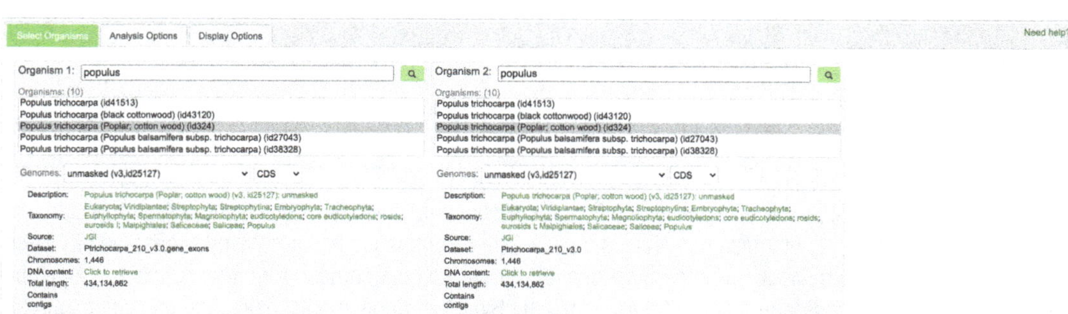

Fig. 4 Step 1—specify genomes

Fig. 5 Step 2—specify Ks in CodeML menu

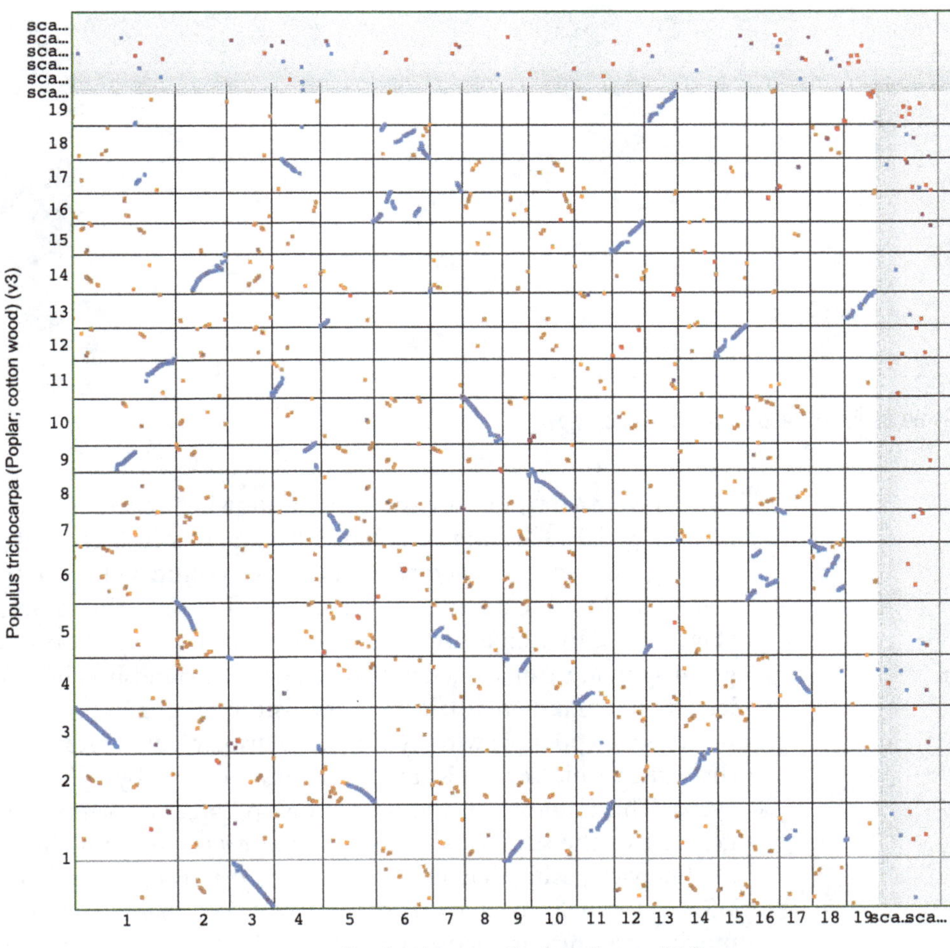

Fig. 6 Syntenic dotplot in SynMap2 with Ks coloration for *Populus* against itself. The blue syntelog chains represent the Salicaceae-specific WGD, while orange represents pairs descending from the *gamma* hexaploidy

Fig. 7 Same as Fig. 6, above, but in legacy view

SynMap2 dotplots are interactive, allowing zooming and panning (Fig. 10). Clicking on a syntenic gene pair in legacy SynMap viewing calls up a zoomed view of the two contigs under comparison (Fig. 11), including a Ks histogram for only this block pair (Fig. 12). GEvo links are called up in SynMap2 by clicking on a given syntelog pair to generate a red circle around it (Fig. 10); in legacy view, a red crosshairs appears instead (Fig. 13).

It is also possible with SynMap2 to display only syntenic chains that exist within certain Ks ranges, in other words, by a given WGD event. This is shown in the next SynMap2 analysis comparing the genome of the sucker fish *Myxocyprinus asiaticus* against itself. In the first view, using default settings, the most-recent sucker-specific WGD is visualized as long, orange syntelog chains, whereas the much more ancient teleost-specific polyploidy event, which is much more rearranged and fractionated, appears in green at higher Ks

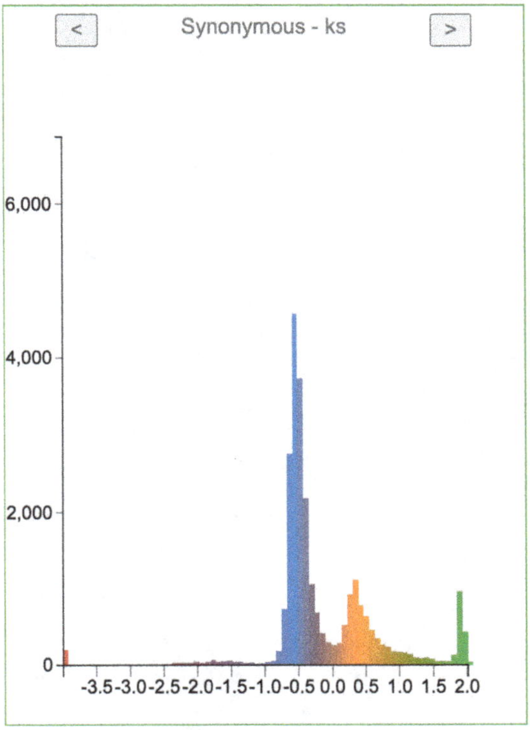

Fig. 8 SynMap2 Ks histogram showing numbers of syntelog pairs (*y*-axis) at particular log[10] values for Ks (*x*-axis). Events matching colorations are as noted in Fig. 6. The green peak to the far right represents erroneous (irrational) Ks calculations resulting from poor syntelog alignments (sometimes also saturated Ks values); such peaks almost always appear for any genomic comparison, for example since gene models are characteristically of varying quality in any annotation. The red peak to the far left, at extremely low log [10] Ks, represents nearly identical haplotigs retained in the otherwise haploid assembly. Note that such red gene pairs detected by SynMap as syntelogs tend to lie near the diagonal, and are therefore directly proximal in the assembly. Genome assemblers often compress haplotigs adjacent to each other, and SynMap can be fooled into detecting these nearly contiguous regions as internal synteny. SynMap can in fact be a useful tool to assess the extent of diploidy in assemblies and annotations. Finally, the very low and broad (purplish) peak to the left of the sharp (blue) Salicaceae WGD likely represents further allelic hits, with the log[10] Ks at the mode representing the haplotype split distance (and time)

values (Fig. 14). Selecting only a particular Ks range in the interactive histogram, i.e., the orange peak, permits recovery of a view restricted to the sucker-specific WGD (Fig. 15).

3.2 FractBias: Visualizing Syntenic Depths and Subgenome Biases

FractBias is part of the SynMap pipeline, as an add-on option available under the Analysis Options tab. FractBias was developed by the CoGe team specifically to visualize fractionation differences—potential biases—between polyploid subgenomes in a

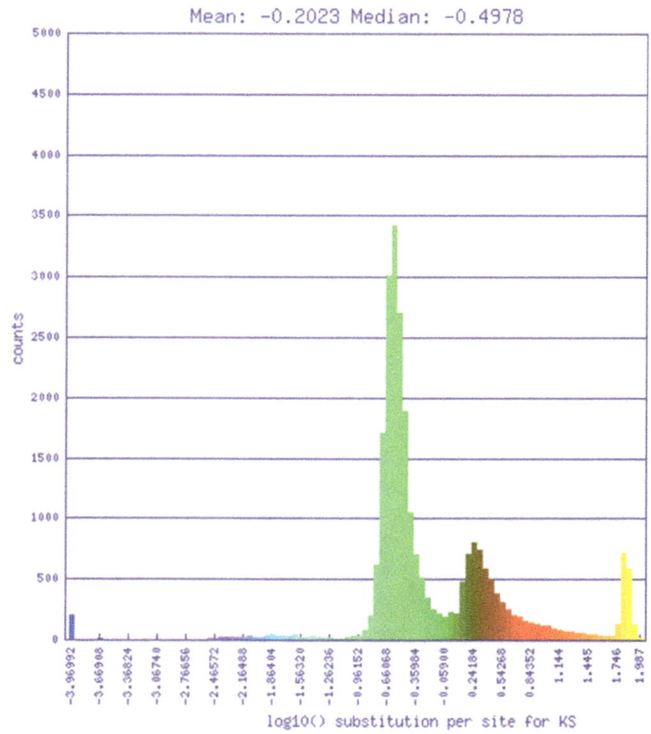

Fig. 9 Same as Fig. 8 above, but in legacy view

query genome, as compared to a reference genome similar enough for syntenic homology to be detected against [12]. This is accomplished through a sliding window analysis of syntelogs between the two genome assemblies, looking for percent retentions per given window of such syntelogs in the query genome relative to the target. FractBias requires that the investigator input a target syntenic depth relationship between the query genome (which need not be annotated) and the reference genome (which must be annotated with CDSs to perform properly); this is executed using the "Quota Align" option [22] under the "Syntenic Depth" menu. Other than modifying the values for "Max query chromosomes" and "Max target chromosomes (to account for chromosome number and contiguity differences in the assemblies under comparison), default settings are usually sufficient to obtain useful views. The default windowing option is "Use all genes in target genome"; this provides the most "life-like" syntenic mappings between query and target, but using "Use only syntenic genes in target genome (increases fractionation signal)" does exactly the latter, while sometimes sacrificing the spatial realism of target genome chromosome sizes and gene numbers. For details on the differences between these two settings, see Joyce et al. (2017), their Supplementary Figures 1, 2, and 3 [12].

Fig. 10 Zoom and pan view of *Populus*:self, chromosomes 1 and 3, showing the same syntenic chain (blue) figured by GEvo in Fig. 3. Small inversions are visible as slope shifts along the diagonal to the right of the vertical line. The small red circle highlights a spot on an ancient, orange *gamma* block where a GEvo view can be called up

As an example, the family containing the sucker fish *Myxocyprinus asiaticus*—the Catostomidae—has experienced two polyploidy events in its evolutionary history since its split with the gars, Lepisosteiformes [23]. The first polyploidization event (3R), which occurred approximately 320 Mya, is shared by all teleost fish [24–27]. Like the salmonid (salmon) [28, 29] and cyprinid (carp) [30, 31] lineages, sucker fishes are known to have undergone an additional ancient WGD [23]. FractBias can be readily used to illustrate both of these events.

Here (Fig. 16), we have adjusted the numbers of query and target chromosome numbers from default settings to account for numbers in the species at hand (suckers have 50 haploid chromosomes [23], gar has 29 [32]). The Quota Align setting used specifies the expected proportions of 4 parts *Myxocyprinus* to 1 part *Lepisosteus* to account for four syntenic sucker blocks resulting from two WGD events in suckers, while no further polyploidies in gars following the split between the two lineages.

Fig. 11 Same as previous, legacy view

The anticipated 4:1 relationship between sucker and gar is immediately apparent as two pairs of sucker chromosomes mapping broadly across a given gar chromosome (Fig. 17). Each pair—originally a single 3R subgenome, now paired due to the later sucker-specific WGD—descends from the ancient 3R event, which has been argued to result from a broad allopolyploidy event [33]. This notion is supported here, given that the two sucker chromosome pairs differ greatly between themselves in terms of fractionation percentage relative to gar (Fig. 17). Similar patterns of extensive fractionation bias, or subgenome dominance [14], have been described across many plants, with one explanation for the phenomenon being (as described above) differential epigenetic transcriptional masking in parental species. The argument is that such biases may be set up extremely early after allopolyploidy events. In contrast, largely unbiased fractionation patterns between subgenomes may reflect autopolyploidy events, allopolyploids that formed between very close relatives, or extreme event recency. As noted above, there are two pairs of sucker chromosomes broadly mapping to the single gar reference chromosome, and the fractionation differences between the two members of each pair are very slight. Therefore, it is possible that the sucker-specific WGD event

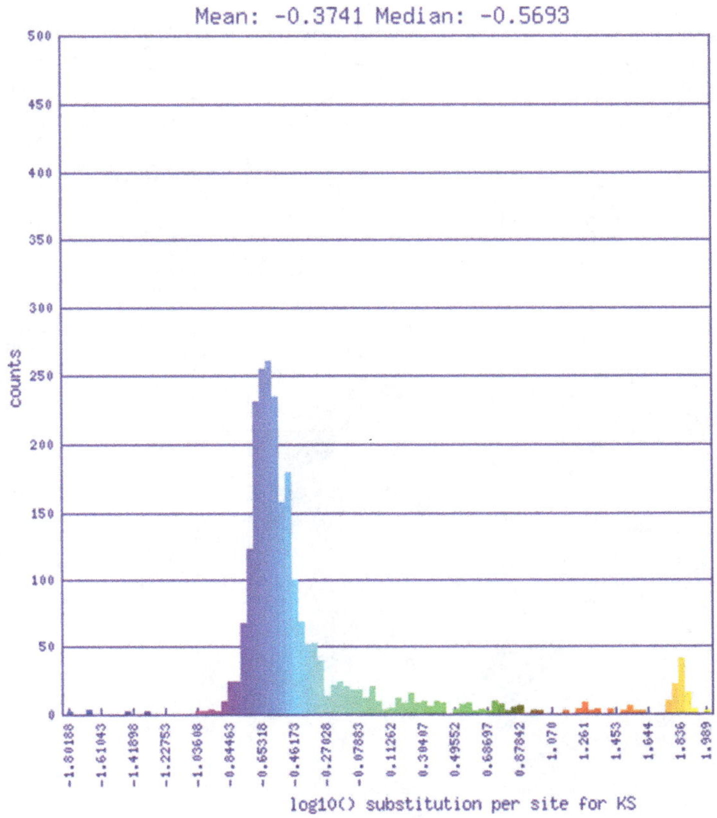

Fig. 12 Ks histogram specific to the zoom-in above for syntelogs in *Populus* chromosomes 1 and 3

was an autopolyploidy or close allopolyploidy event, rather unlike the likely broad cross that gave rise to the doubled genomes (through 3R) of all teleosts.

As noted above, using the windowing option "Use only syntenic genes in target genome (increases fractionation signal)" can provide more strongly syntenic views, but at the cost of scaling the *x*-axis more realistically via windowing based on all genes in the target genome, whether or not found as syntenic between query and target (Fig. 18).

3.3 GEvo: Examining Microsynteny Among Genomic Blocks

GEvo permits close, genome-browser-like views of syntenic blocks within or among species [3, 19]. To call up a pairwise GEvo view, one can click on SynMap results as in Figs. 10 and 13 (Fig. 19). It is also possible to go to GEvo and input one's own genomic feature selections. The clearest way to generate microsyntenic views is to turn off noncoding sequences under the drop down "Mask Sequence:" under the "Sequence Submission" tab.

Note that it is possible to change the size of the view using an entry into "Apply distance to all submissions:" At the upper right, it is possible to add GEvo views together into multiway syntenic

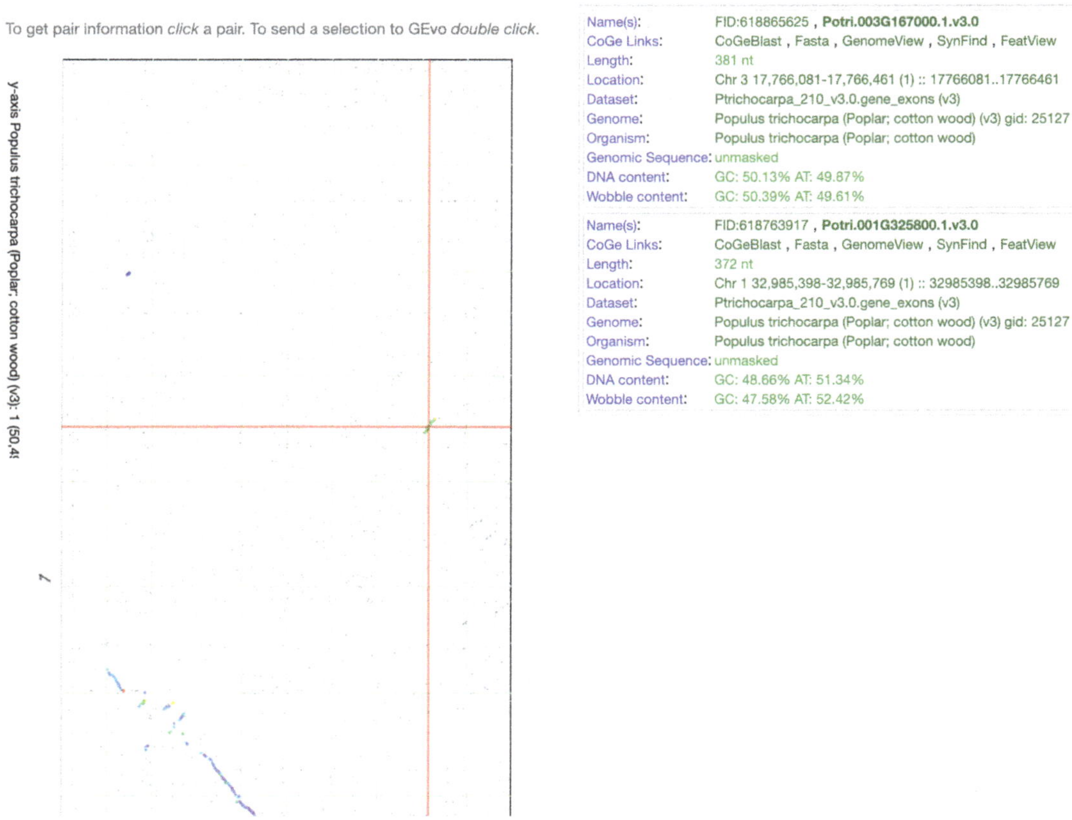

Fig. 13 Calling up a GEvo view in the legacy zoom-in viewer. The red cross-hairs highlight the syntelog pair at upper right

graphics by entering their tiny-URLs into "Merge Previous GEvo Analysis (paste in URL):" A further set of useful options is available under the "Results Visualization Options" tab (Fig. 20).

Here, it is useful to turn off coloration of masked and unsequenced nucleotides to decomplicate microsyntenic views. Additionally, cleaner views of BLAST HSPs are obtained by clicking the "Draw all HSPs on top?" button. In terms of the "Algorithm" tab, default settings are usually appropriate. The view generated from the settings in Figs. 19 and 20 is shown in Fig. 21. To activate connector lines or wedges, one can click on BLAST HSPs individually or hold down shift and click one HSP to highlight all. It is also possible to drag a rectangle around a series of HSPs to highlight them.

A further useful feature for fine-tuning microsyntenic view (in fact already used for Fig. 19; see the "Left sequence:" and "Right sequence:" entries and how they differ from the "Apply distance to all submissions:" entry). Using the slide bars, it is possible to change the size of the view further (Fig. 22), e.g., to provide a zoomed-in depiction (Fig. 23).

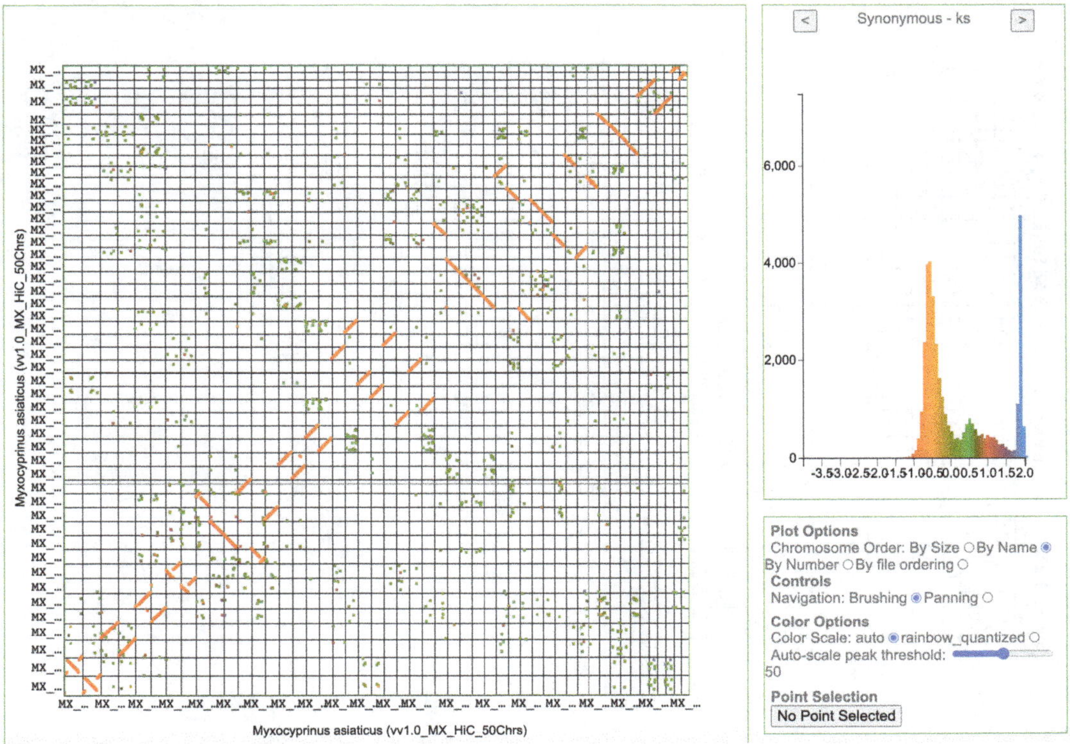

Fig. 14 Self SynMap2 view of the *Myxocyprinus asiaticus* genome. Similar to the poplar genome self-comparison above, irrational Ks calculations for, e.g., poorly aligned syntelogs come into view here as the blue peak to far right

Another useful option is to highlight overlapping features in purple (as opposed to the green default gene model color) between syntenic blocks (Fig. 24). This option nicely shows the extent of fractionation following a WGD event.

3.4 SynFind: Discovery of Blocks of Genomes That Are Syntenic to a Window Containing a Target Gene

SynFind is useful for discovery of syntenic blocks either within or between genomes when an investigator already has a feature of interest to input and compare [20]. SynFind will also simultaneously calculate syntenic depth between genomes compared, based on the given query. Syntenic depth is a numerical measure of the duplication status of a genome that can be useful as an adjunct to other methods in determining polyploid status.

For this exercise, we start with the *Arabidopsis LEAFY* gene, a well-known hub-like regulator in the flowering developmental program [34]. *LEAFY* is typically a single-copy gene in most studied angiosperms; experiments in *Arabidopsis* have shown that inserting different numbers of *LEAFY* copies changes flowering time [35], so it seems likely that selection favoring reproductive success usually limits the number of *LEAFY* copies a species can have [36, 37]. As such, after a WGD event, the likely result would be to rapidly lose extra *LEAFY* copies through the fractionation process (unless sub- or neofunctionalization of copies occurs).

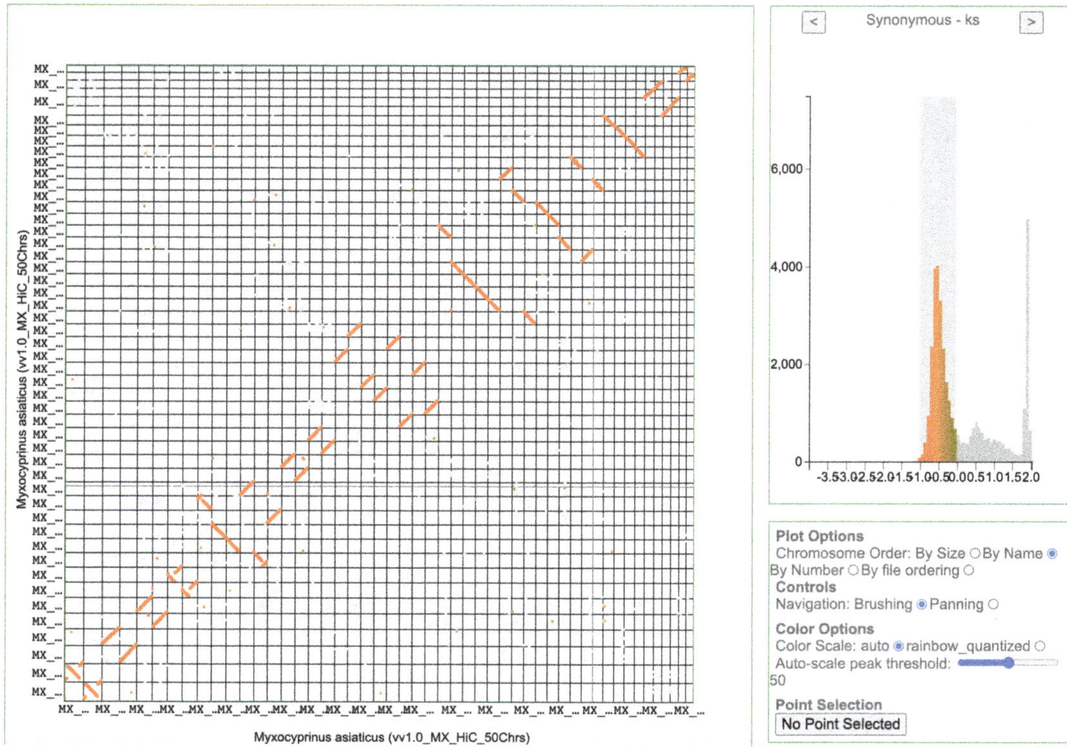

Fig. 15 Self:self SynMap2 view of the *Myxocyprinus asiaticus* genome, selecting only the recent most, sucker-specific WGD for colorized display

Fig. 16 FractBias settings

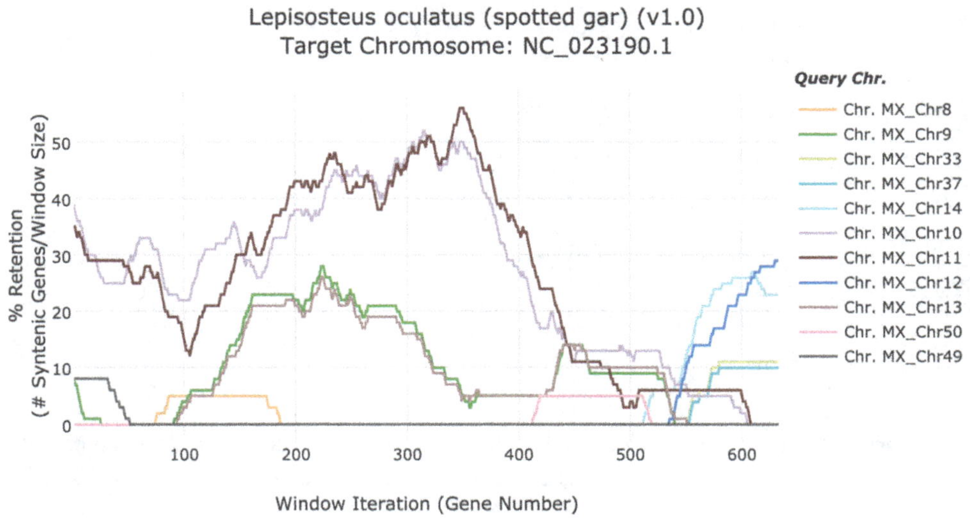

Fig. 17 FractBias view of multiple mappings of *Myxocyprinus asiaticus* chromosomes onto a single spotted gar chromosome, using the default "Use all genes in target genome" option. Sucker chromosomes 10 and 11 represent one-half of the 3R event and are descendants of the sucker-specific WGD. They retain, following the postpolyploidy diploidization (fractionation) process, considerably more genes relative to gar as compared to the chromosome 8 and 13 pair, which are also sucker-WGD descendants, but representing the second 3R subgenome

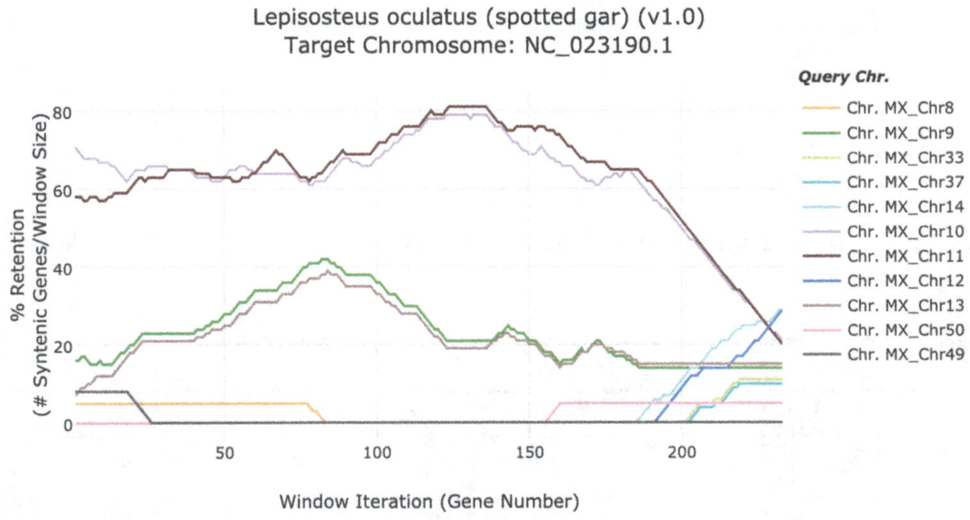

Fig. 18 FractBias view of multiple mappings of *Myxocyprinus asiaticus* chromosomes onto a single spotted gar chromosome, using the "Use only syntenic genes in target genome (increases fractionation signal)" windowing option. In comparison to Fig. 17, the *x*-axis is scaled by windows that include only genes syntenic between both query and target genomes

Using CoGe's FeatView, one can call up a stored feature, which can be a gene name or a gene ID (Fig. 25). Under "CoGe Links:" in the bottom field, one can then press "SynFind" to start the block-searching process across multiple genomes. A default SynFind search considers LAST searches of windows of 40 genes

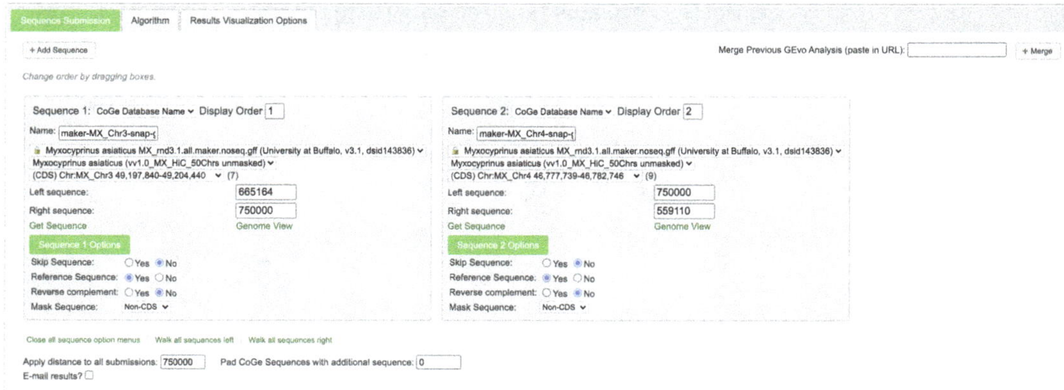

Fig. 19 GEvo view for 2 homeologous blocks of the sucker-specific WGD, as observable in the *Myxocyprinus* genome. This page stems from clicking on one of the highly syntenic orange chromosome pairs in Fig. 15

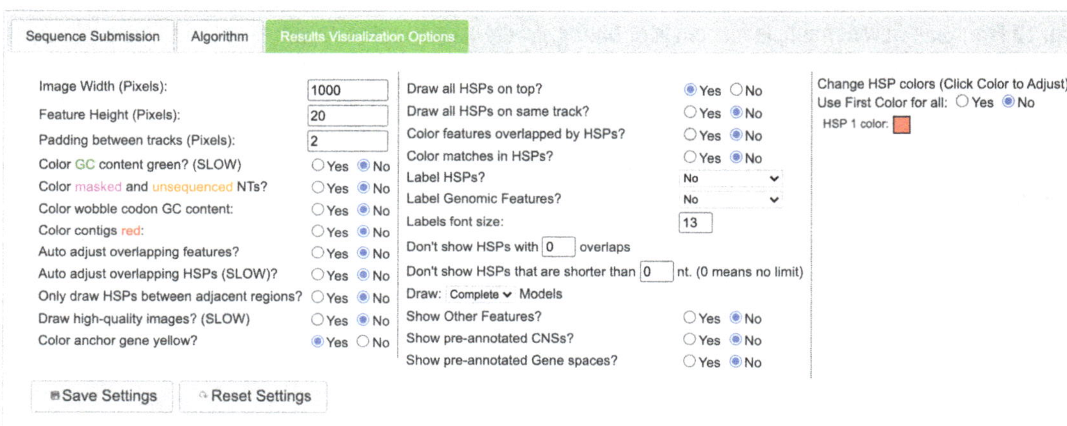

Fig. 20 GEvo view for 2 homeologous blocks of the sucker-specific WGD; the "Results Visualization Options" tab

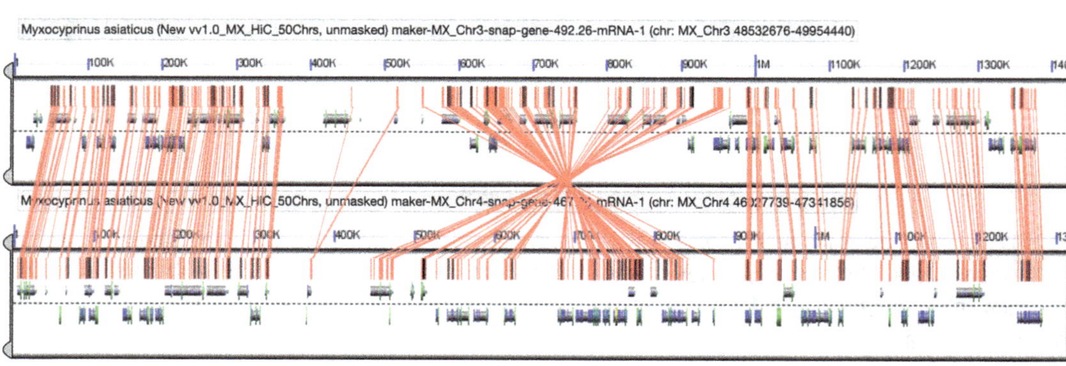

Fig. 21 GEvo view for 2 homeologous blocks of the sucker-specific WGD—note the large inversion at center distinguishing the two homeologous chromosomes

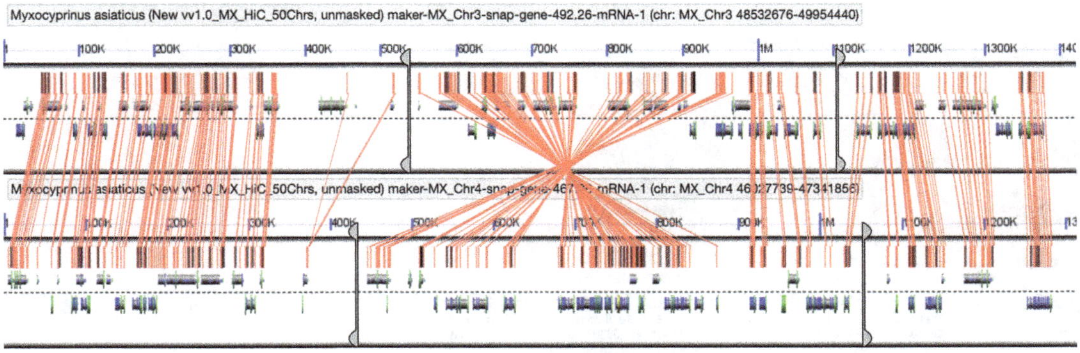

Fig. 22 GEvo view for 2 homeologous blocks of the sucker-specific WGD—the slide bars have been pulled into place to generate the next view, delimited by them

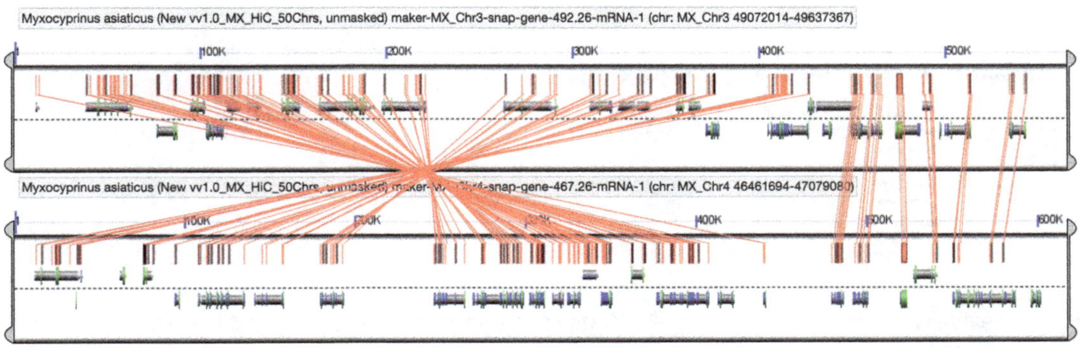

Fig. 23 GEvo view for 2 homeologous blocks of the sucker-specific WGD—new, zoomed-in view after using the slide bars and restarting the analysis

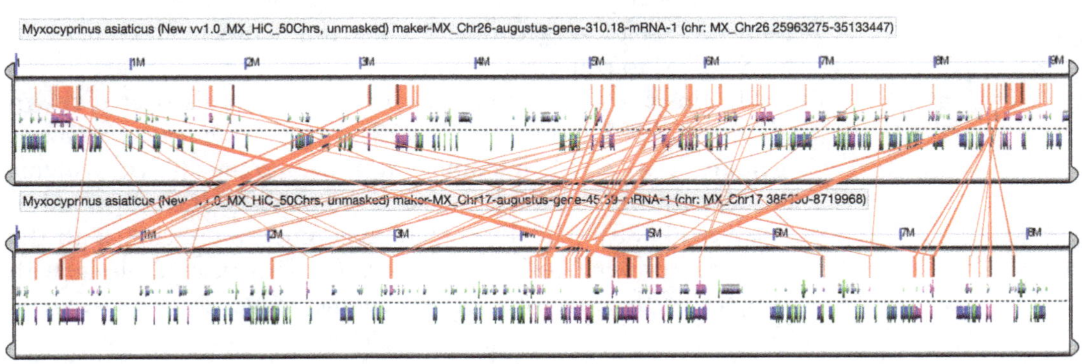

Fig. 24 GEvo view for two homeologous blocks of the ancient 3R WGD in *Myxocyprinus*; this view was called up by clicking on one of the homeolog pairs containing fragmented green syntelog chains in Fig. 14. Clicking on "Color features overlapped by HSPs?" in the "Results Visualization Options" tab colors overlapping gene models purple instead of green. Clearly, many genes in green have moved in or out of this highly fractionated and fragmented block pair since the ancient 3R event that generated it

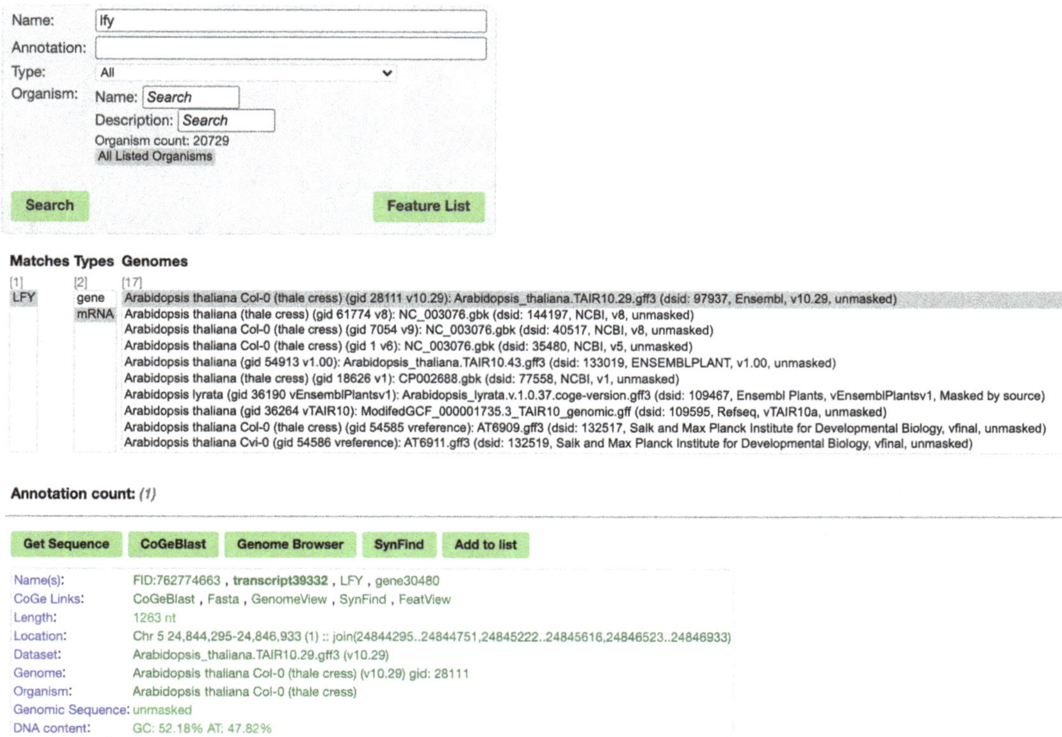

Fig. 25 FeatView window where the query "LFY" (the abbreviation for the *LEAFY* gene) has been entered. We then chose "mRNA" in the drop down

wherein four syntelogs are required for syntenic block discovery. The investigator then must enter the target genomes for the search using "Select Target Genomes"; these must contain CDSs in order for SynFind to run.

In this example, we first use CoGeBlast to discover the *Vitis vinifera LEAFY* homolog (Fig. 27), since *Arabidopsis thaliana*'s genome is so small, fractionated, and rearranged that syntenic blocks between it and other species can be difficult for SynFind to discover [1, 38]. The *Vitis* genome is much more similar to the ancestral core eudicot's in structure (as shown, e.g., along with analyses of the Robusta coffee genome sequence and elsewhere [39, 40]).

We first select the *Vitis* genome for search under "Select Target Genomes." Pressing on "CoGeBlast" under "CoGe Links:" brings up the BLAST field shown in Fig. 26.

Searching the *Vitis* genome with default settings yields 3 BLAST HSPs that hit the same genomic feature (Fig. 27).

Clicking on one of the "Closest Genomic Features" brings up the feature metadata and links window, as shown in Fig. 28.

The investigator may now press the "SynFind" link to initiate a search, first by calling up a SynFind window wherein target genomes can be selected for analysis (Fig. 29).

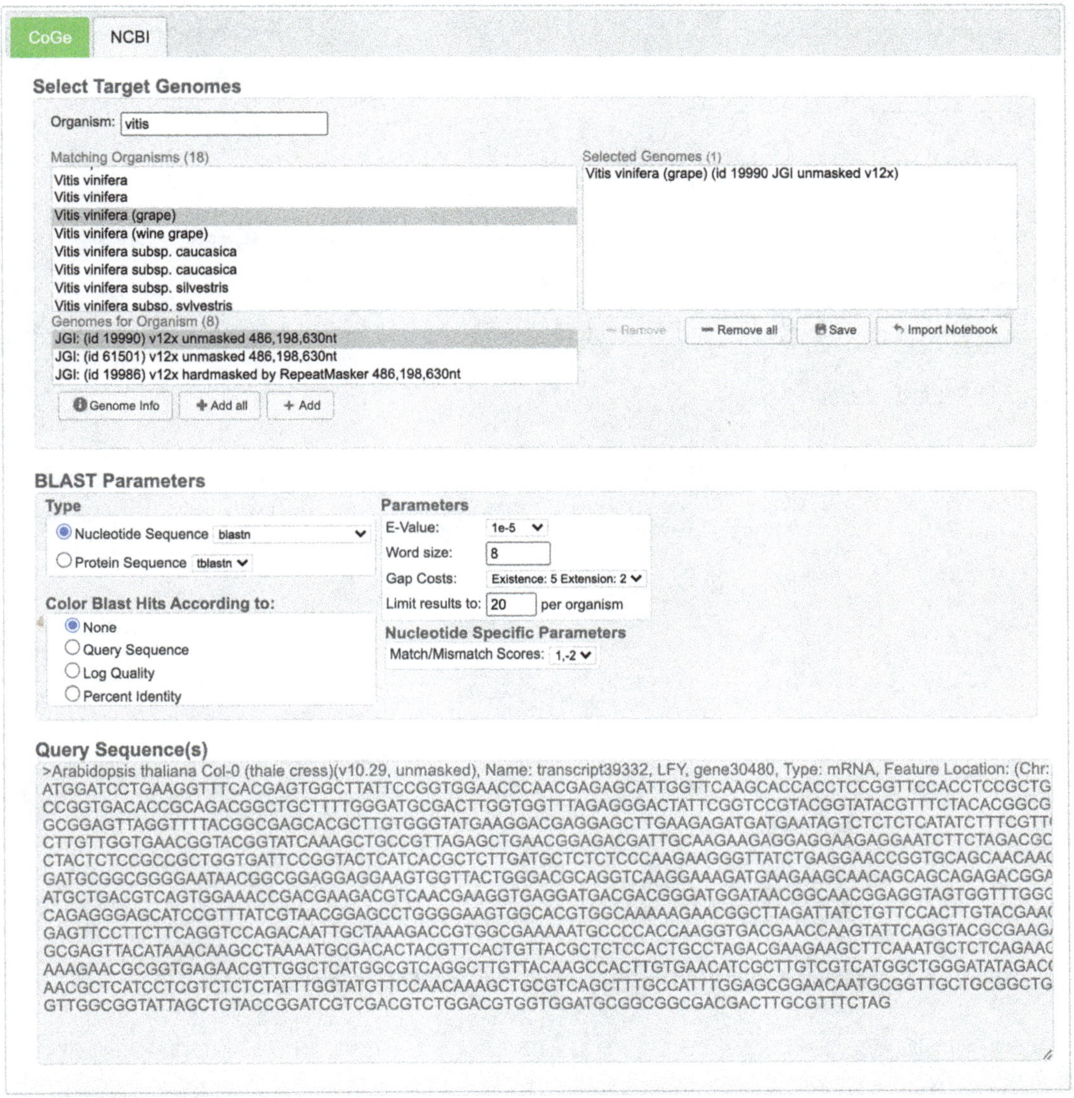

Fig. 26 CoGeBlast search menu including the *LEAFY* query sequence

In this example, we expect to discover only one *LEAFY* gene in *Arabidopsis*, although the diploid and recent *Brassica* polyploids selected [41–46] could contain more, e.g., if the latter have not existed long enough to ensure complete fractionation of extra copies.

In the SynFind results field (Fig. 30), we first see the homologous *Vitis* block used as query. Then, we see that two syntenic *Arabidopsis* blocks were discovered: one containing the *LEAFY* syntelog, and another annotated with a genomic position of type "proxy for region"; this indicates discovery of a syntenic block that is missing a *LEAFY* syntelog (following that particular copy's fractionation), but that is nonetheless homologous enough in terms of

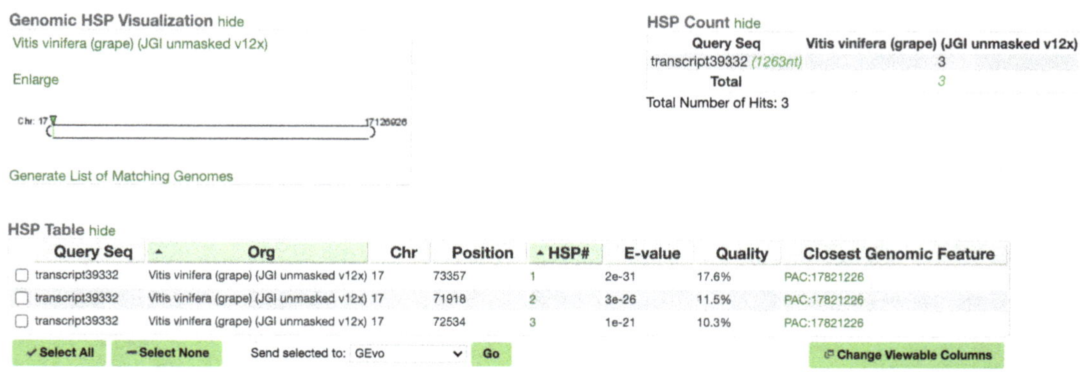

Fig. 27 CoGeBlast results

Fig. 28 Feature metadata and links window for the *Vitis vinifera LEAFY* homolog

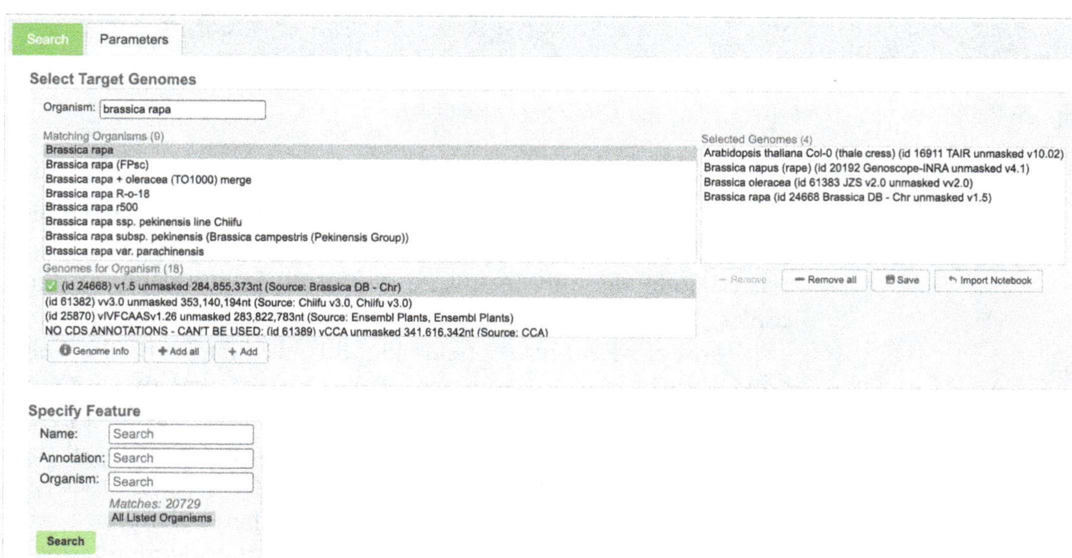

Fig. 29 Brassicaceae genomes entered from "Select Target Genomes" window

Organism	Genome	Type	Name	Chr	Synteny Score	SynMap
Vitis vinifera (grape)	(v12x unmasked)	query	PAC:17821226	17	0	Dotplot
Arabidopsis thaliana Col-0 (thale cress)	(v10.02 unmasked)	syntelog	AT5G61850.1	5	10	Dotplot
Arabidopsis thaliana Col-0 (thale cress)	(v10.02 unmasked)	proxy for region	pos 20739453	5	6	Dotplot
Brassica napus (rape)	(v4.1 unmasked)	proxy for region	pos 14795887	A06	6	Dotplot
Brassica napus (rape)	(v4.1 unmasked)	proxy for region	pos 36944084	C03	5	Dotplot
Brassica napus (rape)	(v4.1 unmasked)	syntelog	GSBRNA2T00008696001	Cnn_random	5	Dotplot
Brassica oleracea	(vv2.0 unmasked)	syntelog	BolC03g059680.2J.m1	C03	10	Dotplot
Brassica oleracea	(vv2.0 unmasked)	syntelog	BolC02g060500.2J.m1	C02	8	Dotplot
Brassica rapa	(v1.5 unmasked)	syntelog	Bra019619	A06	10	Dotplot
Brassica rapa	(v1.5 unmasked)	syntelog	Bra029305	A02	8	Dotplot

Links
Regenerate this analysis: https://genomevolution.org/r/1jezg
Compare and visualize region in GEvo
Generate all dotplots
Generate Feature List
Generate Fasta Sequences
Generate master gene set table
Generate master gene set table (top one syntenlog per organism)

Fig. 30 SynFind results field

surrounding genes to be identified as descending from its most recent lineage-specific WGD event. Three syntenic blocks—one containing a syntelog—were discovered in *Brassica napus*, an allopolyploid [44], but two *LEAFY* syntelogs each were discovered in the *B. oleracea* and *B. rapa* assemblies used.

In terms of syntenic depth, SynFind provides tabular output for each target genome.

Depth 0 indicates no syntenic genes, depth 1 may largely represent orthologous synteny, depth 2 can suggest paralogous synteny descending from a WGD, and so on. In this example (*Brassica napus*, Fig. 31), depth 2 represents a maximum percentage (9.35%) of retained syntelogs, while depth 4 represents a close second place at 8.04% of total unique genes. In between, at depth 3—wherein triplicate syntelogs still occur after the *Brassica napus* allopolyploidy event—there are still 6.67% of total genes discovered by SynFind. Thereafter, syntenic depths possibly suggestive of further WGD events decrease. Some of these likely reflect the ancient *gamma* hexaploidy event underlying all core eudicots, but the principal data are discovery of a Brassicales-specific WGD and the most recent allotetraploidy event underlying *Brassica napus*.

The investigator can then call up a GEvo view of all of the blocks discovered by clicking on the "Compare and visualize region in GEvo" link shown in Fig. 30. Figure 32 shows the resultant GEvo view of the reference block in *Vitis vinifera*; each row of differently-colored BLAST HSPs reflects the search results in Fig. 30 in the same top-to-bottom order. Considerable synteny around the *LEAFY* syntelogs is observed across species; this can be further highlighted using connectors between BLAST HSPs (Fig. 33).

Brassica napus (rape): Science Publication: http://www.sciencemag.org/content/345/6199/950.full (v4.1, id20192): unmasked
Query genome unique genes
Target genome unique genes

Depth 0	8084	of 26346 30.68%
Depth 1	1473	of 26346 5.59%
Depth 2	2464	of 26346 9.35%
Depth 3	1756	of 26346 6.67%
Depth 4	2119	of 26346 8.04%
Depth 5	1517	of 26346 5.76%
Depth 6	1653	of 26346 6.27%
Depth 7	1496	of 26346 5.68%
Depth 8	1339	of 26346 5.08%
Depth 9	1135	of 26346 4.31%
Depth 10	1007	of 26346 3.82%
Depth 11	716	of 26346 2.72%
Depth 12	546	of 26346 2.07%
Depth 13	324	of 26346 1.23%
Depth 14	282	of 26346 1.07%
Depth 15	164	of 26346 0.62%
Depth 16	135	of 26346 0.51%
Depth 17	69	of 26346 0.26%
Depth 18	33	of 26346 0.13%
Depth 19	17	of 26346 0.06%
Depth 20	8	of 26346 0.03%
Depth 21	4	of 26346 0.02%
Depth 22	3	of 26346 0.01%
Depth 23	1	of 26346 0.00%
Depth 24	1	of 26346 0.00%

Fig. 31 Syntenic depth for *Brassica napus*, as determined against *Vitis vinifera* as reference

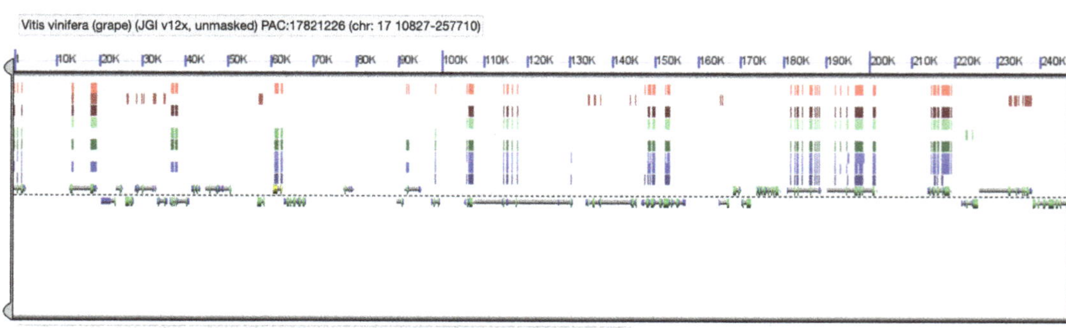

Fig. 32 GEvo view of the *Vitis* reference block used in the *LEAFY* SynFind search

4 Notes

In summary, the CoGe software suite provides a broad series of tools for studying paleopolyploidy in plant and animal genomes. In addition to the primary tools covered here, additional analyses or visual explorations can be conducted on the site but are not covered here for brevity. The genome-browser style visualization tool, EPIC-CoGe, permits exploration of genomes and features from chromosomes to base-pair resolution through a JBrowse application [47]. Custom tracks can be loaded and visualized on the browser. Additional detailed tools are available and can be explored

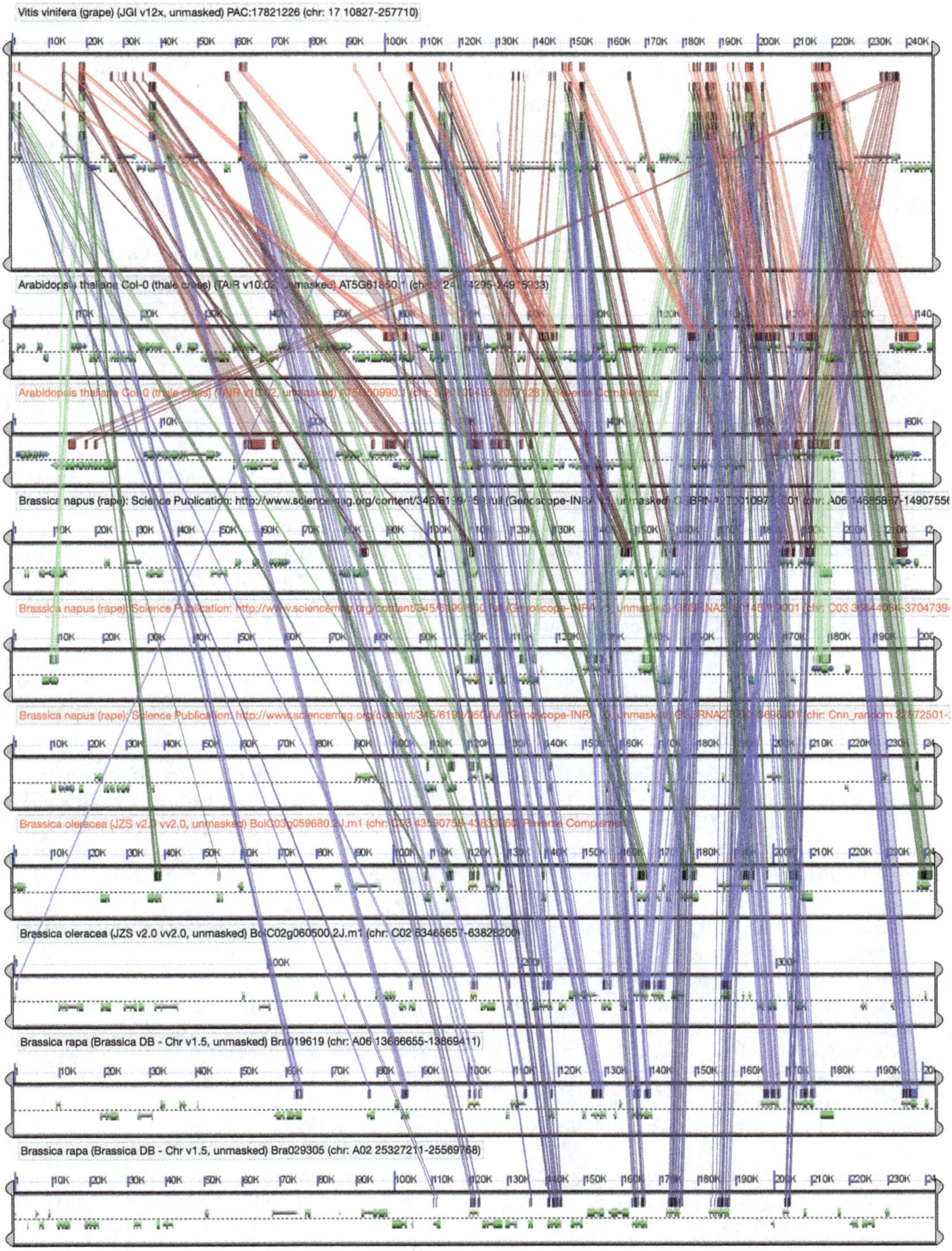

Fig. 33 GEvo view of syntenic blocks around *LEAFY* homologs (or blocks which formerly contained a *LEAFY* homolog prefractionation) in the *Vitis*, *Arabidopsis*, and *Brassica* genomes

on CoGe itself or through the helpful online guide, CoGepedia (https://genomevolution.org/wiki/index.php/). In combination with other tools outlined in this volume, CoGe contributes to a robust understanding of the detection and dynamics of genome evolution following polyploidy.

References

1. Lyons E et al (2008) Finding and comparing syntenic regions among Arabidopsis and the outgroups papaya, poplar, and grape: CoGe with rosids. Plant Physiol 148(4):1772–1781
2. Lyons E et al (2008) The value of nonmodel genomes and an example using SynMap within CoGe to dissect the hexaploidy that predates the rosids. Trop Plant Biol 1(3):181–190
3. Joyce B et al (2017) Comparative genomics using CoGe, hook, line, and sinker. Bioinformatics in aquaculture: principles and methods. Wiley, Hoboken
4. Haug-Baltzell A et al (2017) SynMap2 and SynMap3D: web-based whole-genome synteny browsers. Bioinformatics 33(14):2197–2198
5. Lyons E et al (2011) Using genomic sequencing for classical genetics in E. coli K12. PloS One 6(2):e16717
6. Ibarra-Laclette E et al (2013) Architecture and evolution of a minute plant genome. Nature 498(7452):94–98
7. Tuskan GA et al (2006) The genome of black cottonwood, Populus trichocarpa (Torr. & Gray). Science 313(5793):1596–1604
8. Soltis DE et al (2009) Polyploidy and angiosperm diversification. Am J Bot 96(1): 336–348
9. Chanderbali AS et al (2017) Evolution of floral diversity: genomics, genes and gamma. Philos Trans R Soc B Biol Sci 372(1713):20150509
10. Jiao Y et al (2012) A genome triplication associated with early diversification of the core eudicots. Genome Biol 13(1):1–14
11. Jaillon O et al (2007) The grapevine genome sequence suggests ancestral hexaploidization in major angiosperm phyla. Nature 449(7161): 463–467
12. Joyce BL et al (2017) FractBias: a graphical tool for assessing fractionation bias following polyploidy. Bioinformatics 33(4):552–554
13. Sankoff D, Zheng C (2012) Fractionation, rearrangement and subgenome dominance. Bioinformatics 28(18):i402–i408
14. Alger EI, Edger PP (2020) One subgenome to rule them all: underlying mechanisms of subgenome dominance. Curr Opin Plant Biol 54:108–113
15. Cheng F et al (2018) Gene retention, fractionation and subgenome differences in polyploid plants. Nat Plants 4(5):258–268
16. Garsmeur O et al (2014) Two evolutionarily distinct classes of paleopolyploidy. Mol Biol Evol 31(2):448–454
17. Zhao M et al (2017) Patterns and consequences of subgenome differentiation provide insights into the nature of paleopolyploidy in plants. Plant Cell 29(12):2974–2994
18. Li Q et al (2019) Unbiased subgenome evolution following a recent whole-genome duplication in pear (Pyrus bretschneideri Rehd.). Hortic Res 6(1):1–12
19. Castillo AI et al (2018) A tutorial of diverse genome analysis tools found in the CoGe web-platform using Plasmodium spp. as a model. Database:2018
20. Tang H et al (2015) SynFind: compiling syntenic regions across any set of genomes on demand. Genome Biol Evol 7(12):3286–3298
21. Emms DM, Kelly S (2019) OrthoFinder: phylogenetic orthology inference for comparative genomics. Genome Biol 20(1):1–14
22. Tang H et al (2011) Screening synteny blocks in pairwise genome comparisons through integer programming. BMC Bioinform 12(1): 1–11
23. Krabbenhoft TJ et al (2021) Chromosome-level genome assembly of Chinese Sucker (Myxocyprinus asiaticus) reveals strongly conserved synteny following a catostomid-specific whole-genome duplication. Genome Biol Evol 13(9):evab190
24. Vandepoele K et al (2004) Major events in the genome evolution of vertebrates: paranome age and size differ considerably between ray-finned fishes and land vertebrates. Proc Natl Acad Sci 101(6):1638–1643
25. Meyer A, Van de Peer Y (2005) From 2R to 3R: evidence for a fish-specific genome duplication (FSGD). BioEssays 27(9):937–945

26. Jaillon O et al (2004) Genome duplication in the teleost fish Tetraodon nigroviridis reveals the early vertebrate proto-karyotype. Nature 431(7011):946–957
27. Davesne D et al (2021) Fossilized cell structures identify an ancient origin for the teleost whole-genome duplication. Proc Natl Acad Sci 118(30)
28. Macqueen DJ, Johnston IA (2014) A well-constrained estimate for the timing of the salmonid whole genome duplication reveals major decoupling from species diversification. Proc R Soc B Biol Sci 281(1778):20132881
29. Lien S et al (2016) The Atlantic salmon genome provides insights into rediploidization. Nature 533(7602):200–205
30. Xu P et al (2014) Genome sequence and genetic diversity of the common carp, Cyprinus carpio. Nat Genet 46(11):1212–1219
31. Li J-T et al (2021) Parallel subgenome structure and divergent expression evolution of allo-tetraploid common carp and goldfish. Nat Genet:1–11
32. Braasch I et al (2016) The spotted gar genome illuminates vertebrate evolution and facilitates human-teleost comparisons. Nat Genet 48(4): 427–437
33. Conant GC (2020) The lasting after-effects of an ancient polyploidy on the genomes of teleosts. PLoS One 15(4):e0231356
34. Schultz EA, Haughn GW (1991) LEAFY, a homeotic gene that regulates inflorescence development in Arabidopsis. Plant Cell 3(8): 771–781
35. Blázquez MA et al (1997) LEAFY expression and flower initiation in Arabidopsis. Development 124(19):3835–3844
36. Sayou C et al (2014) A promiscuous intermediate underlies the evolution of LEAFY DNA binding specificity. Science 343(6171): 645–648
37. Albert VA, Oppenheimer DG, Lindqvist C (2002) Pleiotropy, redundancy and the evolution of flowers. Trends Plant Sci 7(7):297–301
38. Kaul S et al (2000) Analysis of the genome sequence of the flowering plant Arabidopsis thaliana. Nature 408(6814):796–815
39. Denoeud F et al (2014) The coffee genome provides insight into the convergent evolution of caffeine biosynthesis. Science 345(6201): 1181–1184
40. Sankoff D, Zheng C (2018) Whole genome duplication in plants: implications for evolutionary analysis. In: Comparative genomics. Springer, Humana Press, New York, NY. pp 291–315
41. Zhang L et al (2018) Improved Brassica rapa reference genome by single-molecule sequencing and chromosome conformation capture technologies. Hortic Res 5(1):1–11
42. Wang X et al (2011) The genome of the mesopolyploid crop species Brassica rapa. Nat Genet 43(10):1035–1039
43. Sun F et al (2017) The high-quality genome of Brassica napus cultivar 'ZS 11'reveals the introgression history in semi-winter morphotype. Plant J 92(3):452–468
44. Liu S, Snowdon R, Chalhoub B (2018) The Brassica napus genome. Springer, Switzerland
45. Liu S et al (2014) The Brassica oleracea genome reveals the asymmetrical evolution of polyploid genomes. Nat Commun 5(1):1–11
46. Liu S, Snowdon R, Kole C (2021) The Brassica oleracea genome. Springer, Switzerland
47. Nelson AD et al (2018) EPIC-CoGe: managing and analyzing genomic data. Bioinformatics 34(15):2651–2653

Chapter 3

Inference of Ancient Polyploidy Using Transcriptome Data

Jia Li, Yves Van de Peer, and Zhen Li

Abstract

Polyploidizations, or whole-genome duplications (WGDs), in plants have increased biological complexity, facilitated evolutionary innovation, and likely enabled adaptation under harsh conditions. Besides genomic data, transcriptome data have been widely employed to detect WGDs, due to their efficient accessibility to the gene space of a species. Age distributions based on synonymous substitutions (so-called K_S age distributions) for paralogs assembled from transcriptome data have identified numerous WGDs in plants, paving the way for further studies on the importance of WGDs for the evolution of seed and flowering plants. However, it is still unclear how transcriptome-based age distributions compare to those based on genomic data. In this chapter, we implemented three different de novo transcriptome assembly pipelines with two popular assemblers, i.e., Trinity and SOAPdenovo-Trans. We selected six plant species with published genomes and transcriptomes to evaluate how assembled transcripts from different pipelines perform when using K_S distributions to detect previously documented WGDs in the six species. Further, using genes predicted in each genome as references, we evaluated the effects of missing genes, gene family clustering, and de novo assembled transcripts on the transcriptome-based K_S distributions. Our results show that, although the transcriptome-based K_S distributions differ from the genome-based ones with respect to their shapes and scales, they are still reasonably reliable for unveiling WGDs, except in species where most duplicates originated from a recent WGD. We also discuss how to overcome some possible pitfalls when using transcriptome data to identify WGDs.

Key words Ancient polyploidy, WGD, Transcriptome assembly, RNA-Seq, K_S age distribution

1 Introduction

It is generally acknowledged that polyploidization, or whole-genome duplication (WGD), has played a significant role in the speciation, evolution, and adaptation of flowering plants, and WGDs have been identified in most angiosperm genomes [1–4]. Indeed, so far, all sequenced angiosperms, except for *Amborella trichopoda* [5] and *Aristolochia fimbriata* [6], seem to have experienced at least one WGD since the divergence of angiosperms [7]. Genome sequences, and especially well-assembled genomes, are great resources for unveiling (ancient) WGDs, because of structural information that can be used for finding synteny and/or

Yves Van de Peer (ed.), *Polyploidy: Methods and Protocols*,
Methods in Molecular Biology, vol. 2545, https://doi.org/10.1007/978-1-0716-2561-3_3,

collinearity (Chen et al. and Victor et al., in this volume). Synteny and/or collinearity analysis compares the location and order of homologous genes within a genome or between genomes. A WGD event can be identified through intragenome synteny or collinearity or by showing so-called double synteny with the genome of another species [8, 9].

It has been recognized that synteny or collinearity analysis is the most reliable approach for identifying ancient WGDs. However, the analysis depends on the continuity of genome assembly, the age of a WGD, and the rate of genome rearrangements after WGDs [10]. Although with the help of new sequencing technologies, a plant genome sequence becomes more accessible than ever before [11, 12], the assembly and annotation of a genome is still limited by computing resources and algorithm deficiency [13, 14]. Moreover, not all species of evolutionary and economic importance have their genomes sequenced (yet), as genome sequencing may be hindered by huge genome sizes and high levels of heterozygosity and ploidy [15–18]. For such species, transcriptome sequencing provides an alternative solution to access gene space. Through de novo transcriptome assembly, the reconstructed gene content can be used to detect WGDs by applying approaches that are entirely independent of the genome sequence, such as standard approaches utilizing age distributions for all the paralogs in a species (or the whole paranome) based on the number of synonymous substitutions per synonymous site (K_S), or, alternatively, the number of transversions per four-fold degenerated site (4dTV) (Chen et al., in this volume). By plotting the number of duplicated genes against the age of the duplication event, the paranome age distribution shows a peak at a specific K_S value if a species has experienced a WGD [19, 20]. Thus, unlike identifying synteny or collinearity using genomes, the paranome age distribution only needs a well-represented gene space, which can be efficiently obtained by transcriptome sequencing these days [21].

Many studies have successfully employed transcriptomic data to build paranome age distributions for the detection of WGDs in various lineages of plants [22–26]. Even before the era of next-generation sequencing, Cui et al. [22] have applied the approach to Expressed Sequence Tag (EST) data and identified WGDs in species from Nymphaeales, Magnoliids, Gnetales, etc. Now, after more than a decade, analysis of fully sequenced genomes of the above species has confirmed most of these previously identified WGDs [10, 27–29]. In addition, paranome age distributions are often combined with phylogenomic approaches involving gene tree—species tree reconciliation to infer WGDs (see more details in Chen et al. in this volume). For instance, using 1124 plant transcriptomes, the one thousand plant (OneKP) transcriptome sequencing project has inferred 244 ancient WGDs across green plants based on paranome age distributions and gene tree—species tree reconciliations. Among the WGDs identified by OneKP,

65 (27%) could be verified by currently published genomes [30]. However, the OneKP project also seems to have missed several WGDs that could be identified through the use of entire genome sequences [31], suggesting that transcriptomes may still have less power than do complete genomes concerning the identification of WGDs.

Indeed, compared with all genes predicted in a genome, the gene space reconstructed by transcriptomes is neither complete nor nonredundant, so that it may affect the correct inference of WGDs. A characteristic of transcriptomes is that many expressed genes are environment (condition) and developmental stage dependent [32] and, as a result, transcriptome sequencing cannot guarantee to cover the complete gene space of a species. For example, even a well-assembled plant transcriptome can just retrieve up to 75% of reference transcripts in a species [33].

Also, transcriptome assembly can be more complicated than genome assembly because sequencing reads from a transcriptome have different abundances at various gene loci, while usually, sequencing reads from a genome show a somewhat uniform coverage [34]. Due to the differences in transcriptome assembly algorithms, some assemblers are reasonably good at assembling transcriptomes of certain species in particular, but no assembler outperforms others in all species. By comparing 20 biological-based and reference-free metrics, five assemblers, namely Trinity, SPAdes, Trans-ABySS, Bridger, and SOAPdenovo-Trans, have been shown to be among the best tools for de novo transcriptome assembly [35].

Finally, transcriptome assembly can assemble different products transcribed from one gene locus, causing redundancy in the reconstructed gene space. On the one hand, a gene locus may produce different transcripts (isoforms) due to alternative splicing. On the other hand, a gene locus may have different alleles and may produce several allelic transcripts, especially in species with high heterozygosity. It has been well acknowledged that high heterozygosity is always an issue for both genome and transcriptome assembly [36, 37]. Thus, when isoforms meet allelic transcripts in transcriptome assembly, they may lead to redundant assembled transcripts that originated from the same gene locus [38]. Furthermore, if the gene locus has a highly similar duplicate, isoforms and allelic transcripts may match with the gene or its duplicate, and it is then difficult to distinguish between sequences from a gene locus and sequences from its duplicate [39], potentially leading to chimeric assembled transcripts [40, 41]. Therefore, selecting an assembled transcript to represent a gene locus is often not a trivial task. Failing to do so may artificially amplify gene family sizes when building paranome age distributions and complicate the identification of signature K_S peaks [42].

Although there are issues with assembling transcriptomes, building paranome age distributions based on transcriptomic datasets has become a widely adopted approach to detect WGDs. However, it is still unclear to what extent the paranome age distributions based on genes from transcriptomes are comparable to those based on genes from genomes. In this chapter, we selected six plant species with published genomes and transcriptomes to study how transcriptomic datasets perform when using paranome K_S distributions to detect the most recent well-documented WGDs in these six species. Using genes predicted in each genome as references, we evaluated the effects of missing reference genes, gene family clustering, and de novo assembled transcripts on the paranome K_S distributions. Our results show that although the transcriptome-based paranome K_S distributions differ from the genome-based ones, they are still reasonably reliable for identifying WGD when using cautiously.

2 Materials

2.1 Plant Genomes and Transcriptomes

We selected six plants with available genomes and RNA-seq datasets (Table 1): two monocot species, i.e., pineapple (*Ananas comosus*) and Phalaenopsis (*Phalaenopsis equestris*), and four eudicot species, i.e., Arabidopsis (*Arabidopsis thaliana*), papaya (*Carica papaya*), soybean (*Glycine max*), and grape (*Vitis vinifera*). The pineapple genome has experienced two ancient WGD events, namely σ and τ, and the most recent WGD has a peak at ~1.2 in the K_S distributions for the whole paranome [43]. The Phalaenopsis genome has one WGD identified with a peak at ~1.1 in the K_S distributions for the whole paranome [44]. In Arabidopsis, two WGDs have been uncovered (since the γ WGD shared by all core eudicots) in the K_S distributions for the whole paranome, and the most recent WGD has a signature K_S peak at ~0.8 [45]. Soybean has experienced a very recent WGD with a K_S peak at ~0.2 and retained more than 75% of the duplicated genes that originated from this WGD [46]. Grape and papaya have experienced no additional WGDs after the ancient hexaploidization (γ) that is shared by all the core eudicots [47]. The ancient hexaploidization (γ) signature K_S peak varies in different K_S distributions (grape ($K_S \approx 1.2$); papaya ($K_S \approx 1.8$)) for the simple reason that different species can have different synonymous substitution rates. RNA-Seq data from leaf, root, and stem were collected for each species except for *C. papaya*, which only has transcriptomes from leaf and root. We also created a "mixed" sample in each species by merging RNA-Seq data from different tissues.

Table 1
Summary of the examined plants in this study

Species[a]	Genome size (Mb)	Gene number	RNA-Seq data		
			Tissue	SRA ID	Size (Gb)
Ananas comosus	316	25,440	Leaf	SRR7663722	4.45
			Root	SRR7663721	4.40
			Stem	SRR7663702	4.23
Phalaenopsis equestris	1086	29,431	Leaf	SRR2080202	1.20
			Root	SRR2080194	4.49
			Stem	SRR2080200	5.95
Arabidopsis thaliana	120	27,655	Leaf	SRR3993754	1.15
			Root	SRR3993762	1.28
			Stem	SRR3993761	1.07
Carica papaya	343	27,768	Leaf	SRR7145703	6.38
			Root	SRR7145705	7.15
Glycine max	978	56,044	Leaf	SRR12744739	6.76
			Root	SRR12744729	7.02
			Stem	SRR12744731	6.75
Vitis vinifera	486	26,346	Leaf	SRR9970452	8.53
			Root	ERR3814249	5.83
			Stem	SRR9970442	8.47

[a]Genome sequences and annotations were downloaded from PLAZA 4.5, except the one of *Ananas comosus*, which was retrieved from NCBI with PRJEB33121

3 Methods

3.1 De Novo Assembly of Transcriptomes

For de novo transcriptome assembly, we implemented three standard pipelines that are widely employed in various evolutionary genomic studies [11, 31, 48–50]. Briefly, Trinity v2.12.0 [34] or SOAPdenovo-Trans v1.03 [51] was first used to de novo assemble cleaned RNA-Seq reads for each sample. After collecting the assembled transcripts, Transdecoder v5.0.2 (https://github.com/TransDecoder/TransDecoder/) was used to predict open reading frames (ORFs). Because redundant assembled transcripts resulted from alternative splicing, allelic transcripts, or highly similar duplicates still exist, the predicted ORFs were clustered by tools like CD-HIT v4.8.1 [52] aiming at selecting a representative sequence at each gene locus. Also, BUSCO (Benchmarking Universal Single-Copy Orthologs) v5.0.0 [53] was integrated into each pipeline for a primary evaluation of the gene space.

3.1.1 Data Preprocessing

Trimmomatic v0.39 [54] with default parameters was used for removing Illumina sequencing adaptors, and low-quality bases in sequencing reads.

```
trimmomatic PE -summary trimmomatic.summary.log fq1.gz fq2.gz
clean.fq1.gz unpaired.fq1.gz clean.fq2.gz unpaired.fq2.gz SLI-
DINGWINDOW:5:20 LEADING:3 TRAILING:3 MINLEN:50
```

- LEADING:3, TRAILING:3. Remove low-quality or N bases (below quality 3) at both ends.
- SLIDINGWINDOW:5:20. Scan the read with a 5-base wide sliding window, cutting when the average quality per base drops below 20.
- MINLEN:50. Discard reads shorter than 50 bases.

3.1.2 Pipeline 1

Pipeline 1 implements a straightforward approach based on Trinity for transcriptome assembly, followed by Transdecoder and CD-HIT for predicting ORFs and removing redundant assemblies, respectively. The pipeline or similar pipelines with different parameters, such as different identity thresholds in CD-HIT, have been widely used in studies like Ren et al. [55] and Cheon et al. [50].

```
Trinity --seqType fq --min_contig_length 150 -left clean.fq1.
gz --right clean.fq2.gz --output trinity_Mixed
```

- -min_contig_length: minimum assembled contig length to report.

```
TransDecoder.LongOrfs -t Mixed.trans.clean.fa
TransDecoder.Predict -t Mixed.trans.clean.fa
cd-hit -i Mixed.transdecoder.pep -c 0.99 -o Mixed.transdeco-
der.cdhit.p1.pep
```

- -c: sequence identity threshold. 0.99 means that sequences with 99% identity are clustered.

3.1.3 Pipeline 2

Pipeline 2 is less commonly adopted than Pipeline 1, but it uses the transcript clustering information embedded within Trinity [49]. The clustering information is defined by shared sequence content when performing de novo transcriptome assembly [34]. Ideally, such a cluster can represent a gene locus, and transcripts in the cluster are considered isoforms derived from the gene locus. Therefore, Pipeline 2. selects the longest ORF from each trinity cluster as the representative ORF for a gene locus.

```
Trinity --seqType fq --min_contig_length 150 -left clean.fq1.
gz --right clean.fq2.gz --output trinity_Mixed
```

- -min_contig_length: minimum assembled contig length to report.

```
TransDecoder.LongOrfs -t Mixed.trans.clean.fa
TransDecoder.Predict -t Mixed.trans.clean.fa
perl Selecting_Trinity_transcript_based_on_Transdecoder_lon-
gest_orfs.pl Mixed.trans.clean.fa Mixed.transdecoder.pep
Mixed.orf_info.xls Mixed.cluster.xls Mixed.longest_orf.uni-
genes.fa Mixed.longest_orf.fa
```

- This is an in-house Perl script to extract the longest ORF in a Trinity cluster (see Code availability).

3.1.4 Pipeline 3

Pipeline 3 is similar to Pipeline 1, but SOAPdenovo-Trans replaces Trinity as the transcriptome assembler in the pipeline. It is the pipeline that has been used in the OneKP project [30].

```
SOAPdenovo-Trans-31mer all -s Mixed.soap.conf -K 25 -F -o
Mixed.soap_trans; GapCloser -a Mixed.soap_trans.scafSeq -b
Mixed.soap.conf -o Mixed.soap_trans.GapCloser.fa
```

- "Mixed.soap.conf": the configuration file for SOAPdenovo-Trans-31mer and GapCloser

```
max_rd_len=150
[LIB]
rd_len_cutoff=150
avg_ins=200
reverse_seq=0
asm_flags=3
map_len=32
q1=clean.fq1.gz
q2=clean.fq2.gz
```

- -K: kmer size
- -F: fill gaps in scaffolds

```
TransDecoder.LongOrfs -t Mixed.trans.clean.fa
TransDecoder.Predict -t Mixed.trans.clean.fa
cd-hit -i Mixed.transdecoder.pep -c 0.99 -o Mixed.transdeco-
der.cdhit.p1.pep
```

- -c: sequence identity threshold. 0.99 means that sequences with 99% identity are clustered.

3.1.5 BUSCO Evaluation

For BUSCO evaluation, the following exemplar command-line was used to infer the completeness of gene space for each sample based on 1614 BUSCOs in the database of embryophyta_odb10.

```
busco -I Mixed.transdecoder.cdhit.pep -l embryophyta_odb10 -m
prot
```

The numbers of predicted ORFs from the three pipelines for different samples are shown in Table 2. In general, Pipeline 1 predicted the most ORFs in all the examined transcriptomes, whereas Pipeline 2 and Pipeline 3 generated different but similar numbers of ORFs (Table 2). Similarly, for a specific sample, Pipeline 1 resulted in the most Complete BUSCOs (including both nonduplicated and

Table 2
The number of predicted ORFs

Species	Samples	Pipeline 1	Pipeline 2	Pipeline 3
Ananas comosus				
	Mixed	116,585	56,267	43,024
	Leaf	58,009	22,781	20,507
	Root	88,084	51,251	38,035
	Stem	53,561	20,986	19,400
Arabidopsis thaliana				
	Mixed	39,126	22,651	21,887
	Leaf	23,951	18,995	17,699
	Root	29,981	21,855	20,280
	Stem	26,922	20,601	19,262
Carica papaya				
	Mixed	71,012	20,606	24,916
	Leaf	46,262	16,515	18,302
	Root	54,506	20,002	21,900
Glycine max				
	Mixed	144,687	38,102	35,400
	Leaf	112,034	26,867	28,361
	Root	83,968	31,817	31,592
	Stem	101,569	29,133	32,485
Phalaenopsis equestris				
	Mixed	70,865	25,982	28,129
	Leaf	11,025	10,510	8624
	Root	49,741	23,205	22,562
	Stem	53,474	22,504	23,313
Vitis vinifera				
	Mixed	83,542	34,044	36,757
	Leaf	65,442	25,969	24,692
	Root	24,634	22,645	25,289
	Stem	59,287	24,866	24,691

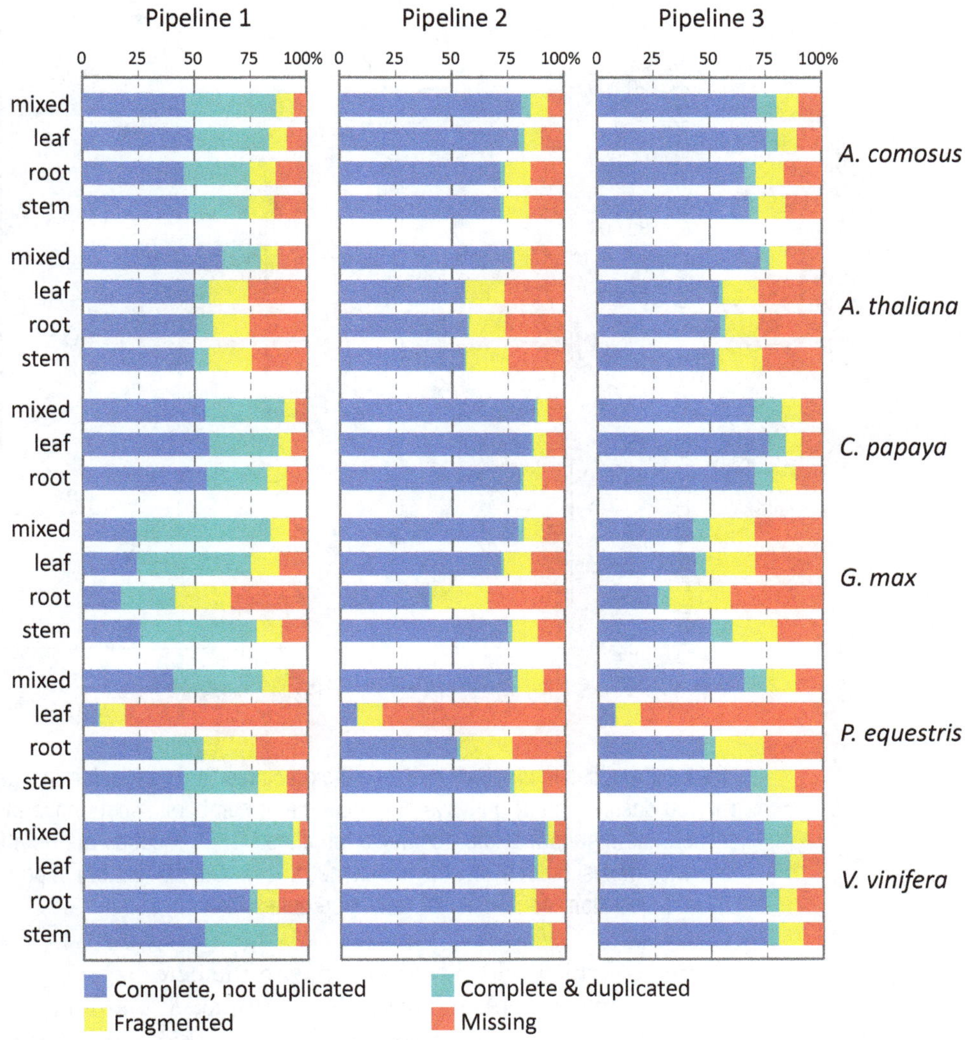

Fig. 1 BUSCO evaluations of predicted ORFs by the three different pipelines for various tissues

duplicated), followed by Pipeline 2 and then Pipeline 3. The fractions of Complete BUSCOs are comparable among different pipelines, except for Pipeline 3, which showed worse performance in assembling the RNA-Seq reads from *G. max* (Fig. 1). Additionally, in all the species and samples, ORFs from Pipeline 1 always have the highest fractions of Duplicated BUSCOs, suggesting that Pipeline 1 produced a certain level of gene space redundancy in different samples.

Further, we directly compared the numbers of predicted ORFs and the numbers of genes in the reference genomes (Fig. 2). Compared to the number of reference genes in the genomes, Pipeline 1 tends to predict many more ORFs. It should be noted that the differences in numbers between the predicted ORFs from Pipeline 1 and those from Pipelines 2 and 3 differ to various extents in the

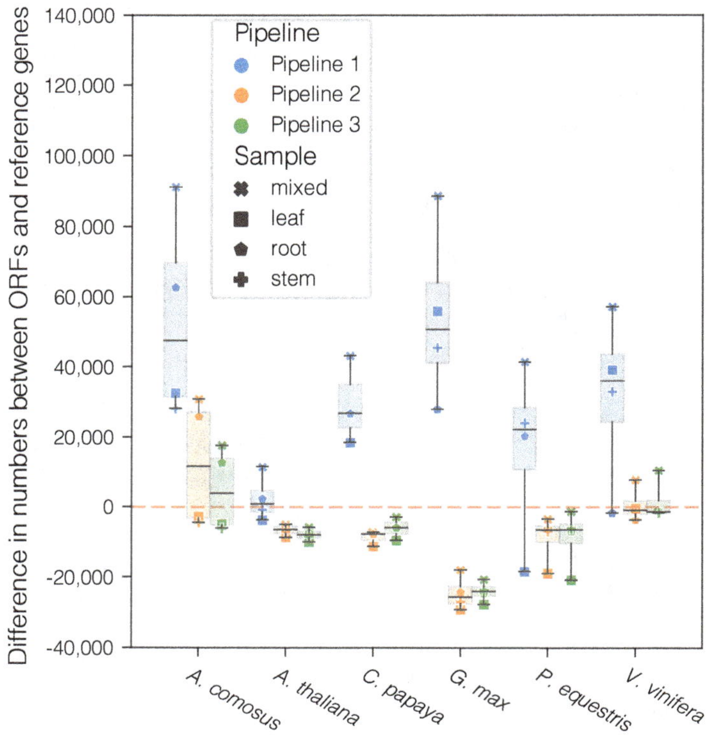

Fig. 2 Differences in numbers between the predicted ORFs and reference genes. The red dashed line at 0 shows no difference in numbers. A dot above the red dashed line means that the number of predicted ORFs exceeds the number of reference genes. A dot below the red dashed line means the number of predicted ORFs is less than the number of reference genes

six species, which is likely correlated with the heterozygous level of sequences in the RNA-Seq reads. For example, *A. thaliana* is a self-pollinated plant with heterozygosity as low as 0.5% [56]. The numbers of predicted ORFs in all three pipelines are close to the number of reference genes. However, compared with other species, Pipeline 1 in *A. comosus* and *G. max* predicted more ORFs than the reference genes (Fig. 2). Both species must have more heterozygous sequences in their RNA-Seq reads because the *A. comosus* genome has a high heterozygosity of 2% [43], while the *G. max* genome still contains many duplicated genes resulted from a very recent WGD [46]. In addition, the BUSCO results for the predicted ORFs from Pipeline 1 in *A. comosus* and *G. max* also show higher fractions of Duplicated BUSCOs than that in other species (Fig. 1), indicating that they still have redundant ORFs resulting from allelic transcripts or duplicated genes. Because both high heterozygosity and recent duplicates with highly similar sequences can cause similar issues in de novo transcriptome assembly, our results suggest that Pipeline 1 may behave suboptimal in dealing with highly heterozygous sequencing reads.

In contrast, compared to the number of genes in the completely sequenced genomes, Pipelines 2 and 3 predicted fewer or sometimes relatively comparable numbers of ORFs. Both pipelines resulted in lower fractions of Duplicated BUSCOs than Pipeline 1 (Fig. 1), suggesting that they removed more ORFs with similar sequences. However, compared with other species, both pipelines predicted much fewer ORFs than the reference genes in *G. max*, but Pipelines 2 and 3 have different performances in the BUSCO evaluations. ORFs from Pipeline 2 have nearly equivalent fractions of Complete BUSCOs but almost no Duplicated BUSCOs. This may be an issue for a species still retaining duplicates from a recent WGD, as we expect more duplicated genes in the genome. Because the duplicated genes retained from the very recent WGD in *G. max* are still quite similar, Trinity could falsely cluster true paralogs with similar sequence content during the assembly process. For the ORFs from Pipeline 3, they have lower fractions of Complete BUSCOs, but higher fractions of Fragmented BUSCOs than do Pipelines 1 and 2 in the samples of *G. max* (Fig. 1), suggesting that Pipeline 3 with SOAPdenovo-Trans handles duplicated genes with highly similar sequences differently from Pipeline 2. In either case, both Pipelines 2 and 3 performed less well in species with a recent WGD than in species without.

3.2 Building K_S Distributions for the Whole Paranomes

Finally, we used the wgd v1.2 program [57] to build K_S distributions for the whole paranomes based on the predicted ORFs from de novo transcriptome assemblies and the reference genes in genomes (Chen et al., in this volume). The wgd suite integrates commonly used K_S and collinearity analysis workflows with Gaussian mixture modeling and result visualization tools, providing researchers with a convenient way to detect WGD events based on genomic or transcriptomic data. Below, we take the predicted ORFs from the mixed sample of *V. vinifera* as an example for using wgd:

```
wgd dmd -I 3 Mixed.selected.transdecoder.p1.cds -o 01.wgd_dmd -
nostrictcds
```

- -I: --inflation FLOAT inflation factor for MCL.
- -nostrictcds: do not enforce proper CDS sequences, which means all the cds, including the complete cds with start codon and stop codon and the incomplete cds, will be used for clustering.

```
wgd ksd 01.wgd_dmd/Mixed.selected.transdecoder.p1.cds.mcl
Mixed.selected.transdecoder.p1.cds -o 02.wgd_ksd
```

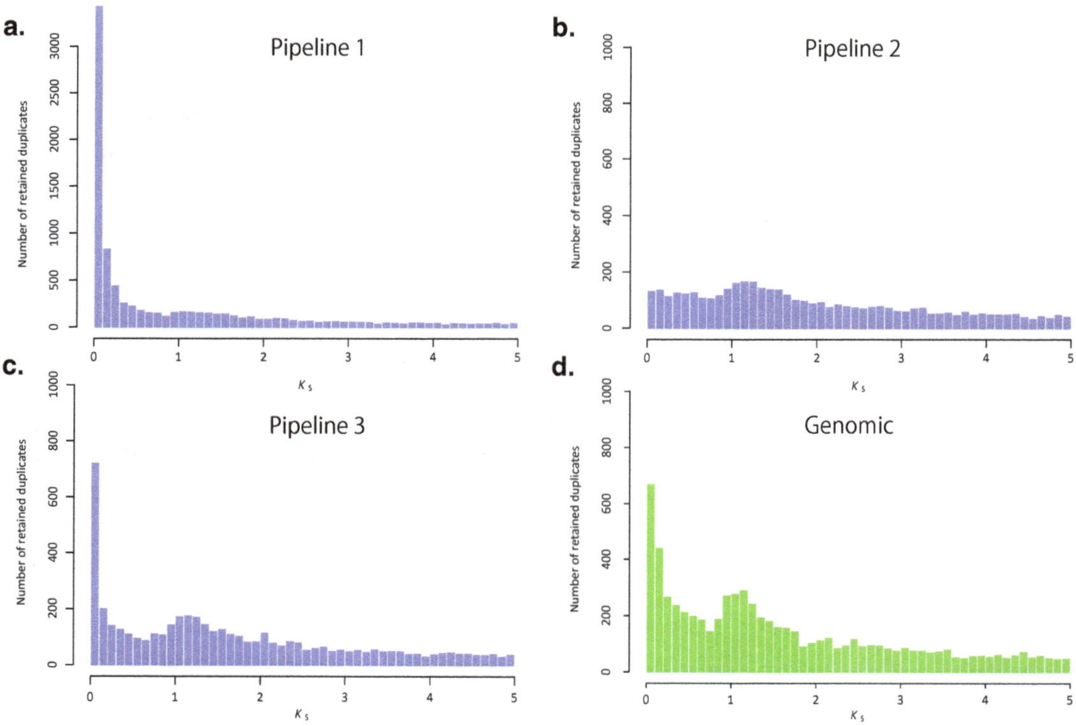

Fig. 3 K_S distributions for the whole paranome of *Vitis vinifera*. (**a–c**) The K_S distributions are based on the ORFs predicted by the three transcriptome assembly pipelines with the mixed sample in *V. vinifera*. (**d**) The K_S distribution based on the reference genes from the *V. vinifera* genome

Using the K_S distributions for the whole paranome of *V. vinifera* as an example (Fig. 3), our results show that K_S distributions based on transcriptomes are different from those based on genes predicted in complete genomes. Also, the transcriptome-based K_S distributions are different from each other depending on the different pipelines used. The peak representing the hexaploidization event in the *V. vinifera* genome at ~1.2 is evident in the K_S distributions based on ORFs from Pipelines 2 and 3, but less so in the K_S distribution based on ORFs from Pipeline 1, the reason being that the transcriptome-based K_S distribution based on ORFs from Pipeline 1 exhibits an abnormally high number of duplicates at low K_S values (0–0.1). Such an abnormally high peak overshadows the WGD signature K_S peak in *V. vinifera*, leading to potential failures in detecting WGDs.

For the rest of the chapter, we further compared the transcriptome-based and genome-based K_S distributions in the six plant species by considering the effects of missing reference genes, gene family clustering, and de novo assembled transcripts. Because the mixed sample in each species combines all the expressed genes from different tissues, it always contains considerably more ORFs than individual tissues, no matter which pipeline was used

(Table 2). In addition, the mixed samples from the six species also have the most Complete BUSCOs in all pipelines (Fig. 1). Therefore, for building K_S distributions and further analyses in the chapter, we only focus on the de novo assemblies based on the mixed samples.

4 Missing Reference Genes and K_S Distributions

The differences in numbers of ORFs and genes indicate that the transcriptome assemblies missed some reference genes in the genome and produced some unknown ORFs. Here, we define the missing reference genes as those that exist in the reference genomes but do not appear in the predicted ORFs in the transcriptome assemblies. Unknown ORFs are defined as the ORFs that only exist in the transcriptomes but are not found in the reference genome. Because missing reference genes must affect K_S distributions, we first determine the gene space that could be reconstructed by the three de novo assembly pipelines, using the predicted genes in the genomes as references.

4.1 Gene Space Reconstructed by Transcriptome Assembly

To obtain an upper bound of the gene space that can be reconstructed by transcriptome sequencing in a species, we first mapped all the RNA-Seq reads of each species to their corresponding genome by Hisat2 v2.1.0 with a parameter "--dta" to only report alignments tailored for transcriptome assemblers [58]. The upper bound of gene space for a sample was then defined as the number of reference genes mapped by at least two RNA-Seq reads. Here, we used a loose cut-off of two RNA-Seq reads for estimating the upper bounds in different species not only because it is the minimum number of RNA-Seq reads used in Trinity [34], but it is also the minimum requirement for doing any assembly. The results show that about 48–69% of the reference genes have the potential to be assembled, and the upper bounds of gene space vary in different species (Fig. 4).

Then, we mapped the ORFs predicted by the three pipelines to their corresponding genomes by BLAT v3.5 [59] to determine how many reference genes could be retrieved from the assembled transcripts. Apparently, not all the reference genes supported by two or more RNA-Seq reads could be assembled. The fractions of reconstructed genes vary from species to species, with the highest in *A. comosus* and the lowest in *C. papaya* and *P. equestris* (Fig. 4). For different pipelines, in general, Pipeline 1 assembled more reference genes than Pipelines 2 and 3, but the fractions are close in each species except for *G. max*. Also, for the predicted ORFs, most are complete ORFs with 100% coverage or nearly complete ORFs with a coverage $\geq$ 90% of the corresponding reference genes. The only exception is that Pipeline 3 assembled most ORFs with a coverage

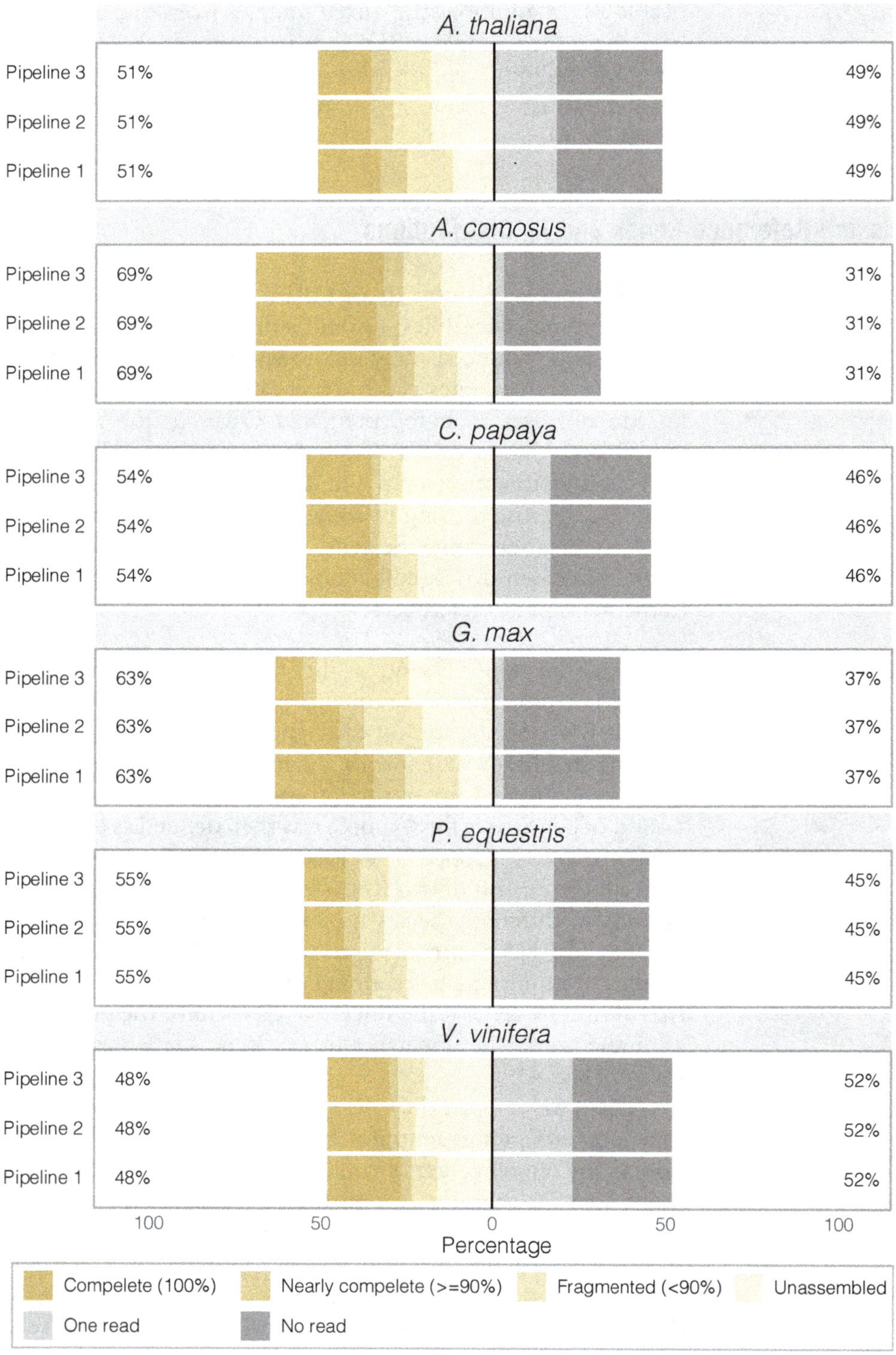

Fig. 4 Reference genes supported by RNA-Seq reads and assembled ORFs, based on the mixed sample in each species. The left part shows the percentages of reference genes supported by at least two RNA-Seq reads (see details in the main text). Assembled ORFs have different coverages for reference genes: "Complete (100%)," "Nearly complete (≥90%)," and "Fragmented (<90%)." The reference genes supported by RNA-Seq reads but without any assembled ORFs are shown as "Unassembled." The right part shows the percentages of reference genes that are not supported "No read," or only supported by "One read" in the mixed samples

less than 90% of the reference genes in *G. max*. As discussed above, this may be related to the poor performance of SOAPdenovo-Trans on species that have undergone a recent WGD.

Because each species has a significant fraction of missing reference genes, we wonder if this would affect identifying WGD signals in K_S distributions. To this end, we marked the reference genes in the genome-based K_S distributions according to their assembly status in the transcriptomes (Fig. 5). We noticed that the potential K_S signals could be recovered by paralogous pairs that are completely or partially assembled in the predicted ORFs. Although the missing reference genes make up a certain proportion of genes at each K_S interval, especially at intervals with small K_S values, they do not really affect the appearance of WGD peaks in the K_S distributions, reassuring the application of transcriptomic dataset in detecting WGDs.

4.2 Redundant ORFs

By mapping the predicted ORFs to the reference genes in each genome, we also evaluated the transcriptome assembly redundancy of the three pipelines. The mapping results demonstrate that Pipeline 1 could produce redundant ORFs, excessing to half of the gene loci having more than one mapped ORFs, especially for *G. max* (Fig. 6). On the contrary, the proportions of gene loci to which only one ORF was mapped increased dramatically for ORFs from Pipelines 2 and 3. Pipeline 2 demonstrates a better capability to remove redundancies because only one representative ORF was kept in each Trinity cluster, but at a risk of falsely clustering paralogous genes into one Trinity cluster. Pipeline 3 seems to be more moderate than the other two pipelines, balancing the number of ORFs and redundancy.

5 Gene Family Clustering and K_S Distributions

Gene families containing paralogous gene pairs are another basis for building K_S distributions. However, in transcriptome assemblies, isoforms and allelic assemblies together with missing genes can confound gene family identification, so we investigated the effects of gene family clustering on the transcriptome-based K_S distributions. We identified 4242, 4683, 3120, 10,979, 3136, and 3680 multigene families in genomes with reference genes in *Ananas comosus Arabidopsis thaliana, Carica papaya, Glycine max, Phalaenopsis equestris,* and *Vitis vinifera*, respectively. For the predicted ORFs, those from Pipeline 1 always have more gene families identified than the reference genes, while the ones from Pipelines 2 and 3 have a comparable number of gene families as the reference genes (Table 3).

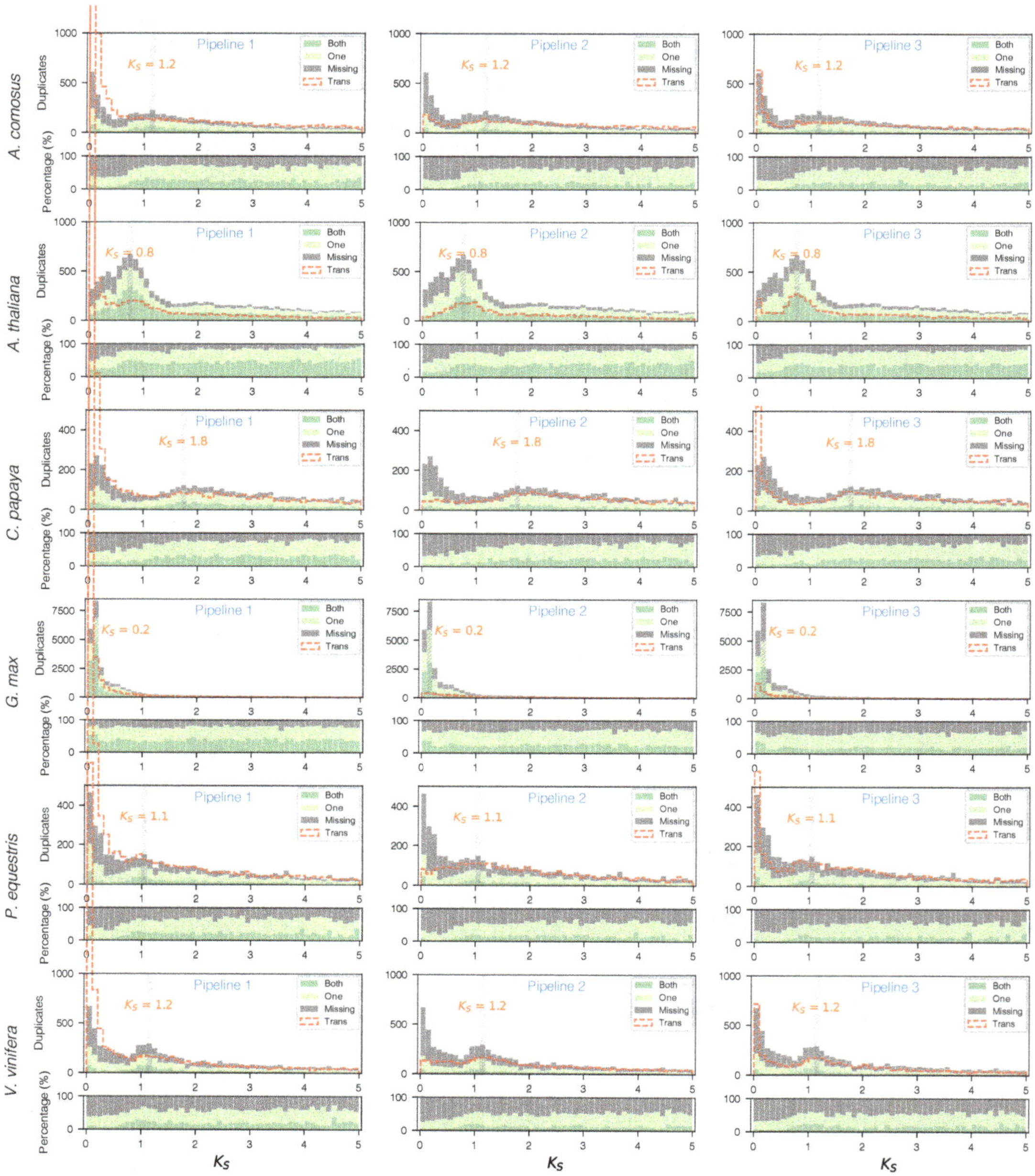

Fig. 5 The impact of the reconstructed gene space on genome-based K_S distributions. "Both" means that the two reference genes of one paralogous pair are present in the predicted ORFs. "One" means that only one of the two reference genes of one paralogous pair is present in the predicted ORFs. "Missing" means neither of the two genes of one paralogous pair is present in the predicted ORFs. The upper part of each subplot shows a genome-based K_S distribution, and the lower part shows the percentages of the three groups at each K_S interval. The red dashed line denotes the transcriptome-based K_S distribution. The gray rectangle denotes the K_S peak of each species

5.1 Presence and Absence of Gene Families

Next, we assessed how well gene families identified on the basis of the reference genes could be predicted based on the ORFs by the three de novo assembly pipelines. To determine the correspondence between gene families of the predicted ORFs and the

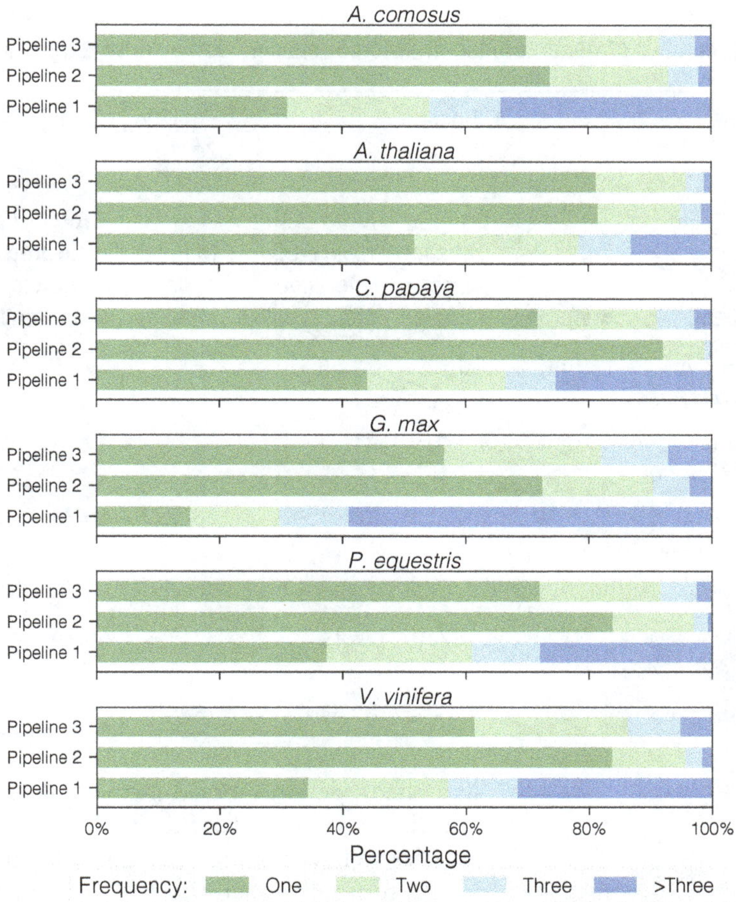

Fig. 6 The redundancy of ORFs at each gene locus. "Frequency" refers to the number of ORFs that could be mapped to each gene locus. "One" means that a gene locus only has one ORF mapped. "Two" means that a gene locus has two ORFs mapped. "Three" means that a gene locus has three ORFs mapped. ">Three" means that a gene locus has more than three ORFs mapped

reference genes, the predicted ORFs were aligned to the reference genes according to Chen et al. [49] using BLAT v3.5 [59]. For subsequent analysis, poor hits with a match length shorter than 100 bp and identity lower than 95% were discarded. The hit with the highest bit-score in BLAT was kept when an ORF had multiple hits to reference genes. If a transcriptome-based gene family only contains ORFs mapped to a single genome-based gene family and vice versa, their correspondence could be precisely determined. However, in many cases, ORFs in transcriptome-based gene families have hits to multiple genome-based gene families and vice versa, so we defined the correspondence of a transcriptome-based gene family and a genome-based gene family if they reciprocally had the most hits to each other. In addition, transcriptome-based gene families that could not match the criteria on correspondence but

Table 3
The number of gene families in different species obtained by different pipelines

Species	Reference	Pipeline 1 Number of gene families	Pipeline 1 Ratio (%) of recovered reference gene families	Pipeline 2 Number of gene families	Pipeline 2 Ratio (%) of recovered reference gene families	Pipeline 3 Number of gene families	Pipeline 3 Ratio (%) of recovered reference gene families
Ananas comosus	4242	12,127	72.63	6399	59.05	6188	61.83
Arabidopsis thaliana	4683	5755	71.34	3358	63.25	3652	64.89
Carica papaya	3120	7512	76.41	2802	66.73	4645	71.38
Glycine max	10,979	13,285	74.46	5048	36.78	5989	37.96
Phalaenopsis equestris	3136	8740	74.14	3211	60.40	4763	66.14
Vitis vinifera	3680	9309	80.54	3880	67.12	5822	73.34

had hits to genome-based gene families were considered problematic families. Gene families with no hits to genome-based gene families were considered unknown families.

It turns out that the ORFs from Pipeline 1 could identify 70–80% of the genome-based gene families. The identification ratio of genome-based gene families decreased to 60–70% for the ORFs from Pipelines 2 and 3 (Table 3). The only exception is *G. max*, which missed nearly half of genome-based gene families in Pipelines 2 and 3, possibly related to the aforementioned assembly issues. The ratios of identified gene families are slightly higher than those of reconstructed gene spaces. As more than half of the gene families of a species exist in the transcriptome assemblies, the transcriptome-based K_S distributions should uncover signature peaks in the K_S distributions for the whole paranome. However, besides the presence and absence of gene families, the shape of K_S distributions also depends on gene family sizes.

5.2 Size Differences of Gene Families

To compare sizes between the transcriptome-based and genome-based gene families, we only selected the corresponding gene families and plotted cumulative distributions for the size differences between a pair of transcriptome- and genome-based gene families. Further, to measure if the size differences are significantly larger or smaller, we sampled the exact number of genes from all the corresponding genome-based gene families 100 times in each

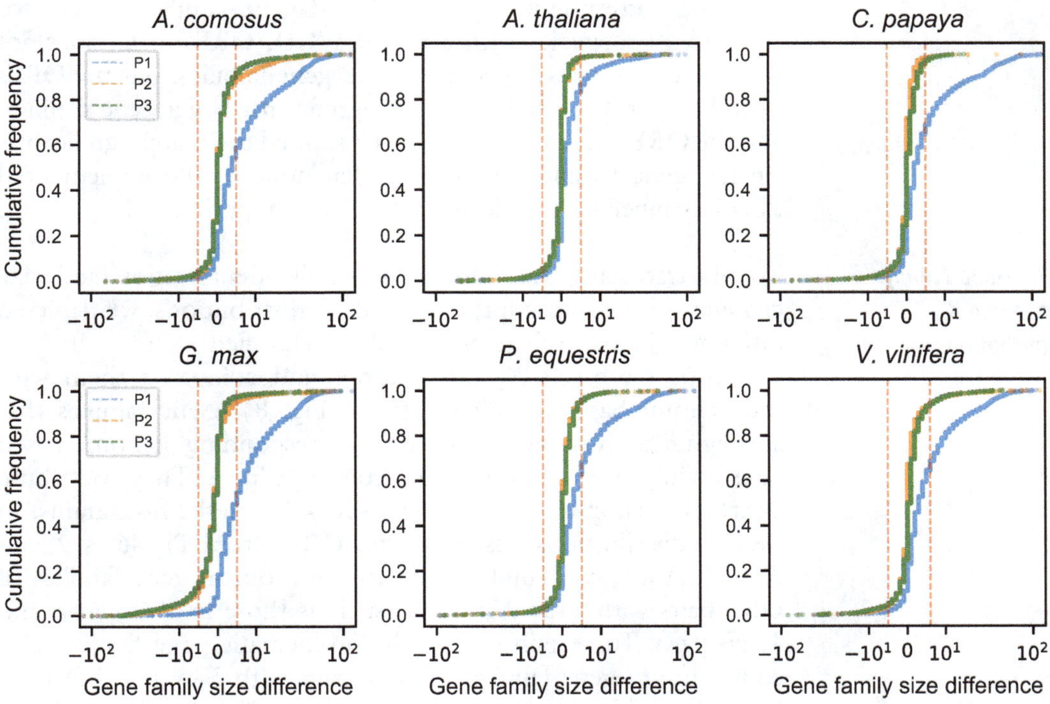

Fig. 7 Cumulative distributions for size differences of gene families between the corresponding transcriptome-based and genome-based gene families. The vertical dashed lines in each subplot correspond to the sizes differences with *z*-scores of −2 and 2, respectively (see details in the main text)

species. Finally, in each resampled set, we calculated z-scores for the size differences between the resampled gene families and the genome-based gene families. We hence could define that a transcriptome-based gene family with a size difference larger and smaller than a *z*-score of 2 and −2, i.e., outside twice standard deviations, is a gene family with significantly different sizes (Fig. 7).

Our results show that, for the transcriptome-based gene families, the ones reconstructed from the ORFs from Pipelines 2 and 3 do vary from the sizes of their corresponding genome-based gene families. However, most of them still have similar sizes as the genome-based gene families. In contrast, the ORFs from Pipeline 1 have more gene families that are significantly larger than their corresponding genome-based gene families, in line with the results that there are more redundant ORFs at the same gene locus in Pipeline 1 (Fig. 6). Heterozygosity also affects gene family clustering of ORFs in transcriptome assembly. Comparing *A. thaliana* with low heterozygosity and *A. comosus* with high heterozygosity, we found that the latter species has more transcriptome-based gene families significantly larger than their corresponding genome-based ones. Interestingly, ORFs from different pipelines produced similar fractions of gene families smaller than the corresponding genome-based gene families, except for *G. max*. As a species with a recent WGD, ORFs in *G. max* not only have much fewer genome-based gene families, but the ORFs from Pipeline 1 formed too many significantly larger gene families, and the ORFs from Pipelines 2 and 3 formed too many significantly smaller gene families, suggesting that none of the implemented assembly pipelines are ideally suited for this species.

5.3 Gene Family Sizes and K_S Distributions

To illustrate the effects of gene family identification and size changes on the transcriptome-based K_S distributions, we depicted different kinds of gene families classified above in the transcriptome-based K_S distributions and compared them with the genome-based K_S distributions (Fig. 8). Gene families that are significantly larger than their corresponding genome-based ones mainly appear in the ORFs from Pipeline 1. They contribute a certain fraction to ORF pairs with $K_S < 0.1$ in the histograms. In the K_S distributions based on the ORFs from Pipelines 2 and 3, such large gene families tend to stand out in gene families of ORF pairs with large K_S values, such as those in *A. comosus* and *V. vinifera*. To certain extent, the problematic gene families also contribute to gene families of ORF pairs with $K_S < 0.1$, but their fractions seem to have no preference toward K_S values. In addition, the fractions of unknown gene families are also different from species to species, but in many species, they tend to be present in gene families of ORF pairs with low K_S values ($K_S < 0.5$).

Our results would suggest that gene families that are larger than their corresponding genome-based gene families and the unknown gene families that do not exist in the "true" genomes may inflate the number of ORF pairs at low K_S values in the K_S distributions, which would affect K_S peaks for WGDs, especially for ORFs from Pipelines 1 and 2. For the K_S distributions based on ORFs from Pipeline 1, the WGD K_S peaks, except the one for the WGD in *A. thaliana* expected at $K_S \approx 0.8$, are hidden in the tail of the histograms, simply because there are too many ORF pairs with small K_S values resulting from large and unknown gene families.

For the K_S distributions based on ORFs from Pipeline 2, these have in general fewer ORF pairs than the K_S distributions based on the reference genes, especially for the ORF pairs with small K_S values (Fig. 8), suggesting that Pipeline 2 may collapse many recent duplicates that would usually compromise the number of duplicates with small K_S values. However, despite somehow obscure, the K_S peaks representing WGD events can still be seen in the K_S distributions for *A. comosus*, *A. thaliana*, *P. equestris*, and *V. vinifera*. Although removing the unknown gene families in the K_S distributions may help increase the visibility of the signature K_S peaks for WGDs, it is impossible to do so when the investigated species has no information on its actual gene space, a typical situation when utilizing transcriptomic data for WGD detection.

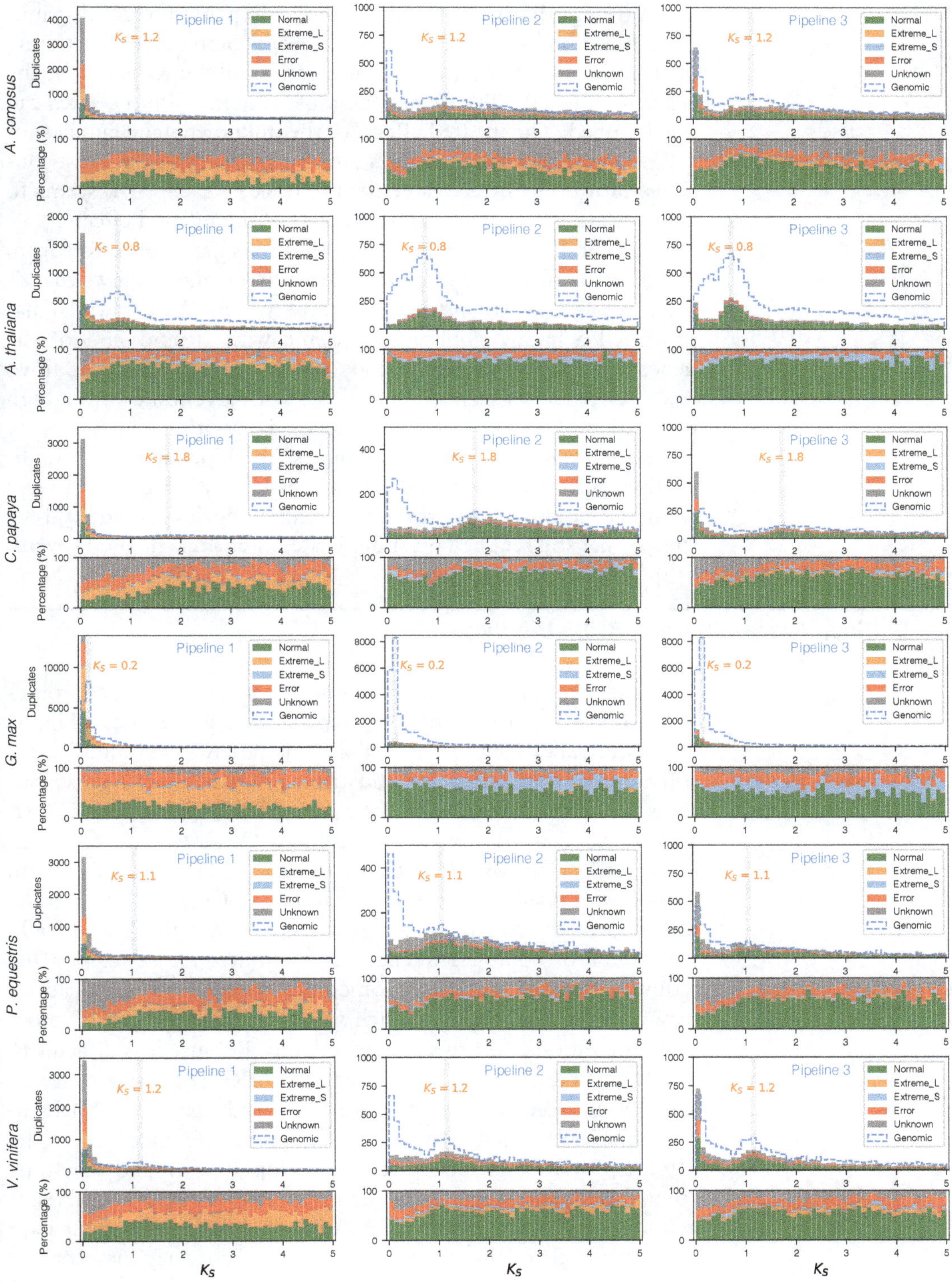

Fig. 8 The impacts of transcriptome-based gene families with extreme sizes on the transcriptome-based K_S distributions. Gene families have extreme sizes if they are significantly larger or smaller than their corresponding genome-based gene families (see details in the main text). The upper part of each subplot shows a transcriptome-based K_S distribution. The lower part shows the percentages of the different kinds of gene families at each K_S interval. The blue dashed line shows the genome-based K_S distribution. The gray rectangle denotes the K_S peak of each species

Although having fewer duplicate gene pairs, the K_S distributions based on ORFs from Pipeline 3 seem comparable with the K_S distributions based on the reference genes. All the K_S peaks for the acknowledged WGDs, except the most recent one in *G. max*, can be identified. The inflated effects of the unknown gene families for ORF pairs with small K_S values are also mild in these K_S distributions. However, the K_S value for the peak in *P. equestris* seems to shift toward a smaller K_S value because of the inflated effects.

Specifically, none of the K_S distributions based on ORFs from the three de novo transcriptome assembly pipelines shows the K_S peak at ~0.2 in *G. max* for its most recent WGD, which has produced a highly duplicated genome with around 75% of the genes still present in a multicopy status [46]. All three de novo assembly pipelines seem to have issues with a genome with a high proportion of recently duplicated genes. They either generate many more significantly larger gene families (for Pipeline 1) or significantly smaller gene families (for Pipelines 2 and 3), suggesting a fine (r)-tuned de novo transcriptome assembly pipeline is required for species that still retained most duplicates after a recent WGD event.

6 De Novo Assemblies and K_S Distributions

Although the three de novo assembly pipelines for transcriptomes show different performances on gene space completeness (Fig. 4) and redundancy (Fig. 6), as well as gene family sizes (Fig. 7), it is still not clear how assembled transcripts or ORFs affect K_S distributions. To this end, we classified the predicted ORFs into five distinct groups based on the mapping results of ORFs to gene loci, i.e., "Correct," "Isoform," "Isoform-like," "Fragmented," and "Unknown" ORFs (Fig. 9). Specifically, a "Correct" ORF is the only best match to a reference gene locus, where the ORF should cover 95% of coding sequences of the reference gene. For the redundant ORFs, although sometimes recent gene duplications may confound redundancy, such ORFs are mainly from different alleles or various isoforms at a gene locus. Because it is difficult to clearly distinguish whether a predicted ORF is from a different allele or an isoform (or both), we here used "Isoform" ORFs to represent ORFs that best match the same reference gene. The "Isoform" ORFs have both start and stop codons, and they only contain sequences from the exons predicted in the reference genome. In contrast, some ORFs with start and stop codons have the best match to the same reference gene, but they may contain extra exons or partial sequences of exons in the reference genome. Because reference gene predictions may also be problematic, we defined such ORFs as "Isoform-like" ORFs. For ORFs that have no start codon and/or stop codons but could be mapped to a part of a reference gene, we classified them as "Fragmented" ORFs. In the

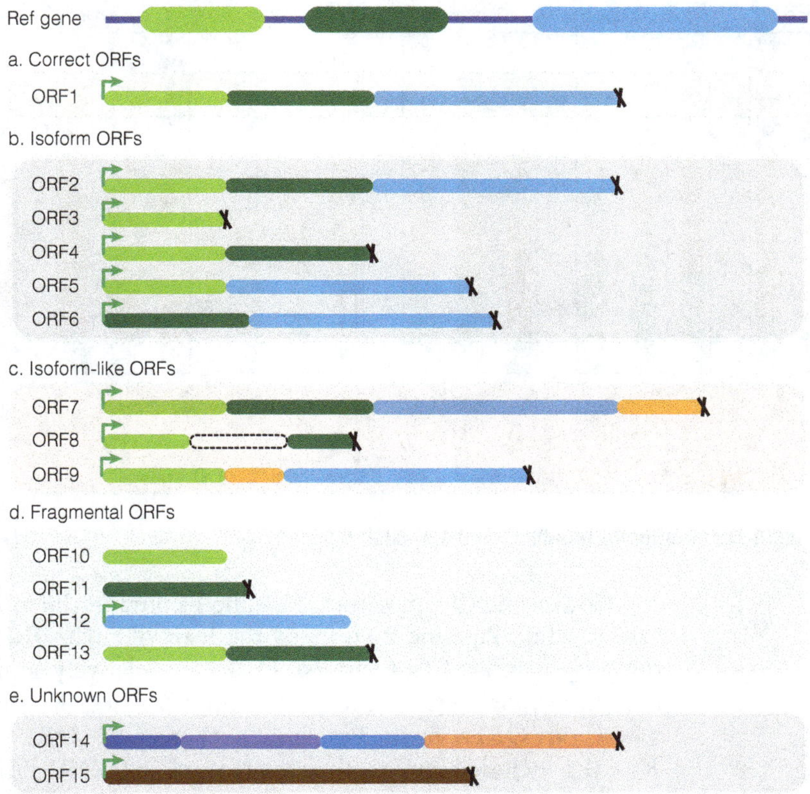

Fig. 9 The classification of assembled ORFs. "ORF1–15" are examples of predicted ORFs mapped to their corresponding reference gene. Five distinct groups of ORFs, including "Correct," "Isoform," "Isoform-like," "Fragmented," and "Unknown" have been depicted. The green arrow denotes a start codon, and the black cross denotes a stop codon

end, the rest are the "Unknown" ORFs that could not be mapped to the reference genomes or any genes thereof.

Among the six plant species investigated, *A. thaliana* has the highest proportion of Correct ORFs, ranging from 23.4% to 50.0% for the three pipelines (Fig. 10). The other species show massive reductions in the proportions of the correct ORFs. *A. thaliana* and *G. max* also have higher proportions of Fragmented ORFs than other investigated species. In addition, *A. thaliana* and *G. max* have the most diminutive proportions of the Unknown ORFs, in line with the gene family analyses where both species have much fewer unknown gene families than other species (Fig. 8). The unknown ORFs may have different sources. They could be de novo assembly artifacts or assembly of contaminated reads in RNA-Seq samples. For example, we found that many Unknown ORFs in *A. comosus* are from microorganisms by annotating these ORFs with the NCBI Nonredundant database, indicating potential contamination when preparing the samples for transcriptome sequencing.

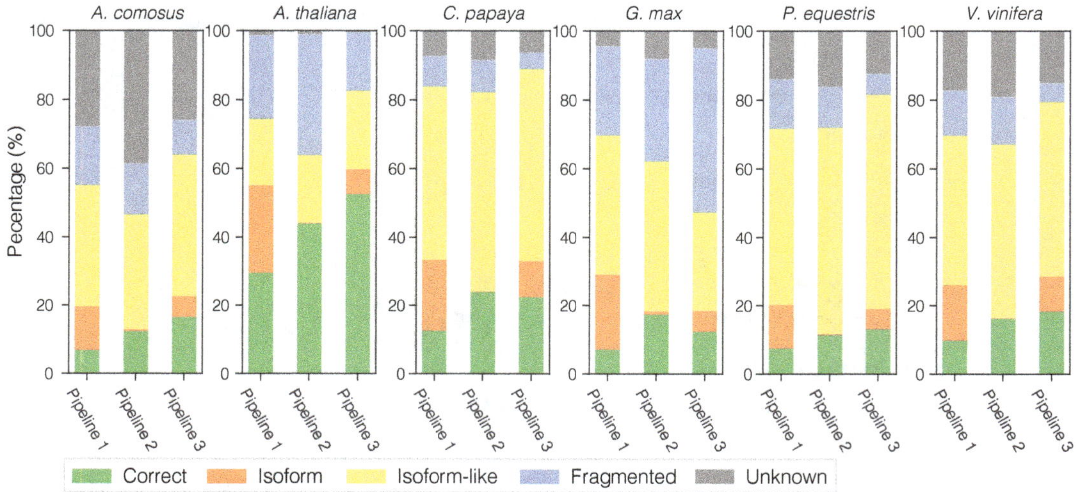

Fig. 10 The percentage of different groups of ORFs in each pipeline

Concerning the pipelines, Pipeline 1 contains the most Isoform ORFs, while Pipeline 2 contains the least of such ORFs, whereas the proportions of Isoform-like ORFs are relatively similar among the three pipelines. In the transcriptome-based K_S distributions based on ORFs from Pipeline 1 (Fig. 11), ORF pairs with $K_S < 0.1$ include many Isoform or Isoform-like ORFs. If an assembly pipeline could not remove the Isoform or Isoform-like ORFs, they would be considered extra members for those gene families. Hence, they result in many gene families with significantly larger sizes than their corresponding genome-based gene families. There is no such pattern in the transcriptome-based K_S distributions based on ORFs from Pipeline 2, while the first bars in the transcriptome-based K_S distributions based on ORFs from Pipeline 3 have more Isoform-like ORFs than Isoform ORFs.

The Fragmented ORFs exist in ORF pairs with all sorts of K_S values. Again, in the transcriptome-based K_S distributions based on ORFs from Pipeline 1, the Fragmented ORFs are mainly present in ORF pairs with $K_S < 0.1$. However, they are also found in ORF pairs with different values, likely forming the problematic gene families in Fig. 8. Because the Fragmented ORFs are relatively short and contain one or a few conserved domains, they might disturb gene family identifications by incorrectly joining different gene families or falsely forming independent gene families.

7 Discussion

Because of the ease of transcriptome sequencing, transcriptomic data have been widely adopted to infer WGDs, and it has become standard practice to examine transcriptomes from dozens, if not hundreds of species, to detect WGDs in a large-scale phylogeny

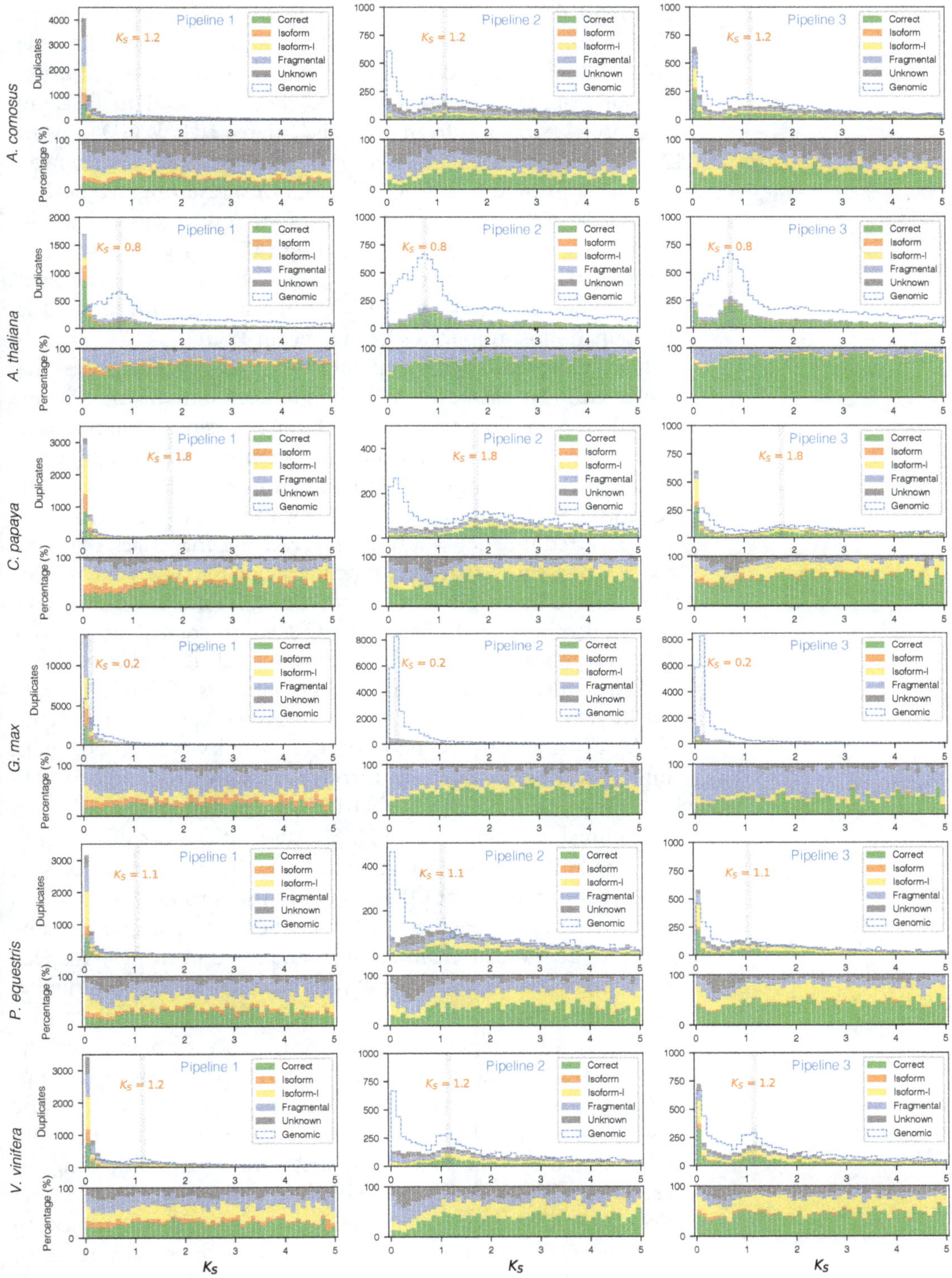

Fig. 11 The impacts of various categories of ORFs on transcriptome-based K_S distributions. Five different groups of ORFs are assigned to a transcriptome-based K_S distribution (the upper part of each subplot). The lower part of each subplot shows the percentages of the five groups of ORFs at each K_S interval. The dark blue dashed line depicts the genome-based K_S distribution. The gray rectangle denotes the K_S peak of each species. "Isoform-I" means the "Isoform-like ORFs" group

[30, 55, 60, 61]. Such studies systematically allow to investigate the importance of WGDs for the evolution of green plants. However, there are some methodological concerns using K_S distributions with transcriptomic datasets because they may lead to fallacious conclusions drawn from falsely detected WGD events [42, 62]. Here, we compared genome-based K_S distributions with transcriptome-based K_S distributions resulting from three different de novo assembly pipelines. Our results show that with proper transcriptome assembly, although the transcriptome-based K_S distributions have different shapes than the genome-based ones, they have the power to identify WGDs but may fail with species that still retained most duplicates from a recent WGD.

Transcriptome assembly pipelines, especially the steps that remove redundant assemblies, have significant impacts on inferring WGDs based on K_S distributions. Despite missing some reference genes and gene families, all the implemented assembly pipelines here could reconstruct enough genes and gene families for WGD detection, if transcriptomes are relatively well sequenced. However, gene space redundancy is a more severe issue than (lack of) completeness. We show that in a pipeline (Pipeline 1) that has been widely applied to various studies (maybe with different parameters, though), Isoform or Isoform-like ORFs are treated as genuine duplicated genes, and most of them have small K_S values less than 0.1. As a result, the extraordinary large numbers of duplicates with small K_S values overshadow the signature WGD peaks in the examined K_S distributions. Some efforts could alleviate the effects of redundant ORFs, for example, through clustering redundant ORFs with a decreasing identity in CD-HIT or by removing ORF pairs with minimal K_S values [42, 62]. However, if the cut-offs used to remove redundancy are too stringent, they have limited effects on the number of redundant ORFs in the transcriptome assembly. On the other hand, if they are too loose, they may collapse genuine paralogous genes with small K_S values and leave an artificial K_S peak slightly larger than 0.1. The peak could then be falsely identified as evidence for a recent WGD event or considered an artifact after removing too many genuine paralogous genes with small(er) K_S values [55]. Although some of the recent WGDs seem to be supported by analyses using genomic data, the K_S peak values from the transcriptome-based K_S distributions are sometimes different from the ones in the genome-based K_S distributions [42], suggesting the inference of WGDs may be still arbitrary and requires further corroboration. Moreover, determining the cut-offs for removing redundant ORFs in transcriptome assembly may require prior knowledge about sequenced species, such as the heterozygosity. It is, of course, helpful to have such information before sequencing a species, but it is also not an easy task for studies including hundreds of species, especially if the cut-offs are required to be fine-tuned from species to species.

Alternatively, relying on the algorithms in de novo transcriptome assemblers is another solution to eliminate redundant ORFs. Trinity and SOAPdenovo-Trans are two de novo transcriptome assemblers based on the *de Bruijn* graph. Both assemblers have been used for detecting WGDs with K_S distributions [63–65]. In Trinity, sequencing reads in the same assembly graph or Trinity cluster are separately assembled, and the final assembled transcripts retain the cluster information. Because a Trinity cluster is often considered to be corresponding to a gene, our Pipeline 2 uses the clustering information to select one representative sequence for each Trinity cluster. As shown in the results of the BUSCO evaluation and K_S distributions, Pipeline 2 removed many duplicated ORFs and scaled-down the overall number of duplicates in the K_S distributions. Nevertheless, the pipeline is inclined to remove more ORF pairs with K_S less than 0.5, leading to failures in detecting very recent WGD events.

Like Trinity, SOAPdenovo-Trans also provides cluster or gene information after transcriptome assembly, but it shows a more reasonable number of gene loci than does Trinity. As shown by us (this chapter) and others [35, 66], Trinity does assemble more Complete BUSCOs than SOAPdenovo-Trans, but it also produces a higher proportion of Duplicated BUSCOs. For instance, in *A. thaliana*, SOAPdenovo-Trans reported 8242 gene loci, one-fourth of which have multiple isoforms with a maximum number of isoforms up to five. On the contrary, Trinity reported 46,364 gene loci, in which 7412 gene loci have multiple isoforms with a maximum number of up to 98 isoforms. Therefore, the assembled results of SOAPdenovo-Trans have much lower redundancy than the results of Trinity, reducing the efforts to further remove redundant ORFs. In addition, our results show that the K_S distributions using ORFs assembled by SOAPdenovo-Trans (Pipeline 3) are more comparable to the genome-based K_S distributions with respect to their shapes and scales. Meanwhile, the running time for SOAPdenovo-Trans is shorter than that of Trinity, which is significantly meaningful when conducting studies with hundreds of species, likely explaining why SOAPdenovo-Trans has been chosen in the OneKP to some extent.

Besides transcriptome assembly, there are other concerns related to RNA-Seq sequencing that may affect using K_S distributions to infer WGDs, such as RNA extraction, library preparation, sequencing depth, and other unexpected events like sample contamination. Compared to genome sequencing and assembly, transcriptome sequencing and assembly are more unstable, and hence challenging to measure their quality. However, if the data volume of the transcriptome is not high enough, the resulting K_S distributions would be less informative and suitable for WGD inference, due to the insufficient gene space. For instance, compared to other samples, the leaf sample of *P. equestris* only produced one-third of total ORFs (Table 1 and Fig. 1). Another issue that should be avoided is sample contamination, as found in the root sample of *A. comosus*, because alien ORFs from other species may have unexpected effects on K_S distributions.

As newly developed sequencing technologies have already been applied to transcriptome sequencing [67], single-molecule and long-read sequencing may help solve some issues related to short-read sequencing. For example, Yue et al. [68] have used PacBio Single Molecule, Real-Time (SMRT) sequencing technology to generate full-length transcriptome data for a sterile triploid *Crocus sativus*, and have successfully identified a recent WGD event in its evolutionary history. In any case, transcriptomic data, increasingly generated as supplements to genomic data, are a great asset for the identification and delineation of ancient polyploid events.

Codes Availability

Commands and scripts used in this study are available at https://github.com/li081766/transcriptome_WGD_project.

Acknowledgments

YVdP acknowledges funding from the FWO (G090919N), from the European Research Council (ERC) under the European Union's Horizon 2020 research and innovation program (No. 833522), and from Ghent University (Methusalem funding, BOF.MET.2021.0005.01).

References

1. Guo J, Xu W, Hu Y et al (2020) Phylotranscriptomics in cucurbitaceae reveal multiple whole-genome duplications and key morphological and molecular innovations. Mol Plant 13: 1117–1133
2. Sheehan H, Feng T, Walker-Hale N et al (2020) Evolution of l-DOPA 4, 5-dioxygenase activity allows for recurrent specialisation to betalain pigmentation in Caryophyllales. New Phytol 227:914–929
3. Xiang Y, Huang C-H, Hu Y et al (2017) Evolution of Rosaceae fruit types based on nuclear phylogeny in the context of geological times and genome duplication. Mol Biol Evol 34: 262–281
4. Van de Peer Y, Ashman T-L, Soltis PS et al (2020) Polyploidy: an evolutionary and ecological force in stressful times. Plant Cell 33: 11–26
5. Albert VA, Barbazuk WB, Depamphilis CW et al (2013) The Amborella genome and the evolution of flowering plants. Science 342: 1241089
6. Qin L, Hu Y, Wang J et al (2021) Insights into angiosperm evolution, floral development and chemical biosynthesis from the Aristolochia fimbriata genome. Nat Plants 7(9):1239–1253
7. Van de Peer Y, Mizrachi E, Marchal K (2017) The evolutionary significance of polyploidy. Nat Rev Genet 18:411–424
8. Van de Peer Y (2004) Computational approaches to unveiling ancient genome duplications. Nat Rev Genet 5:752–763
9. Tang H, Bowers JE, Wang X et al (2008) Synteny and collinearity in plant genomes. Science 320:486–488
10. Wan T, Liu Z, Leitch IJ et al (2021) The Welwitschia genome reveals a unique biology underpinning extreme longevity in deserts. Nat Commun 12:4247
11. Belser C, Istace B, Denis E et al (2018) Chromosome-scale assemblies of plant

genomes using nanopore long reads and optical maps. Nat Plants 4:879–887

12. Michael TP, Jupe F, Bemm F et al (2018) High contiguity Arabidopsis thaliana genome assembly with a single nanopore flow cell. Nat Commun 9:1–8
13. Kersey PJ (2019) Plant genome sequences: past, present, future. Curr Opin Plant Biol 48: 1–8
14. Salzberg SL (2019) Next-generation genome annotation: we still struggle to get it right. Genome Biol 20:92
15. Unamba CIN, Nag A, Sharma RK (2015) Next generation sequencing technologies: the doorway to the unexplored genomics of non-model plants. Front Plant Sci 6:1074
16. Kyriakidou M, Tai HH, Anglin NL et al (2018) Current strategies of polyploid plant genome sequence assembly. Front Plant Sci 9:1660
17. Michael TP, VanBuren R (2020) Building near-complete plant genomes. Curr Opin Plant Biol 54:26–33
18. Voshall A, Moriyama EN (2020) Next-generation transcriptome assembly and analysis: impact of ploidy. Methods 176:14–24
19. Lynch M, Conery JS (2000) The evolutionary fate and consequences of duplicate genes. Science 290:1151–1155
20. Tuskan GA, DiFazio S, Jansson S et al (2006) The genome of black cottonwood, Populus trichocarpa (Torr. & Gray). Science 313: 1596–1604
21. Li Z, Barker MS (2020) Inferring putative ancient whole-genome duplications in the 1000 Plants (1KP) initiative: access to gene family phylogenies and age distributions. GigaScience 9:giaa004
22. Cui L, Wall PK, Leebens-Mack JH et al (2006) Widespread genome duplications throughout the history of flowering plants. Genome Res 16:738–749
23. Cai L, Xi Z, Amorim AM et al (2019) Widespread ancient whole-genome duplications in Malpighiales coincide with Eocene global climatic upheaval. New Phytol 221:565–576
24. Godden GT, Kinser TJ, Soltis PS et al (2019) Phylotranscriptomic analyses reveal asymmetrical gene duplication dynamics and signatures of ancient polyploidy in mints. Genome Biol Evol 11:3393–3408
25. Wang Y, Nie F, Shahid MQ et al (2020) Molecular footprints of selection effects and whole genome duplication (WGD) events in three blueberry species: detected by transcriptome dataset. BMC Plant Biol 20:1–14
26. Vanneste K, Sterck L, Myburg AA et al (2015) Horsetails are ancient polyploids: evidence from Equisetum giganteum. Plant Cell 27: 1567–1578
27. Zhang L, Chen F, Zhang X et al (2020) The water lily genome and the early evolution of flowering plants. Nature 577:79–84
28. Rendón-Anaya M, Ibarra-Laclette E, Méndez-Bravo A et al (2019) The avocado genome informs deep angiosperm phylogeny, highlights introgressive hybridization, and reveals pathogen-influenced gene space adaptation. Proc Natl Acad Sci U S A 116:17081–17089
29. Chen J, Hao Z, Guang X et al (2019) Liriodendron genome sheds light on angiosperm phylogeny and species–pair differentiation. Nat Plants 5:18–25
30. Leebens-Mack JH, Barker MS, Carpenter EJ et al (2019) One thousand plant transcriptomes and the phylogenomics of green plants. Nature 574:679–685
31. Wong GK-S, Soltis DE, Leebens-Mack J et al (2020) Sequencing and analyzing the transcriptomes of a thousand species across the tree of life for green plants. Annu Rev Plant Biol 71:741–765
32. Strickler SR, Bombarely A, Mueller LA (2012) Designing a transcriptome next-generation sequencing project for a nonmodel plant species. Am J Bot 99:257–266
33. Honaas LA, Wafula EK, Wickett NJ et al (2016) Selecting superior de novo transcriptome assemblies: lessons learned by leveraging the best plant genome. PLoS One 11: e0146062
34. Haas BJ, Papanicolaou A, Yassour M et al (2013) De novo transcript sequence reconstruction from RNA-seq using the Trinity platform for reference generation and analysis. Nat Protoc 8:1494–1512
35. Hölzer M, Marz M (2019) De novo transcriptome assembly: a comprehensive cross-species comparison of short-read RNA-Seq assemblers. GigaScience 8:giz039
36. Tigano A, Sackton TB, Friesen VL (2018) Assembly and RNA-free annotation of highly heterozygous genomes: the case of the thick-billed murre (Uria lomvia). Mol Ecol Resour 18:79–90
37. Freedman AH, Clamp M, Sackton TB (2021) Error, noise and bias in de novo transcriptome assemblies. Mol Ecol Resour 21:18–29
38. Yang Y, Smith SA (2013) Optimizing de novo assembly of short-read RNA-seq data for phylogenomics. BMC Genomics 14:328
39. Alkan C, Sajjadian S, Eichler EE (2011) Limitations of next-generation genome sequence assembly. Nat Methods 8:61–65

40. Lander ES, Linton LM, Birren B et al (2001) Initial sequencing and analysis of the human genome. Nature 409:860–921
41. Hahn MW, Zhang SV, Moyle LC (2014) Sequencing, assembling, and correcting draft genomes using recombinant populations. G3 4:669–679
42. Wang H, Guo C, Ma H et al (2019) Reply to Zwaenepoel et al.: Meeting the challenges of detecting polyploidy events from transcriptomic data. Mol Plant 12:137–140
43. Ming R, VanBuren R, Wai CM et al (2015) The pineapple genome and the evolution of CAM photosynthesis. Nat Genet 47:1435–1442
44. Cai J, Liu X, Vanneste K et al (2015) The genome sequence of the orchid Phalaenopsis equestris. Nat Genet 47:65–72
45. Vanneste K, Baele G, Maere S et al (2014) Analysis of 41 plant genomes supports a wave of successful genome duplications in association with the Cretaceous–Paleogene boundary. Genome Res 24:1334–1347
46. Schmutz J, Cannon SB, Schlueter J et al (2010) Genome sequence of the palaeopolyploid soybean. Nature 463:178–183
47. Jaillon O, Aury J-M, Noel B et al (2007) The grapevine genome sequence suggests ancestral hexaploidization in major angiosperm phyla. Nature 449:463–467
48. Zhang G-Q, Liu K-W, Li Z et al (2017) The Apostasia genome and the evolution of orchids. Nature 549:379–383
49. Chen L-Y, Morales-Briones DF, Passow CN et al (2019) Performance of gene expression analyses using de novo assembled transcripts in polyploid species. Bioinformatics 35:4314–4320
50. Cheon S, Zhang J, Park C (2020) Is phylotranscriptomics as reliable as phylogenomics? Mol Biol Evol 37:3672–3683
51. Xie Y, Wu G, Tang J et al (2014) SOAPdenovo-Trans: de novo transcriptome assembly with short RNA-Seq reads. Bioinformatics 30:1660–1666
52. Li W, Godzik A (2006) Cd-hit: a fast program for clustering and comparing large sets of protein or nucleotide sequences. Bioinformatics 22:1658–1659
53. Simão FA, Waterhouse RM, Ioannidis P et al (2015) BUSCO: assessing genome assembly and annotation completeness with single-copy orthologs. Bioinformatics 31:3210–3212
54. Bolger AM, Lohse M, Usadel B (2014) Trimmomatic: a flexible trimmer for Illumina sequence data. Bioinformatics 30:2114–2120
55. Ren R, Wang H, Guo C et al (2018) Widespread whole genome duplications contribute to genome complexity and species diversity in angiosperms. Mol Plant 11:414–428
56. Chin C-S, Peluso P, Sedlazeck FJ et al (2016) Phased diploid genome assembly with single-molecule real-time sequencing. Nat Methods 13:1050–1054
57. Zwaenepoel A, Van de Peer Y (2019) wgd—simple command line tools for the analysis of ancient whole-genome duplications. Bioinformatics 35:2153–2155
58. Kim D, Paggi JM, Park C et al (2019) Graph-based genome alignment and genotyping with HISAT2 and HISAT-genotype. Nat Biotechnol 37:907–915
59. Kent WJ (2002) BLAT—the BLAST-like alignment tool. Genome Res 12:656–664
60. Li Z, Baniaga AE, Sessa EB et al (2015) Early genome duplications in conifers and other seed plants. Sci Adv 1:e1501084
61. Stull GW, Qu X-J, Parins-Fukuchi C et al (2021) Gene duplications and genomic conflict underlie major pulses of phenotypic evolution in gymnosperms. Nat Plants 7(8):1015–1025
62. Zwaenepoel A, Li Z, Lohaus R et al (2019) Finding evidence for whole genome duplications: a reappraisal. Mol Plant 12:133–136
63. Johnson MG, Malley C, Goffinet B et al (2016) A phylotranscriptomic analysis of gene family expansion and evolution in the largest order of pleurocarpous mosses (Hypnales, Bryophyta). Mol Phylogenet Evol 98:29–40
64. Devos N, Szövényi P, Weston DJ et al (2016) Analyses of transcriptome sequences reveal multiple ancient large-scale duplication events in the ancestor of Sphagnopsida (Bryophyta). New Phytol 211:300–318
65. Clark JW, Puttick MN, Donoghue PCJ (2019) Origin of horsetails and the role of whole-genome duplication in plant macroevolution. Proc R Soc B Biol Sci 286:20191662
66. Chopra R, Burow G, Farmer A et al (2014) Comparisons of de novo transcriptome assemblers in diploid and polyploid species using peanut (Arachis spp.) RNA-Seq data. PLoS Data 9:e115055
67. Grabherr MG, Haas BJ, Yassour M et al (2011) Full-length transcriptome assembly from RNA-Seq data without a reference genome. Nat Biotechnol 29:644–652
68. Yue J, Wang R, Ma X et al (2020) Full-length transcriptome sequencing provides insights into the evolution of apocarotenoid biosynthesis in Crocus sativus. Comput Struct Biotechnol J 18:774–783

Chapter 4

POInT: Modeling Polyploidy in the Era of Ubiquitous Genomics

Gavin C. Conant

Abstract

Thirteen years ago, we described an evolutionary modeling tool that could resolve the orthology relationships among the homologous genomic regions created by a whole-genome duplication. This tool, which we subsequently named POInT (the *P*olyploid *O*rthology *In*ference *T*ool), was originally only useful for studying a genome duplication known from bakers' yeast and its relatives. Now, with hundreds of genome sequences that contain the relicts of ancient polyploidy available, POInT can be used to study dozens of different polyploidies, asking both questions about the history of individual events and about the commonalities and differences seen between those events. In this chapter, I give a brief history of the development of POInT as an illustration of the interconnected nature of computational biology research. I then further describe how POInT operates and some of the strengths and drawbacks of its structure. I close with a few examples of discoveries we have made using it.

Key words Polyploidy, Evolutionary model, Synteny

Abbreviations

POInT Polyploid Orthology Inference Tool
WGD Whole-genome duplication

1 Polyploidy and the Advent of Genomics

Very shortly after the rediscovery of Mendel's work [1], geneticists started to consider the role of polyploidy, the doubling (or more) of an organism's chromosome complement, in both genetics and evolution [2–4]. That interest continued, such that, when the first eukaryotic genome was released nearly a century later, it was very quickly shown to have the remnants of an ancient polyploidy encoded within it [5, 6].

Yves Van de Peer (ed.), *Polyploidy: Methods and Protocols*,
Methods in Molecular Biology, vol. 2545, https://doi.org/10.1007/978-1-0716-2561-3_4,

Due to the startlingly rapid improvements in sequencing technologies and the associated tools for assembly and genome comparisons [7], there are now hundreds of available genomes from species that underwent polyploidy at some point in their history [8]. For convenience, these polyploid lineages are often divided based on their age into young neopolyploids and old paleopolyploids [9], with an intermediate category of mesopolyploids used in some cases (e.g., [10, 11]). The precise distinction between these types arguably varies depending on the system and questions in play; for the purposes of this chapter, the most salient feature of the polyploidy is whether it occurred sufficiently long ago that both duplicate gene loss and speciation events have occurred since. Hence, unless otherwise qualified, in what follows polyploidy should be understood to refer to meso- or paleopolyploidy events.

These postpolyploidy gene losses are predicted by evolutionary theory, because, in the most straightforward framework, the genetic redundancy created by the polyploidy protects the organism from the deleterious effects of function-abolishing mutations in one copy of the duplicate pair [12, 13], allowing that copy to be lost through a combination of random mutation and genetic drift. This expectation is largely empirically confirmed by the observation that most duplicate genes are short-lived [14, 15], although it is notable that duplicates produced by polyploidies are longer-lived than are others [16].

2 Gene Loss, Comparative Genomics, and the Need for Models

The story of POInT begins with such duplicate losses, and I think it is instructive to give a brief history of how it developed. I do so less for the intrinsic interest of POInT and more as a reminder that many analysis tools, including POInT, are natural, if unexpected, developments of existing ideas and algorithms.

When a gene duplication of any type is shared among more than a single species, the tools of molecular phylogenetics can be used to model and understand its history [17]. In the special case of a polyploidy, however, there is information beyond the gene sequences themselves that can provide a great deal of assistance in understanding that history: the location of the duplicates in relation to the other genes in the genome. We can refer to groups of homologous genes that occur in the same order in two different genomes as being in synteny (although this conserved order is also sometimes referred to as colinearity, with synteny used instead to describe conserved gene content between genomes). Synteny is also of course useful in comparative genomics more generally (see Chen and Zwaenepoel; Berthelot et al.; in this volume; [18]), but, as I will argue below, it is absolutely essential to understanding the history of a paleopolyploidy event.

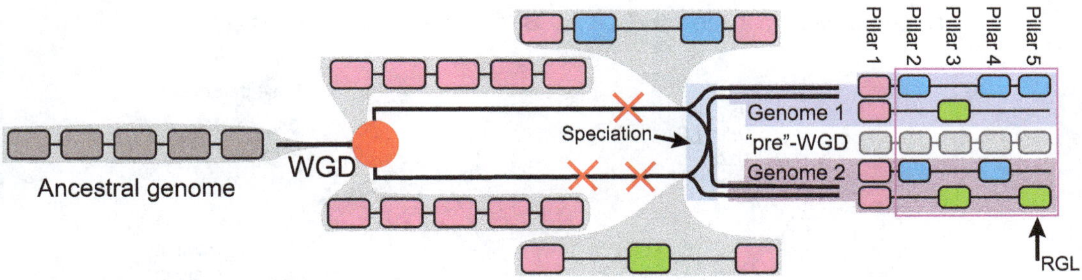

Fig. 1 Schematic of the evolutionary processes modeled with POInT, including gene losses and speciation after a whole-genome duplication. Immediately after the WGD, all five genes are present in two homoeologous copies. Three homoeologous gene losses occur prior to the split of the two species (red "X"s), one in the less fractionated subgenome (Track "0;" yielding the green gene in the lower window) and two from the more fractionated subgenome (Track "1;" yielding the two blue genes in the upper window). After the speciation event, Genome 1 loses a homoeolog from the more fractionated subgenome and Genome 2 loses one from the less fractionated subgenome, resulting in a case of reciprocal gene loss (RGL). The result is five "pillars" of duplicated or lost duplicated genes for the two genomes. The boxed region illustrates the principle that even polyploidies where most or all duplicates have been lost still show detectable patterns of DCS relative to a nonpolyploid outgroup

I was introduced to this argument during my postdoctoral work with Ken Wolfe: Ken and Kevin Byrne had just completed the Yeast Gene Order Browser (YGOB; [19]), a manually-verified set of homologous genes from many yeast genomes, depicted relative to their orders on their chromosomes (http://ygob.ucd.ie). YGOB illustrates the *double-conserved synteny* (DCS) preserved in the yeast genomes after their shared paleopolyploidy by comparing those genomes to other yeast genomes lacking that polyploidy (Fig. 1). A key aspect of DCS is that it is evident not merely among the genes that survive in duplicate from the polyploidy but also among those genes where one of the two duplicate copies has been lost. In principle, with a sufficiently closely related nonpolyploid relative, a polyploidy event could be identified using DCS even if every single duplicate gene pair created by it had lost one of its members (as shown in the boxed region at the extreme right of Fig. 1).

At the time YGOB was created, yeasts were unusual in that genome sequences for many closely related species were available, whereas, in most other groups of eukaryotes, the sequenced genomes were phylogenetically widely spaced. Using YGOB, the Wolfe laboratory made a number of discoveries about the yeast polyploidy and those genomes more generally. They documented the independent loss of alternate duplicate copies in different lineages [15]: These *reciprocal gene losses* (RGLs) could potentially reproductively isolate the lineages in question from each other [20]. They also described a biosyntentic gene cluster that had recently been "born" in bakers' yeast [21] and showed that the earliest phases of duplicate gene losses after the yeast polyploidy had been quite rapid [15].

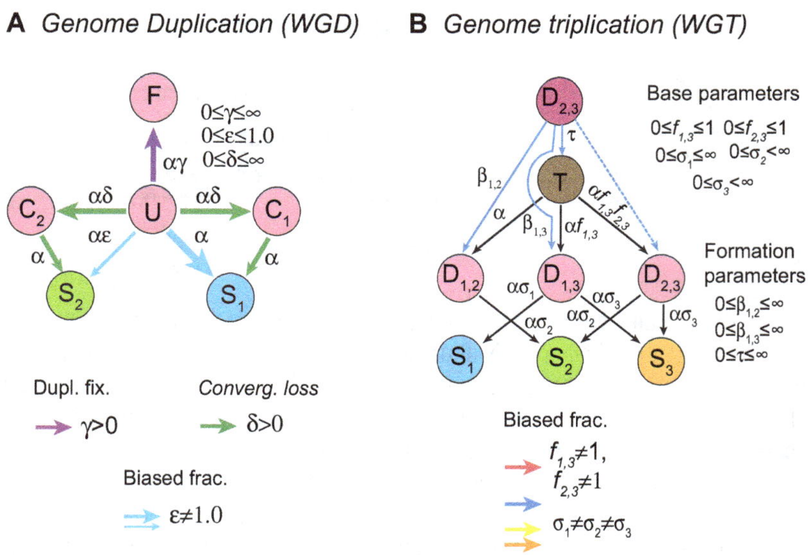

Fig. 2 Models of gene loss after polyploidy for a whole-genome duplication (WGD) and a whole-genome triplication (WGT, hexaploidy). (**a**) WGD: All pillars start in duplicated state **U** (*U*ndifferentiated), from which they can transition to either the three other duplicated states, $\mathbf{C_1}$ (*C*onverging state 1), $\mathbf{C_2}$ (*C*onverging state 2) and **F** (*F*ixed) or to the two single-copy states $\mathbf{S_1}$ (*S*ingle-copy *1*) and $\mathbf{S_2}$ (*S*ingle-copy *2*). We define subgenome 1 ($\mathbf{S_1}$) to be the preferred subgenome with more surviving gene copies: Thus, $\mathbf{C_1}$ and $\mathbf{S_1}$ are states where the genes from the less-fractionated parental subgenome will be or are preserved, and $\mathbf{C_2}$ and $\mathbf{S_2}$ the corresponding states for the more-fractionated parental subgenome. Duplicate fixation is inferred when $\gamma \neq 0$, convergent losses when $\delta \neq 0$, and biased fractionation when $\varepsilon < 1.0$. (**b**) WGT: All pillars start in state **T** (*T*riplicated) and transition first to duplicated states ($\mathbf{D_{x,y}}$) and hence to the single-copy states ($\mathbf{S_x}$). Subgenome 1 is assumed to be favored (fewer losses) and the identity of that subgenome inferred in the POInT computation. Losses from the triplicated state are then increasingly disfavored first to $\mathbf{D_{1,3}}$ (parameter $f_{1,3}$) and $\mathbf{D_{2,3}}$ (parameter $f_{2,3}$). There are also individual rates of loss from the duplicated to single-copy states (σ_x). As described in the text, we add a formation step to this model, with the initial tetraploidy being represented with the $\mathbf{D_{2,3}}$ state, which transitions to state **T** unless losses occur prior to the second allopolyploidy (parameters $\beta_{1,2}$, $\beta_{1,3}$, and τ)

This last question of gene losses reminded me of work by my undergraduate advisor, Paul Lewis, on models for the phylogenetic analysis of nonsequence data [22]. Presented with the comparative data in YGOB, it was not difficult to use its encoded presence/absence data as the states of a phylogenetic model (Fig. 2).

Armed with a simple version of this model, I was able to assist with the next project in the lab: adding a new genome to YGOB. Ken had realized that the genomes that would be most informative as to the history of the yeast polyploidy were not the close relatives of bakers' yeast but rather those like that of *Vanderwaltozyma polyspora* (then known as *Kluyveromyces polysporus*), which had diverged from bakers' yeast shortly after the polyploidy.

It became clear that, because *V. polyspora* and bakers' yeast (*S. cerevisiae*) had diverged so soon after the polyploidy event (Fig. 3), there was a significant challenge in inferring *orthology* for

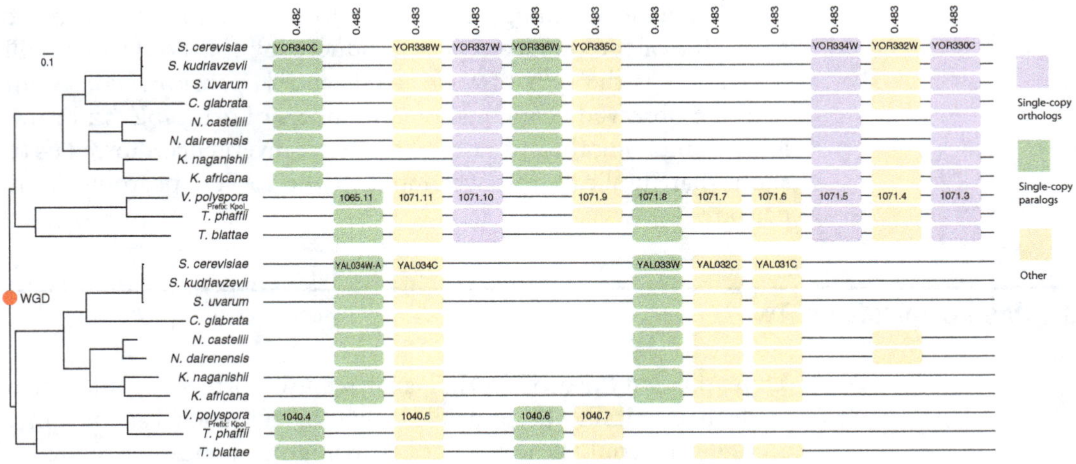

Fig. 3 Inferred orthologous regions from 11 species produced in the yeast WGD event: For clarity, gene names are shown only for *S. cerevisiae* and *V. polyspora*. Pillars shown in purple are cases of single-copy orthologs for these two species: green pillars are single-copy paralogs. Cases where either or both genomes retain homoeologous genes are shown in tan. As discussed in the text, the confidence in the orthology estimate (P_j) is given above each column. Because the yeast WGD has limited evidence for biased fractionation [32], the relationships depicted are degenerate, with the swap of the bottom and top blocks having an identical probability (hence doubling this number gives a sense of the "real" confidence). The mirrored topologies have branch lengths scaled by POInT's estimates of the relative number of duplicate losses ($\alpha \times$ time with a model without biased fractionation, i.e., $\varepsilon = 1.0$ in Fig. 2)

these two genomes. Orthologs, of course, are homologous genes in different species that last shared a common ancestor at the speciation of those species [23]. When two species split almost immediately after a polyploidy, the few *shared* duplicate gene losses that occurred before that split will necessarily produce single-copy orthologs in those species. However, the losses that occur after the speciation will be independent (Fig. 1 illustrates these processes). If we confine ourselves to the genes that are single-copy in both genomes, these independent losses will result in effectively equal numbers of paralogs and orthologs. We show an example of this phenomenon in Fig. 3, where, for the region shown, more of the single-copy genes in *S. cerevisiae* and *V. polyspora* are paralogs (green) than are orthologs (purple).

Under these circumstances, the problem becomes phasing the DCS blocks from *S. cerevisiae* with the orthologous block from *V. polyspora*. YGOB made its orthology phasing by pairing syntenic blocks so as to maximize the number of syntenic positions where all of the polyploid genomes had the same gene status (e.g., present or missing). With this approach, we estimated that about 55% of single-copy genes in *S. cerevisiae* and *V. polyspora* are orthologs and 45% paralogs [24].

This high degree of dissimilarity between the two genomes raises some concerns (to be considered shortly), but we were able to use YGOB's inferences to fit our duplicate loss model and infer the phylogenetic timing of the gene losses. We found that more than 80% of the duplicate pairs produced by the WGD were still duplicated at the split of *S. cerevisiae* and *V. polyspora*, consistent with the observed near equality in numbers of orthologs and paralogs. Comparing the loss rates to sequence divergence allowed us to show that duplicate loss was rapid in the earliest postpolyploidy period.

3 One Polyploidy or Two?

Thus, the first "proto"-POInT analyses used orthology inferences made heuristically. In principle, such an approach can be quite accurate and could be thought of as analogous to multiple sequence alignment in molecular phylogenetics, where appropriately scaled dissimilarity penalties and hill-climbing methods are used to identify the homologous bases/residues that are then analyzed with probabilistic models of sequence evolution [25–27]. However, there were a few potential concerns. Firstly, could we be sure that *S. cerevisiae* and *V. polyspora* were actually products of the same polyploidy event? A heuristic approach will assume a common polyploidy and produce the syntenic blocks most consistent with that hypothesis. But how could we know that independent polyploidies analyzed with such a tool might not produce a similar 55%/45% split of inferred orthologs to paralogs? The heuristic approach would also work best for large synteny blocks, where even those few shared prespeciation losses will overwhelm any chance effects and give correct orthology phasing. But what is "large" in this context and how serious are such chance effects likely to be? Clearly what was needed is a modeling approach that accounted for the uncertainty in the orthology phasing.

The core of this new model came from adapting an unrelated project from my doctoral work. That project was converting the linkage analysis package Genehunter [28, 29] to run on a supercomputer [30]. Genehunter uses the linkage analysis algorithm of Lander and Green [31] to locate a disease-causing genetic variant in the genome by taking a number of genetic markers from several members of a family afflicted with that disease. The algorithm computes the odds of the disease-causing allele being at each of a number of points within the genetic map, given the known genotypes for each individual at each marker and the recombination rates between those markers. What is important about this algorithm is that while the markers for each individual are known, it is not known whether any given marker *allele* is derived from the individual's mother or father. As a result, the calculation carries this parental uncertainty throughout the analysis, meaning that, given n genotyped individuals, at each marker there are (essentially) 2^n inheritance patterns that must be considered.

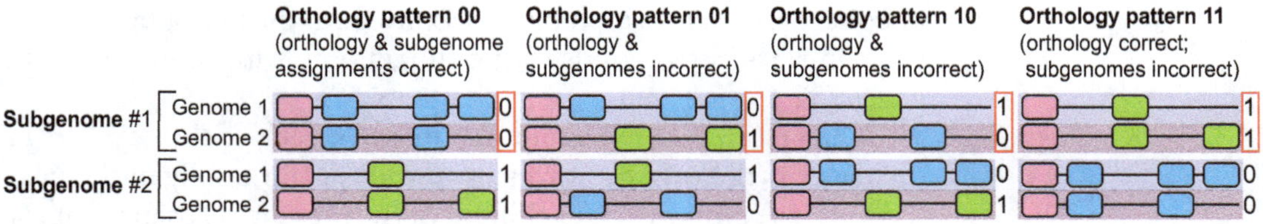

Fig. 4 Four possible orthology relationships (i.e., subgenome assignments) for two genomes sharing a WGD, using the track identifiers (0,1) and colors from Fig. 1, showing how different orthology assignments give rise to different patterns of orthology and paralogy in a column. Hence, the first block depicts the correct orthology taken from Fig. 1, while the remaining three swap subgenome assignments for one or both of the sections of the two extant genomes. POInT computes the likelihood of each and then conditions them on pillars to the left and right (see text)

I realized that the unknown orthology states in the polyploidy model could be treated with the mathematics of the Lander and Green algorithm: It was this insight that made POInT a useful tool. The starting point is the model of duplicate loss along a phylogeny already discussed (Fig. 2). In our original analysis, we assumed that the orthology relationships were known and simply computed the likelihood of the resulting loss data under the model. However, for a given duplicated locus (pillar, Fig. 1), it is perfectly possible to compute a likelihood for any or all possible orthology relationships (Fig. 4). Indeed, for the special case of a duplicate pair fully preserved in each genome, that likelihood is the same for all possible orthology relationships.

4 The POInT Computation

For clarity, let us first assign identifiers 0 and 1 to the two genomic regions produced by the WGD in each of the n genomes considered (Fig. 4). These assignments are purely for bookkeeping purposes: We do not assume that section "0" in one genome is orthologous to section "0" in another. We must compute the likelihood for a set of presence/absence data for every possible combination of orthology relationships: For n genomes, there are 2^n such relationships. POInT can make similar computations for hexaploidies and octoploidies, but the notation is more cumbersome, so I confine my examples here to the case of a genome duplication/tetraploidy. We represent an orthology relationship as a binary vector $\overline{O}$ of length n. For example, for two taxa, the four possible values of $\overline{O}$ are `00` (0 in decimal notation), `01` (1), `10` (2), and `11` (3). The value at the gth position of the vector (0 or 1) indicates which of the two genomic sections is assigned to the "top" track for the gth genome, with the other assigned to the lower track (Fig. 1). When modeling *biased fractionation,* the unequal preservation of duplicate copies from the two subgenomes, this top track corresponds to the subgenome with more surviving genes [32].

The first step of the computation is, for each of the m pillars in the dataset (the origins of which we will explain briefly in the next section), to compute the likelihood of the presence/absence data at that pillar under each of the 2^n orthology relationships, given a phylogeny and set of model parameters. These likelihoods are stored in a vector $\overline{L}$, indexed by the 2^n orthology relationships in binary format.

The insight of the Lander and Green algorithm comes in at this point. We would like to assess if there is an orthology relationship that has relatively high likelihood for a number of pillars. To do so, we take advantage of the synteny relationships between neighboring pillars, if such relationships exist. We therefore design a transition probability matrix $\Theta^{i-i,i}$ that describes the synteny transition between pillars $i-1$ and pillar i. This matrix gives the probability of orthology relationship j at pillar i given that pillar $i-1$ has orthology relationship k, where both j and k represent orthology relationship vectors when read in binary format. The elements of this matrix $\Theta^{i-1,i}_{j,k}$ have the form:

$$\Theta^{i-1,i}_{j,k} = \prod_{l=0}^{n-1} \theta_{i,l}^{((j\wedge k)\gg l)\&1} \cdot (1-\theta_{i,l})^{1-(((j\wedge k)\gg l)\&1)} \tag{1}$$

where "^" represents the bit-wise *exclusive-or* operator, "≫" the bit-wise shift operator, and " & " a bit-wise *and*. Hence, as horrifying as this equation looks, it simply implies that we consider the genomes $l = [0\ldots n-1]$ and construct a product that uses $\theta_{i,l}$ at positions where $\overline{O_j}$ and $\overline{O_k}$ differ and $(1-\theta_{i,l})$ at positions where they are the same. Here, $\theta_{i,l}$ is the probability that the orthology assignments change between a pair of pillars for genome l. Its value, however, differs depending on the synteny information for the two pillars. If, between $i-1$ and i, synteny is maintained in either the upper or lower track or both, $\theta_{i,l} = \theta$, a global constant estimated from the data by maximum likelihood. Otherwise, $\theta_{i,l} = 0.5$ for genome l, meaning position $i-1$ conveys no information on the orthology relationships at i. We will return to the interpretation of θ shortly.

We can now construct expressions for our orthology estimates at a position, given the information at one or more other neighboring positions. Recall that $\overline{L}^i$ contains the likelihood of the data at i for each possible orthology state. We now define $\overline{L}^{i-1|D_0\ldots D_{i-1}}$, the likelihood of each possible orthology state at $i-1$, given the data at pillars $0\ldots i-1$. We can write a recurrence equation for $\overline{L}^{i|D_0\ldots D_i}$ using $\overline{L}^{i-1|D_0\ldots D_{i-1}}$, $\overline{L}^i$, and the transition probability matrix $\Theta^{i-1,i}$:

$$\overline{L}^{i|D_0\ldots D_i} = \overline{L}^i \odot \left(\Theta^{i-1,i} \cdot \overline{L}^{i-1|D_0\ldots D_{i-1}}\right) \tag{2}$$

where "⊙" represents an element-wise vector product. When we define the base-case $\overline{L}^{0|D_0} = \overline{L}^0$, we can apply this equation sequentially to the m pillars, with the total likelihood of the data L being

$$L = \sum_{l=0}^{2^n-1} L_l^{m-1|D_0 \ldots D_{m-1}} \tag{3}$$

In other words, the total likelihood is just the sum of the likelihoods of ending up in each of the 2^n orthology states at the last pillar (pillar $m - 1$). Using eq. 3, we optimize the model parameters by maximum likelihood using standard numerical approaches [33]. POInT can either perform this optimization for a user-supplied phylogenetic topology or search across all possible rooted topologies and return the one with the highest likelihood.

To obtain the orthology inferences themselves, as well as associated confidence estimates, we need to add what is known in the HMM literature as a posterior decoding step. This step is analogous to obtaining the odds ratios for the disease allele placement in the Lander and Green algorithm; a similar approach has been applied to estimating correlated rate variation across sites in a sequence alignment [34]. For a given pillar i, we compute, as before $\overline{L}^{i-1|D_0 \ldots D_{i-1}}$. We then compute the analogous probabilities working from the last pillar $m - 1$ down to pillar $i + 1$:$\overline{L}^{i+1|D_{i+1} \ldots D_{m-1}}$ Conceptually, this step corresponds to applying eq. 2 to the reversed sequence of pillars. We then obtain $\overline{L}^{i|D}$, a vector of conditional likelihoods for all of the 2^n orthology states at pillar i, given all of the observed gene presence/absence data, denoted as$D$:

$$\overline{L}^{i|D} = \left(\overline{L}^i \odot \left(\Theta^{i-1,i} \cdot \overline{L}^{i-1|D_0 \ldots D_{i-1}}\right)\right) \odot \left(\Theta^{i,i+1} \cdot \overline{L}^{i+1|D_{i+1} \ldots D_{m-1}}\right) \tag{4}$$

The elements of $\overline{L}^{i|D}$ sum to L, and the conditional probability P_j of any given state j is just the jth element of that vector divided by that total likelihood L:

$$P_j = \frac{L_j^{i|D}}{L} \tag{5}$$

These conditional probabilities give us estimates of our confidence in a given orthology inference. In Fig. 3, the values at the top are such conditional probabilities. Because the yeast WGD is not marked by biased fractionation, the model is completely symmetrical, with states C_1 and C_2 and S_1 and S_2 being equivalent (Fig. 2). As a result, the orthology relationships shown have the same conditional probabilities as those where the top and bottom tracks are swapped, which accounts for the fact that P_j has values near to ½ rather than to 1 for these examples.

4.1 Comments on the POInT Computation

In the next section, I will give some examples that I hope illustrate the value of POInT for analyzing polyploidies. But I think it is first appropriate to note a few weaknesses.

First, POInT is only as accurate as the presence/absence data used as input. We have described a pipeline for producing these pillars for cases when manual sets like those of YGOB are not available [32, 35]. Briefly, each polyploid genome is searched against an outgroup genome lacking the event using BLAST. The resulting homologs are then placed in DCS blocks relative to that outgroup using simulated annealing [36]. The DCS blocks for individual genomes are merged, and a global pillar order that minimizes synteny breaks is inferred and used by POInT. This approach works especially well for more recent polyploidies with relatively few rearrangements observed between the genomes. We have shown that even for the ancient teleost genome duplication, it is adequate [35] but have found it to be problematic for the very old vertebrate 2R events (data not shown).

A second concern is with the θ parameter from eq. 1. Unlike the model parameters in Fig. 2, θ is an error or nuisance parameter in the model without a biological interpretation. Hence, unlike the Lander and Green algorithm, where the corresponding parameter is a recombination rate, θ only indicates how often our model "changes its mind" about the orthology assignments from pillar to pillar. Fortunately, in practice, θ tends to be quite small: In a recent analysis of seven tetraploidies, we found $0.002 \leq \theta \leq 0.009$, meaning that the inferred orthology was almost invariably maintained between adjacent pillars. It is hence a useful diagnostic tool: If a large (>0.1) value of θ is inferred, the resulting orthology inferences are of low quality.

4.2 Example Uses of POInT

Testing for a shared polyploidy. One of the first uses we made of POInT was testing whether the polyploidies found in *V. polyspora* and *S. cerevisiae* were in fact a single shared event. In our modeling framework, two independent polyploidies can be easily represented as a phylogeny like that in Fig. 3 where the root branch has zero length. In that case, the clades on either side of that root have no common losses from their respective polyploidies. The question of how many *apparently* shared losses we would infer in the case of independent polyploidies can be addressed with simulation: giving POInT a phylogeny with a zero-length root and set of model parameters and asking it to simulate new polyploid genomes under that model. For each simulation, we ask how much smaller the likelihood seen under a model with a forced zero length root (an independent polyploidy model) is than is the likelihood of a model where the root length can take on any value (e.g., a model of a shared polyploidy; [37]). We then ask whether the difference in these two likelihoods for the real data is much greater than the differences seen among the simulations. A large improvement in likelihood for the shared polyploidy model relative to the independent one when analyzing the real data as compared to analyzing the simulations suggests significant evidence of a shared polyploidy between these lineages, and was indeed what we found [37]. We conducted an identical analysis more recently to show that three species of parasitic nematodes also descend from a common triploid ancestor [38].

Confirming the existence of biased fractionation. As with the question of the shared *S. cerevisiae/V. polyspora* genome duplication, inferences of biased fractionation have the potential to be influenced by the methods used to detect them. Hence, while biased fractionation is inferred when heuristic approaches for orthology inference are used on polyploid genomes [39, 40], there remained the slight concern that such methods, in seeking to maximize genomic similarity, might infer a bias in gene losses where none exists. Our models incorporate bias in fractionation with model parameters such as ε in Fig. 2a. Using likelihood ratio tests [41], we initially showed [32] that models with biased fractionation ($\varepsilon < 1.0$) were preferred over those without it ($\varepsilon = 1.0$). However, one might still argue that the maximization approach inherent in making these parameter estimates was giving rise to spurious inferences of bias. In the end, I again adopted a simulation approach, creating simulated polyploidies without biased fractionation and showing that analyzing those simulated genomes using models that included biased fractionation did not give estimates of the ε parameter that were anything like as small (e.g., high bias) as was observed for real data [35]. We can therefore be quite certain that biased fractionation is a common pattern in paleopolyploid genomes [32, 35, 38, 42].

Modeling hexaploidy. We recently completed an analysis of the hexaploidy shared by the *Brassica* crops *B. oleracea* (broccoli, cabbage, cauliflower) and *B. rapa* (kale) and their relatives [42]. The 3: 1 pattern of conserved synteny between these genomes and that of *Arabidopsis* is clear [11], but there is not an obvious single mutational step that could produce such hexaploids. Instead, it is believed that they were formed when two diploids hybridized to form an allotetraploid, which subsequently underwent an allopolyploid hybridization with another diploid to form an allohexaploid [43]. The extant genomes show signs of strong biased fractionation, such that one of the three subgenomes shows many more surviving genes (the LF, or "least fractionated" subgenome) than the other two (MF1 and MF2, "more" and "most" fractionated, respectively). As a result, it was argued that the progenitor of the LF genome was most likely the "last arriving," i.e., the diploid progenitor contributing to the second allopolyploidization. We developed a new type of model for a two-step hexaploidy (Fig. 2b) that includes the transition from a genome duplication (state $D_{2,3}$) to a triplication (state T) prior to gene losses and speciation events after the second allopolyploidy [42]. If the two allopolyploidies were separated in time, not all of the initially duplicated genes would have survived in duplicate to become triplicated. Denoting LF as subgenome 1, we can model this effect with transitions from state $D_{2,3}$ to $D_{1,2}$ (loss of a duplicate from MF2 prior to LF arriving) or from $D_{2,3}$ to $D_{1,3}$ (loss from MF1). Because all of these transitions occurred prior to the first speciation in our data, we cannot estimate whether any given pillar underwent a loss prior to the arrival of LF. However, treating the pillars as a whole allows us to estimate the frequency of such losses, as well as to compare our inferences to models where the LF subgenome was one of the original progenitors, rather than the last arriving. The POInT framework gives us strong confidence both in there having been a number of gene losses prior to the arrival of LF and that LF was indeed the last arriving subgenome, confirming the two-step model for this event.

5 Future Directions and Concluding Remarks

POInT is a complex and slightly idiosyncratic program, and running a full analysis with it is unnecessary for most users. Instead, we have made our inferences from POInT publicly available in a graphical format with $\text{POInT}_{\text{browse}}$ (wgd.statgen.ncsu.edu). Here, users can first visually explore the DCS blocks and orthology inferences and then download the associated coding regions and inferred gene trees, as well as gain access to larger datasets from our analyses. Our main future goal for this portal is to add a "batch download" feature that will allow users to request sets of pillars that meet specified criteria, such as orthology confidence and number of surviving homoeologs. The synteny visualizations can also be downloaded in publication quality: such was the origin of Fig. 3.

We also plan to further extend POInT to handle nested polyploidies, using the framework of Fig. 2b. While computationally expensive, such models could allow the analysis of, for instance, the recent soybean polyploidy simultaneously with the more ancient shared legume WGD. With such an addition, as well as a few other polyploidies we have not yet considered, such as that in the Solanaceae [44], we will be able to provide something close to a comprehensive reference for paleopolyploidies in the sequenced genomes. From there, questions such as the degree of convergent evolution inherent to polyploidy, the sources of biases in duplicate losses and the timing of losses after hybridization can be addressed in detail using POInT's inferences as a starting point (pun intended).

Acknowledgments

I would like to thank J. Thorne and A. Zwaenepoel for assistance with clarifying the mathematical description of POInT's computations.

References

1. Simunek M, Hoßfeld U, Wissemann V (2011) "Rediscovery" revised–the cooperation of Erich and Armin von Tschermak-Seysenegg in the context of the "rediscovery" of Mendel's laws in 1899–1901. Plant Biol 13(6):835–841
2. Clausen R, Goodspeed T (1925) Interspecific hybridization in Nicotiana. II. A tetraploid glutinosa-tabacum hybrid, an experimental verification of Winge's hypothesis. Genetics 10(3):278
3. Kuwada Y (1911) Maiosis in the pollen mother cells of Zea Mays L. (with plate V.). 植物学雑誌 25(294):163–181
4. Taylor JS, Raes J (2004) Duplication and divergence: the evolution of new genes and old ideas. Annu Rev Genet 38:615–643
5. Goffeau A, Barrell BG, Bussey H, Davis RW, Dujon B, Feldmann H, Galibert F, Hoheisel JD, Jacq C, Johnston M, Louis EJ, Mewes HW, Murakami Y, Philippsen P, Tettelin H, Oliver SG (1996) Life with 6000 genes. Science 274(5287):546, 563–567
6. Wolfe KH, Shields DC (1997) Molecular evidence for an ancient duplication of the entire yeast genome. Nature 387(6634):708–713
7. Goodwin S, McPherson JD, McCombie WR (2016) Coming of age: ten years of next-generation sequencing technologies. Nat Rev Genet 17(6):333
8. Van de Peer Y, Mizrachi E, Marchal K (2017) The evolutionary significance of polyploidy. Nat Rev Genet 18(7):411–424
9. Hilu K (1993) Polyploidy and the evolution of domesticated plants. Am J Bot 80(12): 1494–1499
10. Mandáková T, Joly S, Krzywinski M, Mummenhoff K, Lysak MA (2010) Fast diploidization in close mesopolyploid relatives of Arabidopsis. Plant Cell 22(7):2277–2290
11. Wang X, Wang H, Wang J, Sun R, Wu J, Liu S, Bai Y, Mun JH, Bancroft I, Cheng F, Huang S, Li X, Hua W, Freeling M, Pires JC, Paterson AH, Chalhoub B, Wang B, Hayward A, Sharpe AG, Park BS, Weisshaar B, Liu B, Li B, Tong C, Song C, Duran C, Peng C, Geng C, Koh C, Lin C, Edwards D, Mu D, Shen D, Soumpourou E, Li F, Fraser F, Conant G, Lassalle G, King GJ, Bonnema G, Tang H, Belcram H, Zhou H, Hirakawa H, Abe H, Guo H, Jin H, Parkin IA, Batley J, Kim JS, Just J, Li J, Xu J, Deng J, Kim JA, Yu J, Meng J, Min J, Poulain J, Hatakeyama K, Wu K, Wang L, Fang L, Trick M, Links MG, Zhao M, Jin M, Ramchiary N, Drou N, Berkman PJ, Cai Q, Huang Q, Li R, Tabata S, Cheng S, Zhang S, Sato S, Sun S, Kwon SJ, Choi SR, Lee TH, Fan W, Zhao X, Tan X, Xu X, Wang Y, Qiu Y, Yin Y, Li Y, Du Y, Liao Y, Lim Y, Narusaka Y, Wang Z, Li Z, Xiong Z, Zhang Z (2011) The genome of the mesopolyploid crop species Brassica rapa. Nat Genet 43(10):1035–1039
12. Li W-H (1980) Rate of gene silencing at duplicate loci: a theoretical study and interpretation of data from tetraploid fish. Genetics 95: 237–258
13. Nei M, Roychoudhury AK (1973) Probability of fixation of nonfunctional genes at duplicate loci. Am Nat 107:362–372
14. Lynch M, Conery JS (2000) The evolutionary fate and consequences of duplicate genes. Science 290:1151–1155
15. Scannell DR, Byrne KP, Gordon JL, Wong S, Wolfe KH (2006) Multiple rounds of speciation associated with reciprocal gene loss in polyploid yeasts. Nature 440:341–345
16. Walsh B (2003) Population-genetic models of the fates of duplicate genes. Genetica 118(2–3):279–294
17. Li W-H (1997) *Molecular evolution*. Sinauer Associates, Sunderland
18. Kristensen DM, Wolf YI, Mushegian AR, Koonin EV (2011) Computational methods for Gene Orthology inference. Brief Bioinform 12(5):379–391
19. Byrne KP, Wolfe KH (2005) The Yeast Gene Order Browser: combining curated homology and syntenic context reveals gene fate in polyploid species. Genome Res 15(10):1456–1461
20. Werth CR, Windham MD (1991) A model for divergent, allopatric speciation of polyploid pteridophytes resulting from silencing of duplicate-gene expression. Am Nat 137(4): 515–526

21. Wong S, Wolfe KH (2005) Birth of a metabolic gene cluster in yeast by adaptive gene relocation. Nat Genet 37(7):777
22. Lewis PO (2001) A likelihood approach to estimating phylogeny from discrete morphological character data. Syst Biol 50:913–925
23. Koonin EV (2005) Orthologs, paralogs, and evolutionary genomics. Annu Rev Genet 39: 309–338
24. Scannell DR, Frank AC, Conant GC, Byrne KP, Woolfit M, Wolfe KH (2007) Independent sorting-out of thousands of duplicated gene pairs in two yeast species descended from a whole-genome duplication. Proc Natl Acad Sci U S A 104:8397–8402
25. Henikoff S, Henikoff JG (1992) Amino-acid substitution matrices from protein blocks. Proc Natl Acad Sci U S A 89(22): 10915–10919
26. Notredame C, Higgins DG, Heringa J (2000) T-coffee: a novel method for fast and accurate multiple sequence alignment. J Mol Biol 302(1):205–217
27. Felsenstein J (1981) Evolutionary trees from DNA sequences: a maximum likelihood approach. J Mol Evol 17:368–376
28. Kruglyak L, Daly MJ, Reeve-Daly MP, Lander ES (1996) Parametric and nonparametric linkage analysis: a unified multipoint approach. Am J Hum Genet 58:1347–1363
29. Kruglyak L, Lander ES (1998) Faster multipoint linkage analysis using Fourier transforms. J Comput Biol 5(1):1–7
30. Conant GC, Plimpton SJ, Old W, Wagner A, Fain PR, Pacheco TR, Heffelfinger G (2003) Parallel Genehunter: implementation of a linkage analysis package for distributed-memory architectures. J Parallel Distrib Comput 63: 674–682
31. Lander ES, Green P (1987) Construction of multilocus genetic linkage maps in humans. Proc Natl Acad Sci U S A 84:2363–2367
32. Emery M, Willis MMS, Hao Y, Barry K, Oakgrove K, Peng Y, Schmutz J, Lyons E, Pires JC, Edger PP, Conant GC (2018) Preferential retention of genes from one parental genome after polyploidy illustrates the nature and scope of the genomic conflicts induced by hybridization. PLoS Genet 14(3): e1007267em
33. Press WH, Teukolsky SA, Vetterling WA, Flannery BP (1992) Numerical recipes in C. Cambridge University Press, New York
34. Felsenstein J, Churchill GA (1996) A hidden Markov model approach to variation among sites in rate of evolution. Mol Biol Evol 13(1): 93–104
35. Conant GC (2020) The lasting after-effects of an ancient polyploidy on the genomes of teleosts. PLoS One 15(4):e0231356
36. Kirkpatrick S, Gelatt CDJ, Vecchi MP (1983) Optimization by simulated annealing. Science 220(4598):671–680
37. Conant GC, Wolfe KH (2008) Probabilistic cross-species inference of orthologous genomic regions created by whole-genome duplication in yeast. Genetics 179:1681–1692
38. Schoonmaker A, Hao Y, Bird D, Conant GC (2020) A single, shared triploidy in three species of parasitic nematodes. G3 10:225–233
39. Freeling M, Thomas BC (2006) Gene-balanced duplications, like tetraploidy, provide predictable drive to increase morphological complexity. Genome Res 16:805–814
40. Schnable JC, Springer NM, Freeling M (2011) Differentiation of the maize subgenomes by genome dominance and both ancient and ongoing gene loss. Proc Natl Acad Sci U S A 108(10):4069–4074
41. Sokal RR, Rohlf FJ (1995) Biometry, 3rd edn. W. H. Freeman and Company, New York
42. Hao Y, Mabry ME, Edger P, Freeling M, Zheng C, Jin L, VanBuren R, Colle M, An H, Abrahams RS, Washburn JD, Qi X, Barry K, Daum C, Shu S, Schmutz J, Sankoff D, Barker MS, Lyons E, Pires JC, Conant GC (2021) The contributions of the allopolyploid parents of the mesopolyploid Brassiceae are evolutionarily distinct but functionally compatible. Genome Res 31:799–810
43. Tang H, Woodhouse MR, Cheng F, Schnable JC, Pedersen BS, Conant G, Wang X, Freeling M, Pires JC (2012) Altered patterns of fractionation and exon deletions in Brassica rapa support a two-step model of paleohexaploidy. Genetics 190(4):1563–1574
44. Bombarely A, Moser M, Amrad A, Bao M, Bapaume L, Barry CS, Bliek M, Boersma MR, Borghi L, Bruggmann R (2016) Insight into the evolution of the Solanaceae from the parental genomes of Petunia hybrida. Nat Plants 2(6):1–9

Chapter 5

Applying Machine Learning to Classify the Origins of Gene Duplications

Michael T. W. McKibben and Michael S. Barker

Abstract

Nearly all lineages of land plants have experienced at least one whole-genome duplication (WGD) in their history. The legacy of these ancient WGDs is still observable in the diploidized genomes of extant plants. Genes originating from WGD—paleologs—can be maintained in diploidized genomes for millions of years. These paleologs have the potential to shape plant evolution through sub- and neofunctionalization, increased genetic diversity, and reciprocal gene loss among lineages. Current methods for classifying paleologs often rely on only a subset of potential genomic features, have varying levels of accuracy, and often require significant data and/or computational time. Here, we developed a supervised machine learning approach to classify paleologs from a target WGD in diploidized genomes across a broad range of different duplication histories. We collected empirical data on syntenic block sizes and other genomic features from 27 plant species each with a different history of paleopolyploidy. Features from these genomes were used to develop simulations of syntenic blocks and paleologs to train a gradient boosted decision tree. Using this approach, Frackify (Fractionation Classify), we were able to accurately identify and classify paleologs across a broad range of parameter space, including cases with multiple overlapping WGDs. We then compared Frackify with other paleolog inference approaches in six species with paleotetraploid and paleohexaploid ancestries. Frackify provides a way to combine multiple genomic features to quickly classify paleologs while providing a high degree of consistency with existing approaches.

Key words Polyploidy, Whole-genome duplication, Paleologs, Ohnologs, Fractionation, Machine learning, Ks, Synteny

1 Introduction

Polyploidy, or whole-genome duplication (WGD), is a common phenomenon among some lineages of eukaryotes, particularly flowering plants where nearly 30% of extant species are estimated to be recent polyploids [1–3]. Although many of these polyploid species ultimately go extinct [4–7], many return to the diploid state through a process known as diploidization [8]. This is a ubiquitous

Yves Van de Peer (ed.), *Polyploidy: Methods and Protocols*,
Methods in Molecular Biology, vol. 2545, https://doi.org/10.1007/978-1-0716-2561-3_5,

process as most land plants are ancient polyploids that have diploidized, with several lineages containing multiple WGD in their recent and ancient history [9]. WGD-derived genes, also known as paleologs or ohnologs, are rapidly lost during diploidization in a process termed fractionation [8, 10–12]. The paleologs that are not lost during this process can persist in diploidized genomes for millions of years [13, 14]. Paleologs are thought to play important roles in a variety of evolutionary processes including speciation through reciprocal gene loss or lineage-specific rediploidization [15–21] as well as adaptation and the evolution of novelty by sub- and neofunctionalization [22–26]. Recent analysis has shown that paleologs have higher levels of genetic diversity than nonpaleologs and can significantly contribute to domestication [14]. Further exploration into the legacy of ancient polyploidy requires the rapid and accurate identification of paleologs in the genomes of extant diploid descendants.

Current methods for the identification of paleologs utilize sequence comparisons, genome structure information, or a combination of the two. Arguably, the most common and simplest method used to detect paleologs is to identify peaks of gene duplication by examining gene age distributions [27–31]. WGD instantly doubles the number of genes in a genome, and this large, abrupt birth event is visible in gene age distributions. These age distributions are often called Ks plots (Fig. 1) because they are usually made by plotting the synonymous site divergences (Ks) among paralogs in a single genome. Mixture models can be used to identify and place bounds on peaks of gene duplication in Ks plots, and homologous sets of genes within these Ks bounds are assumed to be paleologs [27–29]. Although simple and applicable to a wide range of data, this approach suffers from potentially high

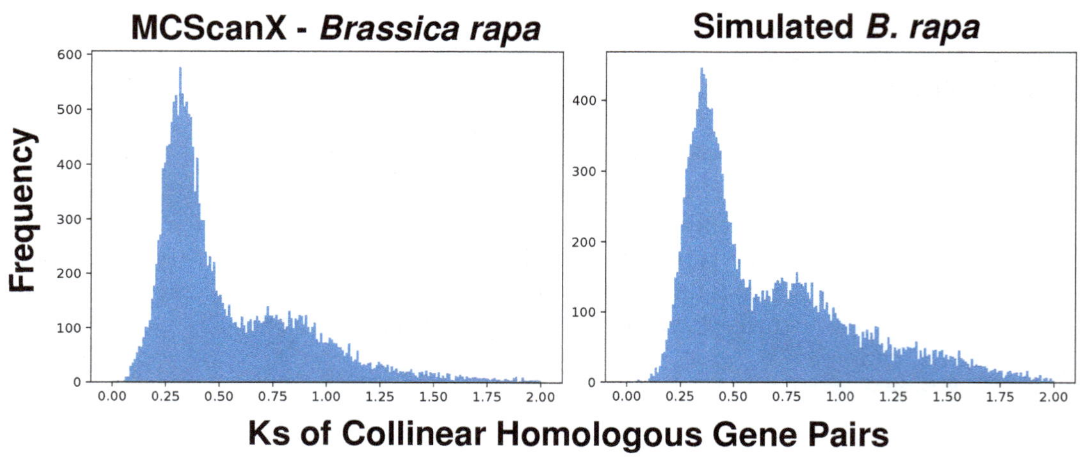

Fig. 1 Comparison of the distribution of collinear homologous gene pair synonymous distance (Ks_{pair}) for syntenic block pairs in MCScanX results from *Brassica rapa* versus simulated MCScanX results

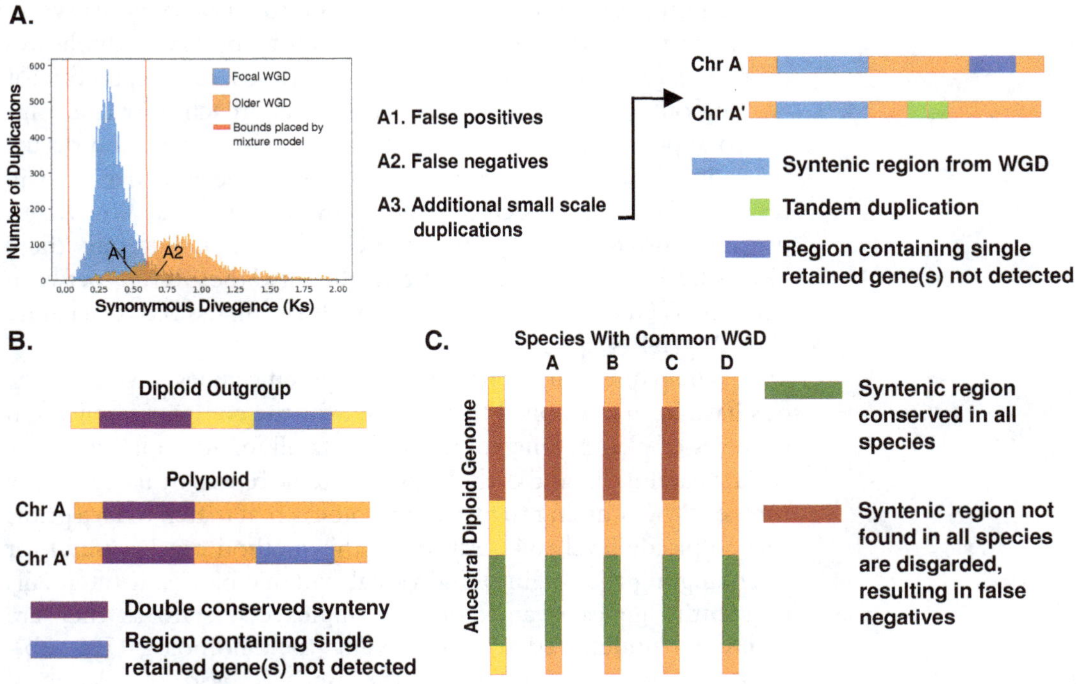

Fig. 2 Examples of major sources of false positives and negatives in a variety of paleolog identification approaches. (**a**) Hard bounds placed by mixture models can produce false positives and negatives (**a**.1, **a**.2) while also misidentifying paralogs from small-scale duplications as paleologs. By using syntenic information some paralogs from small-scale duplications can be excluded at the cost of missing single retained paleologs (**a**3). (**b**) Using DCS blocks improves identification of paleologs that are still present in two or more copies, but can lead to exclusion of single retained paleologs that are missing their other paralog(s). (**c**) Approaches such as POInT discard regions that are not conserved across all species used in the analysis, resulting in false negatives

rates of false positives and negatives. For example, not all genes within a Ks peak were necessarily duplicated by a WGD, and the hard bounds placed by mixture models are not perfectly accurate (Fig. 2a). Another method for identifying paleologs is to leverage genome structure rather than only comparing unordered sequences. This approach uses conserved gene order either within a species or between an outgroup to identify the duplicated regions in genomes from WGDs [32–36]. For the purposes of this chapter, we will be referring to conserved gene order as synteny, although the definition of synteny may vary from author to author [37, 38]. Syntenic regions can also arise from other small-scale duplications or additional layers of older WGD [23, 39, 40].

Several methods have been employed to resolve the origins of syntenic regions. The simplest of these approaches is to compare the overlap of genes within syntenic regions with those identified by Ks-based methods [14, 34, 41–43]. As an alternative to Ks, some studies have used Double Conserved Synteny (DCS) with a diploid outgroup that does not share the WGD. In DCS analyses, intragenomic syntenic regions that match to a homologous syntenic block in the outgroup are considered to be WGD duplicates (Fig. 2b) [35, 36]. Although these methods can be useful for filtering out small-scale duplications, both approaches are challenging to apply when multiple WGDs are present, a common occurrence in plants [9]. Relying solely on syntenic information will result in DCS being detected not only for the focal WGD, but possibly also older or younger WGD [34, 44]. Filtering these blocks by Ks can also be insufficient because the overlap between adjacent WGDs can make them difficult or impossible to separate (Fig. 2a) [34, 45].

To help disentangle the origins of syntenic genes, several methods have been developed that account for the complex duplication histories of plants. One approach, quota-align, attempts to differentiate paralogous and orthologous syntenic blocks by using Binary Integer Programming to fit syntenic blocks that match to an a priori chosen ploidy level [34]. Although this method can be useful for disentangling the origin of individual syntenic blocks, it often fails to identify genes retained in the single copy state as they are inherently undetected due to missing their homolog (Fig. 2b). This method is also limited by the user's a priori knowledge of the evolutionary history of their study organism [34]. As an alternative, POInT (Polyploid Orthology Inference Tool, see Conant, this volume) provides a powerful framework for reconstructing the evolutionary history of ancient WGDs and is one of the few methods to formally model genome evolution after WGD. Although paleolog classification is not the primary goal of POInT, its inferences can be used to identify paleologs [11, 46]. An inherent drawback to POInT's approach is that inferences are only made for regions with conserved synteny across all species examined, resulting in many paleologs being missed, especially those in potentially interesting areas of genomes that have diverged in retention or experienced high rates of rearrangement since a WGD (Fig. 2c) [11, 46]. For sets of paleologs that are detected, the predicted retention status of each copy can be assessed by the model confidence, limiting error in downstream analysis [11, 46]. The tradeoff of these sophisticated models is that it requires nontrivial computational effort to prepare data and run POInT; including the identification of high confidence DCS blocks within species using simulated annealing and the estimation of ancestral gene order using DCS blocks conserved across all species of interest. Overall, each of the existing approaches to identify paleologs has strengths and weaknesses that can be leveraged to provide a specialized tool for paleolog identification and classification that is computationally efficient.

Many large-scale and complex genomic questions have eluded evolutionary biologists due to computational limitations or gaps in theory. Machine learning-based approaches have recently been used to overcome some of these limitations, especially in cases where reasonable models for the biological processes of interest can be simulated [47]. Recent examples include the detection of selective sweeps [48, 49], describing past demographic history and recombination [50–53], landscape connectivity [54], inferring phylogenetic structure [55], species delimitation [56], and hybrid speciation and admixture [57]. The benefit of supervised machine learning in these cases is that they often achieve accurate inferences with much less computational time and effort than traditional inference methods [55, 58]. In the context of identifying paleologs, machine learning allows for an ensemble approach that can leverage the strengths of various methods such as quota-align, DCS analysis, and Ks divergence analysis, yet also allows for the easy integration of updated methods in the future. Here, we introduce a gradient boosted classification approach to identify paleologs using multiple features from synteny and Ks information. This method, Frackify (Fractionation-Classify), is able to make inferences with a high degree of overlap with POInT classifications in conserved syntenic regions while also identifying putative paleologs in regions not analyzed by POInT. We begin with exploring biological data in order to make informed simulations and the caveats of such an approach. We then explore methods for training, testing, and validation a gradient boosted decision tree. Finally, we analyze six paleohexaploid and paleotetraploid plant genomes to compare Frackify with other paleolog inference approaches.

2 Materials

2.1 Dependencies

Frackify is a python 3-based pipeline that currently requires the following python libraries: numpy (1.18.1), matplotlib (3.1.3), sys, statistics, pandas (1.0.1), random, itertools, csv (1.0), os, scipy (1.4.1), xgboost (1.3.3), and pickle [59–65].

2.2 Pipeline Input

1. CDS file for focal species being analyzed.
2. MCScanX or SynMap output with Ks/Kn information.
   (a) Self–Self (intragenomic) comparison.
   (b) Self–Outgroup comparison.
3. Mean Ks for each focal WGD in the species.

Examples and documentation can be found in the Frackify GitLab repo: https://gitlab.com/barker-lab/frackify or Docker: https://hub.docker.com/repository/docker/mmckibben/frackify/general

3 Methods

3.1 Biologically Informed Approach

Frackify, like other supervised machine learning-based approaches, makes predictions using a series of "rules" which it infers from a training dataset where the "answers" are known [66, 67]. Currently, no such dataset exists for paleologs as methods for paleolog identification are not necessarily accurate. Insufficient training data is a common issue in bioinformatics, and many researchers supplement training data using simulations [50, 57]. However, simulating genome evolution post WGD is challenging, and even if models were available for single WGD events, the complex duplication histories of many plants would present considerable challenges to tractiably producing training data across parameter space [9, 11, 68–70]. A prime example is the ongoing exploration of how cytological and genetic diploidization interplay on evolutionary timescales [8]. A more tractable approach is to simulate the detectable signatures of WGD in genomes according to reasonable empirical distributions derived from well-studied examples. In our case, we simulated the output of popular synteny inference tools such as MCScanX and CoGe SynMap [41, 71]. Both of these tools produce output files containing inferred syntenic block pairs either within or between species. Additional settings also allow for Ks values to be produced for homologous gene pairs in the detected syntenic blocks [41, 71]. To help inform our simulations using empirical datasets we used MCScanX to identify syntenic blocks in the genomes of 27 species (Table 1). These species were selected because they have well-studied and assembled genomes across a variety of taxa, including 10 orders and 22 genera that have experienced varying numbers of paleotetraploidies or paleohexaploidies with different levels of divergence. Several key properties of these syntenic blocks were summarized and used to develop simulations of WGD to train Frackify.

A fundamental property of syntenic blocks is their length. In closely related taxa, syntenic blocks can be nearly as long as chromosomes, but over time they are broken up through fractionation and genomic rearrangements [8, 72, 73]. To understand the lengths of syntenic blocks given the level of divergence of the WGD, we explored the distributions of self-self syntenic block lengths from MCScanX in our 27 exemplar species. Across the entire dataset, block lengths of less than 30 genes were predominant with a mean of 17.6 collinear gene pairs; however, there was a long tail of much larger blocks with a maximum of 1285 genes in *Brassica juncea* (Fig. 3). Although the approximate shape of a log-normal distribution was consistent across individual species, more fractionated genomes were skewed toward shorter block lengths. To test this, we fit a log-normal distribution to the distribution of syntenic block lengths in each species using the fitdistr()

Table 1
List of 27 species used for empirical analysis of syntenic block properties, including the version and source for each genome used in this study

Species	Genome accessed	Source
Aethionema arabicum	CoGe id23428	Nguyen et al. [114]
Ananas comosus	CoGe id25735	Ming et al. [115]
Arabidopsis lyrata	CoGe id3068	Rawat et al. [116]
Arabidopsis thaliana	CoGe id1	Swarbreck et al. [117]
Brachypodium distachyon	CoGe id25040	TIBI [118]
Brassica rapa	CoGe id24668	Wang et al. [119]
Brassica juncea	Brassica Database Bju15	Yang et al. [120]
Capsella rubella	CoGe id16754	Slotte et al. [121]
Cleome violacea	CoGe id23822	Emery et al. [11]
Cucumis melo	MELONOMICS CM3.5.1	Testolin et al. [122]
Eutrema parvulum	CoGe id12384	Dassanayake et al. [123]
Glycine max	Phytozome 12 Wm82.a2.v1	Chang et al. [124]
Gossypium hirsutum	NCBI Sequence Read Archive SRX202873	Zhang et al. [125]
Malus domestica	Genome Database for Rosacea	Jung et al. [126]
Oropetium thomaeum	CoGe id25799	VanBuren et al. [127]
Oryza sativa	Rice Genome Annotation Project V7.0	3,000 rice genomes project [128]
Phaseolus vulgaris	Phytozome 12 version 1.0	Schmutz et al. [129]
Prunus persica	Phytozome 12 version Prunus persica 2.1	Verde et al. [130]
Raphanus Sativus	Radish Genome Database RS1.0	Yu et al. [131]
Ricinus communis	Hale, http://castorbean.jcvi.org	http://castorbean.jcvi.org
Setaria italica	Phytozome 12 Setaria italica v2.2	Bennetzen et al. [132]
Sinapus alba	CoGe id33284	Zhang et al. [133]
Solanum lycopersicum	Phytozome 12 iTAG2.4	The Tomato Genome Consortium [134]
Solanum tuberosum	Phytozome 12 Solanum tuberosum v4.03	Sharma et al. [135]
Sorghum bicolor	Phytozome 12 Sorghum bicolor v3.1.1	McCormick et al. [136]
Thellungiella halophila	CoGe id19492	Yang et al. [137]
Vitis vinifera	Ensembl Plants release 51	Jaillon et al. [138]

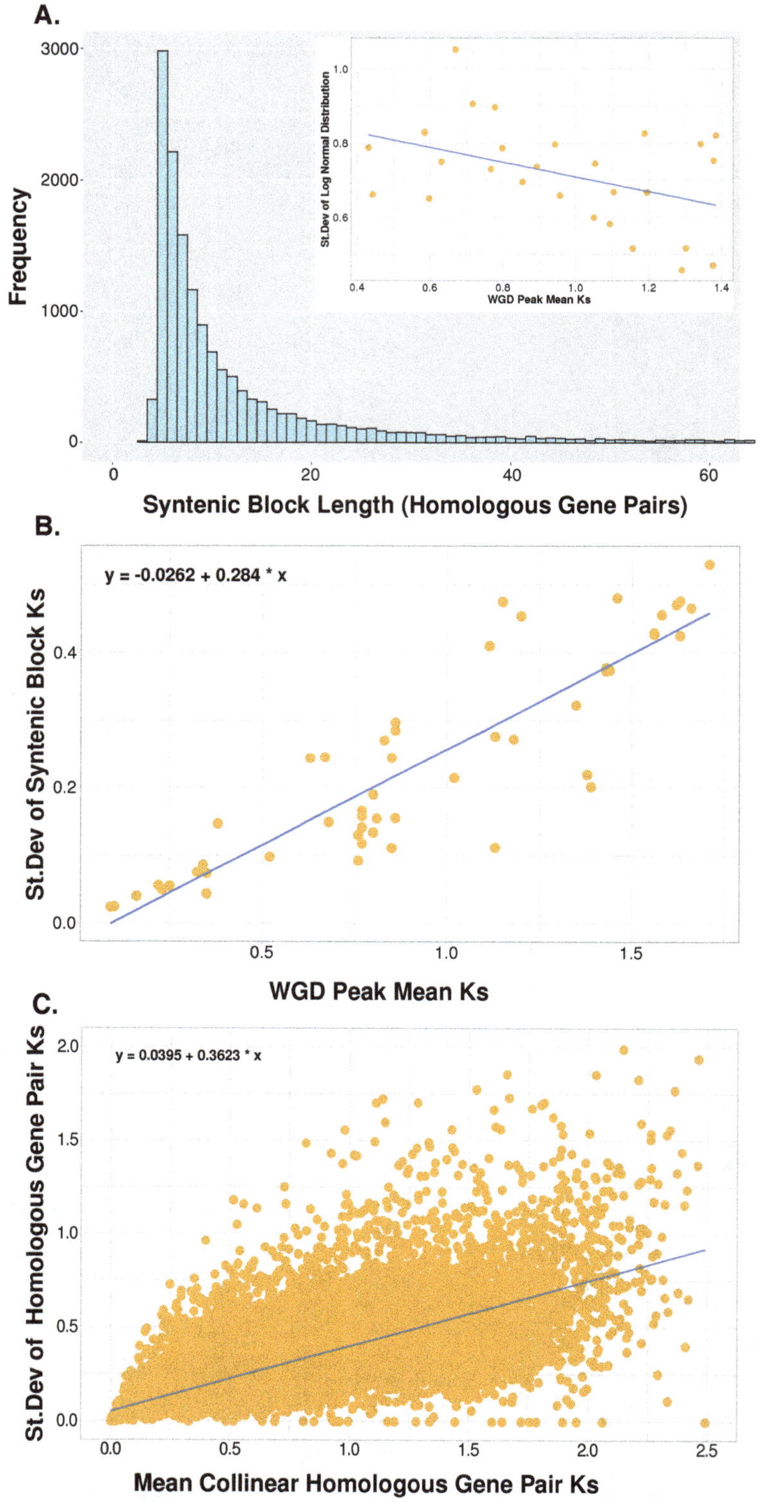

Fig. 3 Comparison of the distribution of syntenic block lengths in terms of gene pairs across the entire dataset of 27 species against the distributions within each individual species. (**a**) The histogram shows syntenic block lengths for gene pairs from the entire dataset of 27 species. The inset shows the standard deviation of log-normal distributions fit to the syntenic block length distributions of each species against Ks_{WGD} of the

function from the MASS R library [74]. We then plotted the mean and standard deviation of those distributions against the mean synonymous divergence (Ks_{WGD}) of collinear homologous gene pairs (Ks_{pair}) of the most recent WGD in each species (Fig. 3a). Using a linear model, we found that although the mean length of blocks did not vary significantly with Ks_{WGD} ($F_{1,27} = -1.909$, $p = 0.067$, $R^2 = 0.123$), it was a significant predictor of the standard deviation (Fig. 3a) ($F_{1,27} = -2.242$, $p = 0.034$, $R^2 = 0.162$). Using this model, we simulated the expected length of syntenic blocks for a given WGD by randomly selecting block lengths from a log-normal distribution with a mean of 17.6 collinear genes pairs (2.25 mean-log) and a standard deviation based on the Ks_{WGD} being simulated.

As syntenic blocks become broken up through time, the surviving collinear homologs also accumulate substitutions [29, 34, 45]. Using the sequence divergence of these homologs, it is possible to screen which syntenic blocks originate from different duplication events, such as the WGD in question or older WGDs and small-scale duplications [29, 34, 45]. Similar to the use of Ks plots to visualize the age distribution of paralogs in a genome, the mean synonymous divergence (Ks) of the collinear homologous gene pairs (Ks_{block}) in each syntenic block pair can be plotted in a frequency distribution. As we could not a priori know the exact WGD each block originates from in our empirical data, we used EMMIX to fit a series of mixture models to a distribution of duplication node Ks values in gene family clusters calculated using DupPipe for each of the 27 species [75, 76]. We then selected the model that fit the distribution for each species best based on the Bayesian Information Criterion (BIC) score and estimated a Ks boundary for each WGD peak by considering the Ks range of those gene pairs with posterior assignment probability exceeding 0.5 for the relevant duplication peak. Syntenic blocks with a Ks_{block} within the bounds of these estimates were considered to have originated from the WGD. We then calculated the standard deviation of Ks_{block} values for each WGD peak and plotted it against Ks_{WGD} of each WGD. We fit a linear model to the data and found that Ks_{WGD} was a significant predictor of the Ks_{block} standard deviation (Fig. 3b) ($F_{1,48} = 11.490$, $p < 0.001$, $R^2 = 0.733$). Using this model, we then simulated the distribution of synonymous divergence among syntenic blocks based on the mean level of divergence for the WGD.

Fig. 3 (continued) youngest WGD in each. A linear model was fit to the data as described in Subheading 3.1. (**b**) The standard deviation of the Ks_{block} for syntenic blocks plotted against the Ks_{WGD} for each WGD detected in all of the 27 species. (**c**) The standard deviation of the Ks_{pair} plotted against the Ks_{block} of each syntenic block pair detected in all of the 27 species

Not only do WGD differ in the variation of synonymous divergence among block pairs, but pairs of collinear homologs within those blocks can also accumulate different levels of synonymous divergence [34, 45, 77]. To account for this variation, we calculated the standard deviation of Ks_{pair} within block pairs and plotted them against the Ks_{block} of each block. We fit a linear model controlling for block length and found that the standard deviation of Ks_{pair} within syntenic block pairs was significantly predicted by Ks_{block} (Fig. 3c) ($F_{1,15,798} = 109.557$, $p < 0.001$, $R^2 = 0.474$). Using this model, we simulated the level of synonymous divergence in collinear homologous gene pairs given the Ks_{block} of the syntenic block pair they are within.

3.2 Simulation

Using the models and distributions estimated from our 27 species, we simulated syntenic blocks from a variety of duplication histories in a format similar to the output of MCScanX (Fig. 1). This simulated training data was then used to train Frackify to classify paleologs in syntenic blocks detected by MCScanX or CoGe SynMap. The simplest of these training data simulated only a single WGD event. In single WGD simulations, we began by specifying the Ks_{WGD} and producing a list of featureless syntenic block pairs (Fig. 4a.1). The number of these block pairs can be determined by the user, allowing for flexibility in the scale of the simulation. Each block pair within the initial list was a different length randomly selected from a log-normal distribution described by Ks_{WGD} and the model of block lengths for our 27 species (Fig. 3a). We then simulated synonymous divergence in each block pair by assigning a random Ks_{block} from a normal distribution with a mean equal to the Ks_{WGD} and standard deviation specified by our model of Ks_{block} in Subheading 3.1 (Figs. 3b and 4a.2). Finally, we incorporated the synonymous divergence in each collinear homologous gene pair within each block pair based on the Ks_{block} (Fig. 4a.3). This was done by selecting from a normal distribution with a mean equal to Ks_{block} and standard deviation from the model of Ks_{pair} within blocks from our 27 species (Fig. 3c). At the end of the three previous steps, we had a list of syntenic blocks similar to those found in MCScanX results for an individual WGD. If additional WGDs are desired, the user can specify to repeat this process for each. Using this approach, we were able to produce simulated syntenic blocks for a wide variety of duplication histories similar to empirical examples (Fig. 1).

Within syntenic blocks produced by WGDs, there are many classes of genes produced by different patterns of retention and additional small-scale duplications [12, 23, 33]. Pairs of homologous genes on syntenic blocks can represent paleologs retained as singletons, duplicates or triplicates, and nonpaleologs produced by tandem duplication or gene conversion [11, 71, 78]. To train Frackify to identify these classes of homologous gene pairs, we

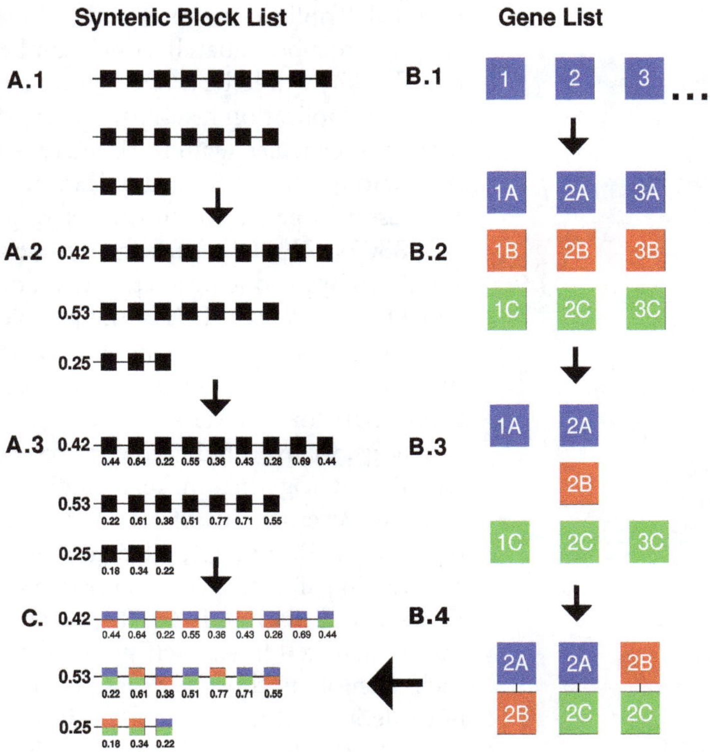

Fig. 4 Flow chart for the basic steps used to simulate syntenic blocks. (**a**) Syntenic block simulation begins by (**a**.1) making a list of blocks, (**a**.2) assigning a Ks_{block} to each block, and (**a**.3) finally assigning individual Ks_{pair} to collinear homologous genes within each block. (**b**) To populate the syntenic blocks with genes, we first (**b**.1) generated a list of gene labels, (**b**.2) duplicated or triplicated the list depending on the nature of the polyploidy being simulated, (**b**.3) removed genes to simulate fractionation, and then (**b**.4) grouped the remaining genes into collinear homologous gene pairs. (**c**) Finally, the two lists were merged by randomly assigning homologous gene pairs to the appropriate syntenic blocks for the WGD of origin

populated our simulated syntenic blocks with genes that reflect these different classes of gene duplications (Fig. 4b). In order to produce genes with varying duplication histories, we began by producing a list of unduplicated genes. The length of this list was based on the number of empty collinear gene pairs in the simulated syntenic blocks from the most divergent WGD so that after WGD and fractionation we would have enough syntenic WGD-derived genes—syntelogs—to populate each simulated syntenic block (Fig. 4b.1). We then produced WGD-derived syntelogs by duplicating or triplicating each gene in the list depending on whether a paleohexaploidy or paleotetraploidy was being simulated (Fig. 4b.2) [79]. Next, we simulated fractionation of these genes after WGD by selecting a random number of genes to remove from each syntelog set using a discrete uniform distribution between

0 and 3 (triplication) or 0 and 2 (duplication) (Fig. 4b.3). This resulted in approximately equal numbers of each paleolog class so that Frackify would be trained with a large sample of each. If a more complex duplication history was desired, we duplicated and fractionated the gene list again to simulate repeated duplication events in the genome. Small-scale duplications can also occur in genomes and result in genes which do not originate from our focal WGD [23, 33, 34, 71, 80]. Genes from these additional duplications were simulated by producing a separate list of nonpaleologs with a length equal to an arbitrary 15% of the paleolog gene list.

To produce the final simulated synteny dataset, we merged our simulated list of paleologs and nonpaleologs into syntenic blocks appropriate for each class (Fig. 4b.4). In the case of double retained paleologs, we expected each copy to be in WGD-derived regions A and B that together make a single syntenic block pair [11, 34, 81]. However, triples of retained genes would be found in three regions (A, B, and C), all being syntenic with each other. This resulted in pairwise block comparisons among all regions (e.g., A–B, A–C, and B–C) [11, 40, 82]. Most single retained genes would not be detected in self–self syntenic block pairs due to having lost their homologs; however, a small number may be aligned with nonpaleologs due to tandem duplications or gene conversion [40, 71, 78, 83, 84]. To simulate this, 10% of the single retained genes were randomly selected and paired with a nonpaleolog to form a misaligned homologous pair, consistent with previous estimates of ortholog family loss in several eukaryotic lineages [85–87]. Once all pairs of collinear homologous genes were produced, they were randomly assigned to syntenic blocks from the WGD of origin to produce the final simulated self-self syntenic dataset (Fig. 4c).

Although self–self synteny data is useful for detecting WGD-derived syntelog sets, relying on it solely could lead to single retained genes being misclassified because they have lost their other paralogs [40, 83, 84]. Previous methods have often identified single retained genes using gene tree information or syntenic comparisons to an outgroup [11, 40, 46, 83, 88]. For Frackify, we identified single retained genes using syntenic information from a related outgroup that the focal species diverged from prior to experiencing the WGD of interest. Similar to WGD, this speciation event can be detected in Ks plots due to a large peak in the synonymous divergence distribution for orthologous genes [28, 89–92]. Genes in the focal species that were not already recognized as paleologs and were orthologous to an outgroup gene were considered to be single retained paleologs. This is because genes that have an ortholog in a preduplication outgroup—identified by both synteny and divergence—were likely present in the genome prior to WGD, and the other copies were presumably lost during fractionation [88, 93]. To simulate orthology with an outgroup, we first simulated the shared duplication history prior to speciation by duplicating the self–self comparison list and removing syntenic blocks from the youngest WGD. Natural variation in synonymous divergence of the shared WGD in the outgroup and ingroup species was simulated by assigning new Ks_{pair} to each syntenic homolog pair using the same methods as the self–self comparison (Fig. 4a.2–3). The ortholog peak was then added by duplicating the gene list once more and assigning new Ks_{pair} to each ortholog pair with a mean Ks (Ks_{orth}) between the focal and the next oldest Ks_{WGD}. Although this final list represents the self–outgroup syntenic comparison, it is possible that some genes may be differentially lost between the focal species and the outgroup [11, 37, 45]. To capture differential gene loss, we randomly removed 10% of the genes from the outgroup, but users can specify different amounts of gene loss if desired [85–87].

Using our simulation framework, we produced self–self and self–outgroup syntenic information for a variety of duplication histories. The simplest histories involved a single WGD event, which can have different levels of synonymous divergence. To represent these cases, we produced 210 simulations with single WGD at Ks increments of 0.10 between 0 and 2 Ks. Some genomes have more complex duplication histories with multiple WGDs, a common occurrence in some land plant lineages [9, 13, 94, 95]. For these complex duplication histories, we also conducted simulations with two and three WGDs. Similar to the single WGD simulations, we placed the youngest WGD at increments of 0.10 Ks starting at zero. For each Ks increment of the youngest WGD, we then simulated one or two additional WGD at distances in increments of 0.10 Ks away from the youngest WGD until the most divergent WGD had a Ks of 2.0. Across the 1, 2, and 3 WGD scenarios, we ran a total of 1540 simulations. To avoid producing an unnecessarily large amount of data, we limited the scale of these simulations to produce an average of 10 syntenic block pairs per WGD compared to the average of 352 in the empirical dataset. These smaller scaled genomes contained an average of 634 genes per simulation. Across all 40,903 syntenic blocks, we simulated a total of 976,896 paleologs and nonpaleologs to train Frackify.

3.3 Training

The Frackify classifier was developed by training a boost decision tree using the XGBoost and Sklearn python libraries [63, 96]. To prepare data for XGBoost, we parsed multiple features from the MCScanX-like simulated data for each individual gene. The same feature extraction approach was also used for MCScanX empirical data. The first of these features included information pertaining to the duplication history of the organism or simulation, such as the Ks_{WGD} of each WGD and Ks_{orth} in the self-outgroup comparison. We also provided the difference in Ks between the focal WGD and possible neighboring WGD to help inform Frackify the possible degree of overlap in their Ks_{block} distributions. Next, we focused on features of the genes themselves, beginning with self–self syntenic information. First, we counted how many syntenic block pairs included the syntenic block each gene was found in. As some of these syntenic regions may have originated from older duplications, we then provided the number of syntenic block pairs which had Ks_{block} within three standard deviations of the focal Ks_{WGD}. From these possible WGD-derived syntenic blocks, we counted the number of putative WGD-derived syntelogs. As an example, we would expect a triple retained paleolog (gene A) to have a syntelog set size of three (A, B, and C) and be found in two syntenic block pairs (A–B and A–C). In the case of multiple overlapping WGD, it is possible this syntelog set may include genes that originated from an older duplication (gene O). To distinguish paleologs in cases like this, we then counted the total number of syntenic block pairs the syntelog set was involved in with Ks_{block} within 3 standard deviations of the Ks_{WGD}. Using our triple retained paleolog example, if all syntelogs in the set are true paleologs, we would expect three syntelogs (A, B, and C) and three syntenic block pairs (A–B, A–C, B–C). If one of the genes is from an older duplication (A, B, and O), we would have three syntelogs but only two syntenic block pairs (A–B, A–O). Finally, we averaged the Ks_{block} of each syntelog and reported how many standard deviations away from Ks_{WGD} the three closest syntelogs were. This initial set of summary statistics was chosen to summarize both the syntenic information for each individual gene and the syntenic relationships of its syntelogs.

In addition to features from the self–self syntenic data, we also included self–outgroup syntenic information to both detect single retained genes and help parse out which genes in a syntelog set may originate from overlapping WGD. First, we searched if each gene was found in syntenic blocks pairs with the outgroup at all. If a gene was in more than one syntenic block pair, we then filtered out only the syntenic blocks with Ks_{block} within three standard deviations of the Ks_{orth}. This process was repeated for all syntelogs in each set to gain a list of all the syntenic blocks that may be from the ortholog peak. We then filtered the outgroup genes in this list of syntenic blocks to find those that were orthologous to multiple syntelogs. If more than one orthologous outgroup gene was found, we calculated the mean Ks_{block} for all the syntenic block pairs that contained the outgroup genes. We selected the outgroup gene with the closest mean Ks_{block} to Ks_{orth} as it was most likely to be the ortholog of the WGD-derived syntelog set. Finally, we listed the number of syntelogs in the set that matched the chosen orthologous outgroup gene. All of the features of the duplication history, self–self, and self–outgroup syntenic comparisons were formulated into a single row of a data frame along with the known class of the given gene.

With our data formatted, we then trained our boosted decision tree while attempting to minimize overfitting [97, 98]. First, we assessed if the portion of the data used for testing and training could bias the model and lead to artificially inflated estimates of model accuracy while also searching for optimized hyperparameters using a Bayesian approach [99]. This was done using a 5-cross fold validation, in which the dataset is repeatedly split into training and testing data, and then an average accuracy of the model's predictions is assessed. We repeated the 5-cross fold validation 200 times with different hyperparameters to minimize the negative mean absolute error [100–102]. Once the final set of hyperparameters was selected, we performed a final 5-cross fold validation, producing predictions for gene classes in the testing data with a mean accuracy of 96.05% varying by 2.32%. Given the low level of variance in how accurately the model identified gene classes, it is unlikely the model would be sensitive to the proportion of the dataset that is used for training and testing. Based on this information, we split our simulated dataset into 70% training and 30% testing data and trained the final model.

3.4 Testing and Validation

Using our simulated test data described above, we used a variety of standard approaches to evaluate the performance of Frackify's predictions for the class of each gene. First, we tested how well the thresholds for each class of gene were defined within the model using a Receiver Operating Characteristic curve (ROC) [103, 104]. For all classes, the false-positive rate (FPR) remained low compared to the true-positive discovery rate (TPR), suggesting the model was optimized correctly to avoid biasing toward one class of gene (Fig. 5). To evaluate the performance of Frackify further, we made a confusion matrix which compared the classifications predicted by Frackify for each gene against the actual class, visualized as a heat map (Fig. 6) [105]. Across the 1540 simulations,

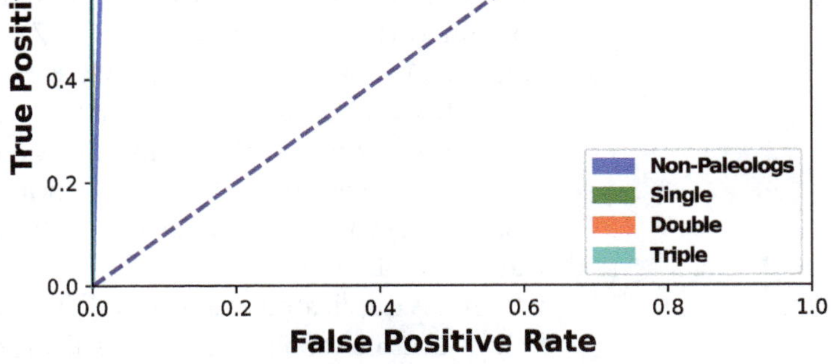

Fig. 5 ROC curves for each of the paleolog classifications predicted by the trained model

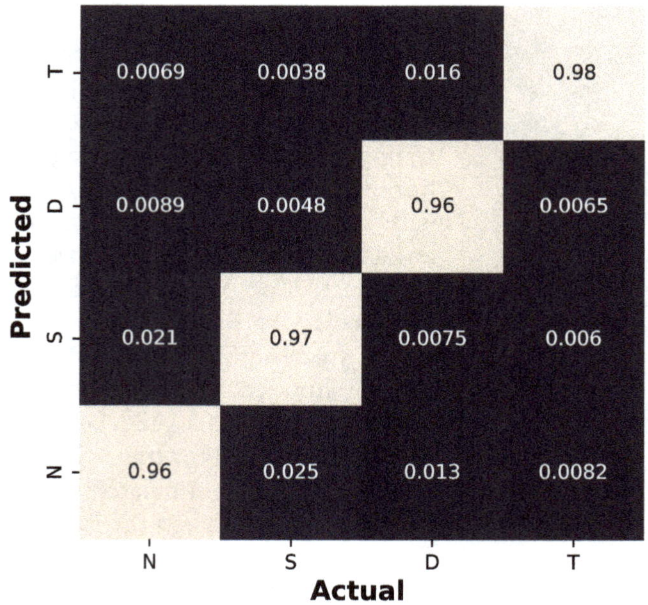

Fig. 6 Heatmap of the true labels for genes from the youngest WGD in the testing data against the classes predicted by the trained model. N, S, D, and T represent nonpaleologs, single, double, and triple retained paleologs, respectively. The scale is the ratio of the genes in the actual class (columns) which were predicted by the trained model to be in each class (rows)

Frackify predicted the majority (~96–98%) of each classification category correctly (Fig. 6). Similarly, there was a low error rate between all classifications with no clear patterns of particular gene classes being mistaken for others. From these observations, Frackify appears to perform well across a large set of simulated duplication histories; however, it is possible that certain scenarios are contributing the majority of the 2–4% of each class identified incorrectly.

To further explore the small fraction of paleologs that were misclassified by Frackify, we considered several cases where paleolog classification might be challenging. For example, if multiple WGD peaks heavily overlap or speciation with the outgroup occurred soon after WGD [34, 88, 106]. To evaluate Frackify's performance in these cases, we assessed predictions for simulations that contained three WGD events (Fig. 7). Despite the multiple overlapping WGDs, nonpaleologs were rarely misidentified as paleologs. However, Frackify less accurately identified different classes of paleologs in cases with highly divergent and overlapping WGDs. There are several potential explanations for why this might be the case. As an example, we considered the scenario where a given gene duplicates through a WGD, followed quickly by speciation, after which it is duplicated in a second WGD. This scenario results in two sets of double retained paleologs from the younger WGD that are orthologous to two genes from the older WGD shared with the outgroup. In this case, Frackify does not misclassify these older

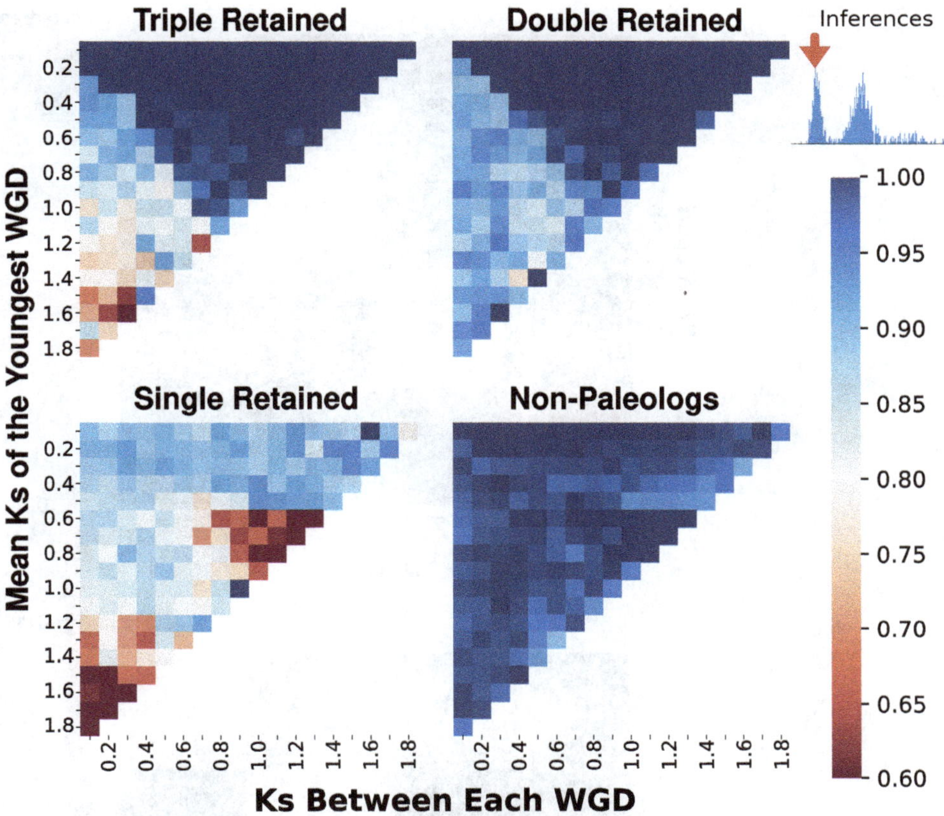

Fig. 7 Heatmaps for the accuracy of single, double, triple retained paleologs and nonpaleologs predicted by Frackify across parameter space for simulations containing three WGD. The *y* axis represents the Ks_{WGD} for the youngest WGD, the *x* axis represents the distance in Ks from the second and third oldest WGD to the focal WGD. Blank space represents parameter space where the oldest WGD occurs beyond the bounds of parameter space. The scale is the accuracy of predictions in terms of the ratio of correctly identified genes in each class out of all genes in that class

duplicate genes as paleologs from the youngest WGD. However, it may misidentify which ortholog each of the syntelogs matches, resulting in a triple retained paleolog set and a single retained paleolog. This issue could be compounded if one or more of the orthologs or paleologs are lost, such as in highly divergent WGD. Across most parameter space, Frackify was able to discern these complex scenarios with a high degree of accuracy (~85% or higher), but classification accuracy declined for Ks_{WGD} greater than 1.00 (Fig. 7).

Frackify is able to maintain a high degree of accuracy across parameter space when making inferences for the youngest WGD in a species; however, it is possible inferences may be complicated when considering the older WGD. To assess Frackify in such scenarios, we made additional paleolog inferences for the second oldest WGD in each simulation with three WGD. For these additional inferences, Frackify maintained a high degree of accuracy for paleologs (~89% or higher) (Fig. 8a). However, paleologs from the

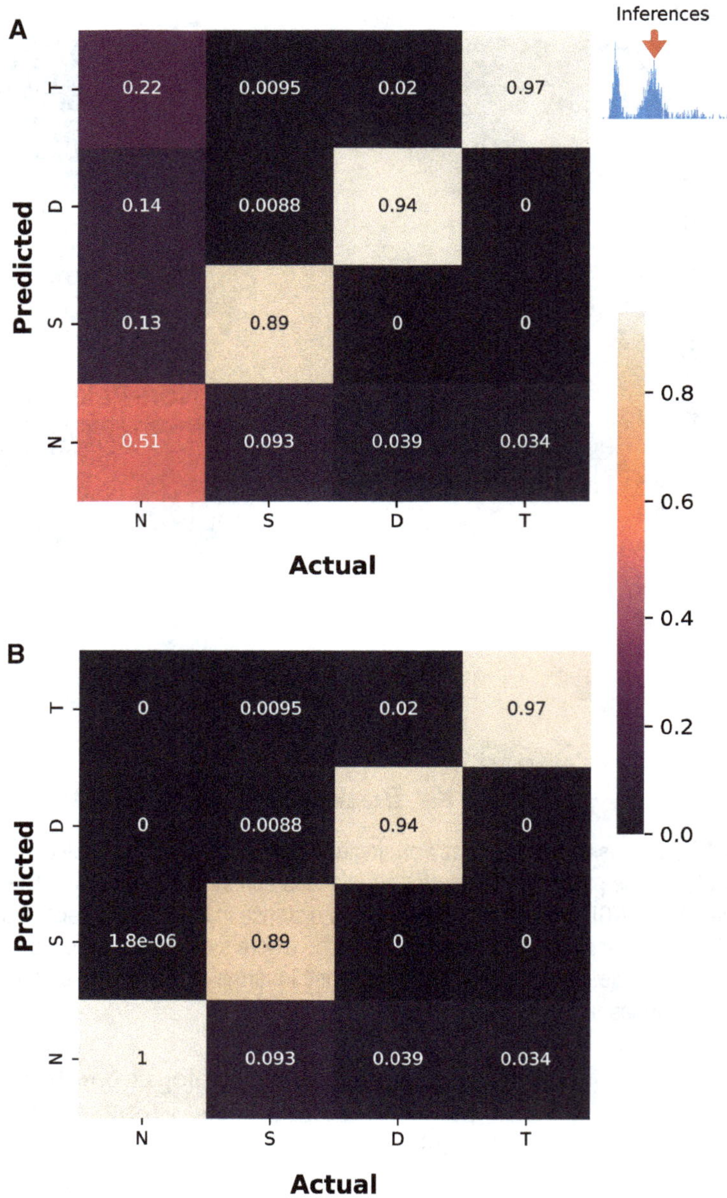

Fig. 8 (**a**) Heatmap for simulations with three WGD where paleolog inferences were made in the second oldest WGD. (**b**) Final paleolog inferences from part A after contamination from the younger duplications is removed through a two part filtering step. Both heatmaps contain the labels for genes in the testing dataset compared against the classes predicted by the trained model. N, S, D, and T represent nonpaleologs, single, double, and triple retained paleologs, respectively. The scale is the ratio of the genes in the actual class (columns) which were predicted by the trained model to be in each class (rows)

younger WGD were frequently misidentified as originating from the second youngest WGD (nonpaleologs in Fig. 8a). Users can remove these false positives by making additional paleolog inferences for the youngest WGD, and removing these paleologs from the inferences for the second oldest WGD. The results of this additional two-part filtering step produce inferences for all classes of genes with a high degree of accuracy (~89% or greater) (Fig. 8b). In summary, Frackify is able to maintain a high degree of accuracy across a wide range of parameter space and in genomes with multiple WGD.

To help understand why Frackify might sometimes misclassify paleologs, we explored the boosted decision tree itself. We used SHapley Additive exPlanation (SHAP) values [65, 107, 108] to help understand how XGBoost was making predictions for each class of gene. SHAP values are a useful data visualization tool to help interpret which features are having a significant impact on a machine learning model [107]. In our case, we noted several important features that had consistently high values for several classes of genes (Fig. 9). In particular, we found that the number of syntelogs in a set and how many match to the same orthologous outgroup gene had high SHAP values across multiple gene classes. These features are likely major drivers of paleolog classification and in combination may explain why nonpaleologs are rarely misclassified. For example, it is possible nonpaleologs from more recent small-scale duplications could be found in the focal WGD Ks peak, but they are not likely to be identified as paleologs because they do not have a clear ortholog in the outgroup. Similarly, these primary features may be used in combination with other features to help discern more complex scenarios. For example, if a set of paleologs contains a paralog that actually originates from an older, shared WGD, that paralog will be syntenic to multiple regions of the outgroup genome. These additional syntenic regions will result in an inflated number of putative orthologous outgroup genes, lending evidence that the paleolog set is contaminated with paralogs. These secondary features can be highly gene class specific, evident by their SHAP values. For example, the most important feature for the classification of double retained paleologs was the number of standard deviations away from Ks_{WGD} the mean Ks_{block} of each syntelog (Fig. 9). These features are likely how the model discerns if syntenic block pairs in the self–self comparison originated from the focal or overlapping WGD. Although we do not know the exact way these features are used to make final predictions, their importance to the model is useful to understand how we may improve Frackify's classifications in the future and for comparing Frackify to other approaches.

3.5 Empirical Examples

Although Frackify performed well when validated with simulated test data, we wanted to test the model's performance on empirical datasets. To accomplish this, we compared Frackify's classifications with paleolog inferences from POInT for six plant species [11, 14, 82]. Many of these species are of agronomic importance and have experienced multiple WGDs, making them a useful testing ground for paleolog inference tools [109–112]. In *Thellungiella halophila*,

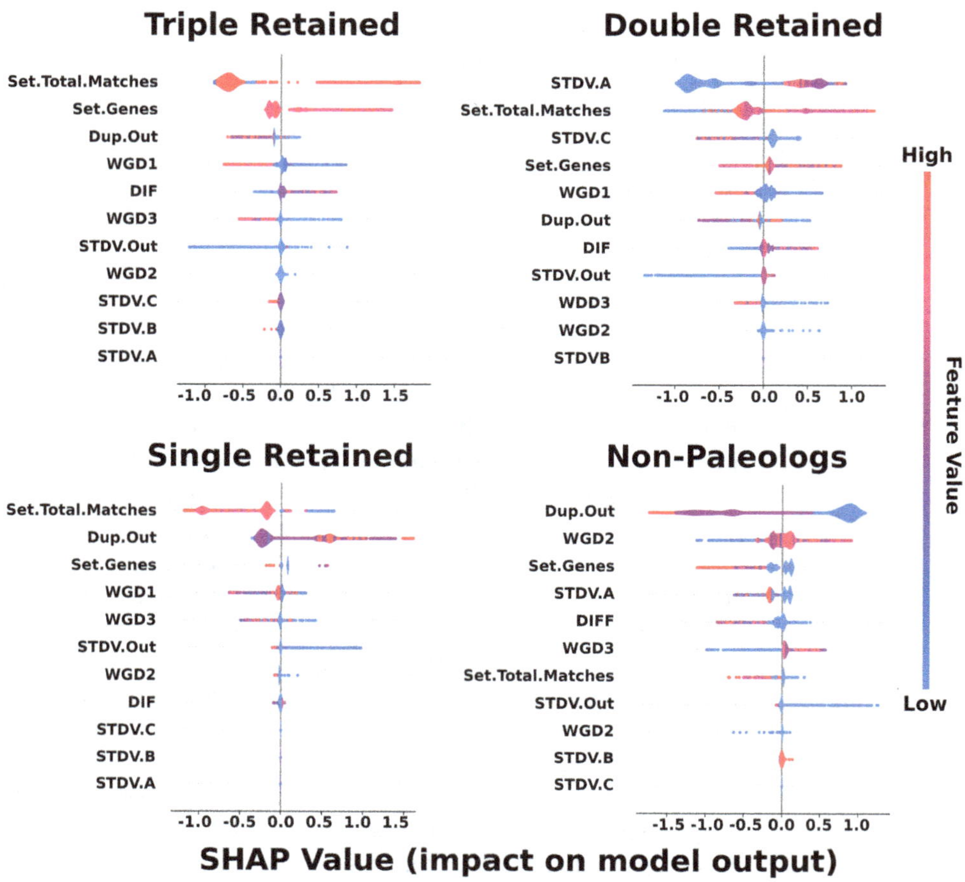

Fig. 9 SHAP values for each class of genes predicted by Frackify. Violin plots represent the distribution of SHAP values for a given class of gene with outliers represented as dots. The color of the violin plot indicates the feature value for the model's output, while the x-axis represents the SHAP value for a given point. Each feature describes the following properties; WGD (1–3) is the K_{WGD} of each WGD in the species, DIF is the difference in Ks between the two youngest WGD, Syn.Genes describes how many genes are in the WGD derived syntelog set, STDV(A–C) describes how many standard deviations away from K_{WGD} each gene in the set is, Dup.Out represents if the a gene matches to an outgroup gene in the orthology peak, Syn.Matches described how many of the genes in the WGD derived slytelog set match to the outgroup gene, and STDV.Out is how many standard deviations away from the orthology peak the outgroup gene is

Capsella rubella, and *Eutrema parvulum,* we analyzed the *At*-α Brassicaceae duplication event (*ARTHα*), estimated to have occurred 14.5–86 Ma with a median WGD peak Ks near 0.70 (Fig. 10) [3, 68]. For *Brassica rapa, B. oleracea,* and *Crambe hispanica,* we analyzed the Brassicaceae whole-genome triplication (*BRNIα*), with a median WGD peak Ks near 0.30 (Fig. 10) [3, 14]. Similar to POInT, we used *Cleome violacea* and *Arabidopsis thaliana* as outgroups for the Brassicaceae whole-genome triplication and *At*-α WGD, respectively [11]. We compared the POInT paleolog inferences in these six genomes to the less rigorous Ks + Synteny overlap-based approach discussed in the introduction and new classifications from Frackify.

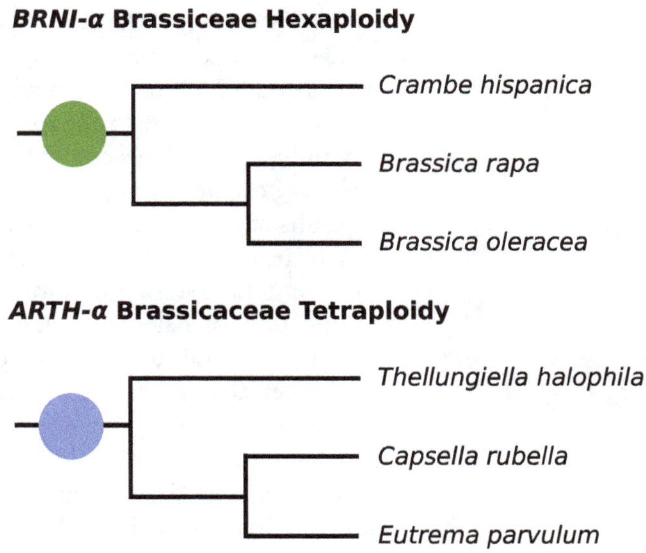

Fig. 10 Basic phylogenetic tree for species sharing the At-α Brassicaceae tetraploidy (*ARTHα*) and Brassicaceae Mesohexaploidy (*BRNIα*) included in this analysis [3]

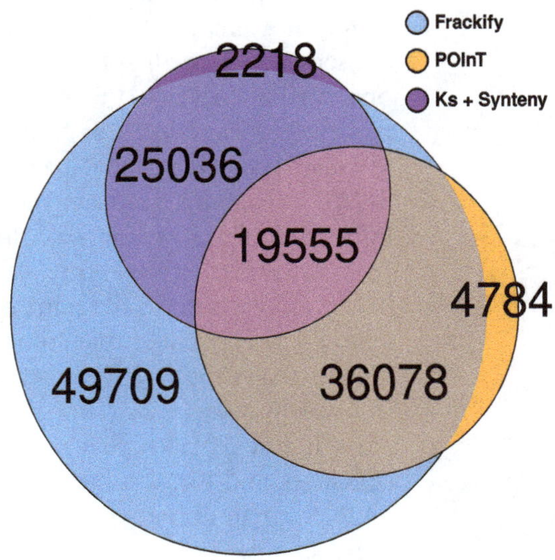

Fig. 11 Euler plot for paleolog identifications across all six species used in the empirical dataset comparison. The area of each region is approximately proportional to the percentage of the total dataset it represents

Across all six species, we found that Frackify identified genes as paleologs with a large degree of overlap with other approaches, but also identified many additional paleologs. We visualized this overlap using Euler plots from the eulerr R library (Fig. 11) [113]. Over half of the paleologs identified were shared across more than one method (58.7% total, 53.0–62.1% across species) (Fig. 11). A relatively small proportion of unique paleologs classifications came from POInT and the Ks + Synteny overlap approach (5.1% total, 0.3–5.5% across species) (Fig. 11). Similarly, Frackify identified 95.3% and 92.1% of the paleologs identified by Ks + Synteny and POInT, respectively. In the case of POInT, Frackify also maintained a high degree of agreement in terms of the retention status of each paleolog classification (81% average, 63–89% range) (Fig. 12). These results suggest that Frackify is not only able to identify the majority of paleologs previously identified by other methods, but also their retention status. In addition to these findings, a significant proportion of paleologs identified were unique to Frackify (36.2% total, 33.0–39.0% across species) (Fig. 11).

To further analyze the additional genes identified by Frackify and qualitatively assess if they are generally accurate or false positives, we conducted a more detailed comparison of classification differences between POInT and Frackify that leveraged SHAP values from the model. For all classes of paleologs identified by POInT, we noted a high degree of agreement with Frackify, suggesting the model was able to recognize these classifications and make accurate predictions (81% average, 63–89% range) (Fig. 12). The major source of disagreement between Frackify and POInT occurred in nonpaleologs. It is possible these nonpaleologs were simply not analyzed by POInT due to being in DCS blocks not conserved across all species [11, 46]. This was supported by the observation that 46,195 (62%) of the additional paleologs identified by Frackify were not found in the output files of POInT. To assess if these additional paleologs were false positives, we used SHAP values to compare their features. The most consistently important features for paleolog classification in Frackify were (1) the WGD-derived syntelog set size and (2) how many genes in the set matched to the same outgroup gene in the ortholog peak (Fig. 9). It is possible that some of the genes identified using these metrics may originate from other WGDs in the genomes. To test if this was the case, we examined how many paleologs were found in any syntenic blocks with a mean Ks within three standard deviations of Ks_{WGD} and Ks_{orth}. Filtering by this metric, we found that out of 74,745 total unique paleologs Frackify identified across all six species, only 389 were found outside our expected Ks_{WGD} and Ks_{orth} distributions. Although some of these paleologs (14,453) were also located in syntenic blocks pairs originating from older WGD, our analyses suggest that they were duplicated in the youngest WGDs rather than being false-positive classifications. Thus, many of the paleologs uniquely recovered by Frackify are not clearly false positives and are likely paleologs missed by POInT.

3.6 Summary

Frackify demonstrates how machine learning approaches can be used to explore the evolution of complex, duplicated genomes. By combining multiple empirical features of genome synteny and divergence, Frackify is able to relatively rapidly identify paleologs in

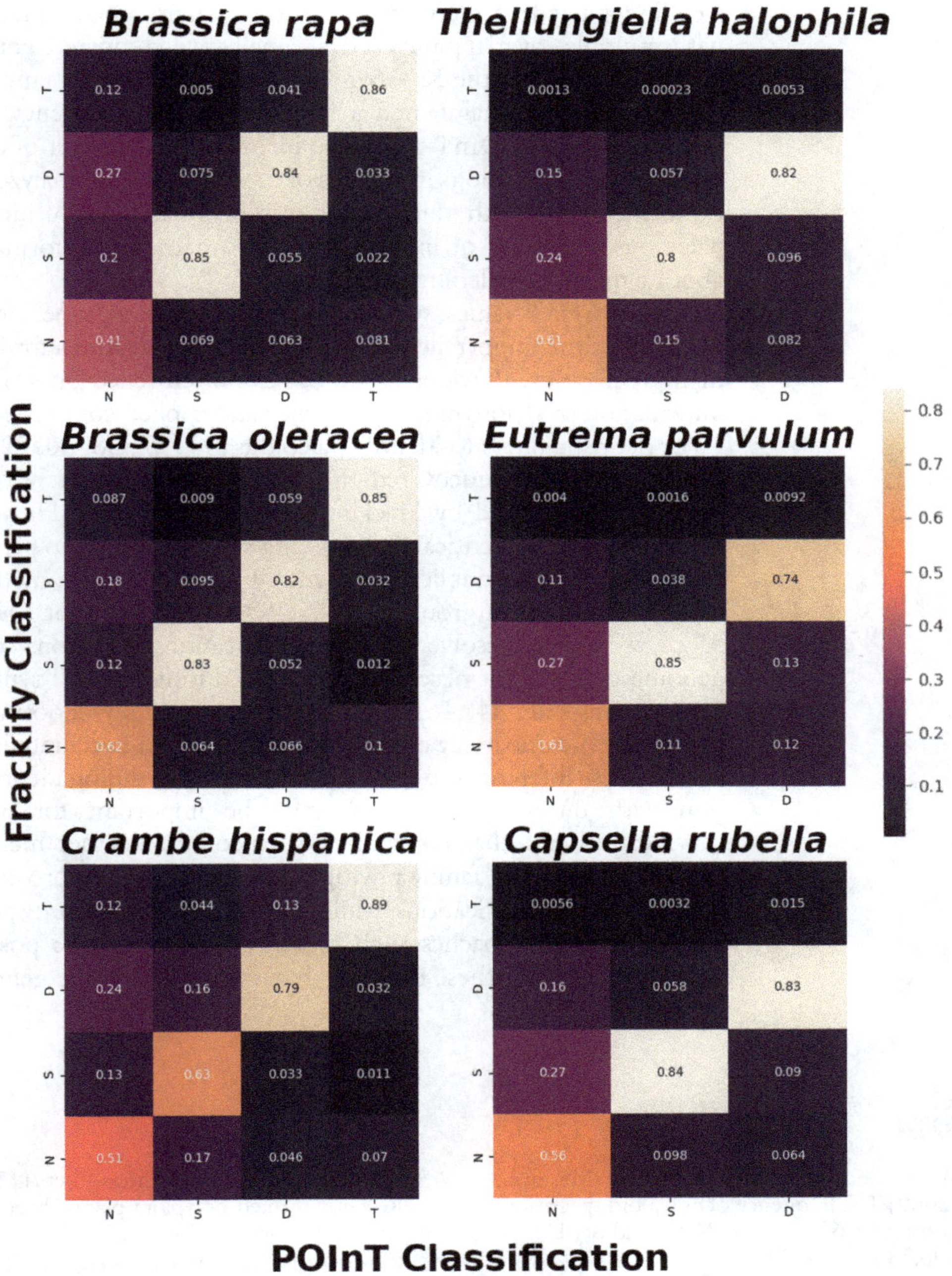

Fig. 12 Heatmaps for predicted class of each gene made by Frackify compared against inferences made by POInT in the six species. On the left hand column are paleohexaploids with up to three gene copies retained, while the right hand column contains paleotetraploids that only retain up to two copies of each gene. Color indicates the level of agreement between the two methods scaled as the ratio of the total number of genes in each class inferred by POInT (columns) compared to Frackify's predictions (rows)

diploidized species with both simple and complex duplication histories. Frackify uses a gradient boosted decision tree trained with simulated syntenic blocks developed from the empirical data of 27 paleopolyploid plant species. In testing, Frackify performed well under a wider variety of duplication scenarios, including those with multiple, overlapping WGDs. Finally, we compared Frackify against POInT and the Ks + Synteny method. Across six empirical datasets, Frackify maintained a high degree of consistency with methods such as POInT while also identifying a large set of additional putative paleologs in regions of the genomes not analyzed by these methods. With this tool, researchers can quickly identify paleologs in species of interest while also gathering information about patterns of paleolog retention.

Using SHAP values, we were also able to gain a deeper understanding of the importance of different genomic features for identifying paleologs. Previous methods have often relied on syntenic information or synonymous divergence and applied complex models or heuristic filters to identify paleologs [11, 27, 28, 30, 32–36, 82]. SHAP values uncovered that syntenic information was the primary feature used by Frackify to classify paleologs. However, Ks information was critical for assigning syntenic blocks to particular WGDs or other duplication events (Fig. 9). Similarly, methods which utilize an outgroup, such as POInT, Frackify, or Quota-Align, are able to resolve paleolog classifications in regions of the genome that are not necessarily detected during self–self syntenic comparisons [11, 34, 46, 82]. As a result, carefully choosing an outgroup to compare against is critical for marking accurate inferences. These inferences suggest that improving ortholog identification with outgroup genomes could be important for better classification and characterization of paleologs. Further merging of synteny and gene family phylogenetic data should improve current paleolog classifications while also allowing for multispecies comparisons. Approaches such as Frackify provide one possible path to combining these disparate types of data to make genomic inferences.

References

1. Wood TE, Takebayashi N, Barker MS et al (2009) The frequency of polyploid speciation in vascular plants. Proc Natl Acad Sci U S A 106:13875–13879
2. Barker MS, Arrigo N, Baniaga AE et al (2016) On the relative abundance of autopolyploids and allopolyploids. New Phytol 210:391–398
3. Li Z, Barker MS (2020) Inferring putative ancient whole-genome duplications in the 1000 Plants (1KP) initiative: access to gene family phylogenies and age distributions. Gigascience, vol 9, p giaa004
4. Arrigo N, Barker MS (2012) Rarely successful polyploids and their legacy in plant genomes. Curr Opin Plant Biol 15:140–146
5. Mayrose I, Zhan SH, Rothfels CJ et al (2011) Recently formed polyploid plants diversify at lower rates. Science 333:1257
6. Román-Palacios C, Molina-Henao YF, Barker MS (2020) Polyploids increase overall diversity despite higher turnover than diploids in the Brassicaceae. Proc R Soc B Biol Sci 287: 20200962
7. Zenil-Ferguson R, Burleigh JG, Freyman WA et al (2019) Interaction among ploidy, breeding system and lineage diversification. New Phytol 224:1252–1265
8. Li Z, McKibben MTW, Finch GS et al (2021) Patterns and processes of diploidization in land plants. Annu Rev Plant Biol 72:387–410

9. One Thousand Plant Transcriptomes Initiative (2019) One thousand plant transcriptomes and the phylogenomics of green plants. Nature 574:679–685
10. Cheng F, Wu J, Fang L et al (2012) Biased gene fractionation and dominant gene expression among the subgenomes of *Brassica rapa*. PLoS One 7:e36442
11. Emery M, Willis MMS, Hao Y et al (2018) Preferential retention of genes from one parental genome after polyploidy illustrates the nature and scope of the genomic conflicts induced by hybridization. PLoS Genet 14: e1007267
12. Cheng F, Wu J, Cai X et al (2018) Gene retention, fractionation and subgenome differences in polyploid plants. Nat Plants 4: 258–268
13. Qiao X, Li Q, Yin H et al (2019) Gene duplication and evolution in recurring polyploidization-diploidization cycles in plants. Genome Biol 20:38
14. Qi X, An H, Hall TE et al (2021) Genes derived from ancient polyploidy have higher genetic diversity and are associated with domestication in *Brassica rapa*. New Phytol 230:372–386
15. Scannell DR, Byrne KP, Gordon JL et al (2006) Multiple rounds of speciation associated with reciprocal gene loss in polyploid yeasts. Nature 440:341–345
16. Muir CD, Hahn MW (2015) The limited contribution of reciprocal gene loss to increased speciation rates following whole-genome duplication. Am Nat 185:70–86
17. Robertson FM, Gundappa MK, Grammes F et al (2017) Lineage-specific rediploidization is a mechanism to explain time-lags between genome duplication and evolutionary diversification. Genome Biol 18:111
18. Schnable JC, Freeling M, Lyons E (2012) Genome-wide analysis of syntenic gene deletion in the grasses. Genome Biol Evol 4: 265–277
19. Werth CR, Windham MD (1991) A model for divergent, allopatric speciation of polyploid pteridophytes resulting from silencing of duplicate-gene expression. Am Nat 137: 515–526
20. Lynch M, Force AG (2000) The origin of interspecific genomic incompatibility via gene duplication. Am Nat 156:590–605
21. Edger PP, Pires JC (2009) Gene and genome duplications: the impact of dosage-sensitivity on the fate of nuclear genes. Chromosom Res 17:699–717
22. Scannell DR, Wolfe KH (2008) A burst of protein sequence evolution and a prolonged period of asymmetric evolution follow gene duplication in yeast. Genome Res 18: 137–147
23. Freeling M (2009) Bias in plant gene content following different sorts of duplication: tandem, whole-genome, segmental, or by transposition. Annu Rev Plant Biol 60:433–453
24. Fajkus P, Peška V, Fajkus J et al (2021) Origin and fates of TERT gene copies in polyploid plants. Int J Mol Sci 22:1783
25. Ohno S (2013) Evolution by gene duplication. Springer Science & Business Media, Berlin
26. Schilling S, Kennedy A, Pan S et al (2020) Genome-wide analysis of MIKC-type MADS-box genes in wheat: pervasive duplications, functional conservation and putative neofunctionalization. New Phytol 225: 511–529
27. Tiley GP, Barker MS, Burleigh JG (2018) Assessing the performance of *Ks* plots for detecting ancient whole genome duplications. Genome Biol Evol 10:2882–2898
28. Barker MS, Kane NC, Matvienko M et al (2008) Multiple paleopolyploidizations during the evolution of the Compositae reveal parallel patterns of duplicate gene retention after millions of years. Mol Biol Evol 25: 2445–2455
29. Cui L, Wall PK, Leebens-Mack JH et al (2006) Widespread genome duplications throughout the history of flowering plants. Genome Res 16:738–749
30. Blanc G, Wolfe KH (2004) Functional divergence of duplicated genes formed by polyploidy during *Arabidopsis* evolution. Plant Cell 16:1679–1691
31. Maere S, De Bodt S, Raes J et al (2005) Modeling gene and genome duplications in eukaryotes. Proc Natl Acad Sci U S A 102: 5454–5459
32. Glover N, Sheppard S, Dessimoz C (2021) Homoeolog inference methods requiring bidirectional best hits or synteny miss many pairs. Genome Biol Evol 13:evab077
33. Singh PP, Arora J, Isambert H (2015) Identification of ohnolog genes originating from whole genome duplication in early vertebrates, based on synteny comparison across multiple genomes. PLoS Comput Biol 11: e1004394
34. Tang H, Lyons E, Pedersen B et al (2011) Screening synteny blocks in pairwise genome comparisons through integer programming. BMC Bioinform 12:102

35. Wolfe KH, Shields DC (1997) Molecular evidence for an ancient duplication of the entire yeast genome. Nature 387:708–713
36. Kellis M, Birren BW, Lander ES (2004) Proof and evolutionary analysis of ancient genome duplication in the yeast *Saccharomyces cerevisiae*. Nature 428:617–624
37. Tang H, Bowers JE, Wang X et al (2008) Synteny and collinearity in plant genomes. Science 320:486–488
38. Passarge E, Horsthemke B, Farber RA (1999) Incorrect use of the term synteny. Nat Genet 23:387
39. Barker MS, Baute GJ, Liu S-L (2012) Duplications and turnover in plant genomes. In: Plant genome diversity, vol 1. Springer, Vienna, pp 155–169
40. Tang H, Wang X, Bowers JE et al (2008) Unraveling ancient hexaploidy through multiply-aligned angiosperm gene maps. Genome Res 18:1944–1954
41. Haug-Baltzell A, Stephens SA, Davey S et al (2017) SynMap2 and SynMap3D: web-based whole-genome synteny browsers. Bioinformatics 33:2197–2198
42. Shi T, Chen J (2020) A reappraisal of the phylogenetic placement of the *Aquilegia* whole-genome duplication. Genome Biol 21(1):295
43. Cannon SB, Sterck L, Rombauts S et al (2006) Legume genome evolution viewed through the *Medicago truncatula* and *Lotus japonicus* genomes. Proc Natl Acad Sci U S A 103:14959–14964
44. Pham SK, Pevzner PA (2010) DRIMM-Synteny: decomposing genomes into evolutionary conserved segments. Bioinformatics 26:2509–2516
45. Simillion C, Vandepoele K, Van Montagu MCE et al (2002) The hidden duplication past of *Arabidopsis thaliana*. Proc Natl Acad Sci U S A 99:13627–13632
46. Conant GC, Wolfe KH (2008) Probabilistic cross-species inference of orthologous genomic regions created by whole-genome duplication in yeast. Genetics 179:1681–1692
47. Schrider DR, Kern AD (2018) Supervised machine learning for population genetics: a new paradigm. Trends Genet 34:301–312
48. Schrider DR, Kern AD (2016) S/HIC: robust identification of soft and hard sweeps using machine learning. PLoS Genet 12: e1005928
49. Kern AD, Schrider DR (2018) diploS/HIC: An Updated approach to classifying selective sweeps. G3 8:1959–1970
50. Flagel L, Brandvain Y, Schrider DR (2019) The unreasonable effectiveness of convolutional neural networks in population genetic inference. Mol Biol Evol 36:220–238
51. Adrian AB, Corchado JC, Comeron JM (2016) Predictive models of recombination rate variation across the *Drosophila melanogaster* genome. Genome Biol Evol 8:2597–2612
52. Gao F, Ming C, Hu W et al (2016) New software for the fast estimation of population recombination rates (FastEPRR) in the Genomic Era. G3 (Bethesda) 6:1563–1571. https://doi.org/10.1534/g3.116.028233
53. Lin K, Futschik A, Li H (2013) A fast estimate for the population recombination rate based on regression. Genetics 194:473–484
54. Pless E, Saarman NP, Powell JR et al (2021) A machine-learning approach to map landscape connectivity in *Aedes aegypti* with genetic and environmental data. Proc Natl Acad Sci U S A 118:e2003201118
55. Azouri D, Abadi S, Mansour Y et al (2021) Harnessing machine learning to guide phylogenetic-tree search algorithms. Nat Commun 12:1983
56. Derkarabetian S, Castillo S, Koo PK et al (2019) A demonstration of unsupervised machine learning in species delimitation. Mol Phylogenet Evol 139:106562
57. Blischak PD, Barker MS, Gutenkunst RN (2021) Chromosome-scale inference of hybrid speciation and admixture with convolutional neural networks. Mol Ecol Resour 21(8):2676–2688
58. Larrañaga P, Calvo B, Santana R et al (2006) Machine learning in bioinformatics. Brief Bioinform 7:86–112
59. Harris CR, Millman KJ, van der Walt SJ et al (2020) Array programming with NumPy. Nature 585:357–362
60. Hunter JD (2007) Matplotlib: a 2D graphics environment. Comput Sci Eng 9:90–95
61. Van Rossum G, Drake FL (2009) Python 3 reference manual CreateSpace
62. Virtanen P, Gommers R, Oliphant TE et al (2020) SciPy 1.0: fundamental algorithms for scientific computing in Python. Nat Methods 17:261–272
63. Chen T, Guestrin C (2016) XGBoost: a scalable tree boosting system. In: Proceedings of the 22nd ACM SIGKDD international conference on knowledge discovery and data mining. Association for Computing Machinery, New York, NY, USA, pp 785–794
64. Van Rossum G (2020) The Python library reference, release 3.8. 2. 16

65. Lundberg SM, Lee SI (2017) A unified approach to interpreting model predictions
66. Breiman L (2001) Statistical modeling: the two cultures (with comments and a rejoinder by the author). SSO Schweiz Monatsschr Zahnheilkd 16:199–231
67. Bzdok D, Altman N, Krzywinski M (2018) Statistics versus machine learning. Nat Methods 15:233–234
68. Barker MS, Vogel H, Schranz ME (2009) Paleopolyploidy in the Brassicales: analyses of the *Cleome* transcriptome elucidate the history of genome duplications in *Arabidopsis* and other Brassicales. Genome Biol Evol 1: 391–399
69. Van de Peer Y, Mizrachi E, Marchal K (2017) The evolutionary significance of polyploidy. Nat Rev Genet 18:411–424
70. Sankoff D, Zheng C, Zhu Q (2010) The collapse of gene complement following whole genome duplication. BMC Genomics 11:313
71. Wang Y, Tang H, Debarry JD et al (2012) MCScanX: a toolkit for detection and evolutionary analysis of gene synteny and collinearity. Nucleic Acids Res 40:e49
72. Vakirlis N, Carvunis A-R, McLysaght A (2020) Synteny-based analyses indicate that sequence divergence is not the main source of orphan genes. elife 9:e53500
73. Bhutkar A, Schaeffer SW, Russo SM et al (2008) Chromosomal rearrangement inferred from comparisons of 12 Drosophila genomes. Genetics 179:1657–1680
74. Venables WN, Ripley BD (2002) Modern applied statistics with S-PLUS. Springer Science & Business Media. https://doi.org/10.1007/978-0-387-21706-2
75. Barker MS, Dlugosch KM, Dinh L et al (2010) EvoPipes.net: bioinformatic tools for ecological and evolutionary genomics. Evol Bioinformatics Online 6:143–149
76. McLachlan GJ, Peel D, Basford KE et al (1999) The EMMIX software for the fitting of mixtures of normal and t-components. J Stat Softw 4
77. Jiao Y, Li J, Tang H et al (2014) Integrated syntenic and phylogenomic analyses reveal an ancient genome duplication in monocots. Plant Cell. 26:2792–2802. https://doi.org/10.1105/tpc.114.127597
78. Wang X-Y, Paterson AH (2011) Gene conversion in angiosperm genomes with an emphasis on genes duplicated by polyploidization. Genes 2:1–20
79. Lyons E, Pedersen B, Kane J et al (2008) Finding and comparing syntenic regions among *Arabidopsis* and the outgroups papaya, poplar, and grape: CoGe with rosids. Plant Physiol 148:1772–1781
80. Haas BJ, Delcher AL, Wortman JR et al (2004) DAGchainer: a tool for mining segmental genome duplications and synteny. Bioinformatics 20:3643–3646
81. Kent WJ, Baertsch R, Hinrichs A et al (2003) Evolution's cauldron: duplication, deletion, and rearrangement in the mouse and human genomes. Proc Natl Acad Sci U S A 100: 11484–11489
82. Hao Y, Mabry ME, Edger PP et al (2021) The contributions from the progenitor genomes of the mesopolyploid Brassiceae are evolutionarily distinct but functionally compatible. Genome Res 31:799–810
83. Zhang Y, Yu Z, Zheng C et al (2021) Integrated synteny- and similarity-based inference on the polyploidization-fractionation cycle. Interface Focus 11:20200059
84. Bhutkar A, Russo S, Smith TF et al (2006) Techniques for multi-genome synteny analysis to overcome assembly limitations. Genome Inform 17:152–161
85. Wyder S, Kriventseva EV, Schröder R et al (2007) Quantification of ortholog losses in insects and vertebrates. Genome Biol 8:R242
86. Dalquen DA, Dessimoz C (2013) Bidirectional best hits miss many orthologs in duplication-rich clades such as plants and animals. Genome Biol Evol 5:1800–1806
87. Zhang J (2003) Evolution by gene duplication: an update. Trends Ecol Evol 18: 292–298
88. Parey E, Louis A, Cabau C et al (2020) Synteny-guided resolution of gene trees clarifies the functional impact of whole-genome duplications. Mol Biol Evol 37:3324–3337
89. Wang J-T, Li J-T, Zhang X-F et al (2012) Transcriptome analysis reveals the time of the fourth round of genome duplication in common carp (*Cyprinus carpio*). BMC Genomics 13:96
90. Li F-W, Brouwer P, Carretero-Paulet L et al (2018) Fern genomes elucidate land plant evolution and cyanobacterial symbioses. Nat Plants 4:460–472
91. Zhang G-Q, Liu K-W, Li Z et al (2017) The *Apostasia* genome and the evolution of orchids. Nature 549:379–383
92. Lai Z, Kane NC, Kozik A et al (2012) Genomics of Compositae weeds: EST libraries,

microarrays, and evidence of introgression. Am J Bot 99:209–218

93. Byrne KP, Wolfe KH (2005) The Yeast Gene Order Browser: combining curated homology and syntenic context reveals gene fate in polyploid species. Genome Res 15:1456–1461
94. Li Z, Barker MS (2019) Inferring putative ancient whole genome duplications in the 1000 Plants (1KP) initiative: access to gene family phylogenies and age distributions. https://www.biorxiv.org/content/10.1101/735076v1
95. Barker MS, Husband BC, Pires JC (2016) Spreading Winge and flying high: the evolutionary importance of polyploidy after a century of study. Am J Bot 103:1139–1145
96. Pedregosa F, Varoquaux G, Gramfort A et al (2011) Scikit-learn: machine learning in python. Mach Learn Python 6
97. Vabalas A, Gowen E, Poliakoff E et al (2019) Machine learning algorithm validation with a limited sample size. PLoS One 14:e0224365
98. Belkin M, Hsu D, Ma S et al (2019) Reconciling modern machine-learning practice and the classical bias-variance trade-off. Proc Natl Acad Sci U S A 116:15849–15854
99. Wu J, Chen X-Y, Zhang H et al (2019) Hyperparameter optimization for machine learning models based on bayesian optimization. Dianzi Keji Daxue Xuebao 17:26–40
100. Kuhn M, Johnson K (2013) Applied predictive modeling. Springer, New York
101. Willmott CJ, Matsuura K (2005) Advantages of the mean absolute error (MAE) over the root mean square error (RMSE) in assessing average model performance. Clim Res 30:79–82
102. Chai T, Draxler RR (2014) Root mean square error (RMSE) or mean absolute error (MAE)? – arguments against avoiding RMSE in the literature. Geosci Model Dev 7:1247–1250
103. Bradley AP (1997) The use of the area under the ROC curve in the evaluation of machine learning algorithms. Pattern Recogn 30:1145–1159
104. Hand DJ, Till RJ (2001) A simple generalisation of the area under the roc curve for multiple class classification problems. Mach Learn 45:171–186
105. Waskom M (2021) seaborn: statistical data visualization. J Open Source Softw 6:3021
106. Tang H, Bomhoff MD, Briones E et al (2015) SynFind: compiling syntenic regions across any set of genomes on demand. Genome Biol Evol 7:3286–3298
107. Štrumbelj E, Kononenko I (2014) Explaining prediction models and individual predictions with feature contributions. Knowl Inf Syst 41:647–665
108. Ribeiro MT, Singh S, Guestrin C (2016) "Why should I trust you?": explaining the predictions of any classifier. In: Proceedings of the 22nd ACM SIGKDD international conference on knowledge discovery and data mining. Association for Computing Machinery, New York, NY, USA, pp 1135–1144
109. Meyer RS, DuVal AE, Jensen HR (2012) Patterns and processes in crop domestication: an historical review and quantitative analysis of 203 global food crops. New Phytol 196:29–48
110. Yu K, Feng M, Yang G et al (2020) Changes in alternative splicing in response to domestication and polyploidization in wheat. Plant Physiol 184:1955–1968
111. Franzke A, Lysak MA, Al-Shehbaz IA et al (2011) Cabbage family affairs: the evolutionary history of Brassicaceae. Trends Plant Sci 16:108–116
112. Mabry ME, Brose JM, Blischak PD et al (2020) Phylogeny and multiple independent whole-genome duplication events in the Brassicales. Am J Bot 107:1148–1164
113. Larsson J (2020) Area-proportional Euler and Venn diagrams with ellipses_. R package version 6.1. 0
114. Nguyen T-P, Mühlich C, Mohammadin S et al (2019) Genome improvement and genetic map construction for *Aethionema arabicum*, the first divergent branch in the brassicaceae family, vol G3 9, pp 3521–3530
115. Ming R, VanBuren R, Wai CM et al (2015) The pineapple genome and the evolution of CAM photosynthesis. Nat Genet 47:1435–1442
116. Rawat V, Abdelsamad A, Pietzenuk B et al (2015) Improving the annotation of *Arabidopsis lyrata* Using RNA-Seq Data. PLoS One 10:e0137391
117. Swarbreck D, Wilks C, Lamesch P et al (2008) The Arabidopsis Information Resource (TAIR): gene structure and function annotation. Nucleic Acids Res 36:D1009–D1014
118. TIBI (2010) Genome sequencing and analysis of the model grass *Brachypodium distachyon*. Nature 463. https://doi.org/10.1038/nature08747
119. Wang X, Wang H, Wang J et al (2011) The genome of the mesopolyploid crop species *Brassica rapa*. Nat Genet 43:1035–1039

120. Yang J, Liu D, Wang X et al (2016) The genome sequence of allopolyploid *Brassica juncea* and analysis of differential homoeolog gene expression influencing selection. Nat Genet 48:1225–1232
121. Slotte T, Hazzouri KM, Ågren JA et al (2013) The *Capsella rubella* genome and the genomic consequences of rapid mating system evolution. Nat Genet 45:831–835
122. Testolin R, Huang H-W, Ferguson AR (2016) The Kiwifruit genome. Springer, Cham
123. Dassanayake M, Oh D-H, Haas JS et al (2011) The genome of the extremophile crucifer *Thellungiella parvula*. Nat Genet 43: 913–918
124. Chang S, Wang Y, Lu J et al (2013) The mitochondrial genome of soybean reveals complex genome structures and gene evolution at intercellular and phylogenetic levels. PLoS One 8:e56502
125. Zhang T, Hu Y, Jiang W et al (2015) Sequencing of allotetraploid cotton (*Gossypium hirsutum* L. acc. TM-1) provides a resource for fiber improvement. Nat Biotechnol 33:531–537
126. Jung S, Lee T, Cheng C-H et al (2019) 15 years of GDR: new data and functionality in the Genome Database for Rosaceae. Nucleic Acids Res 47:D1137–D1145
127. VanBuren R, Bryant D, Edger PP et al (2015) Single-molecule sequencing of the desiccation-tolerant grass *Oropetium thomaeum*. Nature 527:508–511
128. 3,000 rice genomes project (2014) The 3,000 rice genomes project. Gigascience 3:7
129. Schmutz J, McClean PE, Mamidi S et al (2014) A reference genome for common bean and genome-wide analysis of dual domestications. Nat Genet 46:707–713
130. Verde I, The International Peach Genome Initiative, Abbott AG et al (2013) The high-quality draft genome of peach *(Prunus persica)* identifies unique patterns of genetic diversity, domestication and genome evolution. Nat Genet. https://doi.org/10.1038/ng.2586
131. Yu H-J, Baek S, Lee Y-J et al (2019) The radish genome database (RadishGD): an integrated information resource for radish genomics. Database 2019
132. Bennetzen JL, Schmutz J, Wang H et al (2012) Reference genome sequence of the model plant *Setaria*. Nat Biotechnol 30: 555–561
133. Zhang G, Liu X, Quan Z et al (2012) Genome sequence of foxtail millet (Setaria italica) provides insights into grass evolution and biofuel potential. Nat Biotechnol 30: 549–554
134. The Tomato Genome Consortium (2012) The tomato genome sequence provides insights into fleshy fruit evolution. Nature. https://doi.org/10.1038/nature11119
135. Sharma SK, Bolser D, de Boer J et al (2013) Construction of reference chromosome-scale pseudomolecules for potato: integrating the potato genome with genetic and physical maps. Genetics. https://doi.org/10.1534/g3.113.007153
136. McCormick RF, Truong SK, Sreedasyam A et al (2018) The *Sorghum bicolor* reference genome: improved assembly, gene annotations, a transcriptome atlas, and signatures of genome organization. Plant J. https://doi.org/10.1111/tpj.13781
137. Yang R, Jarvis DE, Chen H et al (2013) The reference genome of the halophytic plant *Eutrema salsugineum*. Front Plant Sci 4:46
138. Jaillon O, Aury J-M, Noel B et al (2007) The grapevine genome sequence suggests ancestral hexaploidization in major angiosperm phyla. Nature 449:463–467

Part II

Phylogenetics to Study Polyploidy

Chapter 6

Phasing Gene Copies into Polyploid Subgenomes Using a Bayesian Phylogenetic Approach

William A. Freyman and Carl J. Rothfels

Abstract

This chapter describes the usage of `homologizer` to phase gene copies into polyploid subgenomes. Allopolyploids contain multiple copies of each genetic locus, where each copy potentially belongs to a different subgenome with its own distinct evolutionary history. If gene copies across different loci are incorrectly phased (i.e., assigned to the wrong subgenome), then the bifurcating tree assumption underlying multilocus phylogenetic inference and related analyses will be violated, leading to unsound results. `homologizer` is a highly flexible Bayesian method that uses a phylogenetic framework to infer the posterior probabilities of the phasing of gene copies into subgenomes. We describe how to prepare input data and other considerations needed to perform `homologizer` analyses and demonstrate how to visualize and interpret the results. We first walk through a basic example using `homologizer` to phase gene copies into polyploid subgenomes and then demonstrate how `homologizer` can be used as a hypothesis-testing tool to detect non-homeologous sequences such as hidden paralogs or allelic variation through the tools of Bayesian model comparison.

Key words Allopolyploidy, Homeologs, Hybridization, Phasing, Polyploid phylogenetics, Reticulate evolution, RevBayes

1 Introduction

1.1 Overview

In this chapter we demonstrate `homologizer`, a highly flexible Bayesian method that infers the phasing of gene copies into the subgenomes of hybrid organisms [1]. Allopolyploids and F1 hybrids contain distinct subgenomes each with their own evolutionary history. When sequencing loci from such organisms, unique copies of each locus may be recovered from each subgenome. If each copy of each locus is not assigned to the correct subgenome, then multilocus phylogenetic inference can be erroneous since the assumptions of a single bifurcating tree have been violated [2].

`homologizer` uses a phylogenetic framework to infer the posterior probabilities of the phasing of gene copies across loci and is implemented in the open-source phylogenetic inference

Yves Van de Peer (ed.), *Polyploidy: Methods and Protocols*,
Methods in Molecular Biology, vol. 2545, https://doi.org/10.1007/978-1-0716-2561-3_6,

software RevBayes [3]. RevBayes is a rich and complex phylogenetic inference tool; it can accommodate a vast array of models outside of the scope of this manuscript (see http://revbayes.com). The benefit of homologizer being implemented within RevBayes is that phasing functionality can be added in a simple and modular fashion to all the other types of phylogenetic models already implemented within RevBayes. For example, the phasing inference can be performed on a fixed phylogeny or while simultaneously estimating the phylogeny under a wide range of phylogenetic models, including the multi-species coalescent. Thus homologizer can be used to infer the phylogeny while treating the phasing as a nuisance parameter, or for inferring the phasing while integrating out the phylogeny, or for a combination of these goals. Additionally, this flexibility allows different homologizer phasing models to be compared using standard statistical model comparison approaches such as Bayes factors [4]. This functionality is particularly useful when using homologizer as a hypothesis-testing tool to detect non-homeologous sequences such as hidden paralogs or allelic variation.

Below we will briefly describe the homologizer model, discuss some preliminary concerns, and then walk through two example analyses in detail. The first example analysis uses homologizer to phase gene copies into the subgenomes of a set of allopolyploids. The second example analysis uses homologizer to test whether the observed gene copies of one of the allopolyploids are homeologs from distinct subgenomes or allelic variants arising from the same subgenome. We provide the data and the full scripts required to run these examples at http://github.com/wf8/homologizer. In this chapter we will focus on the practical considerations of preparing input data, setting up and running the analyses, as well as visualizing and interpreting the results. For more detailed and theoretical discussions about the performance of the homologizer model under different evolutionary scenarios we refer the reader to Freyman et al. [1].

1.2 homologizer Model and Assumptions

In a homologizer analysis phylogenies containing hybrid organisms are represented as multilabeled trees ("mul-trees"), where each hybrid accession is present multiple times, once for each subgenome (each distinct evolutionary history; [5]). Gene copies are phased into subgenomes for each locus by swapping which gene copy is assigned to each of that accession's tips in the mul-tree (Fig. 1). The homologizer probabilistic model consists of a set of latent variables that assign gene copies to the various tips of a hybrid accession's mul-tree tips. The prior probabilities of all possible gene copy assignments are assumed to be equal. Given any one assignment of gene copies to the mul-tree tips, a phylogenetic model can then be used to calculate the likelihood of the phasing assignments. The posterior distribution of the phasing assignments

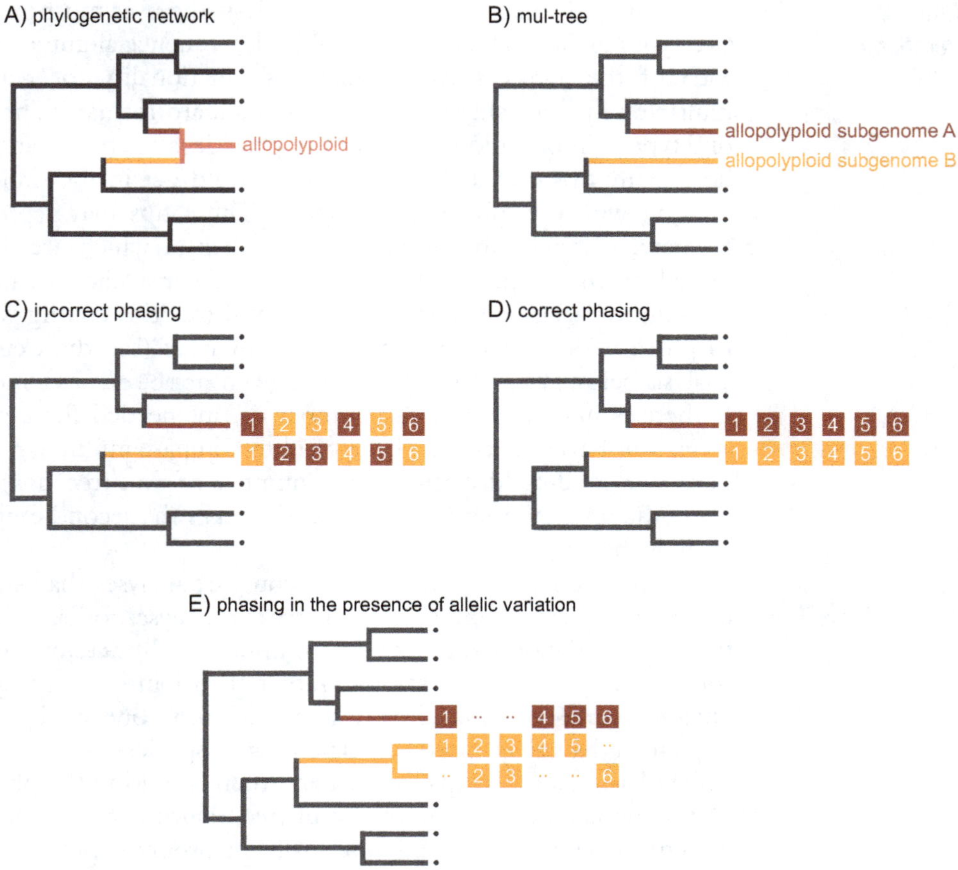

Fig. 1 The `homologizer` model phases gene copies into polyploid subgenomes over a mul-tree. **a**: A phylogenetic network with a single a single hybridization (reticulation) giving rise to an allopolyploid. **b**: The mul-tree representation of the phylogenetic network has two tips (red and orange) representing the two subgenomes of the allopolyploid. **c**: Six loci were sequenced from the allopolyploid (red and orange squares). Two copies (red and orange) of each locus were recovered. Loci 2, 3, and 5 are incorrectly phased; that is they are incorrectly assigned to the wrong subgenome. **d**: After phasing, the gene copies of each locus are now assigned to the correct subgenome. **e**: An alternative phasing model allows for the two gene copies to be allelic variants from the same subgenome. In this example, both copies of locus 2 and 3 are allelic variants from the orange subgenome and no copies of these loci were recovered from the red subgenome

is estimated by drawing sample from the posterior using Markov chain Monte Carlo (MCMC; [6]) algorithms with proposals that swap gene copies among mul-tree tips.

Any phylogenetic model that `RevBayes` offers can be extended to include the specialized `homologizer` MCMC proposals and monitors. This suite of models includes those in which the topology is fixed or jointly estimated, models that are time calibrated, and models that assume a shared gene tree for all loci or those that allow unlinked gene trees such as the multi-species coalescent.

1.3 Practical Considerations

To set up and run a `homologizer` analysis the user must provide two types of data. The first is the multiple-sequence alignments for each of the loci. `RevBayes` includes functionality for reading multiple-sequence alignments in most standard formats. The second type of information the user must provide is a determination of how many tips each hybrid accession should get in the mul-tree (among which to phase gene copies). These tips may represent separate, distinct subgenomes (the number of which are determined by the ploidy of the organism) and/or allelic variants of the same subgenome. The number of mul-tree tips among which to phase is defined within the `Rev` script itself (see the example analysis below) and is thus fixed for a given `homologizer` model. If there is uncertainty in the number of tips needed for a given accession, Bayes factors can be applied to compare multiple `homologizer` models (differing in the number of mul-tree tips) and select the optimal number of tips (as shown in the second example analysis, below).

Some datasets will require more complex analyses than shown in our examples. Scaling up `homologizer` analyses for large high-throughput datasets may not be computationally feasible unless some care is taken. For example, rather than jointly inferring the phylogeny and phasing all loci simultaneously, one could use a sequential Bayesian approach: first infer a species-level mul-tree while phasing a subsample of loci, and then condition the phasing of the remaining loci on that mul-tree. During the secondary phasing analyses one could optionally incorporate phylogenetic uncertainty in the mul-tree by integrating over the posterior distribution of mul-trees, as inferred in the first step. Additionally, if a user expects high levels of incomplete lineage sorting to be present in their dataset, we recommend using `homologizer` within the full multi-species coalescent (MSC) framework, allowing for unlinked gene trees. This model, however, is computationally demanding (like all full MSC models) and can feasibly be applied to only relatively small datasets.

1.4 Getting Started with `RevBayes`

Probabilistic graphical models are commonly used in statistics and machine learning, and their use is growing in all fields of science in which mathematical models are used to make inference, particularly computational biology [7, 8]. The graphical model framework provides a precise formalism for complex, parameter-rich models that enable them to be constructed and utilized effectively. All probabilistic models can be represented as graphs, where each node in the graph is a random variable, and the edges of the graph represent the dependency structure among variables.

There are a number of probabilistic programming languages that are designed to specify graphical models such as `Stan` [9] and `PyMC3` [10]. However, phylogenetic models require inference machinery and distributions not available in those other tools. The software `RevBayes` [3] provides the R-like scripting language `Rev` which enables users to specify complex phylogenetic graphical models [11]. In `RevBayes` there are no default "black-box" models; the user must fully detail the analysis they want to perform. This means there is a great deal of flexibility in setting up custom analyses uniquely tailed to answer the researcher's questions, including `homologizer` analyses.

1.4.1 Installing RevBayes

`RevBayes` can be downloaded from http://revbayes.com. Executables are available for MacOS, MS Windows, and Linux. If you are using MacOS or Linux operating system, we recommend that you add the `RevBayes` executable to your path. The software can also be compiled from the source code, available from https://github.com/revbayes/revbayes.

1.4.2 Running RevBayes

In `RevBayes` the `Rev` language is used to specify models. It is outside of the scope of this paper to describe the full potential of the `Rev` language, but extensive tutorials are available at http://revbayes.com. Like in `R` [12], the commands can be either typed directly into `RevBayes` interactively or written in scripts that are then loaded.

Start `RevBayes` by typing `rb` in the terminal window. The command prompt > will appear. Type

```
5 + 5
```

and hit enter. Now try

```
x = 5 + 5
x
```

If you want to load a `Rev` script from an external file type

```
source("my_script.Rev")
```

You will get an error message because the file `my_script.Rev` does not exist. To quit RevBayes type `q()`.

For complex models and analyses, it is best to create `Rev` script files that contain all of the model parameters, moves, and functions. In the following sections we will provide code to run `homologizer` analyses that should be saved into a single `Rev` script (Fig. 2).

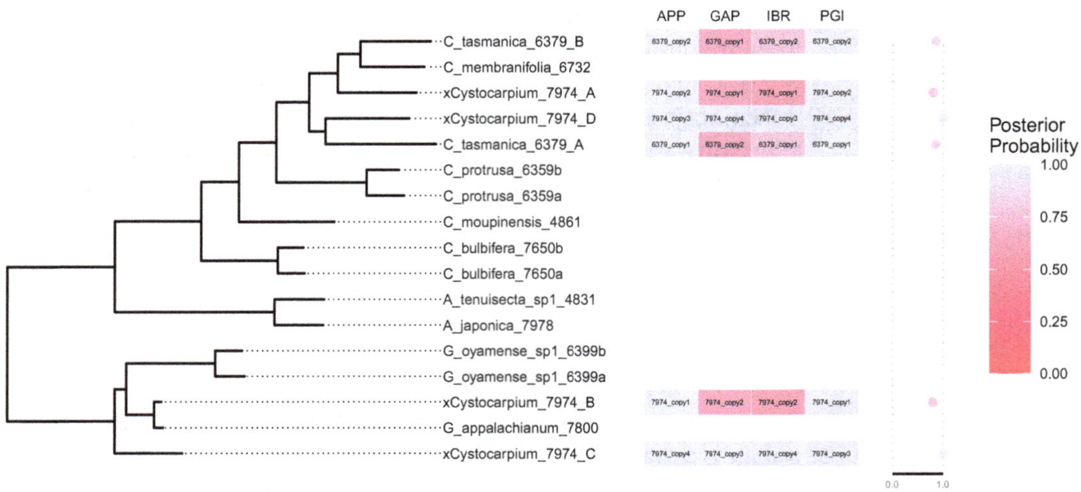

Fig. 2 Final results of a `homologizer` phasing analysis. Shown is the inferred phasing of gene copies into subgenomes summarized on the maximum *a posteriori* (MAP) phylogeny for the Cystopteridaceae dataset. The phase is estimated for the two polyploid accessions `xCystocarpium 7974` and `C tasmanica 6379`. To the right of the tree, each column represents a locus, and the joint MAP phase assignment is shown as text within each box. Each box is colored by the marginal posterior probability of the phase assignment. Adjacent to the heatmap is a column that shows the mean marginal probability across loci of the phasing assignment per tip, which summarizes the model's overall confidence in the phasing of that tip

2 Phasing Gene Copies

2.1 Overview

Our first example analysis uses `homologizer` to phase gene copies into the subgenomes of a set of allopolyploids. The output of the analysis is the posterior distribution of phased homeologs, i.e., the posterior distribution of the assignments of each gene copy, for each locus, into each of the subgenomes of the polyploids. Since we perform joint inference of the phasing and phylogeny, the posterior distribution of the multilocus phylogeny is also inferred, along with all other parameters of the model.

In this example analysis we use a reduced version of the dataset from the fern family Cystopteridaceae previously analyzed in Rothfels et al. [13] and Freyman et al. [1] (reduced to increase the speed of the analyses). The data consist of four single-copy nuclear loci (*ApPEFP_C, gapCpSh, IBR3,* and *pgiC*) for a sample of 11 diploids and two tetraploids.

Here we gloss over many of the details in the phylogenetic model (e.g., substitution models) so that we can focus on the phasing aspect of the analysis. Detailed tutorials on these other aspects of `RevBayes` can be found at http://revbayes.com. All the data and the full scripts required to run this analysis can be found at http://github.com/wf8/homologizer.

2.1.1 Names in the Sequence Alignment Files

One of the trickier aspects of setting up a `homologizer` analysis is the naming of gene copy sequences. In a standard phylogenetic analysis the names used in the sequence alignment are also the names that appear on the tips of the inferred tree. By necessity, `homologizer` adds an extra layer of abstraction between the names in the sequence alignment and the tree's tip names. For example, imagine a polyploid with two subgenomes "subgenome_A" and "subgenome_B". The polyploid has two gene sequences for a given locus named "copy_1" and "copy_2". The objective of the `homologizer` analysis is to figure out which copy of the locus belongs to each of the subgenomes. The inferred mul-tree will have tip names "subgenome_A" and "subgenome_B" but the sequence alignment for the locus should have sequence names "copy_1" and "copy_2".

To keep this straight, we recommend standardizing the naming scheme for gene copy sequences across loci and polyploid subgenomes. For example, the Cystopteridaceae dataset has an accession, ×*Cystocarpium* #7974, that has four sequences for each locus, labelled "7974_copy1" through "7974_copy4", and we wish to phase those copies among four mul-tree tips, labelled "xCystocarpium_7974_A" through "xCystocarpium_7974_D".

2.1.2 Setting up the `Rev` File

For this analysis you could cut and paste the following commands directly into the `Rev` prompt and run each command one by one. However, we recommend you save the `Rev` code into a file (e.g., `cystopteridaceae.Rev`) and then run the analysis by typing `rb cystopteridaceae.Rev` in your command line. The full scripts can be found at http://github.com/wf8/homologizer.

Our first step is to define a vector that holds the input sequence alignment files, one for each locus.

```
alignments = ["data/APP.nex",
              "data/GAP.nex",
              "data/IBR.nex",
              "data/PGI.nex"]
```

We will now loop through and read in each alignment, saving them to the vector `data`.

```
num_loci = alignments.size()
for (i in 1:num_loci) {
    data[i] = readDiscreteCharacterData(alignments[i])
}
```

Next we set the *initial* phase assignments for the polyploid accession `xCystocarpium_7974`. We need to set the phase assignment here to enable the MCMC to initialize. We can randomly assign gene copies to subgenomes; the assignment should not affect the final outcome of the analysis assuming the MCMC is allowed to converge.

We do this by calling the function `setHomeologPhase` on each of the alignments. In the alignments the sequences for this accession are named 7974_copy1 through 7974_copy4. We wish to phase those copies among four mul-tree tips, xCystocarpium_7974_A through xCystocarpium_7974_D.

```
for (i in 1:num_loci) {
    data[i].setHomeologPhase("7974_copy1", "xCystocarpium_7974_A")
    data[i].setHomeologPhase("7974_copy2", "xCystocarpium_7974_B")
    data[i].setHomeologPhase("7974_copy3", "xCystocarpium_7974_C")
    data[i].setHomeologPhase("7974_copy4", "xCystocarpium_7974_D")
}
```

This data set contains a second polyploid C_tasmanica_6379. This accession, though, only has two subgenomes and two gene copies of each locus. However, it is missing a sequence for the gene IBR; for IBR there is only a single copy: 6379_copy1. Recalling that IBR is the third sequence alignment we read in, we can add a blank second IBR gene copy for C_tasmanica_6379:

```
data[3].addMissingTaxa("6379_copy2")
```

Now we again loop through the alignments, this time setting the initial phase for C_tasmanica_6379.

```
for (i in 1:num_loci) {
    data[i].setHomeologPhase("6379_copy1", "C_tasmanica_6379_A")
    data[i].setHomeologPhase("6379_copy2", "C_tasmanica_6379_B")
}
```

The next few sections of code are fairly standard for Rev phylogenetic analyses, and not unique to a homologizer analysis. Since some of the diploid accessions are also missing sequences for some loci, we now add any blank sequences needed so all the alignments contain all the accessions:

```
for (i in 1:num_loci) {
    for (j in 1:num_loci) {
        data[i].addMissingTaxa(data[j].taxa())
    }
}
```

We will need some useful information from the alignments:

```
num_tips = data[1].ntaxa()
n_branches = 2 * num_tips - 3
```

Now create a vector of branch lengths. We will draw each branch length from an exponential distribution. We will also add MCMC scaling moves for each branch length (which we will store in a "moves" vector, indexed by a "mvi" counter).

```
mvi = 0
for (i in 1:n_branches) {
    branch_lengths[i] ~ dnExponential(10)
    moves[++mvi] = mvScale(branch_lengths[i], weight=1.0)
}
```

We will use a uniform topology prior that puts equal probability on all unrooted, fully resolved topologies. Additionally, we will add MCMC moves for the topology, the nearest-neighbor interchange

(NNI) and subtree pruning and regrafting (SPR) tree arrangement moves.

```
topology ~ dnUniformTopology(data[1].taxa())
moves[++mvi] = mvNNI(topology, weight=40.0)
moves[++mvi] = mvSPR(topology, weight=40.0)
```

Finally, we combine the topology and the branch length vector into a deterministic node that represents our phylogeny:

```
tree := treeAssembly(topology, branch_lengths)
```

For the nucleotide substitution models we will specify a general time-reversible (GTR; [14]) model for each locus. We will use an uninformative Dirichlet distribution as prior on the stationary frequencies (`pi`), and for the six exchangeability rates `er`. To estimate `pi` and `er` we use the MCMC move `mvSimplexElementScale`, which randomly changes one element of the simplex and then rescales the other elements so that they sum to one again. For each locus we construct the GTR rate matrix `Q` using the function `fnGTR` which puts together `pi` and `er`.

```
for (i in 1:num_loci) {

    er_prior <- v(1,1,1,1,1,1)
    er[i] ~ dnDirichlet(er_prior)
    er[i].setValue(simplex(v(1,1,1,1,1,1)))
    moves[++mvi] = mvSimplexElementScale(er[i], weight=5)

    pi_prior <- v(1,1,1,1)
    pi[i] ~ dnDirichlet(pi_prior)
    pi[i].setValue(simplex(v(1,1,1,1)))
    moves[++mvi] = mvSimplexElementScale(pi[i], weight=5)

    Q[i] := fnGTR(er[i], pi[i])
}
```

Additionally, we estimate a substitution rate multiplier for each of the alignments except the first one. We draw the rate multipliers from an exponential distribution:

```
for (i in 1:num_loci) {
    if (i == 1) {
        rate_multiplier[i] <- 1.0
    } else {
        rate_multiplier[i] ~ dnExponential(1)
        moves[++mvi] = mvScale(rate_multiplier[i], weight=5)
    }
}
```

Our sequence evolution models are continuous-time Markov chains (CTMC) over the phylogeny. So we pass a GTR rate matrices `Q`, a `rate_multiplier`, and the `tree` into a phylogenetic CTMC distribution, one for each locus. We fix the value of the CTMC to our observed sequence data using the `clamp` function.COMP: Please set the code text within the margin width.

```
for (i in 1:num_loci) {
    ctmc[i] ~ dnPhyloCTMC(tree=tree, Q=Q[i], branchRates=rate_multiplier[i], t
    ctmc[i].clamp(data[i])
}
```

We now have fully defined our phylogenetic model, so we wrap it up and declare it complete:

```
mymodel = model(Q)
```

To infer the phasing, though, we wish to add MCMC phasing proposals. We use the function `mvHomeologPhase` to define a phasing proposal that swaps the sequences between any two mul-tree tips for a given locus. Since our polyploid accession `C_tasmanica_6379` has only two subgenomes A and B, this means we need one `mvHomeologPhase` per locus:

```
for (i in 1:num_loci) {
    moves[++mvi] = mvHomeologPhase(ctmc[i], "C_tasmanica_6379_A", "C_tasmanica
}
```

Note that the `weight` of the move is set to 2. The weight specifies how often this particular MCMC move will be proposed relative to all other moves in our MCMC. If the phasing analysis is not converging, one can try increasing the weight of these moves.

The other polyploid accession `xCystocarpium_7974` has four subgenomes. To enable gene copies to swap among all four subgenomes we need $\binom{4}{2} = 6$ moves for each locus:

```
for (i in 1:num_loci) {
    moves[++mvi] = mvHomeologPhase(ctmc[i], "xCystocarpium_7974_A", "xCystocar
    moves[++mvi] = mvHomeologPhase(ctmc[i], "xCystocarpium_7974_A", "xCystocar
    moves[++mvi] = mvHomeologPhase(ctmc[i], "xCystocarpium_7974_A", "xCystocar
    moves[++mvi] = mvHomeologPhase(ctmc[i], "xCystocarpium_7974_B", "xCystocar
    moves[++mvi] = mvHomeologPhase(ctmc[i], "xCystocarpium_7974_B", "xCystocar
    moves[++mvi] = mvHomeologPhase(ctmc[i], "xCystocarpium_7974_C", "xCystocar
}
```

Finally, we need to set up some monitors to draw samples from the chain. We will set up three monitors used in standard phylogenetic analyses: one that writes a log file for most of the model parameters, another that writes the sampled trees to file, and also a screen monitor so we can view progress on our screen:

```
mni = 0
output_file = "output/homologizer"
monitors[++mni] = mnModel(filename=output_file + ".log", printgen=1)
monitors[++mni] = mnFile(filename=output_file + ".trees", printgen=1, tree)
monitors[++mni] = mnScreen(printgen=1)
```

Additionally we need to define special monitors for logging samples of the phase of each locus. These are defined using `mnHomeologPhase`. We must specify one of these for each of the loci being phased.

```
for (i in 1:num_loci){
    log_file = output_file + "_locus_" + i + "_phase.log"
    monitors[++mni] = mnHomeologPhase(filename=log_file, printgen=1, ctmc[i])
}
```

2.1.3 Running the MCMC

Finally, let us set up our MCMC object and run it. To do this, we pass our model object `mymodel`, the vector of monitors, and the vector of MCMC moves into the `mcmc` function. For this example exercise we will run the analysis for 2000 iterations. For an actual analysis the MCMC should be run much longer.

```
mymcmc = mcmc(mymodel, monitors, moves)
mymcmc.run(generations=2000)
```

This will execute the analysis and you should see output similar to this:

```
Running MCMC simulation
This simulation runs 1 independent replicate.
The simulator uses 73 different moves in a random move schedule with 219.8
moves per iteration

Iter    |   Posterior   |  Likelihood   |     Prior    |    elapsed    |        E
-----------------------------------------------------------------------------------
0       |    -14739.8   |      -14848   |   108.217    |   00:00:00    |   --:--:
1       |    -11255.8   |    -11363.8   |   107.962    |   00:00:01    |   --:--:
2       |    -10664.5   |    -10773.7   |   109.236    |   00:00:01    |   00:16:
3       |    -10440.4   |    -10550.9   |   110.498    |   00:00:01    |   00:11:
4       |    -10405.3   |    -10515.3   |   110.053    |   00:00:02    |   00:08:
...
```

When the analysis is complete, you will have a new directory called `output` that will contain all of the files you specified with the monitors. To check whether the MCMC has converged we can plot the trace of the model parameters found in `output/homologi-zer.log` (Fig. 3). To further assess convergence, this file can be opened in Tracer [15] or analyzed using CODA [16].

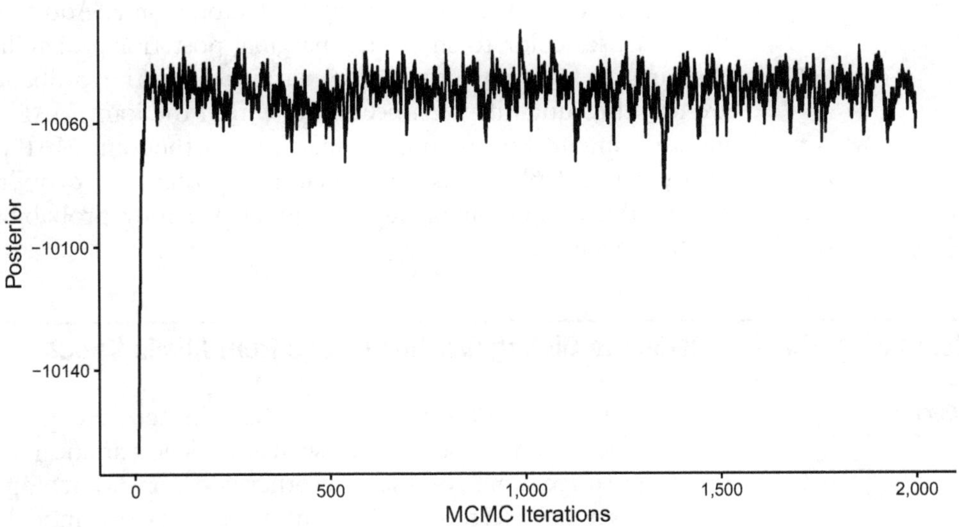

Fig. 3 MCMC trace from the Cystopteridaceae `homologizer` analysis. From this trace the MCMC appears to converge after approximately 100 iterations. For an actual analysis the MCMC should be run much longer

2.1.4 Summarizing the Posterior Distribution

The inferred phasing of gene copies into subgenomes is best summarized in the context of the phylogeny (Fig. 2). So our first step is to summarize the trees sampled by the MCMC. We read in the tree samples:

```
treetrace = readTreeTrace(output_file + ".trees", treetype="non-clock", burnin
```

and summarize the trees into a single maximum *a posteriori* (MAP) tree:

```
map_tree = mapTree(treetrace, output_file + "_map.tree")
```

This command creates the tree file `output/homologizer_-map.tree` that you can plot in APE [17] or FigTree [18]. Since we estimated an unrooted tree, you should use one of these tools to root the tree correctly and save a copy of the rooted tree that we can use to visualize the inferred phasing (as in Fig. 2).

Once we have the rooted MAP tree, the phasing estimates can be summarized and plotted using `R` [12]. For this tutorial we provide a script `plot_phase.R` to generate Fig. 2 (available with the other scripts from these examples at http://github.com/wf8/homologizer). This script can be easily adapted to work for other datasets; see the comments within the script. This plotting functionality will soon be more widely available as part of the `RevGadgets` [19] `R` package.

Figure 2 shows the joint maximum *a posteriori* (MAP) phasing assignment across the set of a polyploid's subgenomes for each locus. This joint MAP assignment is the highest probability assignment of each gene uniquely to a subgenome. Additionally, the plot uses color to show the marginal posterior probability of the phasing for each copy. These marginal posterior probabilities are useful to quantify the uncertainty within the joint MAP phasing assignment. For example, it may be that the joint MAP phase of a given polyploid has a low marginal posterior probability in some subgenomes but a high marginal posterior probability in other subgenomes.

3 Comparing Phasing Models to Distinguish Homeologs from Allelic Variation

3.1 Overview

To distinguish gene copies that evolved in separate polyploid subgenomes from those that arose from allelic variation within the same subgenome (or that are otherwise non-homeologous), one can set up a series of different `homologizer` model that differ in the number of mul-tree tips available to phase. The statistical fit of these models can then be compared using Bayes factors [4].

Consider the example in Fig. 1d, e. In panel d the allopolyploid has two subgenomes (red and orange). Two copies of six loci are correctly phased into the two subgenomes. However, in panel e, both copies of loci two and three are actually allelic variants from the orange subgenome, and no copies of these loci were recovered from the red subgenome. In real datasets it can be hard to distinguish between these two scenarios. With `homologizer`, however, the researcher can set up two models: one that allows phasing among two mul-tree tips for the allopolyploid, and another that allows phasing among three mul-tree tips. Bayes factors are then used to compare to two models and determine how many tips should be used.

For our second example analysis, we will return to the Cystopteridaceae dataset and test whether the polyploid accession `C_tasmanica_6379` should be phased into two mul-tree tips (as done in example 1 above) or whether allelic variation is present and three mul-tree tips are needed for phasing.

3.2 Computing Marginal Likelihoods

To compare the two mul-tree tip and three mul-tree tip `homologizer` phasing models using Bayes factors, we first need to ensure that the data used under both models are the same. For example, when adding a third tip for the polyploid accession `C_tasmanica_6379`, we will add a new set of blank sequences to this tip, so this set of blank sequences needs to be added to both models (the two mul-tree tip and three mul-tree tip phasing models). However, only in the three-tip model will we allow gene copies to be phased into this third tip (in the two-tip model the third tip will exist, but will only be associated with blank sequences). The code from our first example above specifies the two-tip phasing model. To add the third blank tip, we need to modify the code where we set the initial phase of each gene copy to:

```
for (i in 1:num_loci) {
    data[i].setHomeologPhase("6379_copy1", "C_tasmanica_6379_A")
    data[i].setHomeologPhase("6379_copy2", "C_tasmanica_6379_B")
    data[i].addMissingTaxa("6379_BLANK")
    data[i].setHomeologPhase("6379_BLANK", "C_tasmanica_6379_C")
}
```

This code adds a blank sequence (which we have labeled "6379_BLANK") to each locus and associates those blank sequences with a mul-tree tip called "C_tasmanica_6379_C". Since we have not created any moves associated with this third tip (there is no associated `mvHomeologPhase` command), no gene copies will be phased into it. To ease the interpretation of the results after phasing, we recommend that the names for blank sequences include "BLANK", e.g., "6379_BLANK" or "6379_BLANK1" and "6379_BLANK2" if a locus is missing more than one sequence.

For the Bayes factors, we will compute the marginal likelihood of each model using a stepping-stone analysis [20]. To do so for the two-tip phasing model that we now have in hand, we can simply swap out this section of code:

```
mymcmc = mcmc(mymodel, monitors, moves)
mymcmc.run(generations=2000)
```

for this section of code:

```
pow_p = powerPosterior(mymodel, moves, monitors, output_file + ".out",
                       cats=50, sampleFreq=1)
pow_p.burnin(generations=200, tuningInterval=50)
pow_p.run(generations=1000)
ss = steppingStoneSampler(file=output_file + ".out",
                          powerColumnName="power", likelihoodColumnName="likel
print(ss.marginal())
```

This code sets up the stepping-stone sampler that uses 50 stepping stones, sampling 1000 states from each step. Once it is complete, the code prints the marginal likelihood to the screen. When run, it looks like this:

```
Running burn-in phase of Power Posterior sampler for 200 iterations.
The simulator uses 72 different moves in a random move schedule with 222 moves

Progress:
0---------------25---------------50---------------75--------------100
********************************************************************

Running power posterior analysis ...
Step  1 / 51          ****************************************
Step  2 / 51          ****************************************
Step  3 / 51          ****************************************
...
Step 49 / 51          ****************************************
Step 50 / 51          ****************************************
Step 51 / 51          ****************************************
-10313.19
```

The final number (`-10313.19`) is the log marginal likelihood of the two-tip `homologizer` phasing model.

3.3 Setting up the Alternative `homologizer` Model

We can now modify the two-tip phasing model specified above so that it allows phasing among all three mul-tree tips. We will change the code where we define the MCMC proposals that allow different phasing assignments to be explored. Previously, since `C_tasmanica_6379` only had two mul-tree tips to phase among, a single `mvHomeologPhase` per locus was sufficient:

```
for (i in 1:4) {
    moves[++mvi] = mvHomeologPhase(ctmc[i], "C_tasmanica_6379_A", "C_tasmanica
}
```

Now the above code must be changed to:

```
for (i in 1:4) {
    moves[++mvi] = mvHomeologPhase(ctmc[i], "C_tasmanica_6379_A", "C_tasmanica
    moves[++mvi] = mvHomeologPhase(ctmc[i], "C_tasmanica_6379_A", "C_tasmanica
    moves[++mvi] = mvHomeologPhase(ctmc[i], "C_tasmanica_6379_B", "C_tasmanica
}
```

Now we can compute the marginal likelihood of the three-tip model using a stepping-stone analysis just as we did for the two-tip model: this analysis results in a marginal likelihood calculation of -`10244.20`.

3.4 Comparing the Two `homologizer` Models

Bayes factors are the ratio of the marginal likelihoods of the two models being compared. In this case, we have computed −10244.2 for the three mul-tree tip phasing model and −10313.19 for the two mul-tree tip phasing model. Since these are log marginal likelihoods, we subtract them—- 10244.2 − (−10313.19) = 68.99— to compute the Bayes factor. Here, the Bayes factor of the three mul-tree tip model compared to the two mul-tree tip model is 68.99, which is "strong" support for the three-tip model [4]. Note that if you run this example your marginal likelihood estimates may differ slightly. The marginal likelihoods will converge more closely if we ran longer stepping-stone analyses, for example, increasing the number of states sampled from each stone to 5000 rather than 1000.

This ability to use `homologizer` as a data-exploration and hypothesis-testing tool may result in key insights about the polyploid accession that can significantly impact downstream interpretations. In this example, the three-tip model strongly out-performed the two-tip model for the polyploid `C_tasmanica_6379`. Indeed, in Freyman et al. [1] we use `homologizer` to show that the two *gapCpSh* copies from this polyploid are allelic variants from the same subgenome (sister to one another in the phylogeny), with a blank copy phased with high posterior probability to the other subgenome. Recognizing these copies as alleles rather than homeologs resulted in significantly altered downstream inferences about the evolutionary history of this species, including very different inferences of its parentage.

4 Conclusion

Due to the complexity of using `RevBayes`, `homologizer` is not an easy "off-the-shelf" tool to use. However, we hope that this chapter introduces researchers to the potential of this powerful and flexible inference framework and helps them decide whether `homologizer` is an appropriate tool for their data and questions. It is our hope that the examples provided here will help interested users get started with their analyses.

References

1. Freyman WA, Johnson MG, Rothfels CJ (2020) Homologizer: phylogenetic phasing of gene copies into polyploid subgenomes. bioRxiv
2. Rothfels CJ (2021) Polyploid phylogenetics. New Phytol 230(1):66–72
3. Höhna S, Landis MJ, Heath TA, Boussau B, Lartillot N, Moore BR, Huelsenbeck JP, Ronquist F (2016) RevBayes: Bayesian phylogenetic inference using graphical models and an interactive model-specification language. Syst Biol 65(4):726–736
4. Kass RE, Raftery AE (1995) Bayes factors. J Am Stat Assoc 90(430):773–795
5. Huber KT, Oxelman B, Lott M, Moulton V (2006) Reconstructing the evolutionary history of polyploids from multilabeled trees. Mol Biol Evol 23(9):1784–1791. [Online] http://mbe.oxfordjournals.org/cgi/content/abstract/23/9/1784
6. Metropolis N, Rosenbluth AW, Rosenbluth MN, Teller AH, Teller E (1953) Equation of state calculations by fast computing machines. J Chem Phys 21(6):1087–1092
7. Jordan MI (2004) Graphical models. Stat Sci 19(1):140–155
8. Airoldi EM (2007) Getting started in probabilistic graphical models. PLoS Comput Biol 3(12):e252
9. Gelman A, Lee D, Guo J (2015) Stan: a probabilistic programming language for Bayesian inference and optimization. J Educ Behav Stat 40(5):530–543
10. Salvatier J, Wiecki TV, Fonnesbeck C (2016) Probabilistic programming in Python using PyMC3. Peer J Comput Sci 2:e55
11. Höhna S, Heath TA, Boussau B, Landis MJ, Ronquist F, Huelsenbeck JP (2014) Probabilistic graphical model representation in phylogenetics. Syst Biol 63(5):753–771
12. R Core Team (2013) R: a language and environment for statistical computing. R Foundation for Statistical Computing, Vienna. http://www.R-project.org/
13. Rothfels CJ, Pryer K, Li F-W (2017) Next-generation polyploid phylogenetics: rapid resolution of hybrid polyploid complexes using PacBio single-molecule sequencing. New Phytol 213(1):413–429
14. Tavaré S (1986) Some probabilistic and statistical problems in the analysis of DNA sequences. Lect Math Life Sci 17(2):57–86
15. Rambaut A, Drummond AJ, Xie D, Baele G, Suchard MA (2018) Posterior summarization in Bayesian phylogenetics using Tracer 1.7. Syst Biol 67(5):901
16. Plummer M, Best N, Cowles K, Vines K (2006) CODA: convergence diagnosis and output analysis for MCMC. R News 6(1):7–11
17. Paradis E, Claude J, Strimmer K (2004) APE: analyses of phylogenetics and evolution in R language. Bioinformatics 20(2):289–290
18. Rambaut A (2009) Figtree. http://tree.bio.ed.ac.uk/software/figtree/
19. Tribble C et al. (2020) RevGadgets: an R Package for visualizing Bayesian phylogenetic analyses from RevBayes. Methods Ecol Evol
20. Xie W, Lewis PO, Fan Y, Kuo L, Chen M-H (2011) Improving marginal likelihood estimation for Bayesian phylogenetic model selection. Syst Biol 60(2):150–160

Chapter 7

Constraining Whole-Genome Duplication Events in Geological Time

James W. Clark and Philip C. J. Donoghue

Abstract

The timing of whole-genome duplication (WGD) events is crucial to understanding their role in evolution and underpins many hypotheses linking WGD to increased diversity and complexity. As such, means of estimating the timing of the WGD events relative to their macroevolutionary outcomes are of considerable importance. Molecular clock methods facilitate direct estimation of the absolute timing of WGD events, integrating information on the rate of sequence evolution between species while accommodating the uncertainty inherent to the fossil record. We present an explanation of the best practice for constructing fossil calibrations and estimating the age of WGD events via molecular clock methods in the program MCMCtree, with an example dataset based on a well-characterized WGD event within the flowering dogwoods (*Cornus*). The approach presented herein allows for the estimation of the age of WGD events and subsequent speciation events, allowing the relationship between WGD and the macroevolutionary outcomes to be explored. In our example, we show that in the case of flowering dogwoods, the WGD event long predates the end-Cretaceous mass extinction and that the two events may be independent.

Key words Molecular clock, Polyploidy, Whole-genome duplication, Fossil calibration, Cornus

1 Introduction

Whole-genome duplication (WGD; Polyploidy) has occurred in the evolutionary history of several major land plant lineages, as well as in fungi and animals. These events are often invoked as agents of macroevolutionary change [1], and instances of WGD have been linked to morphological innovations [2], biogeographic shifts [3], lineage longevity [4], and increased rates of species diversification [5]. Each of these evolutionary hypotheses depends on an estimate of the timing of the WGD event to justify a correlation, let alone causation. The timing of WGD events can be considered in both relative and absolute terms. The relative (or phylogenetic) timing of a WGD event identifies the lineage (branch on a phylogenetic tree) in which the WGD event occurred, based on identifying which species do and do not exhibit genomic evidence of that event.

Yves Van de Peer (ed.), *Polyploidy: Methods and Protocols*,
Methods in Molecular Biology, vol. 2545, https://doi.org/10.1007/978-1-0716-2561-3_7,

However, many hypotheses that relate macroevolutionary consequences to WGD events, such as increased rates of diversification or morphological innovation, are also dependent on the absolute (geological) timing of the event. Following WGD, the process of diploidization or fractionation is believed to result in a time lag between the duplication itself and any proposed macroevolutionary outcomes [6]. The extent of this lag is important, since a drawn-out period of diploidization may explain the disparate outcomes of WGD across sister lineages [7].

Methods of dating WGD events can be categorized as phylogenetic and nonphylogenetic. The primary nonphylogenetic method is to derive the rate of nonsynonymous substitutions (*Ks*) among paralogous gene pairs [8, 9]. Across all paralogous pairs, the distribution of *Ks* values should exhibit a peak if multiple pairs duplicated at the same time, as would be the case in a WGD event. Once identified, this peak can be converted into units of geological time by selecting an external calibration. However, converting the substitution rate using a single-point calibration is problematic since it a) assumes a strict rate of molecular evolution among loci and b) places excessive confidence in the calibration. Further, these methods are known to fail when dating increasingly ancient WGD events since saturation in the rate of substitutions can obscure signal [9, 10].

Phylogenetic approaches rely on the reconciliation of the evolutionary history of genes and species. The most straightforward phylogenetic approach is to bracket the age of gene duplication events between relevant species divergences. The WGD must be older than all of the lineages that underwent the event, but younger than the divergence of those that did not (Fig. 1). Thus, the simplest way of estimating the age of a WGD event is to provide a range between these two ages. However, in order to provide a reasonable estimate of the age of the WGD, this approach relies on a few assumptions. First, that the ages of the species divergences are known and reliably estimated. When characterizing novel WGD events in lineages that previously lacked genomic resources, it is possible that the evolutionary timeline will be poorly understood. Second, that the time between the two species divergences is small relative to their geologic age. Since this method does not directly estimate the timing of WGD, species that sit on long evolutionary branches will only be able to constrain the age of WGD events to unhelpfully broad intervals. Given these considerable shortfalls, we outline a means of directly estimating the age of WGD events. This approach incorporates both phylogenomic data and information from the fossil record in a molecular clock analysis, capable of precisely and accurately coestimating the timing of WGD and subsequent species divergence [11, 12]. To demonstrate this approach, we provide theoretical and practical examples, including the estimation of the timing of a WGD event shared by extant

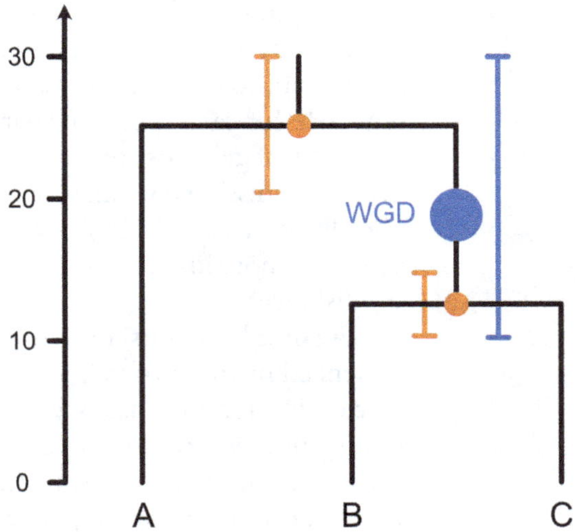

Fig. 1 Constraining the age of whole-genome duplication (WGD) events. Species B and C have undergone a WGD event (blue dot) following their divergence from species A. The timing of species divergence (orange dots) between species B and C and B+C and A is shown as orange bars representing the confidence interval. The WGD event is thus constrained by the minimum divergence time between B and C (10 Ma) and also the maximum divergence time between B+C and A (30 Ma)

members of the flowering dogwood genus, *Cornus* [13]. To perform the analyses described in this chapter, a dataset and set of control files can be found at https://doi.org/10.6084/m9.figshare.16867108.

2 Materials

2.1 Required Data

Molecular clock methods integrate the rate of molecular evolution, measured in substitutions per site of amino acid or nucleotide sequences, with an external calibration. The external calibration may vary, but it is typically provided in the form of a temporal constraint informed by fossil evidence [14]. Therefore, a minimum requirement is molecular sequence data from all species in question that contain a signal of the WGD event and a set of fossil calibrations that inform the divergence times of said species.

Molecular sequence data comes in the form of individual gene families. The most important criterion in selecting a gene family for the analysis is that it contains a clear signal of the WGD event and that the species relationships it predicts do not differ overly from the known species tree. This can be investigated using high-throughput bioinformatic methods, such as gene tree–species tree reconciliation programs, or by manually reconstructing the gene tree and visually inspecting the results.

Beyond this, the quality of the molecular data present in a gene family is measured using the same criterion that is typical of phylogenetic approaches. Each gene family should ideally contain one sequence for each species or paralog in the analysis, though due to processes of gene loss or transcriptomes with lower coverage, some degree of missing data may be acceptable. The length of the gene is not always a reliable indicator of quality, but, in general, longer genes are more likely to contain useful signal over longer evolutionary distances.

Fossil calibrations are established based on the oldest robustly evidenced member of a clade (*see* Subheading 2.3). The selection of species with molecular sequence data and suitable calibrations should not be considered in isolation; taxonomic sampling that maximizes information from the fossil record will result in more accurate estimates of divergence times and, in turn, WGD. Finally, a phylogeny of species relationships is required. This should represent the current best hypothesis of relationships among all species as it is not estimated during these analyses.

2.2 Annotating Sequences

Unlike in a typical phylogenomic or molecular clock analysis, each species may be represented more than once in the tree as a result of a duplication event, forming two or more paralogy groups. Species need to be labeled so as to identify the species or taxon but also the paralogy group to which it belongs (Fig. 2). Paralogy groups must be consistently labeled *within* gene families, but if multiple gene families are being used, then the assignment to either paralogy group *between* gene families is arbitrary (Fig. 2).

2.3 Constructing Calibrations

Several criteria are required to construct a fossil calibration following best practice [15]. These are outlined in the example below:

1. *Node:* Cornaceae + Alangiaceae – Curtisiaceae.
2. *Fossil taxon: Eydeia jerseyensis* [CUPC-1601, Cornell University Palaeobotanical Collection, Cornell University, Ithaca] from the South Amboy Fire Clay Member of the Raritan Formation (Turonian), New Jersey, USA [16].
3. *Phylogenetic justification:* Phylogenetic analysis of morphological characters placed the extinct genus *Eydeia* on the stem of the NMD Group (Nyssaceae, Davidiaceae, Mastixiaceae), in turn sister to Cornaceae + Alangiaceae [16].
4. *Minimum age:* 89.37 Ma.
5. *Maximum age:* 115 Ma.
6. *Age justification:* The age of the Raritan formation has been derived from palynology [17]. The South Amboy Fire Clay member correlates with the *Complexipollenites exigua–Santalacites minor* palynological zone [18, 19], which is considered middle to late Turonian. We follow Atkinson et al. [16] and

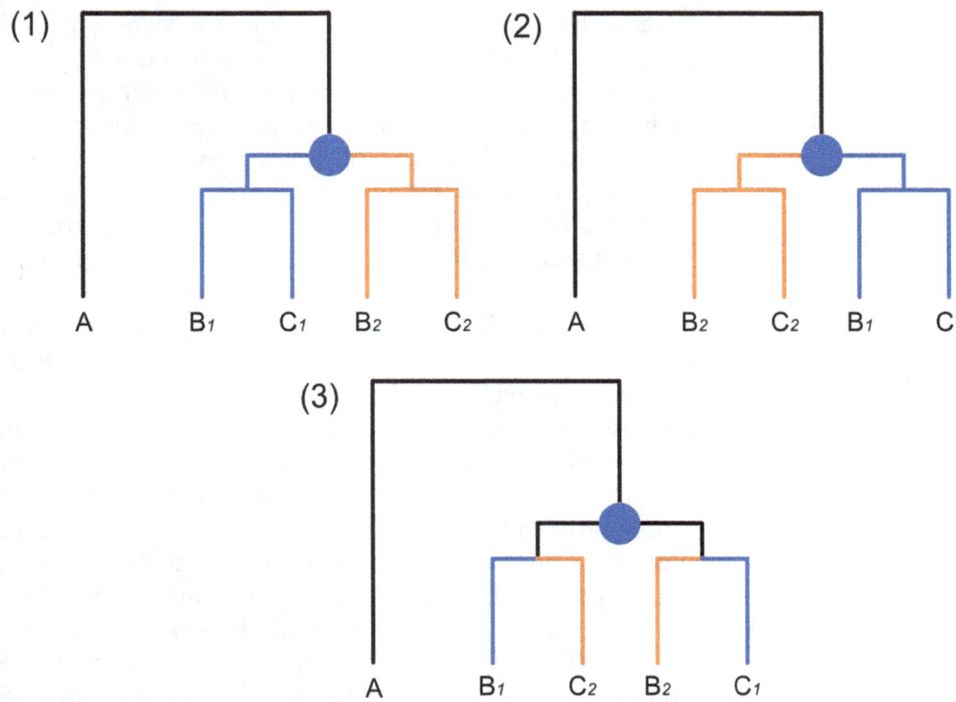

Fig. 2 Annotating sequences across paralogy groups. After duplication (blue dot), each species is represented more than once in the gene tree. Species should be consistently labeled *within* paralogy groups (1 and 2), though between gene families the assignment to either paralogy group is not important (1 vs. 2). However, inconsistent labeling within paralogy groups (3) is incorrect

conservatively consider it latest Turonian, and thus establish a minimum age based on the age of the Turonian-Coniacian boundary of the Turonian, 89.75 ± 0.38 [20].

The maximum age is derived from the most comprehensive analysis of angiosperm divergence times to date [21], which estimated a maximum age for crown group Cornales of 115 Ma.

First, the node on the phylogeny which the fossil is calibrating is clearly stated, in this case, the divergence of Cornaceae and Alangiaceae from Curtisiaceae. Second, the specimen on which the calibration is based, along with where it was collected and where it is housed, is provided. Next, the fossil's inclusion within the stated clade is justified, often based on a formal phylogenetic analysis or the presence of unambiguous synapomorphies established in a previous phylogenetic analysis considering phenotypic data. Fossils are usually used to inform clade-age minimum constraints. Their ages are rarely known directly but, rather, established through correlations between the rock strata in which they are found and strata that have been dated directly. This indirect approach leads to minimum and maximum age interpretations of the fossil, the youngest of which is taken to establish minimum clade-age constraints [14]. These are not sufficient and must be supplemented by maximum clade-age constraints that are commonly established by fitting an arbitrary mathematical distribution to the minimum clade-age constraint to express some visceral perception of how well the fossil on which it is based approximates the true clade age [22]. Alternatively, maximum clade-age constraints can be informed based on evidence of the absence of fossil representatives of a clade constraints [23]. In either instance, justification must be provided for how these ages were derived. Fossil calibrations should be carefully evaluated, and if a full justification is not presented, then a citation to a relevant paper with such a justification should be provided [15].

Different software packages have different means of implementing fossil calibrations, but they are commonly specified as a probability distribution. Different probability distributions reflect different interpretations of the fossil record. A uniform distribution between a minimum and maximum provides a conservative constraint, since no greater probability is assigned to any age [24]. Other probability distributions, such as exponential or Cauchy, may specify a greater probability of a certain node age. Such distributions are commonly applied, yet there is often little justification for weighting the probability toward a particular age and so, though less informative, a uniform distribution is preferable [24]. The R package "MCMCTreeR" allows the specification and visualization of multiple different probability distributions and can be used to produce an input phylogeny with fossil calibrations annotated to the relevant nodes [25].

When assigning fossil calibrations, it is important to also consider the uncertainty of our assessment of the fossil record. Though many fossils have been assigned as members of extant lineages, there is always the possibility of error, especially as hypotheses of clade membership are not always tested. In MCMCTree, this uncertainty can be modeled in the form of "soft" constraints, where a probability tail reflects the possibility that a node may be younger or older than the provided minimum or maximum [26]. These are specified as follows:

```
B(0.8937,1.15,0.01,0.01)
```

The "B" specifies that this calibration contains both a minimum and a maximum constraint. When both a minimum and a maximum are provided, the distribution between them is always uniform. The first number, 0.8937, specifies the minimum constrain, in hundreds of millions of years, followed by the maximum. The final two numbers represent the probability that either the minimum and maximum age can be exceeded, in this case 1%. These soft bounds are especially important when specifying maximum ages.

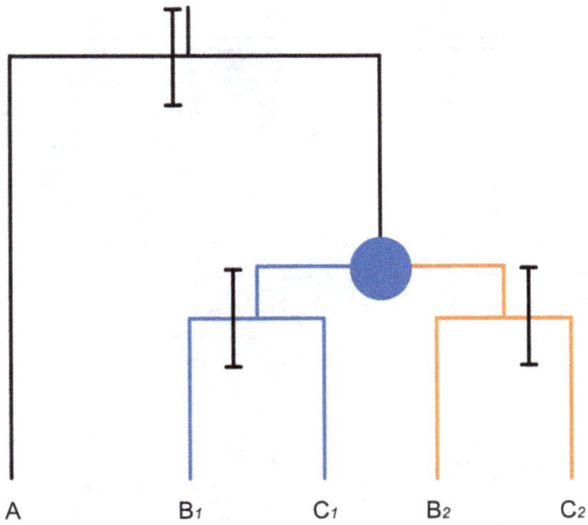

Fig. 3 Cross-calibration. Black bars represent fossil calibrations that can constrain the divergence of B from C and B+C from A. The calibration that constrains the divergence of B from C is present twice in the gene tree after the duplication (blue dot). If this calibration is modeled identically in each paralogy group, then they are cross-calibrated

When dating duplications within gene families, the same speciation node can be represented more than once within a tree in different paralogy groups (Fig. 3). In these instances, "cross-calibration" should be employed, where the same calibration with the same probability distribution is applied to the equivalent speciation nodes in both paralogy groups [27]. A more powerful implementation of this approach, "cross-bracing," has also been proposed, whereby the equivalent speciation nodes are constrained to be exactly the same age [27]. To date, this has only been implemented in the software package BEAST2 [27, 28].

2.4 Examining the Prior

Within MCMCTree, it is possible to examine the combined prior probabilities of the underlying tree and the fossil constraints. This is known as the effective, or joint, time prior, and it can be estimated by running the molecular clock analysis without the sequence data. This is a fast and important step in the analysis, since the tree and fossil priors can interact. For example, the tree topology can truncate the prior for the node ages of some nodes to achieve the expectation that ancestral nodes are older than their descendants. It is important to see the effective prior and to make sure that it does not conflict with the priors that you specified [29]. To run this analysis, we can run the prior control file (`prior.ctl`), where MCMCTree is told to ignore the sequence data:

```
useData = 0
```

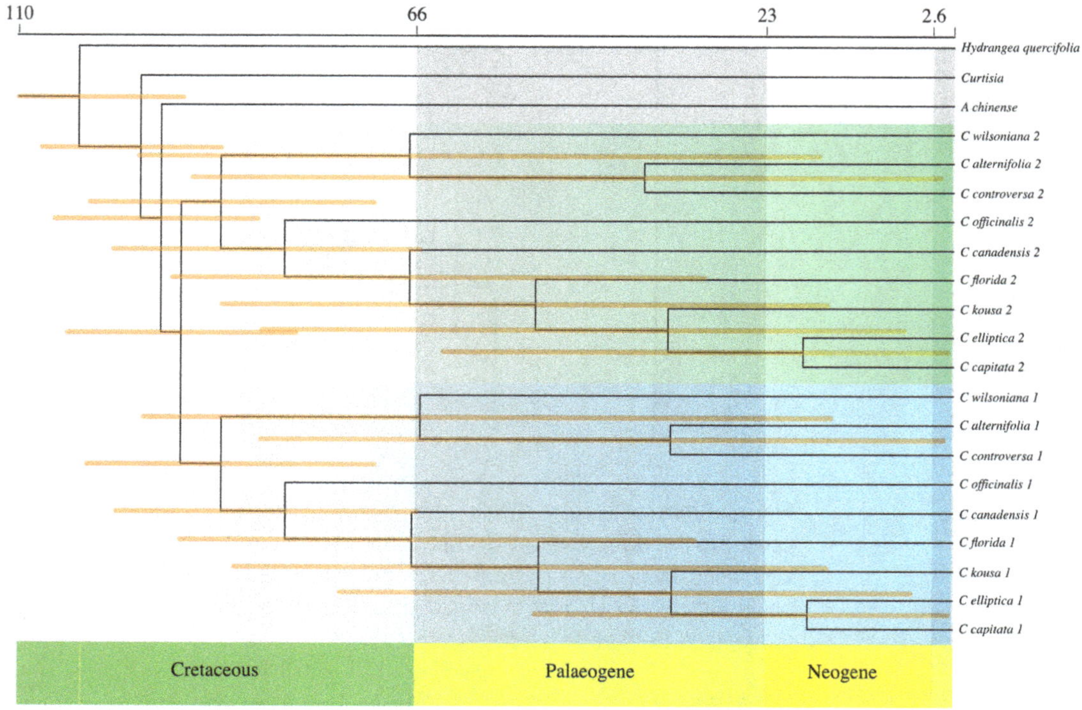

Fig. 4 The effective prior, estimated by running the analysis without molecular sequence data. Divergence times are in millions of years before the presence, as indicated by the top bar. Orange bars represent the 95% highest posterior density (HPD) for each node, with each paralogy group colored blue or green

You will often observe a truncation of the prior that was specified. In our example, the node age prior assigned to the Cornaceae +Alangiaceae node was between 89.37 and 115 Ma, yet the 95% highest posterior density (HPD) for the effective prior is between 90 and 112 Ma (Fig. 4). This is caused by the interaction between the node age prior and the underlying tree model. This is also an opportunity to assess the effect of cross calibration (*see* Subheading 2.3). On each side of the duplication event, the effective priors on divergence times should be very similar; otherwise, a calibration may have been incorrectly specified.

2.5 Running an Analysis

We will describe a clock analysis using the normal approximation method in MCMCTree [30–32]. This is a two-step dating approach, that first estimates branch lengths and allows an approximation of the likelihood surface in the program codeml or baseml [30, 32], and then runs the clock model in MCMCTree. This method is relatively fast and tractable for large datasets. It also allows a high degree of control over all model parameters, all specified in a control file. The sequence data is contained in a Phylip file, consisting of 15 individual gene families that have been concatenated.

The first step in running the normal approximation method is to generate a set of temporary files for the program *codeml* to work with. This is done in the MCMCTree control file (`step_one.ctl`), where the usedata = 3 tells the program to begin the normal approximation method.

```
seed = -1
seqfile = alignment.phy
treefile = calibration_tree.txt
mcmcfile = step_1.txt
outfile = step_1.out
ndata = 1
seqtype = 2
usedata = 3
clock = 2
RootAge = U(1.15,0.001)
```

This will generate a set of files named "`tmp0001.*`". These are the temporary files that are provided to codeml. The most important is `tmp001.ctl`, which is a control file. By default, the file contains instructions to estimate branch lengths according to only a simple model of molecular evolution. Instead, we must change it to describe the appropriate model:

```
seqfile = tmp0001.txt
treefile = tmp0001.trees
outfile = tmp0001.out
noisy = 3
seqtype = 2
aaRatefile = jones.dat
fix_alpha = 0
alpha = 0.5
ncatG = 4
Small_Diff = 0.1e-6
getSE = 2
method = 1
```

Here, we have specified a model file "`jones.dat`" (the JTT model [33]) and have allowed the rate to vary across sites, according to a discrete gamma distribution with four categories and a shape parameter of alpha = 0.5 (a JTT+G4 model). Running codeml will produce a file (named "`rst2`") containing branch length information and the Hessian matrix required for the normal approximation method. To run the normal approximation, we must first rename "`rst2`" to "`in.BV`." Then, in the MCMCtree control file "`step_two.ctl`" change:

```
usedata = 3 to usedata = 2
```

This is now set up to run the analysis with the branch length information estimated from the sequence data. We must also specify some properties of the MCMC, such as the length, sampling frequency, and amount of burn-in.

```
burnin = 2000
sampfreq = 100
nsample = 15000
```

The overall length of the chain is the product of the sample frequency and the number of samples. So, above a sample frequency of every 100 and a total number of samples of 15,000 would specify a chain of length 1,500,000. The burn-in line specifies the number of samples to run the chain for before being discarded. Here, 2000 samples would require 200,000 generations, so the total chain length, including burn-in, is 1,700,000.

2.6 Interpreting and Visualizing Results

The output of the analysis is contained in two files. The treefile ("`FigTree.tre`) contains a Nexus format, and the file "mcmc.txt" contains information from each sampled generation of the Markov chain Monte Carlo (MCMC). First, we can check that the MCMC has run for a sufficient number of generations. The "mcmc.txt" file can be loaded directly into Tracer [34]. This provides the effective sample size (ESS) for each parameter in the analysis. Generally, ESS values greater than 200 are sufficient. The MCMC files from multiple independent runs can be loaded to further compare the posteriors. If each run has converged, the posterior distributions between runs should be the same.

A more intuitive way of looking at the results is to plot the time-scaled phylogeny. This can be done quickly in any tree visualizing program. Publication-ready figures can be produced directly from the MCMCtree output in MCMCtreeR with a set of simple commands [25]:

```
dated.tree <- MCMCtreeR::readMCMCtree("FigTree.tre")
```

```
MCMC.tree.plot(dated.tree, analysis.type = "MCMCtree", time.
correction = 100, plot.type = "phylogram",lwd.bar=5, scale.res
= c("Period"), node.method = "bar")
```

Here, the nodes within the tree are scaled to the *mean* estimates, with the option of plotting bars to represent the 95% HPD. Note that while mean ages are useful for visualizing the tree (the nodes must be positioned somewhere), they may be a misleading (inaccurately precise) interpretation of the results that is likely to exclude the true divergence time [35]. Instead, it is essential to always report the 95% HPD intervals when presenting divergence

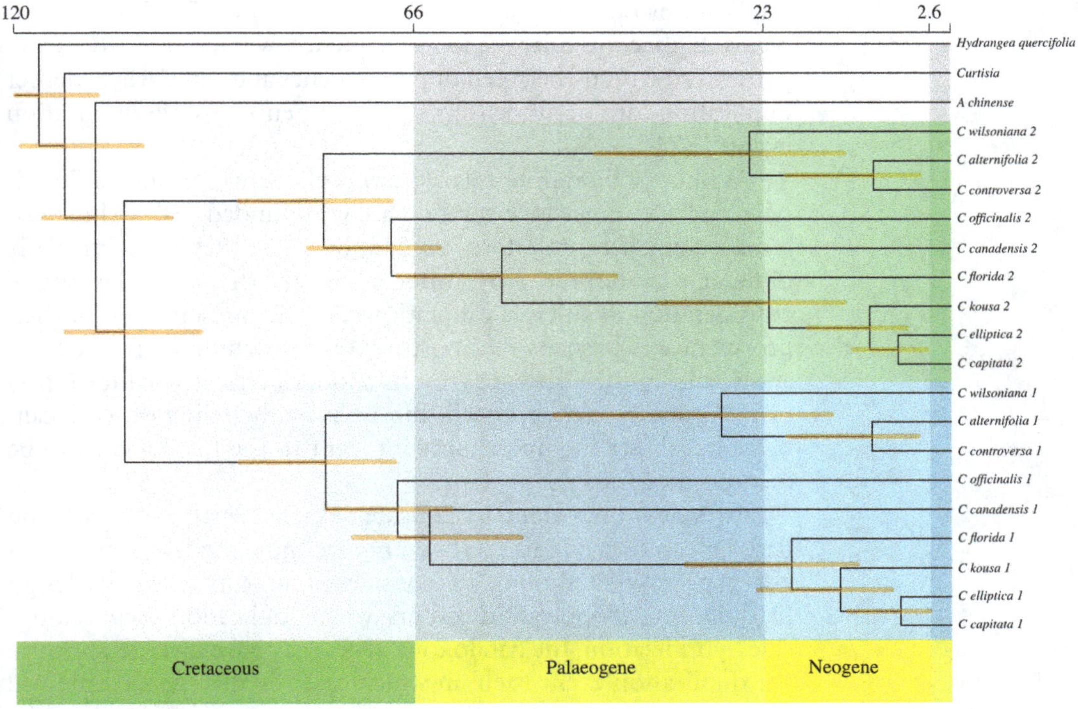

Fig. 5 The posterior estimates of divergence and duplication times derived from the molecular clock analysis. Divergence times are in millions of years before the presence, as indicated by the top bar. Orange bars represent the 95% highest posterior density (HPD) for each node, with each paralogy group colored blue or green

time estimates and to interpret evolutionary history within the context of this uncertainty. In the case of *Cornus*, having run the molecular clock analysis, we can see that the 95% HPD estimates for each node are considerably more precise (Fig. 5), showing the effect of the sequence data on the model. We estimate the duplication event to have occurred within a 109 to 92 Ma (Albian-Cenomanian; middle Cretaceous) interval (Fig. 5).

3 Discussion

We have demonstrated a phylogenetic approach to directly infer the age of WGD events that integrates information from molecular sequence data and a fossil record. Like any analysis, this approach is constrained by the quality of the information provided, relying on a true signal of WGD in the gene families and sufficient information from the fossil record. A major assumption is that gene families containing the signal of a duplication event are derived from WGD rather than local small-scale or chromosomal duplication events. Previous studies have found instances of high-throughput methods identifying WGD events that have more likely been local duplications occurring at a high frequency [36, 37]. Where possible, a high contiguity genome sequence will be able to detect patterns of synteny between duplicate genes and can clarify whether duplications are derived from small- or genome-scale duplication events [37].

Without calibration, relaxed molecular clocks cannot differentiate rate and time since the two are confounded [38]. Thus, systems lacking a reliable fossil record, or those that are particularly difficult to interpret, may suffer a loss of accuracy. The careful consideration of suitable calibrations and the presentation of their provenance is perhaps the most important component of a clock study, and in their absence, results should be treated with caution [15]. Crucially, our approach, through the modeling of fossil calibrations, allows the uncertainty inherent to the fossil record to be incorporated into the analysis.

It is also important to consider exactly what event is being dated given that not all WGD events are the same and their different expectations should be considered in experimental design [10]. During allopolyploidy events, when duplication occurs alongside hybridization, the paralogous genes may have diverged prior to the duplication event itself, in which case the divergence time will reflect the divergence of the two parent species rather than the duplication. Likewise, autopolyploidy events can be followed by a period of tetrasomal inheritance, and paralogs only begin to segregate after a period of fractionation [12]. In this case, the dating will identify the period at which the paralogs diverge, rather than the duplication itself. The extent to which these discrepancies will impact the estimation of the timing of WGD events is unclear. Typically, allopolyploidy occurs between closely related species (though see [39]), and so the difference between the parent divergence and the duplication will be minimal. Similarly, in salmonids, it has been shown that post-WGD, the genome retained a state of tetrasomic inheritance lasting for 17 to 39 million years [40]. The relative contribution of auto- and allopolyploidy in plant evolution is an area of active research [41–43], though the outcomes of the two are predicted to be different [42, 44].

In spite of these caveats, molecular clock methods provide the best means of estimating the time of WGD and in turn facilitate the testing of the most fascinating hypotheses linked to WGD. One of the most prevalent hypotheses surrounding WGD in plants is the apparent clustering of events around the K-Pg boundary [45, 46]. It is proposed that WGD could have provided a selective advantage during the ecological changes in the wake of the mass extinction event, allowing polyploid lineages to survive [47]. The WGD event in *Cornus* is one event that clusters here, originally estimated to have occurred between 85 and 66 Ma and linked to mass extinction event [13]. Instead, we find that the WGD event occurred 109–92 Ma, predating the K-Pg mass extinction event by

43 to 29 million years. We also find that the crown age of *Cornus* also predates the K-Pg mass extinction, as supported by the fossil species *C. piggae* from the Maastrichtian/Campanian [48]. While our results find that the WGD event long predated the K-Pg boundary, this is still compatible with a model wherein polyploid lineages showed greater resilience through the end-Cretaceous mass extinction event.

Another advantage of directly estimating the age of WGD events using the methods outlined in this chapter is that they allow the estimation of both the timing of the WGD event and subsequent speciation events. WGD has been linked to increased rates of diversification in several plant lineages, though often after a "lag" period [5, 49]. This lag has been predicted to reflect the period of diploidization post-WGD, where gene loss, rearrangement, and sub- and neofunctionalization occur [6]. In estimating both WGD and speciation events, this lag can be directly measured. In the case of *Cornus*, we can see that the WGD event predates the divergence of the crown lineage by between 40 and 5 million years. The length of this lag period may be important for hypotheses of causality—estimates of the age of the angiosperm-specific WGD event suggest that it predates the divergence of crown angiosperms by 27 to 65 Ma [11, 50], making a causal relationship between WGD and the success of angiosperms tenuous.

4 Concluding Remarks

We hope to have demonstrated that molecular clock methods provide a means of more accurately and precisely estimating the timing of WGD events. The precision they provide scales with the amount of sequence data available for analysis, and this can be a limiting factor for ancient WGD event, the phylogenetic footprint of which has been eroded through biased gene loss, leading to very few gene families available for dating formative events like the Zeta spermatophyte WGD event [11]. Nevertheless, even a small number of gene families can be sufficient to obtain clade-age estimates with sufficient precision to test macroevolutionary hypotheses. These include, for example, the role of WGD in the origin of angiosperms, the relationship between ploidy and the K-Pg mass extinction, and, indeed, whether the macroevolutionary effect of WGD events lies, if anywhere, with the co-option of redundant duplication genes or their differential loss. Hence, we anticipate that absolute dating methods can support research into the nature, causes, and consequences of WGD events as a generalized phenomenon.

References

1. Clark JW, Donoghue PC (2018) Whole-genome duplication and plant macroevolution. Trends Plant Sci 23(10):933–945
2. Zhang Z, Coenen H, Ruelens P, Hazarika RR, Al HT, Oguis GK, Vandeperre A, van Noort V, Geuten K (2018) Resurrected protein interaction networks reveal the innovation potential of ancient whole-genome duplication. Plant Cell 30(11):2741–2760. https://doi.org/10.1105/tpc.18.00409
3. Barker MS, Li Z, Kidder TI, Reardon CR, Lai Z, Oliveira LO, Scascitelli M, Rieseberg LH (2016) Most Compositae (Asteraceae) are descendants of a paleohexaploid and all share a paleotetraploid ancestor with the Calyceraceae. Am J Bot 103(7):1203–1211
4. Vanneste K, Sterck L, Myburg AA, Van de Peer Y, Mizrachi E (2015) Horsetails are ancient polyploids: evidence from Equisetum giganteum. Plant Cell 27(6):1567–1578. https://doi.org/10.1105/tpc.15.00157
5. Landis JB, Soltis DE, Li Z, Marx HE, Barker MS, Tank DC, Soltis PS (2018) Impact of whole-genome duplication events on diversification rates in angiosperms. Am J Bot 105(3): 348–363. https://doi.org/10.1002/ajb2.1060
6. Schranz ME, Mohammadin S, Edger PP (2012) Ancient whole genome duplications, novelty and diversification: the WGD Radiation Lag-Time Model. Curr Opin Plant Biol 15(2):147–153
7. Dodsworth S, Chase MW, Leitch AR (2016) Is post-polyploidization diploidization the key to the evolutionary success of angiosperms? Bot J Linn Soc 180(1):1–5
8. Lynch M, Conery JS (2000) The evolutionary fate and consequences of duplicate genes. Science 290(5494):1151–1155
9. Vanneste K, Van de Peer Y, Maere S (2013) Inference of genome duplications from age distributions revisited. Mol Biol Evol 30(1): 177–190
10. Doyle JJ, Egan AN (2010) Dating the origins of polyploidy events. New Phytol 186(1): 73–85
11. Clark JW, Donoghue PCJ (1858) 2017 Constraining the timing of whole genome duplication in plant evolutionary history. Proc R Soc B Biol Sci 284:20170912. https://doi.org/10.1098/rspb.2017.0912
12. Macqueen DJ, Johnston IA (2014) A well-constrained estimate for the timing of the salmonid whole genome duplication reveals major decoupling from species diversification. Proc R Soc B Biol Sci 281(1778):20132881. https://doi.org/10.1098/rspb.2013.2881
13. Yu Y, Xiang Q, Manos PS, Soltis DE, Soltis PS, Song B-H, Cheng S, Liu X, Wong G (2017) Whole-genome duplication and molecular evolution in Cornus L. (Cornaceae) – Insights from transcriptome sequences. PLoS One 12(2):e0171361. https://doi.org/10.1371/journal.pone.0171361
14. Donoghue PC, Benton MJ (2007) Rocks and clocks: calibrating the Tree of Life using fossils and molecules. Trends Ecol Evol 22(8): 424–431
15. Parham JF, Donoghue PC, Bell CJ, Calway TD, Head JJ, Holroyd PA, Inoue JG, Irmis RB, Joyce WG, Ksepka DT (2012) Best practices for justifying fossil calibrations. Syst Biol 61(2):346–359
16. Atkinson BA, Martínez C, Crepet WL (2018) Cretaceous asterid evolution: fruits of Eydeia jerseyensis sp. nov. (Cornales) from the upper Turonian of eastern North America. Ann Bot 123(3):451–460. https://doi.org/10.1093/aob/mcy170
17. Christopher RA (1982) The occurrence of the Complexiopollis-Atlantopollis zone (Palynomorphs) in the Eagle Ford Group (Upper Cretaceous) of Texas. J Paleontol 56:525–541
18. Doyle JA, Robbins EI (1977) Angiosperm pollen zonation of the continental Cretaceous of the Atlantic Coastal Plain and its application to deep wells in the Salisbury Embayment. Palynology 1:41
19. Christopher RA (1979) Normapolles and triporate pollen assemblages from the Raritan and Magothy Formations (Upper Cretaceous) of New Jersey. Palynology 3(1):73–121
20. Gale A, Mutterlose J, Batenburg S, Gradstein F, Agterberg F, Ogg J, Petrizzo M (2020) The Cretaceous period. In: Geologic time scale 2020. Elsevier, pp 1023–1086
21. Barba-Montoya J, dos Reis M, Schneider H, Donoghue PCJ, Yang Z (2018) Constraining uncertainty in the timescale of angiosperm evolution and the veracity of a Cretaceous Terrestrial Revolution. New Phytol 218(2):819–834. https://doi.org/10.1111/nph.15011
22. Ho SYW, Phillips MJ (2009) Accounting for calibration uncertainty in phylogenetic estimation of evolutionary divergence times. Syst Biol 58(3):367–380. https://doi.org/10.1093/sysbio/syp035
23. Benton M, Donoghue P, Asher R (2009) Calibrating and constraining molecular clocks. The Timetree of Life 35:86

24. Warnock RC, Yang Z, Donoghue PC (2012) Exploring uncertainty in the calibration of the molecular clock. Biol Lett 8(1):156–159
25. Puttick MN (2019) MCMCtreeR: functions to prepare MCMCtree analyses and visualize posterior ages on trees. Bioinformatics 35(24): 5321–5322
26. Yang Z, Rannala B (2006) Bayesian estimation of species divergence times under a molecular clock using multiple fossil calibrations with soft bounds. Mol Biol Evol 23(1):212–226
27. Shih PM, Matzke NJ (2013) Primary endosymbiosis events date to the later Proterozoic with cross-calibrated phylogenetic dating of duplicated ATPase proteins. Proc Natl Acad Sci 201305813. https://doi.org/10.1073/pnas.1305813110
28. Bouckaert R, Vaughan TG, Barido-Sottani J, Duchêne S, Fourment M, Gavryushkina A, Heled J, Jones G, Kühnert D, De Maio N et al (2019) BEAST 2.5: An advanced software platform for Bayesian evolutionary analysis. PLoS Comput Biol 15(4):e1006650. https://doi.org/10.1371/journal.pcbi.1006650
29. Warnock RC, Parham JF, Joyce WG, Lyson TR, Donoghue PC (2015) Calibration uncertainty in molecular dating analyses: there is no substitute for the prior evaluation of time priors. Proc R Soc B Biol Sci 282(1798): 20141013
30. Yang Z (2007) PAML 4: Phylogenetic analysis by maximum likelihood. Mol Biol Evol 24(8): 1586–1591. https://doi.org/10.1093/molbev/msm088
31. Thorne JL, Kishino H, Painter IS (1998) Estimating the rate of evolution of the rate of molecular evolution. Mol Biol Evol 15(12): 1647–1657
32. Reis MD, Yang Z (2011) Approximate likelihood calculation on a phylogeny for Bayesian estimation of divergence times. Mol Biol Evol 28(7):2161–2172
33. Jones DT, Taylor WR, Thornton JM (1992) The rapid generation of mutation data matrices from protein sequences. Bioinformatics 8(3): 275–282
34. Rambaut A, Drummond AJ, Xie D, Baele G, Suchard MA (2018) Posterior summarization in Bayesian phylogenetics using Tracer 1.7. Syst Biol 67(5):901
35. Warnock RCM, Yang Z, Donoghue PCJ (1857) 2017 Testing the molecular clock using mechanistic models of fossil preservation and molecular evolution. Proc R Soc B Biol Sci 284:20170227. https://doi.org/10.1098/rspb.2017.0227
36. Li Z, Tiley GP, Galuska SR, Reardon CR, Kidder TI, Rundell RJ, Barker MS (2018) Multiple large-scale gene and genome duplications during the evolution of hexapods. Proc Natl Acad Sci 115(18):4713–4718
37. Nakatani Y, Mc Lysaght A (2019) Macrosynteny analysis shows the absence of ancient whole-genome duplication in lepidopteran insects. Proc Natl Acad Sci 116(6):1816–1818
38. dos Reis M, Donoghue PCJ, Yang Z (2016) Bayesian molecular clock dating of species divergences in the genomics era. Nat Rev Genet 17(2):71–80. https://doi.org/10.1038/nrg.2015.8
39. Rothfels CJ, Johnson AK, Hovenkamp PH, Swofford DL, Roskam HC, Fraser-Jenkins CR, Windham MD, Pryer KM Natural History Editor: Mark A.M. 2015 natural hybridization between genera that diverged from each other approximately 60 million years ago. Am Nat 185(3):433–442. https://doi.org/10.1086/679662
40. Gundappa MK, To T-H, Grønvold L, Martin SAM, Lien S, Geist J, Hazlerigg D, Sandve SR, Macqueen DJ (2021) Genome-wide reconstruction of rediploidization following autopolyploidization across one hundred million years of salmonid evolution. bioRxiv:2021.2006.2005.447185. https://doi.org/10.1101/2021.06.05.447185
41. Barker MS, Arrigo N, Baniaga AE, Li Z, Levin DA (2016) On the relative abundance of autopolyploids and allopolyploids. New Phytol 210(2):391–398. https://doi.org/10.1111/nph.13698
42. Spoelhof JP, Soltis PS, Soltis DE (2017) Pure polyploidy: closing the gaps in autopolyploid research. J Syst Evol 55(4):340–352
43. Doyle JJ, Sherman-Broyles S (2017) Double trouble: taxonomy and definitions of polyploidy. New Phytol 213(2):487–493
44. Parisod C, Holderegger R, Brochmann C (2010) Evolutionary consequences of autopolyploidy. New Phytol 186(1):5–17
45. Lohaus R, Van de Peer Y (2016) Of dups and dinos: evolution at the K/Pg boundary. Curr Opin Plant Biol 30:62–69. https://doi.org/10.1016/j.pbi.2016.01.006
46. Vanneste K, Maere S, Van de Peer Y (2014) Tangled up in two: a burst of genome duplications at the end of the Cretaceous and the consequences for plant evolution. Philos Trans Royal Soc B Biol Sciences 369(1648): 20130353. https://doi.org/10.1098/rstb.2013.0353

47. Levin DA (2020) Did dysploid waves follow the pulses of whole genome duplications? Plant Syst Evol 306(5):1–4
48. Atkinson BA, Stockey RA, Rothwell GW (2016) Cretaceous origin of dogwoods: an anatomically preserved Cornus (Cornaceae) fruit from the Campanian of Vancouver Island. PeerJ 4:e2808–e2808. https://doi.org/10.7717/peerj.2808
49. Tank DC, Eastman JM, Pennell MW, Soltis PS, Soltis DE, Hinchliff CE, Brown JW, Sessa EB, Harmon LJ (2015) Nested radiations and the pulse of angiosperm diversification: increased diversification rates often follow whole genome duplications. New Phytol 207(2):454–467. https://doi.org/10.1111/nph.13491
50. Jiao Y, Wickett NJ, Ayyampalayam S, Chanderbali AS, Landherr L, Ralph PE, Tomsho LP, Hu Y, Liang H, Soltis PS (2011) Ancestral polyploidy in seed plants and angiosperms. Nature 473(7345):97–100

Chapter 8

SCORPiOs, a Novel Method to Reconstruct Gene Phylogenies in the Context of a Known WGD Event

Elise Parey, Hugues Roest Crollius, and Camille Berthelot

Abstract

Phylogenetic gene trees recapitulate the evolutionary history of genes across species, forming an essential framework for comparative genomic studies. In particular, within the context of whole-genome duplications (WGDs), they serve as a basis to investigate patterns of duplicate gene retention and loss, timing of genome rediploidization, and, more generally, to explore the functional consequences of the duplication in descending species. Yet, despite ever more sophisticated models to describe the evolution of gene sequences, building accurate gene trees remains a challenge in ancient polyploid taxons. WGDs generate complex gene families with many duplicated copies and recurrent gene losses, which complicate this task even more. Here, we describe how to use SCORPiOs, a novel method that leverages synteny conservation to provide more accurate phylogenies in the presence of a known WGD event.

Key words Genome duplication, Synteny, Gene phylogenies, Orthology, Snakemake workflow, Molecular evolution

1 Introduction

A phylogenetic gene tree describes how a gene family has evolved from a single ancestral gene. Gene families diversify as sequences accumulate substitutions, and are duplicated, lost, and separated in different species during speciation. Classically, evolutionary relationships between genes of extant species are inferred by sequence similarity and represented as a binary tree, which is then compared with the species tree to identify duplication and speciation events. Genes that diverge via speciation are orthologs, while genes separated by duplication are paralogs. Orthologous genes usually perform similar functions across species [1, 2], and orthologous relationships are widely used to annotate gene functions in non-model species by transferring annotations from model species. Paralogs, however, tend to be more functionally divergent and will accrue to new functions and expression patterns over evolutionary

Yves Van de Peer (ed.), *Polyploidy: Methods and Protocols*,
Methods in Molecular Biology, vol. 2545, https://doi.org/10.1007/978-1-0716-2561-3_8,

time [3]. Correctly identifying orthologs and paralogs in the context of WGD is crucial before addressing a number of biological questions, and in particular identifying genes similarly or differentially retained across species. As many genes remain in two copies after a WGD event, distinguishing between paralogous and orthologous copies is a challenging task.

In the past 50 years, a variety of methods and models have been developed to systematically leverage gene sequences and reconstruct gene family evolution histories. They include parsimony, distance-based, or maximum likelihood and Bayesian frameworks, which represent current standards. Finding the maximum likelihood (ML) tree corresponds to finding the tree topology, branch lengths, and parameters of the substitution model that is the most likely to have generated the observed sequence data. However, the solution space for tree topologies grows semifactorially with the number of sequences: there are 2,027,025 possible unrooted trees for a set of 10 sequences and 2.21×10^{20} for 20 sequences. Interested readers can have a look at the TreeCounter software (https://cme.h-its.org/exelixis/web/software/tc/index.html), developed by Alexandros Stamatakis, to compute with precision the number of possible tree topologies with n leaves as n grows large. Since exploring all solutions is clearly unfeasible, available software rely on sophisticated heuristics designed to efficiently search for the ML tree [4–7]. When reconstructing phylogenetic gene trees, ML methods can suffer from both insufficient exploration of this solution space and limited phylogenetic signal in the sequences because sequences of individual genes are typically short. Indeed, different gene tree topologies often have a likelihood close to that of the estimated ML tree and cannot be confidently rejected on the basis of the sequence alignment alone. Thus, other criteria are necessary to select the best topology. Recently developed methods use the reconciliation cost, or parsimony of the duplication-loss scenario, to select among competing topologies [8–11].

SCORPiOs (Synteny-guided CORrection of Paralogies and Orthologies) fills this gap by leveraging another source of evolutionary information, which is particularly informative in the context of WGD: patterns of synteny conservation, measured as the similarity in gene organization along chromosomes between different genomes. WGDs leave very specific synteny signatures in descending genomes, even after hundreds of millions of years of evolution. Indeed, as a WGD initially duplicates all chromosomes, extant genomes descending from the WGD are characterized by pairs of duplicated genomic regions. These can be uncovered when comparing a duplicated genome with a related non-duplicated outgroup: a segment of genes in the outgroup matches two homologous segments in the duplicated genome, a genomic pattern called double-conserved synteny (DCS, Fig. 1) [12]. These DCS regions can provide valuable additional information to guide

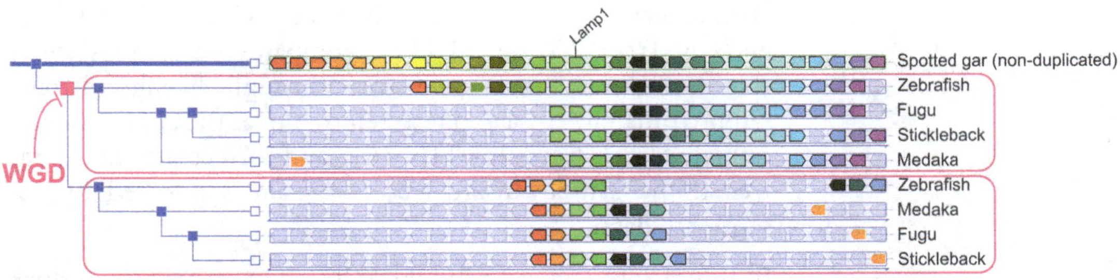

Fig. 1 Double-conserved syntenies. The spotted gar is used as reference, revealing sister duplicated regions in duplicated teleost, i.e., double-conserved syntenies. On the left, the phylogenetic tree for the *lamp1* gene is shown, with the whole-genome duplication (WGD) highlighted. Duplicated segments on each side of the duplication have similar patterns of genes retention (boxes of the same color represent genes belonging to the same gene family)

the tree building process. Indeed, genes that are found in DCS with the outgroup genome are highly likely to be duplicated copies descending from the WGD. Additionally, when comparing DCS among duplicated genomes, orthologous duplicated segments are expected to have more similar gene contents and orders than paralogous segments. Indeed, since orthologous segments diverged more recently (divergence at speciation) than paralogous ones (divergence at the WGD), orthologous segments share a longer ancestral branch punctuated with gene losses and genomic rearrangements. Identifying such orthologous duplicated segments, in turn, can help identify orthologous genes independently from gene sequence evolution. SCORPiOs uses these patterns of double-conserved synteny after a known WGD event to guide gene tree reconstruction in the context of WGD. While SCORPiOs was initially developed to improve gene trees with respect to fish genome duplications, it is relevant to WGDs in other taxons (see "Material: A short guide to data preparation" below).

Briefly, SCORPiOs takes as primary input either a genome-wide set of input gene trees or gene alignments, and outputs a set of WGD-aware, synteny-guided, corrected trees. The main steps of SCORPiOs are as follows:

0. reconstruct starting gene trees with TreeBeST [13] (optional, only if starting from alignments),
1. collect orthologous relationships between a non-duplicated outgroup and duplicated genomes,
2. compare duplicated genomes and match paralogous and orthologous duplicated segments, using the DCS revealed by the orthologs with the outgroup,
3. compile all pairwise synteny predicted orthologs and paralogs for each gene family in a graph and derive synteny-aware tree topologies, with polytomies,

4. correct gene trees: resolve polytomies with ProfileNJ [14] and/or TreeBeST phyml [13], compare the synteny-aware tree with the initial gene tree through likelihood tests (Approximately-Unbiased test, referred to as the approximately unbiased (AU) test below [15, 16]), and correct the initial gene tree if the synteny-aware tree is equally supported by the sequence data.

In this chapter, we describe the SCORPiOs package and show how it runs using real datasets as examples.

2 Material: Software and Input Data

2.1 SCORPiOs on GitHub and ReadTheDocs

SCORPiOs is available on GitHub at https://github.com/DyogenIBENS/SCORPIOS. To download the package as it was at the time of publication [17], which is version v1.1, go to https://github.com/DyogenIBENS/SCORPIOS/tree/v1.1.0. Note, however that minor bug fixes, additional options, as well as usage simplification have been implemented since (for details, see the change history at https://github.com/DyogenIBENS/SCORPIOS/blob/master/CHANGELOG.rst). As such, the latest update is always the recommended choice for any installation.

Along with quick start instructions provided on GitHub, a complete installation and usage instruction manual is also available on ReadTheDocs (https://scorpios.readthedocs.io/en/latest/). Because detailed command-line instructions are already available, we focus here on describing and illustrating the SCORPiOs workflow with examples and refer the users to the documentation for more precise technical instructions.

2.2 Structure of the SCORPiOs Package

SCORPiOs is a pipeline that combines newly developed tools designed to perform synteny comparisons across WGD-descended species, together with state-of-the-art sequence-based tree building software. It is written in Python and implemented as a Snakemake pipeline [18], which enables reproducible, self-contained, and scalable workflows. SCORPiOs can run on MacOS and Linux, and takes advantage of the integration of Snakemake with the Conda package manager to automatically deploy dependencies in virtual environments. Thus, users need not handle dependencies installation directly, which makes SCORPiOs simple to install. The only installation requirement is to have miniconda installed and to download SCORPiOs from GitHub.

The core of the SCORPiOs package is composed of three components: the set of Conda environments (stored in the "env" folder), the custom Python modules (stored in the "scripts" folder), and the

Snakemake files: the master file "Snakefile" defines global configurations, while the *.smk modules contain the code for each pipeline step (`module_build_trees.smk` for **step 0**, `module_orthology_table.smk` for **step 1**, `module_synteny_ortho_para.smk` for **step 2**, `module_graphs_orthogroups.smk` for **step 3** and `module_correct_trees.smk` for **step 4**). As a Snakemake pipeline, SCORPiOs runs within its root directory, where outputs are also generated, and where input data and configuration files should also preferably be stored.

2.3 A Short Guide to Data Preparation

SCORPiOs is designed to reconstruct gene phylogenies from genomes that were subjected to a known WGD event, using at least one closely-related, non-duplicated outgroup. The selected outgroup genome serves to uncover double-conserved synteny (DCS) in the WGD-duplicated genomes (Fig. 1). The construction of DCS by SCORPiOs is a critical step to ensure that as many gene trees as possible can be reliably corrected. It requires a high-contiguity outgroup genome, which diverged as closely as possible prior to the WGD event. Only gene families with at least one copy present in the outgroup genome can be analyzed by SCORPiOs, a requirement that can be problematic if a gene has been lost in the outgroup genome. To alleviate this, SCORPiOs can use more than one outgroup genome. Further, SCORPiOs can handle gene trees containing several WGD events (nested or not), as long as an outgroup genome is available for each duplication. Since SCORPiOs performs pairwise synteny comparisons between duplicated genomes, the minimal number of duplicated genomes for a given WGD is 2. However, SCORPiOs will be more relevant for datasets with many descending duplicated genomes, as pure sequence-based trees are usually more error-prone in this case. Note also that only genome duplication events are currently supported, and not triplication nor any higher polyploidizations levels.

Before running SCORPiOs, users should collect the gene trees, gene alignments, and species tree for all genomes relevant to a study, along with gene coordinate files for genomes to use in synteny analyses (the outgroup(s) used to reveal the DCS and the duplicated genomes). When gene trees are not readily available, users are required to assemble a genome-wide dataset of gene families across all selected study species and align coding sequences. Technical details regarding all required inputs and expected formats can be found in the on-line documentation. Finally, a key prior to running SCORPiOs is the configuration file, which specifies the paths to input files and the parameters. Below is an example of a minimal configuration file, containing only the required entries. Again, we refer to SCORPiOs' documentation for detailed explanation of all other available parameters.

```
#---------------------INPUTS-------------------------#

# INPUT1 - The gene trees to correct, as a single file in New Hampshire Extended (.nhx) format.
trees: data/example/forest.nhx

# ALTERNATIVE INPUT1 – The Gene-to-species mapping file, required if starting from the alignments: a single text file with two columns: gene_name species_name.
#genes_sp_mapping: data/example/genes_sp_mapping.txt

# INPUT2 - The multiple sequence alignments used to build the trees, as a single fasta file.
alis: data/example/ali.fa.gz

# INPUT3 - The genes coordinates for all duplicated species and outgroup(s).
#one file per species, in BED (.bed) format.
genes: data/example/genes/genes.%s.bed

# INPUT4 - The species tree in Newick format with labelled internal nodes (ancestor names).
species_tree: data/example/species_tree.nwk

#---------------------OUTPUTS-----------------------#

# Provide a job name, which will be appended to the output folder name.
jobname: example

#---------------------PARAMS-------------------------#

# Whole-genome duplication(s) and outgroup(s).
# The name of the WGD should be indicated using the name of the first ancestor below the duplication in the species tree (here Clupeocephala).
# Multiple outgroups can be specified, they should be comma-separated (for instance Lepisosteus.oculatus,Amia.calva).
WGDs:
    Clupeocephala: Lepisosteus.oculatus
```

3 Methods: Improved Gene Trees with SCORPiOs

3.1 Running SCORPiOs in Simple and Iterative Modes

Once the input data is prepared, the configuration file is completed, and the Conda environment is installed and activated, SCORPiOs can be run by simply invoking Snakemake within the SCORPiOs folder. The required command-line arguments include specification of the configuration file with `--configfile`, the `--use-conda` flag for correct dependency deployment with Conda, and the `--cores` argument followed by the requested number of CPU cores:

```
snakemake --configfile config.yaml --use-conda --cores 1
```

Further, it is recommended to redirect the standard output and the standard error to files. Indeed, SCORPiOs writes summary statistics to the standard output that users may want to save for inspection after a run:

```
snakemake --configfile config.yaml --use-conda --cores
1 >out_statistics.txt 2>err.txt
```

The above command launches SCORPiOs in the "simple" mode, using configurations specified in `config.yaml`. A second running mode is available: the "iterative" mode. In the iterative mode, SCORPiOs performs several successive rounds of tree corrections. Indeed, because synteny comparisons rely on the orthologous relationships in the gene trees, correcting gene trees can improve the quality of the synteny inferences and thus allow in turn to build better gene trees. The following command allows to run SCORPiOs in the iterative mode:

```
bash iterate_scorpios.sh --snake_args="--configfile config.
yaml" >out_statistics.txt 2>err.txt
```

By default, the iterative correction stops when no additional trees are corrected OR after five iterations. The default can be overridden with the `--max_iter` and `--min_corr` options. For instance, to stop the iterative correction when less than 20 trees are corrected OR after 7 iterations:

```
bash iterate_scorpios.sh --snake_args="--configfile config.
yaml" --min_corr 20 --max_iter 7 >out_statistics.txt 2>err.
txt
```

The iterative correction mode is particularly useful with large datasets of many studied genomes, where initial sequence-based gene trees are most challenging to reconstruct due to the large tree topology space. As an illustration, we present the application of SCORPiOs to two datasets containing duplicated teleost genomes: one with 10 teleosts and a second with 74 teleosts (Fig. 2a, b). The number of corrected gene trees is more than twice higher for the 74 teleost dataset than for the 10 teleosts dataset (Fig. 3a). In addition, while the small set consisting of 10 genomes reaches convergence after two iterations (i.e., 0 corrected trees at iteration 3), the largest 74 species set still shows 74 corrections at iteration 5 (Fig. 3b). It is important to note that the sum of corrections per iteration does not equal the total number of corrected trees, as some trees are often corrected several times. In any new iteration, for computational efficiency, SCORPiOs will only consider gene trees from genes located in the genomic vicinity of genes belonging to families that were corrected at the previous iteration, which correspond to trees with updated synteny information. Because the most computationally expensive steps in SCORPiOs are to

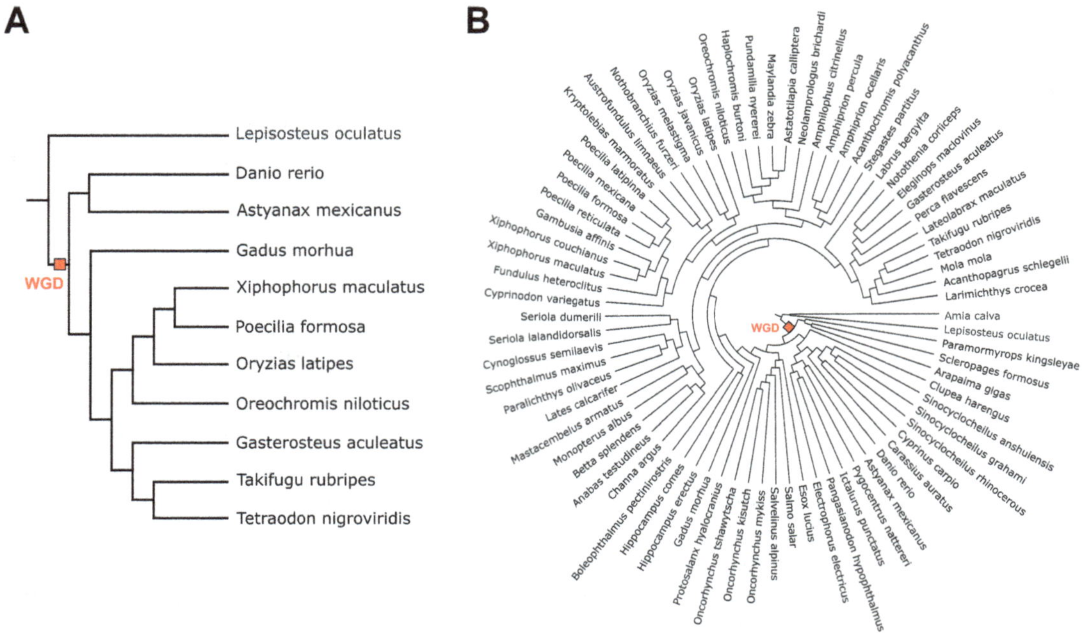

Fig. 2 Species trees for two datasets of teleost genomes. (**a**) Species tree for the 10-teleost dataset, with the spotted gar (*Lepisosteus oculatus*) as non-duplicated outgroup. (**b**) Species tree for the 74-teleost dataset, with the spotted gar (*Lepisosteus oculatus*) and the bowfin (*Amia calva*) as non-duplicated outgroups

resolve gene trees and to perform the AU tests that compare the support given by the sequences to the initial and corrected trees, the runtime significantly decreases at each iteration (Fig. 3c). RAM usage is also reasonable on both sets (Fig. 3d).

The number of corrections per iteration is reported in the output summary statistic file and can be directly extracted with the following command:

```
sed -n '/SUBTREES/,/---/p' out_statistics.txt
```

which outputs 2361, 35, and 0 corrected subtrees for the 3 iterations on the 10-teleosts dataset:

```
----------------- SUBTREES RE-GRAFTING -------------------------
Whole-genome duplication: Clupeocephala
On 7837 total gene trees, 2361 total corrected subtrees in 1592 different trees
1221 corrections required topological changes for other branches in the tree


----------------- SUBTREES RE-GRAFTING -------------------------
Whole-genome duplication: Clupeocephala
On 7837 total gene trees, 35 total corrected subtrees in 35 different trees
22 corrections required topological changes for other branches in the tree
----------------------------------------------------------------


----------------- SUBTREES RE-GRAFTING -------------------------
Whole-genome duplication: Clupeocephala
On 7837 total gene trees, 0 total corrected subtrees in 0 different trees
0 corrections required topological changes for other branches in the tree
----------------------------------------------------------------
```

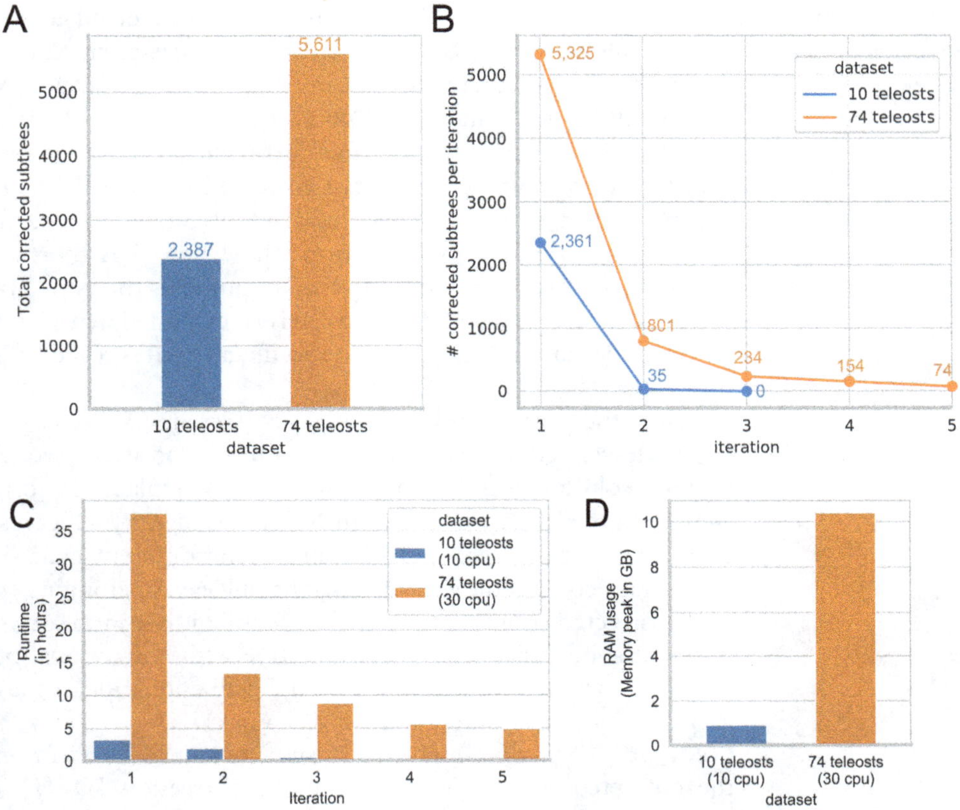

Fig. 3 Iterative SCORPiOs correction accounting for the teleost WGD. (**a**) Total corrected subtrees for two teleost datasets, containing 10 and 74 genomes, respectively. (**b**) Corresponding per-iteration correction numbers. (**c**) By-iteration runtime on the two datasets, given as real elapsed wall-clock time. The 10-teleosts dataset was run on 10 CPU cores and the 74-teleost dataset on 30 CPU cores. In these runs, branch lengths were not recomputed after corrected subtree reinsertion (`brlength: 'n'` in the configuration file), and community detection in graphs was performed with spectral clustering (`spectral: 'y'`). (**d**) RAM usage for runs on each of the two datasets

Similarly, the total number of corrected trees after an iterative correction can be extracted from the set of output gene trees, referred to as "forest" below, since corrected nodes are tagged in the output. The command below gives the total number of corrections for the WGD "Clupeocephala" after two iterations:

```
grep -o CORR_ID_Clupeocephala=Y SCORPiOs_corrected_forest_3_-
with_tags.nhx | wc
```

More details on the format of the corrected gene tree forest and output statistics are given in the next sections.

3.2 Visualizing the Corrected Gene Trees

The primary output of SCORPiOs is the set of corrected gene trees: a single file with the complete set of gene trees after correction. After a run, the corrected forest can be found in the output folder, which will be, for instance, `SCORPiOs_10-teleosts/` for a run with the `jobname` argument set as "10-teleosts" in the configuration file. In the simple mode, the corrected forest will be `SCORPiOs_output_0.nhx`, while in the iterative mode, it is `SCORPiOs_output_5_with_tags.nhx` where 5 is the number of the final correction iteration. The output is in the New Hampshire eXtended format (NHX), which is an extended newick format where nodes and leaves can have specific attributes, called "NHX tags" below.

In the corrected forest, corrected subtrees, i.e., WGD-descended subtrees for which a more consistent sequence-synteny solution was found by SCORPiOs, replace the original subtree in the input tree. Custom NHX tags identify all leaves of a corrected subtree. Because trees can contain different gene subfamilies, tags contain different IDs so that all leaves belonging to the same corrected subtree can easily be identified. For instance, if two different subfamilies are corrected, a tree with corrections for the WGD in the ancestor "Clupeocephala" will contain leaves tagged with `CORR_ID_Clupeocephala=1` and `CORR_ID_Clupeocephala=2`. In such a case, all leaves of one of the corrected subfamilies will have the correction ID "1" (i.e., `CORR_ID_Clupeocephala=1`) and leaves of the second will have the correction ID "2" (`CORR_ID_Clupeocephala=2`). In the iterative mode, the correction tags are extended to include the iteration at which the subtree was first corrected `CORR_ID_Clupeocephala_1=1`; in addition to leaves of corrected subtrees, internal nodes corresponding to a corrected WGD are labeled with `CORR_ID_WGD=Y`.

While corrected trees can be visualized with common tree visualization tools (phylo.io, ete3, ggtree; [19–21]), SCORPiOs comes with a specific tree correction visualization script, based on the ete3 Python module (Fig. 4). To run the script, users should first make sure that ete3 is installed, along with Python (version 3.6 or above), which can be done by simply activating the SCORPiOs Conda environment. Additionally, before running SCORPiOs, users should request that temporary individual tree files are saved during the correction process by specifying `save_tmp_trees: 'y'` in the configuration file, as these files are read by the visualization script to create the images. Available formats for the generated figures include png, pdf, and svg, which can be specified with option -f (png as default).

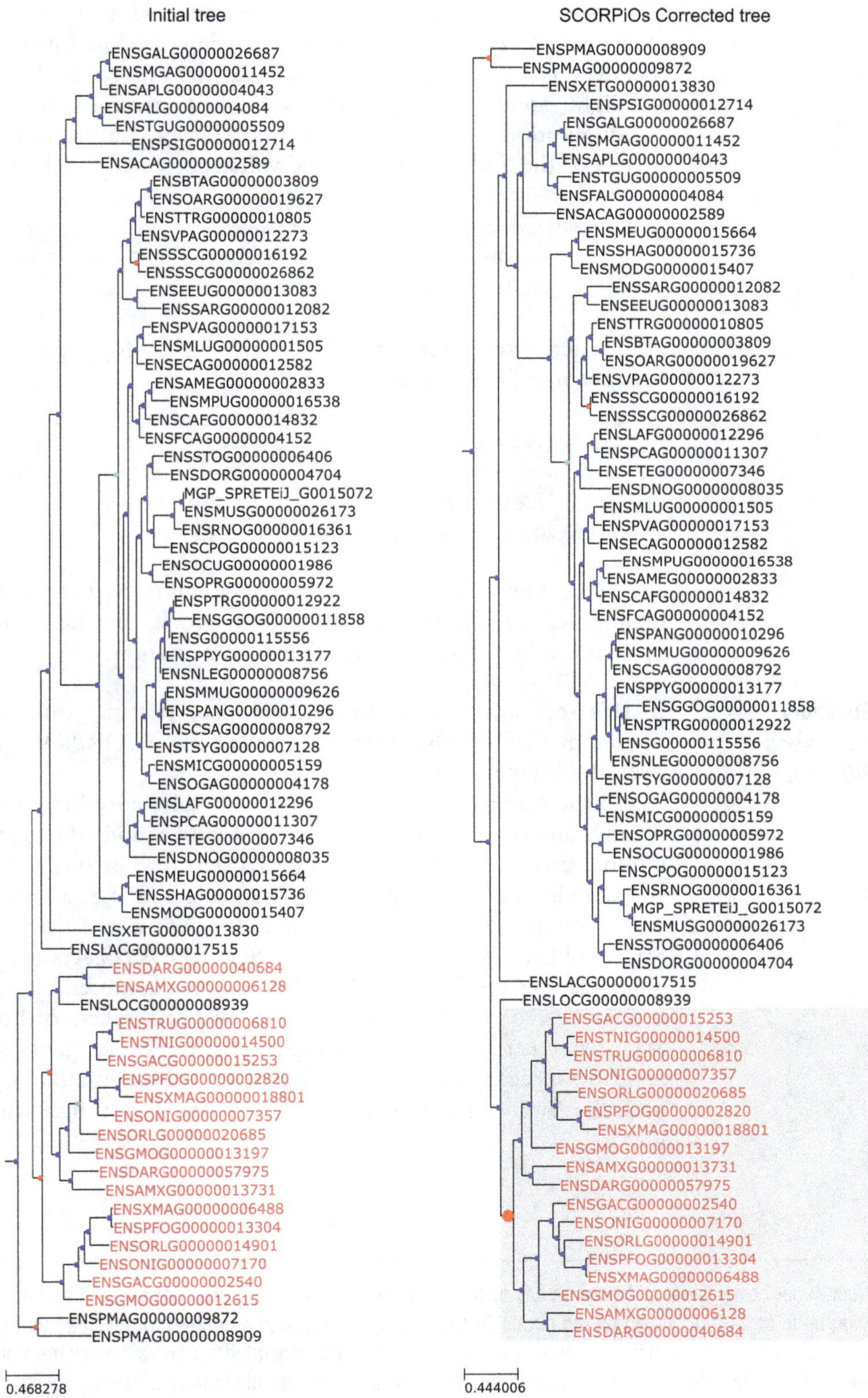

Fig. 4 Images generated by the tree correction visualization script. Before- (left) and after- (right) correction trees, with speciation nodes in blue, duplications in red, and dubious duplications in cyan. In the corrected

To generate "before" and "after" tree images for all corrections after a run, users should launch the following command within the SCORPiOs root folder, which specifies the corrected WGD to highlight (identified by the name of the first ancestor species after the duplication, as in the configuration file), the outgroup species used, and the folder containing individual corrected tree files:

```
python scripts/trees/make_tree_images.py --wgd Clupeocephala
--outgr 'Lepisosteus.oculatus' -i SCORPiOs_teleost-10/Correc-
tions/tmp_whole_trees_1 -o trees_img
```

To generate "before" and "after" tree images for a specific tree (for tree number 1019 here):

```
python scripts/trees/make_tree_images.py --wgd Clupeocephala
--outgr 'Lepisosteus.oculatus' -i SCORPiOs_teleost-10/Correc-
tions/tmp_whole_trees_1/cor_1019 SCORPiOs_teleost-10/Correc-
tions/tmp_whole_trees_1/ori_1019 -o trees_img
```

If the tree contains several corrected subtrees, these will be highlighted in different colors. More details on the available options can be found in the on-line documentation.

3.3 Summary Statistics After a SCORPiOs Run

In this section, we describe the summary statistics output file written by SCORPiOs, using as example a real SCORPiOs run on a set of 74 duplicated teleost genomes.

In the summary statistics file, the only difference between the iterative and simple mode is that in the iterative mode the iteration number precedes each set of computed statistics. Below, we show an example of a summary statistics file generated after an iterative correction run. The first two sections of the output statistics refer to **Step 1** of the workflow. In this step, SCORPiOs collects orthologous relationships between the selected non-duplicated outgroup and duplicated genomes. This is done in two passes: first, orthologs are directly extracted from the input gene trees (i.e., genes separated by speciation nodes, "phylogenetic orthology table"); second, the families are updated to include genes with high synteny

Fig. 4 (continued) tree, the plcd4 gene family in teleost is highlighted with a gray background and the WGD duplication node at its root shown as red circle. Despite being displayed in a different order, the other genes have not been rearranged. Note that the images were cropped from the generated figures as the real tree contains 810 genes (see the full original ensembl tree at http:/may2017.archive.ensembl.org/Lepisosteus_oculatus/Gene/Compara_Tree?g=ENSLOCG00000008939;r=LG12:19269108-19286319;t=ENSLOCT00000010918;collapse=10946801,10947841,10947854,10949312,10949710,10949042,10948637,10949048)

conservation across species, which also strongly suggests orthology. These orthologous relationships are stored in the "Final orthology table," which is used for all synteny comparisons performed in **Step 2**. Reported numbers include the number of orthologous families collected after each pass and the genes they contain in all genomes:

```
----------
Iteration: 1
----------

-------------- PHYLOGENETIC ORTHOLOGY TABLE -------------------
Whole-genome duplication: Osteoglossocephalai (Outgroup Lepisosteus.oculatus)


15517 total families in the phylogenetic orthology table

17366 Paramormyrops.kingsleyae genes in the table
17287 Scleropages.formosus genes in the table
15302 Arapaima.gigas genes in the table
15133 Danio.rerio genes in the table
[...]
13061 Tetraodon.nigroviridis genes in the table
-----------------------------------------------------------------



--------------- FINAL ORTHOLOGY TABLE -------------------------
Whole-genome duplication: Osteoglossocephalai (outgroup Lepisosteus.oculatus)


16000 total families in the final orthology table
19817 Paramormyrops.kingsleyae genes in the final orthology table
19760 Scleropages.formosus genes in the final orthology table
17485 Arapaima.gigas genes in the final orthology table
18390 Danio.rerio genes in the final orthology table
[...]
16348 Tetraodon.nigroviridis genes in the final orthology table
For 15682 families, a synteny orthology graph can potentially be built
(304 with too few genes & 14 in a too short window)
-----------------------------------------------------------------
```

At **Step 2**, the outgroup genome serves as a reference to compare pairs of duplicated genomes. For each window of 15 genes on the outgroup genome (default, can be parametrized), orthologs in the two compared genomes should fall, for the majority, on two different genomic segments in a double-conserved synteny pattern. SCORPiOs uses a scoring system to identify pairs of orthologous duplicated segments, which are expected to be more similar in terms of molecular evolution and gene retention patterns. These pairwise orthologous relationships between duplicated segments are then propagated to the genes they contain, which creates a set of orthologous relationships for families of **Step 1**. These orthologous relationships can be represented as a graph, where nodes are the genes and the edges are the synteny-predicted orthologous relationships. If the process was completely error-free, the graphs would consist of two separated cliques of orthologous genes (orthogroups), each descended from one duplicated gene copy after the WGD. In practice, errors in some of the pairwise synteny-inferred orthologous relations can obscure these orthogroups. SCORPiOs takes advantage of communities detection algorithms to remove the erroneous edges in the graphs and identify the orthogroups. The "Communities Detection in Graphs" section of the summary file reports the number of families where such orthology graphs have been computed, as well as the algorithms used to identify the two communities of orthologous genes. In the example below, for the Osteoglossocephalai (Teleost) WGD, 15,319 graphs were processed from the synteny comparisons that used the spotted gar (*Lepisosteus oculatus*) as reference outgroup:

```
------------- COMMUNITY DETECTION IN GRAPHS --------------------

Whole-genome duplication: Osteoglossocephalai (outgroup Lepisosteus.oculatus)

15388 total orthology graphs effectively built
(Families can fail to produce graphs if gene members cannot be threaded
in the synteny analysis)

15319 processed graphs out of 15388 (69 discarded multigenic families)
3219 (21.01 %) graphs were two separated cliques
11831 (77.23 %) graphs cut with the Girvan-Newman algo.
269 (1.76 %) graphs cut with the Kerningan-Lin algo.
-----------------------------------------------------------------
```

When several outgroup genomes are selected, all synteny graphs are built independently for each outgroup. Then, for each gene family, SCORPiOs selects the most reliable graph to predict the orthologous genes communities, which is the graph with fewer edges to remove to obtain two fully separated orthogroups. In this dataset, the bowfin (*Amia calva*) was used as a second outgroup. Because some gene families are present only in one of the two outgroup genomes, the total number of selected graphs was of 17,070.

The next section presents the confrontation of the synteny-predicted orthogroups to the initial gene trees. A gene tree is defined as synteny-consistent if the two orthogroups identified after community detection in the corresponding synteny graph form two separated clades in the tree, with the gene of the reference outgroup correctly positioned as outgroup. Otherwise, the tree is synteny-inconsistent and requires correction. Note, however, that a subset of the synteny-inconsistent families are not considered for correction by SCORPiOs, because they correspond to complex multigenic families that are difficult to reconstruct accurately. These multigenic families correspond to families with more than 1.5 genes per species on average in at least one orthogroup. A summary of the tree against synteny predictions comparisons is reported in the "Tree vs synteny constraints" section:

```
--------------TREES vs SYNTENY CONSTRAINTS ------------

Whole-genome duplication: Osteoglossocephalai

17070 total subtrees with predicted synteny constraints out of 17493
(423 discarded inconsistent multigenic subtrees)
9349 out of 17070 (54.77 %) synteny-consistent subtrees
7721 out of 17070 (45.23 %) synteny-inconsistent subtrees to correct
--------------------------------------------------------
```

Finally, SCORPiOs attempts to correct synteny-inconsistent trees, starting with the gene tree topology derived from the synteny analysis. This topology contains polytomies for each post-duplication orthogroup, and the outgroup gene branching as an outgroup. The polytomies are first resolved into a binary tree using ProfileNJ [14]. The support given by the sequence alignment to the resolved synteny-aware tree and the initial sequence-based tree is then compared to each other using the AU test [15]. The likelihood AU test ensures that the synteny tree remains consistent with the sequence data. In practice, the correction is accepted if the likelihood of the corrected tree is statistically equivalent or better than the likelihood of the initial tree. For rejected trees, an alternative resolved binary tree is built with TreeBeST PhyML and then similarly tested against the original tree. The ProfileNJ solution is always tested first because it has a lower computational cost. However, ProfileNJ gives a high weight to the parsimony of the duplication and loss scenario, which can sometimes result in a tree that is poorly supported by the sequence alignment. In contrast to ProfileNJ, TreeBeST phyml outputs trees that fit the sequence data better. Combining ProfileNJ and TreeBeST phyml in such a two-steps approach results in a larger number of sequence-synteny consistent solutions. In the example shown below, out of the 7,721 ProfileNJ solutions tested, 4368 were accepted, among which 2,502 solutions are a significantly better fit to the sequences. In the end, with the accepted ProfileNJ (4,368) and TreebeST (957) solutions, 5,325 gene trees were corrected:

```
-------------------- AU-TESTs ----------------------------
Whole-genome duplication: Osteoglossocephalai

Likelihood-tests results for ProfileNJ solutions.
Total : 7721 tested subtrees
Accepted correction (similar or better lk): 4368 (56.6%)
Accepted correction with higher lk: 3493 (45.2%)
Accepted correction with sign. higher lk: 2502 (32.4%)
-----------------------------------------------------------


-------------------- AU-TESTs ---------------------------
Whole-genome duplication: Osteoglossocephalai

Likelihood-tests results for TreeBest solutions.
Total : 3353 tested subtrees
Accepted correction (similar or better lk): 957 (28.5%)
Accepted correction with higher lk: 621 (18.5%)
Accepted correction with sign. higher lk: 390 (11.6%)
4368 subtrees not tested (reason : ProfileNJ solution already accepted)
-----------------------------------------------------------
```

In the final step of the pipeline, SCORPiOs inserts these corrected gene subtrees back into the original gene trees, which may contain species other than the duplicated genomes and outgroup. These original complete trees can also contain several gene subfamilies. Because of this, in the example below, the 5325 corrected subtrees are found in only 3,412 complete trees. In some cases, placing the subtrees back in the trees requires that genes of other species be rearranged, for instance when a gene of a distant species was mistakenly placed within the subtrees with duplicated species. In the example below, for 2,889 of the 5,325 corrections, reinserting the corrected subtree in the tree involved moving genes from other species:

```
---------------- SUBTREES RE-GRAFTING ------------------------
Whole-genome duplication: Osteoglossocephalai

On 26692 total gene trees, 5325 total corrected subtrees in 3412 different trees
2889 corrections required topological changes for other branches in the tree
--------------------------------------------------------------
```

3.4 Tracking the Correction History for a Specific Gene Family

It is often of interest to track the complete correction history for a given gene family, including the pairwise synteny comparisons, the predicted orthogroups from the synteny graph, the synteny-aware tree with polytomies, the resolved binary trees from ProfileNJ and TreeBeST, and the likelihood AU tests results. When setting save_subtrees_lktest: "y" in the configuration file, SCORPiOs saves all of these intermediary outputs to files. The commands below, executed within the output folder, illustrate how to access these results. Throughout a SCORPiOs run, a gene family is identified by the name of the gene in the outgroup genome, which serves to name the intermediary files. Here, we show intermediary files after a run on a 10-teleosts dataset for the family plcd4 which we already described in the SCORPiOs paper [17]. The gene identifier for the plcd4 gene in the spotted gar outgroup, ENSLOCG00000008939, identifies the gene family in the intermediate files, as shown below. We present the intermediate files at iteration 1 (see the "_1" suffix in the paths to the files).

To print all raw pairwise synteny-derived orthologs (before community detection in the graphs):

```
zgrep ENSLOCG00000008939 Synteny/Sorted_SyntenyOrthoPred_Clu-
peocephala_Lepisosteus.oculatus_1.gz
```

To print the list of genes predicted in each "a" and "b" post-duplication group for the family (i.e., after community detection in graphs):

```
grep ENSLOCG00000008939 Graphs/GraphsOrthogroups_Clupeocepha-
la_Lepisosteus.oculatus_1
```

To print the corresponding gene tree topology with synteny-aware polytomies, in the newick format:

```
cat Trees/ctrees_1/Clupeocephala/C_ENSLOCG00000008939.nh
```

To print the binarized synteny-guided profileNJ tree topology:

```
cat Corrections/PolyS_1/Clupeocephala/ENSLOCG00000008939.nh
```

To print the result from the AU test:

```
cat  Corrections/Res_polylk_1/Clupeocephala/Res_EN-
SLOCG00000008939.txt
```

```
# reading SCORPiOs_SCORPiOs_10-
teleosts//Trees/subalis_1/Clupeocephala/ENSLOCG00000008939.pv
# rank item   obs    au     np |   bp     pp     kh     sh    wkh    wsh |
#   1    1  -25.7  0.947  0.949 |  0.950  1.000  0.948  0.948  0.948  0.948 |
#   2    2   25.7  0.053  0.051 |  0.050  7e-12  0.052  0.052  0.052  0.052 |
```

The likelihoods of the trees are compared using the CONSEL software [16], which implements several likelihood tests for tree topologies. SCORPiOs uses the approximately unbiased (AU) test implemented in CONSEL to test for significant difference in likelihoods. The AU test has indeed been demonstrated less biased than other methods [15]. The output above is the raw output from CONSEL, described in more details in the CONSEL manual (http://stat.sys.i.kyoto-u.ac.jp/prog/consel/quick.html). The columns that are of interest for SCORPiOs are "rank," "item," "obs," and "au." The "item" column identifies the tested tree: Item "1" is always the SCORPiOs tree (the ProfileNJ solution in this example), and item "2" is the initial sequence-based tree. The column "rank" ranks the tree with respect to their log-likelihood, while "obs" gives the difference in log-likelihoods. Here, the SCORPiOs tree has the highest log-likelihood (rank 1). The column "au" gives the *p*-value to the AU test, where trees with *p*-value < 0.05 can be rejected. SCORPiOs accepts the solution when it has a higher or statistically equivalent likelihood (lower likelihood, but the null hypothesis of equivalent likelihoods cannot be rejected, i.e. AU *p*-value > 0.05). In this example, the solution is accepted because the likelihood is improved.

Since the ProfileNJ solution was accepted, SCORPiOs did not compute a TreeBeST phyml solution for this family. Otherwise, the TreeBeST phyml resolved tree topology would have been printed to Corrections/TreeB_1/Clupeocephala/ENSLOCG00000008939.nh and the AU test to Corrections/Res_treeBlk_1/Clupeocephala/Res_ENSLOCG00000008939.txt. Again, for further information regarding SCORPiOs intermediary output files, we refer the reader to the on-line documentation.

4 Conclusion

In this chapter, we describe a typical SCORPiOs run aimed at improving gene trees in the context of the teleost WGD event. We illustrate the workflow with interpretations of summary statistics at each of the main steps. SCORPiOs is still in active development and will propose new functionalities in the future. In particular, one feature currently under development will investigate unresolved sequence-synteny disagreements after a SCORPiOs run. Indeed, SCORPiOs only corrects gene trees for which it can find a solution that is consistent with both sequence and synteny evolution. This leaves a fraction of synteny-inconsistent gene trees that SCORPiOs does not correct. One reason for such disagreements between the history of a gene as a sequence and as a locus, in the context of WGD, can be lineage-specific rediploidization [22]. Integrating diagnosis of lineage-specific rediploidization will further broaden the scope of SCORPiOs as a toolkit to better understand how genes and genomes evolve after polyploidization events.

References

1. Altenhoff AM, Studer RA, Robinson-Rechavi-M, Dessimoz C (2012) Resolving the ortholog conjecture: orthologs tend to be weakly, but significantly, more similar in function than paralogs. PLoS Comput Biol 8:e1002514. https://doi.org/10.1371/journal.pcbi.1002514
2. Tatusov RL, Koonin EV, Lipman DJ (1997) A genomic perspective on protein families. Science 278:631–637
3. Ohno S (1970) Evolution by gene duplication. Springer, Berlin, Heidelberg
4. Stamatakis A (2014) RAxML version 8: a tool for phylogenetic analysis and post-analysis of large phylogenies. Bioinformatics 30:1312–1313. https://doi.org/10.1093/bioinformatics/btu033
5. Guindon S, Dufayard J-F, Lefort V et al (2010) New algorithms and methods to estimate maximum-likelihood phylogenies: assessing the performance of PhyML 3.0. Syst Biol 59:307–321. https://doi.org/10.1093/sysbio/syq010
6. Price MN, Dehal PS, Arkin AP (2010) FastTree 2 – approximately maximum-likelihood trees for large alignments. PLoS One 5:e9490. https://doi.org/10.1371/journal.pone.0009490
7. Minh BQ, Schmidt HA, Chernomor O et al (2020) IQ-TREE 2: New models and efficient methods for phylogenetic inference in the genomic era. Mol Biol Evol 37:1530–1534. https://doi.org/10.1093/molbev/msaa015
8. Szöllősi GJ, Rosikiewicz W, Boussau B et al (2013) Efficient exploration of the space of reconciled gene Trees. Syst Biol 62:901–912. https://doi.org/10.1093/sysbio/syt054
9. Morel B, Kozlov AM, Stamatakis A, Szöllősi GJ (2020) GeneRax: a tool for species-Tree-aware maximum likelihood-based gene family Tree inference under gene duplication, transfer, and loss. Mol Biol Evol 37:2763–2774. https://doi.org/10.1093/molbev/msaa141
10. Scornavacca C, Jacox E, Szöllősi GJ (2015) Joint amalgamation of most parsimonious reconciled gene trees. Bioinformatics 31:841–848. https://doi.org/10.1093/bioinformatics/btu728
11. Comte N, Morel B, Hasić D et al (2020) Treerecs: an integrated phylogenetic tool, from sequences to reconciliations. Bioinformatics 36:4822–4824. https://doi.org/10.1093/bioinformatics/btaa615
12. Kellis M, Birren BW, Lander ES (2004) Proof and evolutionary analysis of ancient genome duplication in the yeast Saccharomyces cerevisiae. Nature 428:617–624. https://doi.org/10.1038/nature02424
13. Vilella AJ, Severin J, Ureta-Vidal A et al (2009) EnsemblCompara GeneTrees: complete, duplication-aware phylogenetic trees in vertebrates. Genome Res 19:327–335. https://doi.org/10.1101/gr.073585.107
14. Noutahi E, Semeria M, Lafond M et al (2016) Efficient gene Tree correction guided by genome evolution. PLoS One 11:e0159559. https://doi.org/10.1371/journal.pone.0159559
15. Shimodaira H (2002) An approximately unbiased test of phylogenetic Tree selection. Syst Biol 51:492–508. https://doi.org/10.1080/10635150290069913
16. Shimodaira H, Hasegawa M (2001) CONSEL: for assessing the confidence of phylogenetic tree selection. Bioinformatics 17:1246–1247. https://doi.org/10.1093/bioinformatics/17.12.1246
17. Parey E, Louis A, Cabau C et al (2020) Synteny-guided resolution of gene Trees clarifies the functional impact of whole-genome duplications. Mol Biol Evol 37:3324–3337. https://doi.org/10.1093/molbev/msaa149
18. Köster J, Rahmann S (2012) Snakemake—a scalable bioinformatics workflow engine. Bioinformatics 28:2520–2522. https://doi.org/10.1093/bioinformatics/bts480
19. Robinson O, Dylus D, Dessimoz C (2016) Phylo.io : interactive viewing and comparison of large phylogenetic Trees on the web. Mol Biol Evol 33:2163–2166. https://doi.org/10.1093/molbev/msw080
20. Huerta-Cepas J, Serra F, Bork P (2016) ETE 3: reconstruction, analysis, and visualization of Phylogenomic data. Mol Biol Evol 33:1635–1638. https://doi.org/10.1093/molbev/msw046
21. Yu G (2020) Using ggtree to visualize data on Tree-Like structures. Curr Protoc Bioinformatics 69:e96. https://doi.org/10.1002/cpbi.96
22. Robertson FM, Gundappa MK, Grammes F et al (2017) Lineage-specific rediploidization is a mechanism to explain time-lags between genome duplication and evolutionary diversification. Genome Biol 18:111. https://doi.org/10.1186/s13059-017-1241-z

Chapter 9

Inferring Chromosome Number Changes Along a Phylogeny Using chromEvol

Anna Rice and Itay Mayrose

Abstract

Chromosome numbers have long been used for the identification of key genomic events such as polyploidy and dysploidy. These inferences are often challenging, particularly when applied to large phylogenies, or clades in which more than a few chromosome number transitions had occurred. Here we describe the chromEvol computational framework that infers shifts in chromosome numbers along a phylogeny using probabilistic models of chromosome number change. Given chromosome count data and an associated phylogeny, chromEvol identifies such patterns by fitting probabilistic models of chromosome number evolution to the data. We describe the chromEvol workflow using available online tools, including the specification of the desired models, the examination of model fit to the data, and the inference of ploidy levels. The pipeline can be used by the wide scientific community and requires no previous computational or programming skills.

Key words Chromosome number, Dysploidy, Polyploidy, chromEvol, Phylogenetics, Polyploidy, CCDB, OneTwoTree

1 Introduction

The number of chromosomes within a nucleus is a basic characteristic of the eukaryotic genome. Providing a concise description of the cell karyotype, together with the simplicity by which it can be obtained and its stability across repeated measurements, the use of chromosome counts has increased in popularity in a myriad of biological studies. Changes in chromosome numbers are of crucial impact on key genomic and macroevolutionary processes such as reproductive isolation and lineage diversification and provide important evidence for species determination and phylogenetic relationships [1, 2]. While holding a clear phylogenetic signal (e.g., [3, 4]), chromosome numbers exhibit substantial variations,

Yves Van de Peer (ed.), *Polyploidy: Methods and Protocols*,
Methods in Molecular Biology, vol. 2545, https://doi.org/10.1007/978-1-0716-2561-3_9,

both within and across closely related species, particularly in plants [5–7]. The most dramatic chromosomal change, in both number and genomic content, is whole-genome multiplication, or polyploidization, which consists of the acquisition of one or more complete chromosome sets to the genome. Polyploid species often differ from their diploid progenitors in morphological, physiological, and life history characteristics, and these differences may contribute to the establishment of polyploid species in novel ecological settings, altering diversification opportunities [8–15]. Thus, polyploidy is often seen as one of the main processes that has shaped the evolution of eukaryotic organisms. Another common phenomenon underlying chromosome number variation is dysploidy, involving changes by a single chromosome number, typically without an immediate effect on the genomic content. Dysploidy usually occurs through several types of genome rearrangements, mainly due to chromosome fission or fusion events [1, 16]. Inferring the transitions that occurred along each branch of the phylogeny allows to decode the pathways by which changes in chromosome numbers proceed in the clade of interest and to estimate ancestral chromosome numbers. This in turn allows to categorize taxa at the tips of the tree as diploids or polyploids.

A model that was specifically designed to emulate the process of chromosome number evolution along a phylogeny was first presented by Mayrose et al. [17] and was termed *chromEvol*. This probabilistic framework incorporates a continuous-time Markov process, defined by a rate matrix Q, which describes the instantaneous rate of change from a genome with i haploid chromosomes to a genome with j haploid chromosomes. The entries in this matrix are determined based on a combination of model parameters that defines the rate of change for different types of transitions. For example, the standard chromEvol model assumes three types of transitions: whole-genome duplication (an exact duplication of the number of chromosomes with rate ρ; e.g., $n = 10 \rightarrow n = 20$), a single chromosome number increase (ascending dysploidy with rate λ; e.g., $10 \rightarrow 11$), or a single chromosome number decrease (descending dysploidy with rate δ; e.g., $10 \rightarrow 9$). Other chromosome number transitions included in chromEvol are (1) demi-duplication (with rate μ), which accounts for possible multiplications of the number of chromosomes by 1.5, leading, for example, to triplication events (e.g., $10 \rightarrow 15$), and (2) the addition of the monoploid (base) number β (with rate ν), representing another triplication pathway, as well as putative hybridization scenarios (e.g., assuming $\beta = 9$, transitions from $n = 18$ to $n = 27$, 36, or 45 are allowed). Combining these possibilities, the most general Q matrix is defined as follows:

$$Q_{ij} = \begin{cases} \lambda + \lambda_l \times (i-1) & j = (i+1) \\ \delta + \delta_l \times (i-1) & j = (i-1) \\ \rho & j = 2i \\ \mu & j = 1.5i \\ \nu & (j-i) \text{ is divisible by } \beta \\ 0 & \text{otherwise} \end{cases} \quad (1)$$

The diagonal entries are determined by the constraint that each row in Q sums to zero. In Eq. 1, λ_l and δ_l are rate modifiers that allow for the possibility that the rates of ascending and descending dysploidy depend on the current number of chromosomes.

The rate matrix described above allows the likelihood function to be computed, given a specified phylogeny and assignments of chromosome numbers to the tip taxa. Additionally, from this general rate matrix, several alternative models may be derived, each with a different subset of parameters. The model that best fits the data is chosen by computing the likelihood of each examined model and applying a model selection criterion, such as the Akaike information criterion (AIC; [18]). This procedure chooses the model whose fit is best, relative to other examined models, but does not assess the absolute model fit. Notably, relying on the best model may be vulnerable to incorrect conclusions when the underlying evolutionary process deviates substantially from current modeling assumptions [19].

In the context of the chromEvol model, violations may include deviations from the memorylessness property of the Markovian process (i.e., transition rates are not affected by the sequence of events leading to the number of chromosomes in the genome), time homogeneity (transition rates are identical along each branch and over all branches of the phylogeny), and phylogenetic structure (i.e., ignoring the possibility of reticulate evolutionary events). Thus, the procedure of model selection is followed by a model adequacy procedure, which assesses the absolute fit of a specified model by testing its capability to generate data with similar characteristics as those found in the original data. The model adequacy procedure for the evolution of chromosome numbers is fully described in Rice and Mayrose [20].

Following the completion of a chromEvol run, users can view the inferred chromosome numbers at each ancestral node of the phylogeny and the number and types of inferred transitions that had occurred along the branches of the tree (see all output files in Subheading 2.6 *Interpreting chromEvol Web Server Results*). In addition to the standard output, chromEvol can be used to perform ploidy-level inference. This procedure enables to categorize extant taxa as diploids or polyploids, relative to the ancestral chromosome number of the group in question (i.e., assuming that the root of the phylogeny was in a diploid state). In this procedure, a taxon is inferred as polyploid (diploid) if the expected number of polyploidization events from the root to the tip is above (below) a certain polyploid (diploid) threshold. These thresholds are determined for each dataset using a simulation-based procedure, whose details are described in Glick and Mayrose [21]. Note that this procedure consumes extensive computational time and thus its execution should be specified by the user as it is not performed by default.

2 Methods

In this section we provide detailed instructions on how to use the chromEvol methodology using its online implementation (http://chromevol.tau.ac.il/), including links to additional helpful online tools. Readers will find in the *Notes* section instructions regarding the offline version (*see* **Note 1**).

2.1 Input Data

The necessary inputs for chromEvol are as follows: (1) a phylogeny in a standard NEWICK format and (2) chromosome count data in FASTA format (Fig. 1). The chromosome count for each tip taxa may be either a single integer number (in case no variation exists for that taxon or a summary statistics of an observed chromosome count distribution, such as the median) or, in the case of multiple counts per taxon, a string representing the possible distribution, where each count is followed by its frequency. For example, in the

a

```
(C. malzacianum:0.5, ((C. maritimum:0.2,C. discolor:0.2):0.2,
(C. pulchellum:0.3,(C. mairei:0.2, (C. capense:0.1,
C. centaurioides:0.1) :0.1):0.1):0.1):0.1):0.1;
```

b

```
>C. pulchellum
9=0.09_18=0.72_27=0.19
>C. centaurioides
X
>C. capense
18
>C. mairei
27
>C. discolor
20
>C. maritimum
10
>C. malzacianum
28
```

c

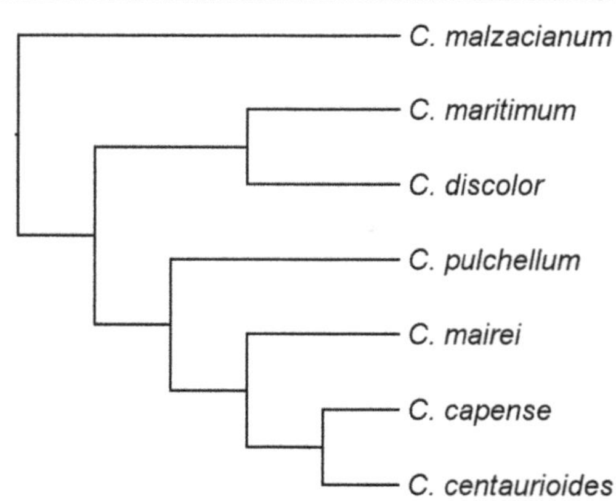

Fig. 1 ChromEvol inputs. (**a**) A phylogeny, represented in a NEWICK text file. (**b**) Chromosome counts in a FASTA format. Note that in this example, *Centaurium centaurioides* has no chromosome count information and is thus represented by *X* and that *Centaurium pulchellum* is represented by a range of possible counts (9 with probability 0.09, 18 with probability 0.72, and 27 with probability 0.19). (**c**) A graphical representation of the NEWICK format in (**a**) as obtained using FigTree

case the recorded counts for a taxon are $n = 20, 20, 30, 30$, the distribution will be written as $20 = 0.5\_30 = 0.5$.

By default, the chromEvol web server assumes that both input data types are available. However, the web server also allows other entry points by automatically retrieving the phylogeny or chromosome counts for the analyzed clade from other online resources. These advanced options are detailed in Subheading 2.5*Missing Input*. ChromEvol requires that the tip labels of the phylogeny and the count data perfectly overlap, such that there are no counts in the data file that are missing from the phylogeny. Furthermore, the phylogeny and count data should match completely in terms of spelling and naming conventions (*see* **Note 2**). However, in case there are tips in the phylogeny that users wish to add to the analysis without their corresponding counts, these taxa should be assigned with *X* in the data file. In the online web server, users can choose to automatically handle discrepancies in the input data, either by assigning missing taxa with *X* or by trimming such taxa from the phylogeny such that all tips have chromosome count information.

2.2 Model Selection

The next step is to select one or more models for hypothesis examination. Each model is represented by a combination of rate parameters (Eq. 1). Users can select from a set of six default models (as detailed in [20]), or design their own model with a specified set of model parameters, by selecting which transition type to include. The web server allows to evaluate up to six models in a single run. The possible transitions rates include rates of ascending dysploidy (gain), descending dysploidy (loss), duplication, demi-duplication, and base number transition. By default, each included rate parameter is modeled by a constant function, meaning, that its transition rate does not depend on the current chromosome number. Alternatively, users can choose to model rates of ascending and descending dysploidy using a linear function. Future developments will allow an elaborated set of functions that will be applied to additional types of transitions.

2.3 Model Adequacy

Optionally, users can choose to perform a model adequacy test. This analysis tests the absolute (rather than the relative) fit of the chosen models to the data. In case the evaluated models include the base number parameter β, another round of optimization is performed, resulting in increased running time (see [20]).

2.4 Ploidy Inference

Optionally, chromEvol can be used to categorize extant species as diploids or polyploids. By clicking the *Perform ploidy inference* checkbox, this analysis will be executed automatically using the results obtained by the chromEvol optimization. The input of this step is the FASTA format count file, together with either a single phylogeny or a set of multiple trees (for instance, a sample from the posterior distribution over tree topologies obtained using a

Bayesian phylogenetic inference tool, such as MrBayes; [22]), in NEWICK format. If the latter is provided, chromEvol will account for uncertainty in the assumed phylogeny. The output of this step is the inferred ploidy level for each taxon (1, polyploid; 0, diploid) or NA (not available) in case this could not be reliably inferred. The inferred ploidy levels for each taxon are accompanied by two reliability scores. The *simulation reliability score* uses simulations that are based on the selected model (and its inferred model parameters) to detect taxa for which ploidy level could not be reliably inferred. These are the taxa that suffer from high false-positive or false-negative rates according to the simulated data (e.g., a taxon was inferred as diploid while it was simulated as polyploid). In many cases, such taxa are placed next to long terminal branches, in which multiple transitions could occur, leading to unreliable inferences (for more details see [21]). The *phylogeny robustness score* is based on comparing the inferred ploidy levels across 100 phylogenies, thus accounting for phylogenetic uncertainties. The *combined score* is the average of the above two scores. By default, in case one of the scores is lower than 0.95, the inference for that taxon is marked as unreliable.

2.5 Missing Input

The input data required for a chromEvol analysis consists of a phylogeny and corresponding chromosome count data. However, often users might be missing either one of the inputs. Here we detail the usage of two online tools that are linked to the chromEvol server and allow users to automatically retrieve chromosome count data or a phylogeny for their taxa of interest.

2.5.1 Missing Chromosome Counts

Chromosome counts can be retrieved in two ways: (1) directly from the Chromosome Counts Database (CCDB) or (2) from the chromEvol website. For the first option, users should enter the CCDB site (http://ccdb.tau.ac.il/) and search for their desired taxa. This can be done either by providing a single name of any taxonomic level (e.g., species, genus) in the search box or by browsing through the major plant groups, families, and genera. Each name in CCDB is classified as accepted scientific name, synonym, or unresolved, following a name resolution procedure that is based on the naming conventions of The Plant List (http://www.theplantlist.org/; V1.1). The counts can be downloaded to a *csv* file format to view the results offline. Additional details about the database can be found in Rice et al. [7]. The chromEvol server enables users to automatically match chromosome counts data to any tip taxa that is present in the input phylogeny. This option is available from the tab *Download CCDB counts* in the chromEvol web server. Users can choose how to handle multiple counts per taxon—either retrieve a single counts statistic (median, minimum, or maximum of counts) or the distribution of all available counts.

2.5.2 Missing Phylogeny

Reconstructing a phylogeny is frequently a tedious procedure, involving marker selection, data cleaning and retrieval, the selection of both the alignment and tree reconstruction methods, and being minded of the various parameters involved in their execution. *OneTwoTree* (http://onetwotree.tau.ac.il) is an online tool that unifies many of the steps involved in phylogeny reconstruction in an easy-to-use manner [23]. Given a list of taxa names, this tool searches the NCBI GenBank for available sequence data, chooses the most appropriate set of nucleotide markers for the available taxa, computes a concatenated multiple sequence alignment, and reconstructs the phylogeny using either maximum likelihood or Bayesian techniques. OneTwoTree is now linked to the chromEvol server, allowing users to automatically reconstruct the phylogeny for all taxa that are present in a provided chromosome counts input file. To use this option, users should enter the *Download tree* tab and provide the chromosome count input file. All taxa names from the file are automatically directed to the OneTwoTree server, allowing users to conveniently modify reconstruction parameters. The tree can be downloaded in a NEWICK format from the Results page. Further details regarding the various methodologies and options of OneTwoTree are described in Drori et al. [23].

2.6 Interpreting chromEvol Web Server Results

Once a chromEvol run has ended, the results are directed to the chromEvol Results page. The first output presented is a summary of the optimized rate parameters for each selected model along with the likelihood scores obtained for each. The model that best fits the data according to the AIC is indicated at the top of this table. Next, users can click to expand the model adequacy results to assess the absolute fit of the chosen models to the data. *P*-values are displayed for each of the test statistic incorporated in this procedure and *p*-values ≤ 0.05 are indicative of model inadequacy for that test statistics. For each chosen model, chromEvol generates multiple results that can be viewed online or saved for further analyses offline. These are available under *Tested models results files* and include the following:

1. Results summary. Run statistics of the analysis, including inferred model parameters, optimized likelihood value, and inferred chromosome number distribution at the root.
2. ML ancestors tree. The most likely chromosome number inferred at each internal node of the tree. These inferences are placed as bootstrap values at each internal node and can be displayed using a tree-viewing program (e.g., FigTree available at http://tree.bio.ed.ac.uk/software/figtree/).
3. Ancestors distribution table. The complete inferred distribution (i.e., the marginal probability of each chromosome number to occur at each ancestral node) is provided in a downloadable comma-separated text file.

4. Inferred transitions. This file provides the inference regarding the expected number of transitions that occur along each branch of the phylogeny. Branches with an expectation above 0.5 of any transition type are first summarized. This is followed by a table that lists the expected number of events for all branches of the phylogeny. A table at the end of this file provides, for each type of transitions, the expected number of events from the root to the tip. The name of the branch is given as the name of the node bellow it (further from the root).
5. All nodes tree. A tree file that specifies the names for all nodes (internal and leaves) of the input tree. These names are given as the bootstrap values and can be viewed in a tree viewer.

Further details about the chromEvol output files can be found in the chromEvol manual (http://chromevol.tau.ac.il/downloads/chromEvol_v2.0_manual.pdf).

3 Working Example: *Centaurium*

To exemplify the usage of the chromEvol web server, we analyzed the chromosome number evolution in the genus *Centaurium* (Gentianaceae). The evolution of chromosome numbers in this genus, using a phylogeny of 26 taxa, was recently analyzed in Maguilla et al. [24], and here we illustrate a similar analysis using the online tools. The following steps were executed:

(i) We reconstructed the phylogeny using sequence data available at the NCBI GenBank by querying the OneTwoTree web server with the name "Centaurium" in the search field. This resulted in a phylogeny that encompasses 27 taxa, including eight taxa whose rank is below the species level (subspecies or varieties) and an outgroup that was identified automatically by OneTwoTree for rooting purposes.

(ii) We used the obtained phylogeny to automatically retrieve chromosome counts from CCDB. This was performed by entering the tab *Download CCDB counts* in the chromEvol web server. In total, 20 taxa with matching chromosome counts were retrieved.

(iii) The above two files were uploaded in their designated input area of the chromEvol web server homepage (marked as #1 and #2 in Fig. 2). To match between the two inputs, we ticked the checkmark next to *Remove tips with no counts data* option (#3 in Fig. 2), which filters from the phylogeny any tip without chromosome count data.

(iv) We selected several models of chromosome number evolution to analyze the data. Each model is defined by a set of transition

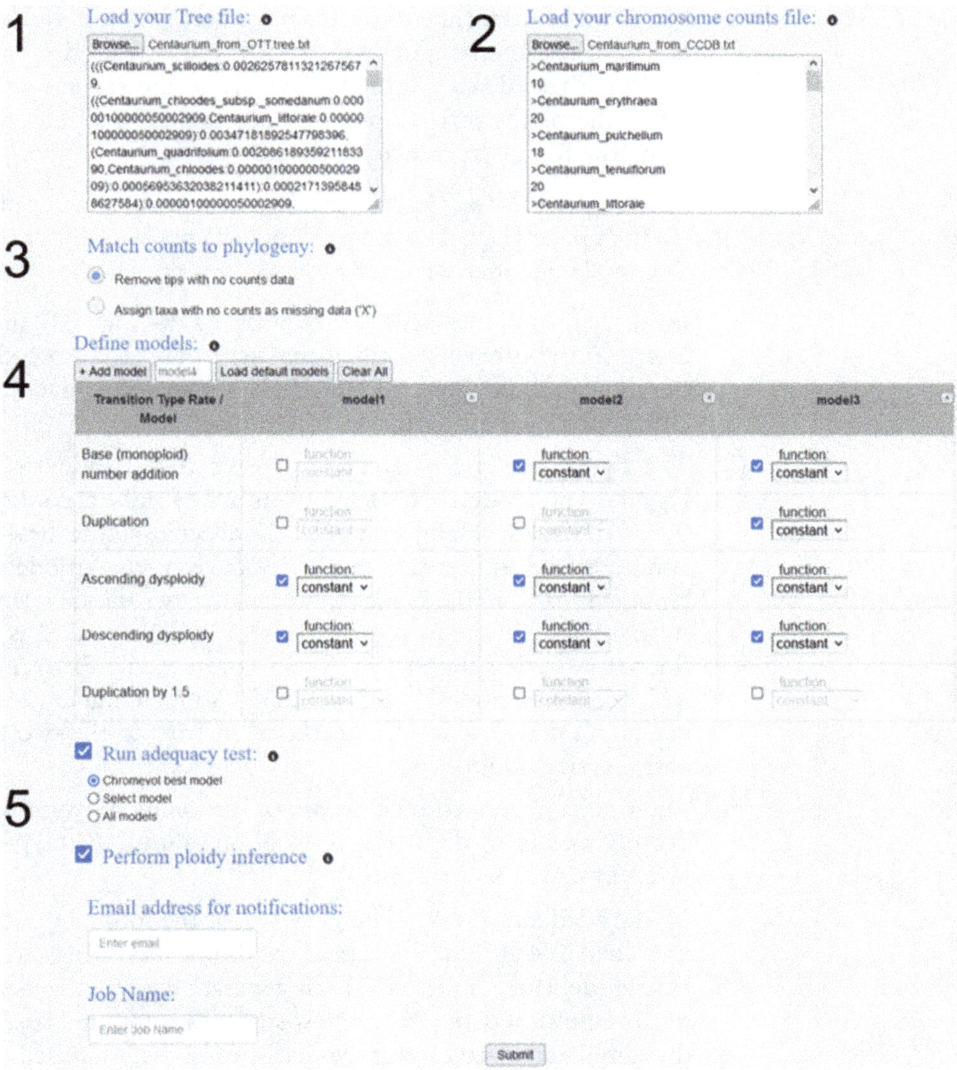

Fig. 2 A screenshot of the chromEvol server homepage, with the respective inputs of the working example

rate parameters and the corresponding function that specifies how to model the dependency between the transition rates and the current number of chromosomes. For illustration purposes, only three models were analyzed in this example (#4 in Fig. 2), although generally it is recommended to examine a more comprehensive set of models (e.g., using the *Load default models* option). In this example, "model1" includes only the possibility of descending and ascending dysploidy transitions without allowing any other ploidy transition, "model2" additionally includes the possibility of transition in an inferred base number, while "model3" is the most general model tested, which further allows duplications in the number of chromosomes.

(v) We chose to perform two optional analyses (#5 in Fig. 2)—the first is to test the absolute adequacy of the best model, among the three tested, to the input data, while the second analysis infers the ploidy level (diploid or polyploid) of all input taxa using the Ploidy Inference Pipeline.

Clicking the *Submit* button executes the analysis and uploads the Results page (Fig. 3), which is updated periodically until the run is completed, after which the following is provided:

(i) Input files and messages (marked as #1 in Fig. 3). In this example, the user was notified that seven taxa were present in the input tree but were filtered due to unavailable data in the chromosome count input file.

(ii) The optimized rate parameters of each tested model (#2 in Fig. 3). The models are sorted from left to right according to their AIC score. Here, "model3" is selected as the best one, although the difference in AIC between this model and "model4" is only 0.58, meaning that the two models are not substantially different (in general, ΔAIC higher than 2 is considered as strong support [25]). In contrast, the difference in AIC between "model1" and "model3" is much higher (26.79), indicating that "model1" is markedly inferior to the two alternative evaluated models.

(iii) The results generated by chromEvol, for online viewing of the downloaded filed (#3 in Fig. 3; *see* Subheading 2.6*Interpreting chromEvol Web Server Results*).

(iv) Model adequacy results. The p-values for the four test statistics examined (#4 in Fig. 3) were above 0.05, indicating that the best model (here, "model3") can generate data that are similar to the empirical data (i.e., the test statistics computed based on the simulated data cannot be statistically distinguished from those computed based on the empirical data at the 0.05 level), and thus, this model is inferred as adequate.

(v) Ploidy-level inference. By executing the ploidy-level inference pipeline (#5 in Fig. 3), 8 taxa were inferred as diploids (all with chromosome numbers equal to 10), while 12 were inferred as polyploids (all with chromosome numbers above 18).

4 Notes

1. *Stand-alone options.* Chromosome counts can also be downloaded using R package *chromer* [26]. Note that using this package does not format the chromosome counts to a FASTA format. Additionally, the tools presented in this chapter—OneTwoTree and chromEvol—are both available in a stand-

User inputs:

1 Counts file | Tree file | Defined models

The following species were found in your phylogeny but are missing in your counts file:
Centaurium chloodes subsp. somedanum, Centaurium cachanlahuen, Centaurium malzacianum, Centaurium uliginosum, Centaurium quadrifolium subsp. barrelieri, Centaurium littorale subsp. compressum, Isertia coccinea

Removing tips with no counts data.

Chromevol results:

Best model is: model3

2

Parameter / Model	model3	model2	model1
Base (monoploid) number addition	0.8862	1.5651	-
Duplication	0.8699	-	-
Ascending dysploidy	0.0000	0.5772	88.3845
Linear ascending dysploidy	-	-	-
Duplication by 1.5	-	-	-
Descending dysploidy	0.7687	1.3494	94.8616
Linear descending dysploidy	-	-	-
Base (monoploid) number	10	10	-
ΔAIC from best model	0.0000	0.5830	26.7891

Tested models results files:

3 model3

Results summary | ML Ancestors Tree | Ancestors distribution Table | Inferred Transitions | All Nodes Tree

model2

model1

Model adequacy results

4

Statistic / Model	model3
Variance	0.10
Entropy	0.19
Parsimony	0.86
Parsimony vs. time	0.38

Ploidy-level results

Download ploidy-level results | View results on tree

5

Taxon	Chromosome count	Ploidy inference	Phylogeny robustness score	Simulation reliability score	Combined score
Centaurium_scilloides	10	Diploid	1.00	1.00	1.00
Centaurium_littorale	20	Polyploid	1.00	1.00	1.00
Centaurium_quadrifolium	10	Diploid	1.00	1.00	1.00

Fig. 3 A screenshot of the chromEvol server Results page, showing the results of the working example

alone version. These stand-alone versions can be executed using Linux OS, while chromEvol can also be used on Windows, and OneTwoTree is also available as a Virtual Box. The stand-alone versions and associated user manuals and documentation are available at http://onetwotree.tau.ac.il/download.html and http://chromevol.tau.ac.il/downloads.html.

2. *Spelling conventions.* Both inputs—the phylogeny and the chromosome counts—should match, not only in their binomial names but also in all other characters. We thus recommend that users avoid using characters such as dots and commas and replace all spaces with underscores.
3. *Modeling the dependency between transition rates and the current number of chromosomes.* At present, the constant and linear functions are available for rates of increasing and decreasing dysploidy, while other transition types are restricted to follow the constant function. In future releases, we plan to expand the list of possible functions and allow other transition types to follow a nonconstant function.

Acknowledgments

We thank Josef Sprinzak for the ongoing developments of the chromEvol web server. This work was supported by a grant from The Israel Science Foundation (961/17 to IM).

References

1. Weiss-Schneeweiss H, Schneeweiss GM (2013) Karyotype diversity and evolutionary trends in angiosperms. In: Greilhuber J, Dolezel J, Wendel JF (eds) Plant genome diversity, vol 2. Springer, Vienna
2. Guerra M (2008) Chromosome numbers in plant cytotaxonomy: concepts and implications. Cytogenet Genome Res 120:339–350
3. Carta A, Bedini G, Peruzzi L (2018) Unscrambling phylogenetic effects and ecological determinants of chromosome number in major angiosperm clades. Sci Rep 8:1–14
4. Vershinina AO, Lukhtanov VA (2017) Evolutionary mechanisms of runaway chromosome number change in Agrodiaetus butterflies. Sci Rep 7:1–9
5. Ruffini Castiglione M, Cremonini R (2012) A fascinating island: 2 n= 4. Plant Biosyst Int J Deal with all Asp Plant Biol 146:711–726
6. Khandelwal S (1990) Chromosome evolution in the genus Ophioglossum L. Bot J Linn Soc 102:205–217
7. Rice A, Glick L, Abadi S et al (2015) The Chromosome Counts Database (CCDB) - a community resource of plant chromosome numbers. New Phytol 206:19–26
8. Levin D (1983) Polyploidy and novelty in flowering plants. Am Nat 122:1–25
9. Ramsey J, Schemske DW (2002) Neopolyploidy in flowering plants. Annu Rev Ecol Syst 33:589–639
10. Rice A, Šmarda P, Novosolov M et al (2019) The global biogeography of polyploid plants. Nat Ecol Evol 3:265–273
11. Leitch AR, Leitch IJ (2008) Genomic plasticity and the diversity of polyploid plants. Science 320:481–483
12. Spoelhof JP, Soltis PS, Soltis DE (2017) Pure polyploidy: closing the gaps in autopolyploid research. J Syst Evol 55:340–352
13. Soltis D, Soltis P, Schemske D (2007) Autopolyploidy in angiosperms: have we grossly underestimated the number of species? Taxon 56:13–30
14. Ramsey J, Ramsey TS (2014) Ecological studies of polyploidy in the 100 years following its discovery. Philos Trans R Soc B Biol Sci 369: 1–20
15. Mayrose I, Zhan SH, Rothfels CJ et al (2011) Recently formed polyploid plants diversify at lower rates. Science 333:1257
16. Mayrose I, Lysak MA (2021) The evolution of chromosome numbers: mechanistic models and experimental approaches. Genome Biol Evol. https://doi.org/10.1093/gbe/evaa220

17. Mayrose I, Barker MS, Otto SP (2010) Probabilistic models of chromosome number evolution and the inference of polyploidy. Syst Biol 59:132–144
18. Akaike H (1974) A new look at the statistical model identification. IEEE Trans Automat Contr 19:716–723
19. Brown JM, Thomson RC (2018) Evaluating model performance in evolutionary biology. Annu Rev Ecol Evol Syst 49:95–114
20. Rice A, Mayrose I (2021) Model adequacy tests for probabilistic models of chromosome-number evolution. New Phytol 229: 3602–3613
21. Glick L, Mayrose I (2014) ChromEvol: assessing the pattern of chromosome number evolution and the inference of polyploidy along a phylogeny. Mol Biol Evol 31:1914
22. Ronquist F, Teslenko M, van der Mark P et al (2012) MrBayes 3.2: efficient Bayesian phylogenetic inference and model choice across a large model space. Syst Biol 61:539–542
23. Drori M, Rice A, Einhorn M et al (2018) OneTwoTree: an online tool for phylogeny reconstruction. Mol Ecol Resour 18: 1492–1499
24. Maguilla E, Escudero M, Jiménez-Lobato V et al (2021) Polyploidy expands the range of Centaurium (Gentianaceae). Front Plant Sci 12. https://doi.org/10.3389/fpls.2021.650551
25. Burnham KP, Anderson DR (1998) Practical use of the information-theoretic approach. In: Model selection and inference: a practical information-theoretic approach. Springer, New York
26. Pennell M, Martinez PA (2016) chromer: interface to chromosome counts database API

Chapter 10

PURC Provides Improved Sequence Inference for Polyploid Phylogenetics and Other Manifestations of the Multiple-Copy Problem

Peter Schafran, Fay-Wei Li, and Carl J. Rothfels

Abstract

Inferring the true biological sequences from amplicon mixtures remains a difficult bioinformatics problem. The traditional approach is to cluster sequencing reads by similarity thresholds and treat the consensus sequence of each cluster as an "operational taxonomic unit" (OTU). Recently, this approach has been improved by model-based methods that correct PCR and sequencing errors in order to infer "amplicon sequence variants" (ASVs). To date, ASV approaches have been used primarily in metagenomics, but they are also useful for determining homeologs in polyploid organisms. To facilitate the usage of ASV methods among polyploidy researchers, we incorporated ASV inference alongside OTU clustering in PURC v2.0, a major update to `PURC` (Pipeline for Untangling Reticulate Complexes). In addition, PURC v2.0 features faster demultiplexing than the original version and has been updated to be compatible with `Python 3`. In this chapter we present results indicating that using the ASV approach is more likely to infer the correct biological sequences in comparison to the earlier OTU-based `PURC` and describe how to prepare sequencing data, run PURC v2.0 under several different modes, and interpret the output.

Key words Allopolyploidy, Amplicon sequence variants, Amplicon sequencing, Moderate data, OTU inference, PacBio, Polyploid phylogenetics, Reticulate evolution

1 Overview

1.1 The Multiple-Copy Problem

There are many situations in biology in which an individual organism may harbor multiple closely related gene copies and where knowing the number of copies and their individual sequences is important for downstream inferences. For example, a major system of self-incompatibility in plants (e.g., the ability of a plant to reject fertilization attempts from its own pollen and thus avoid mating with itself) is based on the presence of particular combinations of "S-alleles," and the characterization of these highly polymorphic S-alleles across individuals is thus essential for understanding the mechanism and consequences of this form of self-incompatibility

Yves Van de Peer (ed.), *Polyploidy: Methods and Protocols*,
Methods in Molecular Biology, vol. 2545, https://doi.org/10.1007/978-1-0716-2561-3_10,

[1, 2]. Similarly, many applications based on the multispecies coalescent, such as delimiting species with BPP [3, 4], rely on including the sequences from both alleles present in a diploid individual. This "multiple-copy problem" [5], however, is particularly pronounced in phylogenetic studies of polyploids. Because many polyploids unite subgenomes from different progenitor species (they are hybrid polyploids, or "allopolyploids"), their true evolutionary history is a network rather than a strictly bifurcating tree, and to infer this reticulate history, the homeologous sequences—the copies from each of the subgenomes—need to be recovered and reconstructed accurately [6].

The classic way of recovering multiple gene copies from a single individual is molecular cloning: the desired marker is amplified by PCR, the amplicon is cloned into plasmid vectors that are used to transform *Escherichia coli*, and multiple colonies of the transformed bacteria are re-amplified and sequenced [e.g. 7, 8]. This approach is labor-intensive and expensive (for a triploid, for example, one would need to sequence 11 colonies to have a 95% chance of getting all three copies), making datasets with many samples or many loci impractical. Short-read next-generation sequencers, such as those of the Illumina platform, offer some relief, in that the reads come from individual molecules (rather than representing a form of majority-rule consensus of the molecules present, as is the case with Sanger sequencing), and sequencing costs are dramatically reduced. However, in order to recover the individual copies, the reads need to be assembled accurately. This assembly step can be bioinformatically prohibitive, especially if there are more than two copies present, and always faces a hard limit: to correctly assemble the full length of each copy of a target sequence, consecutive variable sites have to be separated from each other by no longer than the read length.

1.2 The PURC Approach

To help facilitate the recovery of all homeologous sequences from polyploid accessions (and for other manifestations of the multiple-copy problem), [9] published a molecular lab workflow based on PacBio long-read sequencing, with an associated bioinformatics pipeline (the "Pipeline for Untangling Reticulate Complexes" [PURC]) to infer the individual copies from the PacBio reads. This approach capitalized on PacBio's circular consensus sequencing (CCS) technology to generate contiguous reads for long (>1000 bp) phylogenetically informative regions, thus avoiding the need for assembly and allowing for the accurate retrieval of all copies present in individual polyploids [9–11]. A single PacBio SMRT cell can generate sequences for multiple loci for hundreds of samples, providing an economical means to generate powerful "moderate data" datasets for polyploid phylogenetics.

The wetlab workflow involved standard PCR with barcoded forward primers; the amplicons are then pooled, roughly standardized by amplicon concentration and the anticipated number of copies present in each accession, and sequenced on a single PacBio SMRT cell. PURC can demultiplex the resulting reads by locus, barcode, and phylogenetic affinity (i.e., the same barcode can be used multiple times in a single sequencing run, once for each phylogenetically discernable group, such as a genus). By taking an iterative chimera-removal and clustering approach [12, 13], PURC then infers the underlying biological sequences. The output from PURC is one alignment for each locus, which includes all the copies present in each of the accessions, labelled by their accession ID and coverage (the number of reads that constitute that sequence). Since its release, PURC has been used to investigate allopolyploidy in genus-level datasets in diverse plant lineages [10, 14–16], in studies of cytotype variation within species [11], and for applications where polyploidy itself was incidental to the primary research questions [17–20]. Most notably, Blischak et al. [21] created a program to adapt PURC to data generated by microfluidic PCR, reducing one of the main limitations of amplicon-based approaches (the time and expense associated with PCR itself).

1.3 PURC v2.0

Since the release of PURC, several studies have demonstrated shortcomings of OTU clustering, especially a tendency to overestimate the number of sequences, difficulty in determining appropriate similarity thresholds, and inability to replicate and compare OTUs between analyses [22–25], with the overestimation problem reported for PURC specifically [14, 21]. An alternative approach to identify and separate PCR and sequencing errors from biological sequences is to apply an error model, where read abundance, composition, and quality scores are used to infer whether each unique read is likely to have been derived from another observed sequence [26]. Reads inferred to represent biological sequences by these methods are called amplicon sequence variants (ASVs), exact sequence variants, or zero-radius OTUs. DADA2 [26] is one of the most popular software packages for inferring ASVs and is particularly flexible because it incorporates separate error models for Illumina and PacBio CCS data. Additionally, DADA2 can take sequences as "priors," increasing the sensitivity of the algorithm for sequences that are similar to the priors (https://benjjneb.github.io/dada2/pseudo.html).

In order to take advantage of potential improvements of ASVs, PURC v2.0 incorporates DADA2 alongside Vsearch [27], an open-source alternative to Usearch [12], allowing PURC v2.0 to run in four different ways:

1. OTU clustering alone
2. ASV inference alone
3. OTU clustering and ASV inference in one run
4. OTU clustering followed by ASV inference with the OTUs as priors

To test our incorporation of DADA2, we reproduced OTUs and ASVs generated by [25], where a mock community of five cyanobacteria was created by combining equal quantities of *rbcL-X* amplicons generated from pure cultures. PURC v2.0 was run on four replicates of the five-taxon mock community, performing OTU clustering with default parameter values (clustering thresholds by round: 0.997, 0.995, 0.990, and 0.997; final size threshold = 4) and inferring ASVs (DADA2) with a maximum of five expected errors allowed per read and length outliers removed based on Tukey's outlier test [28]. We also tested data from species of *Isoetes* from [15] using the same parameters.

Our results agree with [25] in showing that ASV inference was more accurate than OTU clustering (Fig. 1). The same five ASVs were recovered from every replicate, whereas OTU clustering identified six to nine OTUs per replicate. Most replicates contained OTUs identical to ASVs, but OTUs with small sequence deviations were common and a few spurious OTUs with 75%–92% identity to ASVs were also generated. Because the *Isoetes* data from [15] are

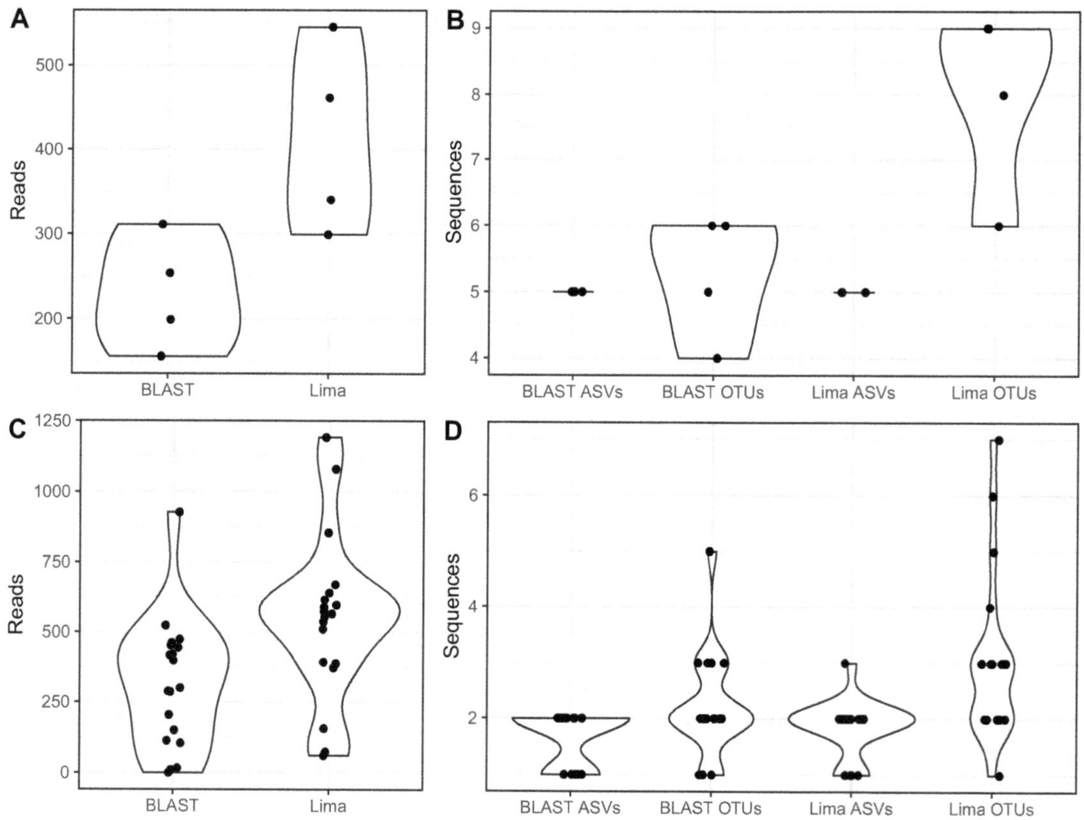

Fig. 1 Comparison of demultiplexing and sequence inference methods using data from [25] and [15]. (**a**) Number of reads assigned to four replicates of the five-taxon mock community. (**b**) Sequences inferred for the four replicates of the five-taxon mock community. (**c**) Number of reads assigned to each *Isoetes* sample. (**d**) Sequences inferred for each *Isoetes* sample

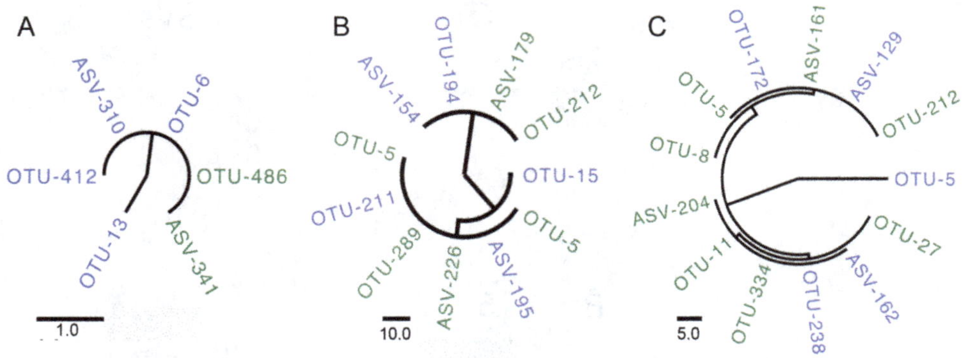

Fig. 2 Maximum-parsimony trees of OTUs and ASVs from two PCR replicates of three *Isoetes* species ((**a**) diploid *I. echinospora* Taylor #6989-1; (**b**) allotetraploid *I. maritima* Taylor #6983; (**c**) allotetraploid *I.* aff. "new hybrid A" Taylor #6988-2). PCR replicate 1 is colored green and replicate 2 is colored blue. Numbers following tip names indicate coverage (the number of reads constituting each OTU/ASV). Trees are rooted at their midpoints. Scale bar represents number of substitutions. (Data are from [15])

empirical and true sequences are unknown, we evaluated ASV- versus OTU-based inferences of these sequences using two independent PCR amplifications of each of three individuals, thought to represent one diploid and two allotetraploids (*I. echinospora* Taylor #6989-1, *I. maritima* Taylor #6983, and *I.* aff. "new hybrid A" Taylor #6988-2, respectively). For these accessions we inferred 20 OTUs versus 10 ASVs in total. Both PCR replicates yielded identical ASV inferences—one sequence for the diploid and two for each of the allotetraploids (Fig. 2)—and these sequences all had high coverage (they represented many individual reads; Fig. 3). However, OTU inference was less consistent. For the diploid, both PCR replicates resulted in a high-coverage OTU that was identical to the one from ASV inference, but one of the replicates inferred two additional low-coverage OTUs (for example, OTU-6 contained five deletions and two ambiguous nucleotides relative to OTU-412 and ASV-310; Fig. 2a.) Similarly, for the tetraploids, for both PCR replicates, OTU inference found the sequences corresponding to the ASVs, but also found additional, presumably spurious, low-coverage sequences (Figs. 2 and 3). One of these sequences—OTU-5 in replicate 2 of the allotetraploid *I.* aff. "new hybrid A" Taylor #6988-2—was uniquely divergent, with 36 SNPs and two deletions totaling 97 bases. The origins of the five reads that constitute this OTU are unclear, but were not due to low-quality basecalls or incorrect assignment by barcodes. In every case, the identical ASVs and OTUs were by far the most abundant (Fig. 3); they likely represent the true biological sequences.

PURC v2.0 also includes PacBio's lima tool (https://lima.how) as a new method for demultiplexing reads on Linux operating systems. Using lima, we could recover an average of 74% more reads

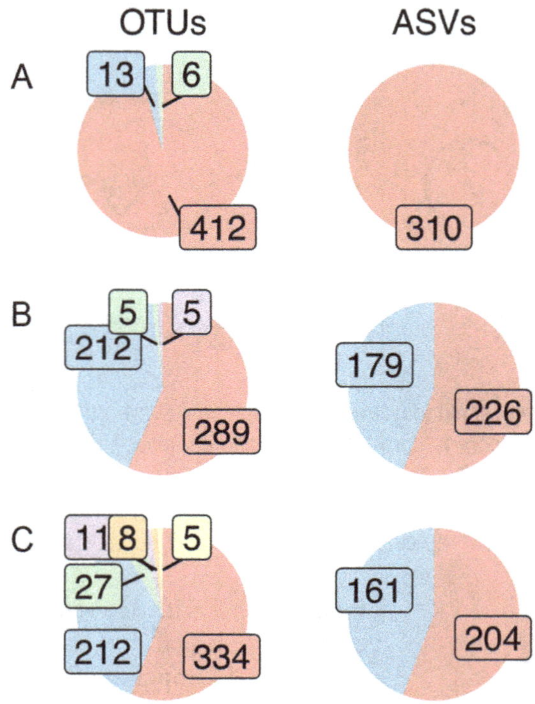

Fig. 3 Proportions of reads contributing to each OTU and ASV for three *Isoetes* species ((**a**) diploid *I. echinospora* Taylor #6989-1; (**b**) allotetraploid *I. maritima* Taylor #6983; (**c**) allotetraploid *I. aff.* "new hybrid A" Taylor #6988-2), from one PCR replicate. Labels indicate number of reads in each OTU/ASV and sequences are colored from largest to smallest within each chart. (Data are from [15])

from the [15] data than with the original BLAST-based method in PURC, including recovering one individual for which no sequences were identified using BLAST; the per-sample increase ranged from 24% to 640% (Fig. 1c). The number of both OTUs and ASVs increased in the lima-demultiplexed dataset but ASVs fluctuated less, with 74% of samples having the same number of ASVs as in the BLAST-demultiplexed data versus only 26% for OTUs.

There were some consistent trends in our analyses of mock community and real polyploid data. lima consistently recovered more reads from each sample, which resulted in an increase in the number of OTUs, possibly by inclusion of more divergent reads rising above the threshold for dropping low-abundance clusters. However, the inclusion of the additional reads had a much smaller effect on ASV inference. The variance in the number of ASVs was always smaller than that for OTUs and especially in the mock community analyses where every replicate was correctly inferred (Fig. 1). While [25] showed that by varying OTU clustering parameters they could more accurately reconstruct the true sample composition, this approach is unreliable in cases where the true sequences are unknown and where it may be tempting to change parameters until the results match expectations. Based on our results, we recommend lima demultiplexing and ASV inference as the primary method for running PURC v2.0. To compare OTU clustering and ASV inference on your own data, PURC v2.0 allows generating both simultaneously, with summary files to examine the number and abundance of sequences for each sample. If it appears that too few ASVs are found, a new analysis can be run with the OTU sequences as priors, increasing the sensitivity of the algorithm and reducing the detection limit for variants. This approach may be particularly useful for samples with little data or where two biological sequences are very similar, although it remains to be tested in polyploids.

2 Materials

2.1 Hardware

A personal computer with a multicore processor and at least 8-GB RAM, though this will vary based on the input data size.

2.2 Software

1. A Unix- or Linux-based operating system (e.g., macOS, Ubuntu), or on Windows, a Linux virtual machine can be used.
2. Conda package manager (https://docs.conda.io).

3 Methods

The following instructions as well as additional information for troubleshooting can be found in the main PURC v2.0 repository at https://bitbucket.org/peter_schafran/purc/.

3.1 Installing PURC v2.0

The most up-to-date version of PURC can be downloaded as a compressed TAR file in the main repository. When uncompressed, a new directory named `purc` will contain executable and installation files.

```
curl -L https://bitbucket.org/peter_schafran/purc/raw/master/
purc_v2.tar.gz -O
tar -xzf purc_v2.tar.gz
cd purc
```

We recommend installing dependencies through Conda. Two YAML files, one for macOS and one for Linux, are included in the repository and can be used to create a new PURC environment containing all necessary dependencies with one of these commands:

```
# macOS
conda env create -n purc --file purc_macos.yaml
conda activate purc
```

```
# Linux
conda env create -n purc --file purc_linux.yaml
conda activate purc
```

3.2 Preparing Input Files

Sequence data are expected to be 99% accurate amplicon sequences containing sample-specific nucleotide sequences (barcodes) on one or both ends of each read, as well as the priming sites used to generate the amplicons. If using PacBio reads generated by their standard protocol (https://ccs.how), they should not need any modification. Either FASTA or FASTQ formatted data are accepted, though FASTQ is required to perform ASV inference. In addition to the data, PURC v2.0 requires four files:

1. Barcode file listing the barcode sequences
2. Reference sequences file, so that reads can be oriented, assigned to the correct locus, and sorted into phylogenetic groups if individual barcodes are used for multiple accessions
3. Map file(s), which link barcode and group IDs to unique accessions (one map file for each locus)
4. Configuration file, which includes information on specific settings, the primer sequences, and other necessary information

3.2.1 Barcode File

Barcode sequences are provided in FASTA format. The sequence ID is the barcode name, and sequences should be oriented in the 5'-3' direction. For example:

```
>BC01
ACTACATATGAGATGA
>BC02
TCATGAGTCGACACTA
>BC03
TATCTATCGTATACGC
```

If using the dual barcode function, the barcode names must be BCF1, BCF2, ... and BCR1, BCR2, ... for barcodes on forward and reverse primers, respectively. This information is used to check for barcodes found in the incorrect orientation. We recommend barcodes that are unique (including reverse complements), especially if using the option to demultiplex with lima. However, if a subset of duplicate barcodes is detected, PURC v2.0 will attempt to run separate analyses through lima with the unique and duplicate barcodes and then merge the output.

3.2.2 Reference Sequence File

Reference sequences must be provided in FASTA format. Each sequence should specify its locus name in the sequence ID (e.g., locus = ApP), even if the data being analyzed represent only one locus. A group name (e.g., group = A) can be provided to demultiplex by BLAST comparison to reference sequences if multiple samples share barcodes. A taxon name (e.g., ref_taxon = Cystopteris_bulbifera) or other descriptor can be included to provide additional information for the user but this information is not used by PURC. Fields are separated by forward slashes, so a complete sequence ID line may look like the following:

```
>locus=ApP/group=A/ref_taxon=Cystopteris_bulbifera
```

Note that the total sequence ID should not exceed 50 characters or else BLAST database construction will fail. All reference sequences for a locus must be in the same orientation to allow proper orientation of the reads.

3.2.3 Map File

The map file indicates which barcodes and/or groups correspond with each sample. This file is a tab-delimited text file with one of three configurations:

1. If each sample contains just one barcode (e.g., only the forward primer is barcoded) and each barcode is unique to one sample, the first column contains barcode names (from the barcode FASTA file) and the second column contains the name of the corresponding sample.

```
BC01 Cystopteris_fragilis_Utah
BC02 Cystopteris_fragilis_Arizona
BC03 Cystopteris_fragilis_Taiwan
```

2. If each sample contains one barcode, but individual barcodes are used for multiple samples, then the first column contains the barcode name, the second column contains a group ID (corresponding to a group specified in a reference sequence ID), and the third column contains the sample name.

```
BC01 A Acystopteris_japonica_Taiwan
BC01 B Gymnocarpium_dryopteris_Ontario
BC02 A Acystopteris_japonica_Japan
```

3. If samples are dual-barcoded, the first column contains the forward primer barcode name, the second column contains the reverse primer barcode name, and the third column the sample name.

```
BCF1 BCR1 Cystopteris_fragilis_Utah
BCF2 BCR1 Cystopteris_fragilis_Arizona
BCF1 BCR2 Cystopteris_fragilis_Taiwan
```

3.2.4 Config File

All PURC v2.0 operation is controlled via the config file, a text file containing information such as file names and run parameters. Paths to the data file, barcode file, reference sequence file, and map files are specified here, as well as primer sequences, run options, and optional parameter settings. Each line item is described by comments in the file (not shown here).

The first section of the config file specifies input files for reads, barcodes, and references and names of the prefix appended to output files, the output folder, and the log file.

```
[Files]
Input_sequence_file = 1-PacBio_seq.fastq
in_Barcode_seq_file = 2-barcodes.fasta
in_RefSeq_seq_file = 3-ref_sequences.fasta
Output_prefix = purc_run
Output_folder = purc_out
Log_file = log
```

The second section provides information about the locus or loci to be processed. The locus name must match that provided in the reference sequence file. If processing multiple loci, their order in `Locus_name` and `Locus-barcode-taxon_map` must match.

```
[Loci]
Locus_name = ApP, GAP
Locus-barcode-taxon_map = 4-map_APP.txt, 4-map_GAP.txt
```

The third section is used to provide primer sequences in the 5'-3' direction: these, too, must be in the same order as, e.g., `Locus_name`. IUPAC ambiguous nucleotide codes are accepted.

```
[Primers]
Forward_primer = GGACCTGGSCTYGCTGARGAGTG, TCTGCMCATGCMATT-
GAAAGAGAG
Reverse_primer = GGAAGVACCTTYCCTACTGCCTG, TAGCTGCTCRAATTC-
CATKSAT
```

The fourth section allows the user to choose between several run modes depending on their data.

- `Mode` controls whether PURC v2.0 checks for interlocus concatemers or not (`Mode` should be set to `1` for single-locus data).
- `Multiplex_per_barcode` indicates if barcodes are reused for multiple individuals within the same locus or if each barcode (for a given locus) is unique to one sample (barcode reuse is not available for dual barcoding).

- `Dual_barcode` is used to specify how the barcodes are arranged on each read. Note that options `1` and `2` are only treated differently if using `lima` to demultiplex.
- `Barcode_detection` describes where to look for barcodes in each read. If set to `0`, the "ErrMidBC" flag is included in the sequence name if barcodes are not at the ends of the sequence. This setting has no effect if using `lima`.
- `Recycle_chimeric_seq` controls whether PURC v2.0 splits interlocus chimeras into their respective loci and includes them in downstream analysis, or discards such chimeras. This setting has no effect if `Mode = 1`.
- `Recycle_no_barcoded_seq` allows for the use of Smith-Waterman local alignment [29] to try to identify barcodes in any sequences that failed to have a significant BLAST match. This setting has no effect if using `lima` to demultiplex.
- `Clustering_method` specifies which method to use for sequence inference. Option `2` does ASV inference and OTU clustering together and is required for `Use_OTU_priors = TRUE`.
- `Align` controls whether each final sequence file produced (one per locus per clustering method) is aligned using `MAFFT` [30].

For example:

```
[PPP_Configuration]
Mode = 0 # 0: Check concatemers and then full run
 # 1: Skip concatemer-checking
Multiplex_per_barcode = 0 # 0: Each barcode contains only one
sample
 # 1: Each barcode contains multiple samples
Dual_barcode = 0 # 0: Barcodes only on one primer
 # 1: Unique barcodes on both primers
 # 2: Same barcode on both primers
Barcode_detection = 1 # 0: Search barcode in entire sequences
 # 1: Search barcode only at the ends of sequences
Recycle_chimeric_seq = 0 # 0: Do not recycle
 # 1: Split chimeric sequences into respective locus
Recycle_no_barcoded_seq = 0 # 0: Do not recycle
 # 1: Use Smith-Waterman algorithm to find barcodes if BLAST
fails
Clustering_method = 0 # 0: Use DADA2 ASV inference
 # 1: Use Vsearch OTU clustering
 # 2: Use both clustering methods
Align = 1 # 0: No aligning attempted.
 # 1: Final consensus sequences will be aligned with MAFFT
```

Other optional parameters follow this section of the config file, including those to change the operation of `DADA2` and `Vsearch`.

DADA2 options include specifying minimum and maximum lengths and maximum number of expected errors for reads to be included and whether to include each sample's OTU sequences as priors. If minimum and/or maximum read length is set to 0, that parameter is calculated using Tukey's equation for outliers [28]:

$$minLen = Q_1 - 1.5(Q_3 - Q_1), maxLen = Q_3 + 1.5(Q_3 - Q_1)$$

where Q_1 is the first quartile and Q_3 the third quartile, and results are rounded to the nearest integer. Outliers are recalculated for each sample. If minLen/maxLen values are user-supplied, they are applied globally. The maximum number of expected errors is estimated by DADA2 based on read quality scores. If putative spurious ASVs are produced, this number can be reduced, while if too many reads are discarded during filtering, it can be increased.

- `minLen` is the minimum length for a read to be included in analysis. Set to 0 to automatically detect short outliers.
- `maxLen` is the maximum length for a read to be included in analysis. Set to 0 to automatically detect long outliers.
- `maxEE` is the maximum number of expected errors estimated by DADA2 based on read quality scores. Reads with more expected errors than `maxEE` are discarded.
- `Use_OTU_priors` determines whether to use the OTU sequences as priors. When activated, the OTUs for each sample are used to infer ASVs for that sample. This can be useful if the number of reads per sample is low, or if well-supported OTUs seem to be lost during ASV inference. However, it can increase the risk of creating spurious ASVs. Requires `Clustering_method = 2`.

For example:

```
[DADA Filtering Parameters]
minLen = 0 # The minimum length to keep a read
maxLen = 0 # The maximum length to keep a read
maxEE = 5 # Reads with greater than maxEE "expected errors" are
discarded
Use_OTU_priors = FALSE # Set to TRUE to use OTU output
sequences as priors for ASV inference
```

Available `Vsearch` options are for the identity levels for each round of clustering, minimum size to retain a cluster, and the abundance skew for detecting chimeric sequences.

- `clustIDn` specifies the identity threshold for clustering reads during the *n*th round of clustering. For example, 0.997 means a read must have 99.7% identity to a cluster's centroid sequence in order to be included in that cluster.

- `sizeThreshold1` sets the minimum number of reads in a cluster for it to be retained for the next round of OTU clustering.
- `sizeThreshold2` is the same as `sizeThreshold1`, but is only applied to the final clustering output.
- `abundance_skew` is the minimum ratio of parent sequences to putative chimeric sequence required to classify a sequence as chimeric. Parent sequences are expected to be at least twice as abundant as their chimeras.

For example:

```
[Clustering Parameters]
clustID1 = 0.997 # The similarity criterion for the initial
VSEARCH clustering
clustID2 = 0.995 # The similarity criterion for the second
clustering
clustID3 = 0.990 # The similarity criterion for the third
clustering
clustID4 = 0.997 # The similarity criterion for the FINAL
clustering
sizeThreshold1 = 1 # The min. number of sequences/cluster for
that cluster to be retained
sizeThreshold2 = 4 # The min. number of sequences/cluster for
that cluster to be retained

[Chimera-killing Parameters]
abundance_skew = 1.9
```

If the user's operating system is detected as Linux-based, PURC v2.0 will default to using `lima` for demultiplexing. To override and use BLAST-based demultiplexing, change the override flag in the config file effect on other operating systems.

```
[Lima Override]
Lima_override = 0 # Set to 1 to use BLAST-based demultiplexing
```

3.3 Running PURC v2.0

3.3.1 Full Run with Demultiplexing and Sequence Inference

Once all input files are complete, `PURC` is initiated by calling the main `PURC` script with the config file as the argument:

```
purc.py config.txt
```

On completion, the output directory will have this structure, where log, output prefix, locus, and clustering method are replaced with those specified by the config file. It will contain a subdirectory for each locus and within each locus directory a subdirectory for each sample.

```
Output_folder
+-- log
+-- Output_prefix_1_bc_trimmed.fa
+-- Output_prefix_2_pr_trimmed.fa
+-- Output_prefix_3_annotated.fa
+-- Output_prefix_4_locus_clustering-method.fa
+-- Output_prefix_4_locus_clustering-method.aligned.fa
+-- Output_prefix_5_counts.xls
+-- Output_prefix_5_proportions.tsv
+-- Output_prefix_5_proportions.pdf
+-- Locus/
| +-- Locus.fa
| +-- Sample_1/
| +-- Sample_1_ASVs.fa
| +-- Sample_1_OTUs.fa
| +-- Sample_1_read_lengths.pdf
+--tmp/
```

- log—log file documenting the PURC run.
- Output_prefix_1_bc_trimmed.fa—all reads containing valid barcodes.
- Output_prefix_2_pr_trimmed.fa—reads with primer sequences removed (OTU clustering only).
- Output_prefix_3_annotated.fa—reads that could be assigned to samples based on barcodes or groups.
- Output_prefix_4_locus_clustering-method.fa—combined output sequences from all samples following OTU clustering and/or ASV inference. One file per locus per clustering method.
- Output_prefix_4_locus_clustering-method.aligned.fa—alignment of output sequence file.
- Output_prefix_5_counts.xls—summary of results containing number of reads surviving at each step in processing and final sequences per sample.
- Output_prefix_5_proportions(.tsv/.pdf)—summary of read coverage for the OTUs/ASVs for each sample. The PDF presents these data as pie charts with labels indicating the read coverage of each slice (as in Fig. 3).
- Locus/Locus.fa—all reads annotated to this locus.
- Locus/Sample_1/Sample_1_ASVs.fa—final ASV sequences for this sample (if applicable).
- Locus/Sample_1/Sample_1_OTUs.fa—final OTU sequences for this sample (if applicable).
- Locus/Sample_1/Sample_1_read_lengths.pdf—histogram of read lengths prior to ASV inference. Dotted lines indicate the limits for discarding too short/too long reads.

If PURC v2.0 is interrupted, it can be resumed by running again with the same config file. As long as all parameters are the same, it will determine the last completed step and continue.

3.3.2 Analyses on Previously Demultiplexed Data

PURC v2.0 includes a secondary script, `purc_recluster.py`, that can be used to perform OTU clustering on prior PURC runs or on data that have been demultiplexed by another method. Unlike the main script, `purc_recluster.py` operates with only command-line arguments.

```
./purc_recluster.py -f annotated_seq_file -o output_folder\
-c clustID1 clustID2 clustID3 clustID4\
-s sizeThreshold1 sizeThreshold2
```

If using PURC demultiplexed data, the input file is the `output_prefix_3_annotated.fa` file. If preparing your own data, the file must be FASTA formatted with name lines structured as follows:

```
>Isoetes_sp_Schafran1|LFY|BCF58^BCR2|
m170705_030709_42153_c10121536_s1_p0/62/ccs
```

The header line has elements separated by "|", where the sample name and locus name are the first and second elements, respectively. Any other elements after the sample and locus names, such as barcode, group, and read IDs, are not used. Four clustering identity thresholds are specified by the `-c/--clustering_identities` flag and two size thresholds specified by the `-s/--size_threshold` function identically to their respective parameters in the config file.

The reclustering script produces output organized similarly to the PURC output detailed above. This output contains two FASTA files for each locus, one of OTU sequences combined from all samples and the other an alignment of those sequences. There are folders for each locus and, within each, separate folders containing working files for each sample. A summary file called `purc_cluster_counts.xls` contains information about the number of reads, OTUs, and chimeras found per sample and per locus.

4 Conclusions

PURC [9], in conjunction with PacBio circular consensus sequencing, introduced an economical and effective alternative to time-consuming cloning and Sanger sequencing for generating broad multilocus datasets for groups containing polyploids. With PURC v2.0 we have significantly improved upon the earlier version, most notably by addressing shortcomings of OTU clustering by the incorporation of ASV inference. In addition to the generally greater accuracy of ASV (versus OTU) inference, our implementation of "straight" ASV inference and ASV inference with priors alongside OTU inference allows users to compare the results from all three approaches. PURC v2.0 can thus function as a data exploration tool, providing users with the opportunity to detect and interrogate unexpected patterns of variation in their amplicon sequencing datasets.

We anticipate that PURC v2.0 will prove to be a valuable component of biologists' toolkits. While amplicon-based data generation does not scale as well as most other reduced-representation techniques, such as Hyb-Seq [e.g. 31, 32], it is a cost-effective way to sequence a greater number of samples for fewer—but more informative—loci, allows for the easy integration of new data with historical datasets, and is a powerful means of supporting "moderate data" approaches [33, 34] (i.e., the production of multilocus datasets that are large enough to be phylogenetically informative yet sufficiently small to allow for thorough curation and model selection; see also [35]). Moderate data are particularly effective for the phylogenetic study of polyploids, where the limiting factor is typically systematic error rather than stochastic error [36]: the accurate recovery and analysis of the full set of homeologous sequences (avoiding chimeras) is more important than the absolute amount of data available per se [6].

The main application of PURC v2.0 is thus likely to be as a component of a "polyploid phylogenetics" workflow [6]. For example, a researcher can use PURC v2.0 to generate a multilocus nuclear dataset for a broad taxon sample, including polyploids, phase the loci (determine, for each locus, which copy of each polyploid comes from which subgenome) with a `homologizer` [37], and use those phased multilocus data for downstream phylogenetic inference such as divergence-time or species-tree estimation. Such a workflow would allow for the investigation of many outstanding questions related to polyploid evolution and would also permit the phylogenetic study of groups that contain polyploids regardless of whether polyploidy itself is of central interest. In addition, PURC v2.0 is not restricted to cases of the multiple-copy problem. For example, to answer questions of hybrid parentage, plastid regions can be amplified and co-sequenced with nuclear loci and processed in the same `PURC` run, and PURC v2.0 is also an effective way to generate sequence data for basic diploids. Finally, while not specifically designed for metabarcoding, the underlying programs in PURC v2.0 are widely used in this field, making `PURC` useful for demultiplexing samples and sequence inference for other downstream analyses (e.g., [38]).

References

1. Ramanauskas K, Igić B (2017) The evolutionary history of plant T2/S-type ribonucleases. PeerJ 5:e3790
2. Goldberg EE, Kohn JR, Lande R et al (2010) Species selection maintains self-incompatibility. Science 330:493–495
3. Yang Z, Rannala B (2010) Bayesian species delimitation using multilocus sequence data. Proc Natl Acad Sci 107:9264–9269
4. Yang Z (2015) The BPP program for species tree estimation and species delimitation. Curr Zool 61:854–865
5. Griffin PC, Robin C, Hoffmann AA (2011) A next-generation sequencing method for overcoming the multiple gene copy problem in polyploid phylogenetics, applied to *Poa* grasses. BMC Biol 9:19. https://doi.org/10.1186/1741-7007-9-19
6. Rothfels CJ (2021) Polyploid phylogenetics. New Phytol 230:66–72
7. Schuettpelz E, Grusz AL, Windham MD, Pryer KM (2008) The utility of nuclear *gapCp* in resolving polyploid fern origins. Syst Bot 33:621–629
8. Li F-W, Pryer KM, Windham MD (2012) *Gaga*, a new fern genus segregated from *Cheilanthes* (Pteridaceae). Syst Bot 37:845–860. https://doi.org/10.1600/036364412X656626
9. Rothfels CJ, Pryer KM, Li F-W (2017) Next-generation polyploid phylogenetics: rapid resolution of hybrid polyploid complexes using PacBio single-molecule sequencing. New Phytol 213. https://doi.org/10.1111/nph.14111
10. Dauphin B, Grant JR, Farrar DR, Rothfels CJ (2018) Rapid allopolyploid radiation of moonwort ferns (*Botrychium*; Ophioglossaceae) revealed by PacBio sequencing of homologous and homeologous nuclear regions. Mol Phylogenet Evol 120:342–353. https://doi.org/10.1016/j.ympev.2017.11.025
11. Kao T-T, Rothfels CJ, Melgoza-Castillo A et al (2020) Infraspecific diversification of the star cloak fern (*Notholaena standleyi*) in the deserts of the United States and Mexico. Am J Bot 107:658–675
12. Edgar RC (2010) Search and clustering orders of magnitude faster than BLAST. Bioinformatics 26:2460–2461
13. Edgar RC, Haas BJ, Clemente JC et al (2011) UCHIME improves sensitivity and speed of chimera detection. Bioinformatics 27:2194–2200
14. Morales-Briones DF, Tank DC (2019) Extensive allopolyploidy in the neotropical genus *Lachemilla* (Rosaceae) revealed by PCR-based target enrichment of the nuclear ribosomal DNA cistron and plastid phylogenomics. Am J Bot 106:415–437. https://doi.org/10.1002/ajb2.1253
15. Suissa JS, Kinosian SP, Schafran PW et al (2022) Homoploid hybrids, allopolyploids, and high ploidy levels characterize the evolutionary history of a western North American quillwort (Isoëtes) complex. Mol Phylogenet Evol 166:107332
16. Blischak PD, Thompson CE, Waight EM et al (2020) Inferring patterns of hybridization and polyploidy in the plant genus *Penstemon* (Plantaginaceae). bioRxiv
17. Kao T-T, Pryer KM, Freund FD et al (2019) Low-copy nuclear sequence data confirm complex patterns of farina evolution in notholaenid ferns (Pteridaceae). Mol Phylogenet Evol 138:139–155. https://doi.org/10.1016/j.ympev.2019.05.016
18. Chery JG, Acevedo-Rodríguez P, Rothfels CJ, Specht CD (2019) Phylogeny of *Paullinia* L. (Paullinieae: Sapindaceae), a diverse genus of lianas with dynamic fruit evolution. Mol Phylogenet Evol 140:106577
19. Wolfe AD, Blischak PD, Kubatko L (2021) Phylogenetics of a rapid, continental radiation: Diversification, biogeography, and circumscription of the beardtongues (*Penstemon*; Plantaginaceae). bioRxiv
20. Frost LA, O'Leary N, Lagomarsino LP et al (2021) Phylogeny, classification, and character evolution of the tribe Citharexyleae (Verbenaceae). Am J Bot 108(10):1982–2001
21. Blischak PD, Latvis M, Morales-Briones DF et al (2018) `Fluidigm2PURC`: automated processing and haplotype inference for double-barcoded PCR amplicons. Appl Plant Sci 6:e01156
22. Callahan BJ, McMurdie PJ, Holmes SP (2017) Exact sequence variants should replace operational taxonomic units in marker-gene data analysis. ISME J 11:2639–2643
23. Barnes CJ, Rasmussen L, Asplund M et al (2020) Comparing DADA2 and OTU clustering approaches in studying the bacterial communities of atopic dermatitis. J Med Microbiol 69:1293–1302
24. Joos L, Beirinckx S, Haegeman A et al (2020) Daring to be differential: Metabarcoding analysis of soil and plant-related microbial

communities using amplicon sequence variants and operational taxonomical units. BMC Genomics 21:733

25. Nelson JM, Hauser DA, Li F-W (2021) The diversity and community structure of symbiotic cyanobacteria in hornworts inferred from long-read amplicon sequencing. Am J Bot 108 (9):1731–1744
26. Callahan BP, McMurdie PJ, Rosen MJ et al (2016) DADA2: high-resolution sample inference from Illumina amplicon data. Nat Methods 13:581–583. https://doi.org/10.1038/nmeth.3869
27. Rognes T, Flouri T, Nichols B et al (2016) VSEARCH: a versatile open source tool for metagenomics. PeerJ 4. https://doi.org/10.7717/peerj.2584
28. Tukey JW (1977) Exploratory data analysis. Addison-Wesley Publishing Company, Reading
29. Smith TF, Waterman MS (1981) Identification of common molecular subsequences. J Mol Biol 147:195–197
30. Katoh K, Standley DM (2013) MAFFT multiple sequence alignment software version 7: improvements in performance and usability. Mol Biol Evol 30:772–780
31. Breinholt JW, Carey SB, Tiley GP et al (2021) A target enrichment probe set for resolving the flagellate land plant tree of life. Appl Plant Sci 9. https://doi.org/10.1002/aps3.11406
32. Johnson MG, Pokorny L, Dodsworth S et al (2019) A universal probe set for targeted sequencing of 353 nuclear genes from any flowering plant designed using k-medoids clustering. Syst Biol 68:594–606
33. Rothfels CJ, Li F-W, Sigel EM et al (2015) The evolutionary history of ferns inferred from 25 low-copy nuclear genes. Am J Bot 102: 1089–1107
34. Rothfels CJ, Larsson A, Kuo L-Y et al (2012) Overcoming deep roots, fast rates, and short internodes to resolve the ancient rapid radiation of eupolypod II ferns. Syst Biol 61:490
35. Frost LA, Lagomarsino LP (2021) More-curated data outperforms more data: Treatment of cryptic and known paralogs improves phylogenomic analysis and resolves a northern Andean origin of *Freziera* (Pentaphylacaceae). bioRxiv
36. Philippe H, Brinkmann H, Lavrov DV et al (2011) Resolving difficult phylogenetic questions: why more sequences are not enough. PLoS Biol 9:e1000602
37. Freyman WA, Johnson MG, Rothfels CJ (2020) `homologizer`: phylogenetic phasing of gene copies into polyploid subgenomes. bioRxiv. https://doi.org/10.1101/2020.10.22.351486
38. Goldberg AR, Conway CJ, Tank DC et al (2020) Diet of a rare herbivore based on DNA metabarcoding of feces: selection, seasonality, and survival. Ecol Evol 10:7627–7643

Part III

Analysis of Gene Expression and Regulation in Polyploids

Chapter 11

Analyses of Genome Regulatory Evolution Following Whole-Genome Duplication Using the Phylogenetic EVE Model

Ksenia Arzumanova, Rori V. Rohlfs, Lars Grønvold, Marius A. Strand, Torgeir R. Hvidsten, and Simen R. Sandve

Abstract

Whole-genome duplications (WGDs) are important in shaping the evolution of complex genomes, including rewiring of genome regulation. To address key questions about how WGDs impact the evolution of genome regulation, we need to understand the relative importance of selection versus drift and temporal evolutionary dynamics. One promising class of statistical models that can help address such questions are phylogenetic Ornstein-Uhlenbeck (OU) models.

Here we present a computational pipeline for the comparative phylogenetic analyses of genome regulation using an OU model. We have implemented this model in R and provide a step-by-step protocol for the use of this model, including example scripts and simulated test data. We provide the nonspecialist a brief overview of how this model works and how to perform tests for signatures of selection on genome regulation as well as power simulations to aid in experimental design and interpretation of results. We believe that these resources could help polyploidy research move forward in an era of rapidly increasing functional genomics data across the tree of life.

Key words Regulatory evolution, Whole-genome duplication, Phylogenetic models, EVE model, R

1 Introduction

Studies of how whole-genome duplications (WGDs) have impacted the evolution of genome regulation are commonly restricted to pairwise comparisons of gene duplicates within a single polyploid genome, with some studies extending these analyses to single non-duplicated outgroup genomes [1]. Solely relying on pairwise comparative approaches severely limits the types of evolutionary questions we can address. Firstly, the lack of an underlying evolutionary model makes it difficult to address key evolutionary questions such as the role of selection versus drift in shaping genome

Yves Van de Peer (ed.), *Polyploidy: Methods and Protocols*,
Methods in Molecular Biology, vol. 2545, https://doi.org/10.1007/978-1-0716-2561-3_11,

regulatory evolution following WGD. Secondly, when WGD events are older and followed by speciation events, we are often interested in evolutionary questions that necessitate the analyses of genome regulatory evolution across species at specific times (i.e., branches) during species evolution. To get to these types of questions, we need a statistical framework that can model genome regulatory phenotypes across the evolutionary history of a gene under different evolutionary scenarios of drift and selection [1].

One promising class of methods that can overcome the shortcomings of pairwise comparisons are phylogenetic Ornstein-Uhlenbeck (OU) models [2, 3]. Within the OU framework, it is possible to model the evolution of any quantitative feature under stabilizing selection, neutral evolution, or directional selection through the evolutionary history of genes and species while taking phylogenetic relatedness into account. In principle, the evolution of any type of functional genomics data can be modelled, e.g., gene expression levels, tissue specificity of gene expression, gene co-expression network centrality, chromatin accessibility, and different histone tail modifications (from now referred to as regulatory phenotypes). OU models have been extensively used to analyze macroevolutionary patterns [4], but as functional omics datasets have exploded, these models have more recently been adopted to analyze the evolution of gene expression [2, 5, 6]. Yet, so far the use of OU models to study the consequences of WGD has been limited [7].

There have been two major obstacles that likely have limited the use of phylogenetic OU models in the context of genome regulation following WGDs: (i) functional genomics data availability from comparable samples across several species and (ii) lack of software and standardized computational pipelines which are accessible for nonspecialists. The recent revolution in low-cost high-throughput functional genomics data generation is now rapidly eroding away the data limitation obstacle. In this chapter, we aim to address the second obstacle by presenting a computational pipeline with example scripts that will equip nonspecialists to use OU models to test for signatures of changes in selection pressure on regulatory phenotypes after WGDs. Furthermore, we also discuss approaches to handle cross-species normalization of RNA-seq data and example scripts to perform power simulations.

2 Overview of Analytical Pipeline

The computational pipeline presented in the following section includes five main steps (Fig. 1): (1) estimating ortholog-ohnolog relationships and gene trees, (2) cross-species normalization of genome regulatory phenotype data, (3) specifying and testing of

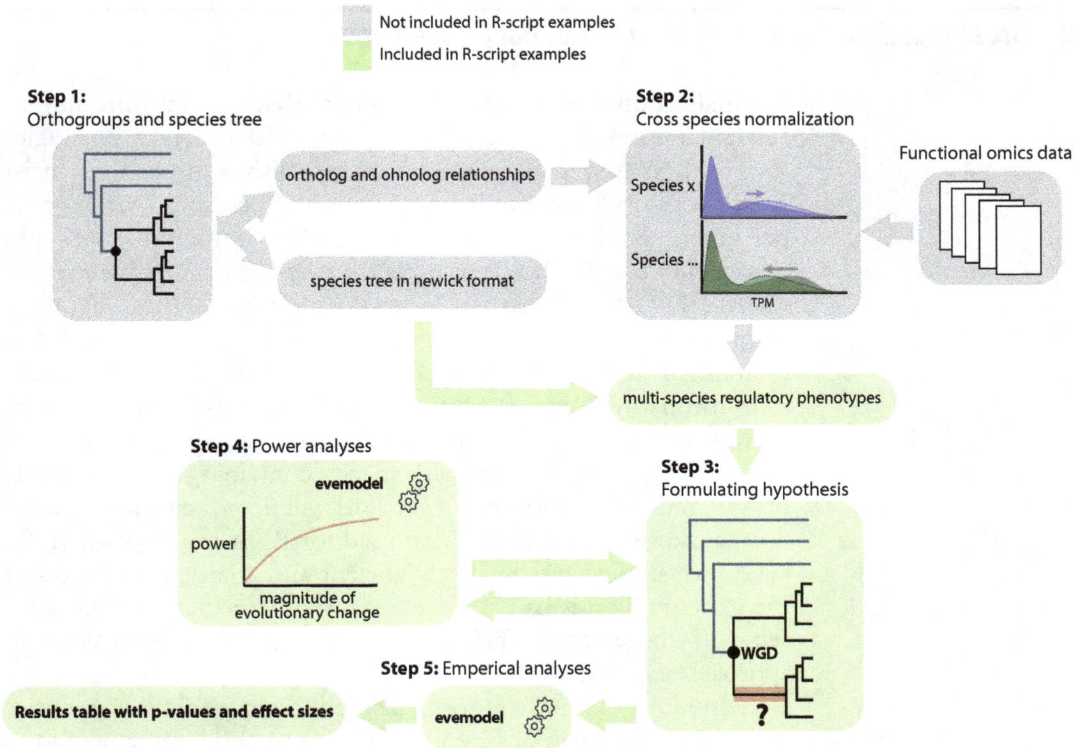

Fig. 1 Analytical pipeline when analyzing the evolution of genome regulatory phenotypes using a phylogenetic OU model

evolutionary hypotheses using *evemodel* in R, (4) exploration of study design and statistical power, and (5) running the OU models and interpreting the results. In this chapter we will provide a brief theoretical background for **steps 2–5**, and for **steps 3–5**. We also make available the R-code. This code will perform each pipeline step using a small simulated dataset of gene expression data from duplicated salmonid genomes and outgroup teleost species.

For **step 1**, however, we will not include a separate section. This is a step that requires running stand-alone ortholog prediction tools (e.g., orthofinder [8] and broccoli [9]) outside R and is not a focus for this chapter. We do however provide an example script (https://gitlab.com/sandve-lab/import-ensembl-gene-trees) which showcases how to generate ortholog-ohnolog relationships directly from ENSEMBL gene trees. This script can act as a convenient starting point when all species in the study have an annotated genome in ENSEMBL. We have also included an example script that can be used to combine the functional omics data and ortholog-ohnolog relationships in a format that is convenient for cross-species normalization in **step 2**.

3 Cross-Species Normalization of Regulatory Phenotypes

Normalization is an essential step in all analyses of high-throughput sequencing-based functional omics data. In the case of single-species RNA-seq data, several well-established procedures exist going from reads via counts to expression values, such as transcripts per million reads (TPM), that are scaled for transcript length and sequencing depth (these methods have been thoroughly reviewed elsewhere [10]). However, normalization procedures allowing comparisons of gene expression levels *across* species are much less established and much more challenging. The added complexity stems primarily from the fact that different species contain different numbers of genes with complex orthology relationships, with an extreme case being datasets with species containing different numbers of WGDs. The following sections will describe a cross-species normalization approach we have used for RNA-seq data when the WGD we studied was relatively ancient and a large proportion of the gene duplicates had returned to a singleton state. Other datasets with more recent WGDs might need different normalization approaches.

A popular method for normalizing differences between samples is the trimmed mean of M values (TMM) method implemented as part of the R package edgeR [11]. Briefly, the TMM method divides read counts by sample-specific scaling factors chosen such that the ratio between the new scaled expression values in one reference sample and the mean values across all other samples (M value), after excluding highly expressed genes (trimmed M values), approach 1.0 for as many genes as possible. Although this method works well within species, there are several challenges with using the TMM method on samples from different species. Firstly, the method requires a one-to-one mapping between genes in different samples, something that is not generally the case when samples come from different species and in particular when these species have undergone WGDs. Secondly, the method assumes that most genes do not significantly change expression across the sample collection (i.e., is not differentially expressed). This assumption is important and can be illustrated with an example: Suppose that we study two related species of which one species has undergone an ancient WGD. Following the WGD, expression levels have reverted to their pre-duplication levels by either pseudogenization of one copy or evolution of lower expression levels for both copies. In this case, the majority of genes in the duplicated species could have seen a 50% downregulation compared to the orthologs in the unduplicated species, and the TMM scaling factors would be close to 0.5 for the samples from the duplicated species. After applying these scaling factors, the downregulation of the duplicated genes would be incorrectly removed while an artificial upregulation would instead be introduced for the single-copy genes.

Faced with this complex picture, we devised a TMM-based method for normalizing data from species with a relatively ancient WGD. This method first normalized expression values within species, using the standard TMM method, and then between species. The between-species normalization calculates scaling factors based on singletons (i.e., ortholog groups containing only one gene from each species) and rely on the assumption that many of these genes will have maintained similar expression levels across species. The procedure includes the following steps:

1. For each species, generate a gene expression matrix with, e.g., TPM values.
2. Within each species, normalize the TPM values across samples (i.e., replicates) using the *calcNormFactors(method = "TMM")* function in edgeR.
3. Identify 1:1 orthologs from the orthogroups and gene trees to use as a baseline for cross-species normalization.
4. Generate a cross-species mean expression matrix containing the mean expression value per species (calculated across all samples from that species). This matrix will have a 1:1 ortholog group as rows, species as columns, and mean expression as values.
5. Calculate normalization factors for each species by applying the TMM method to the cross-species mean expression matrix.
6. Apply the species-specific normalization factors from **step 5** to the expression matrices of each species from **step 1**.

An example code for running this pipeline can be found in our code repository [7].

4 A Biologist's Guide to the EVE Model

To be able to use the EVE model to test evolutionary questions, we will need a basic understanding of the model parameters and how to interpret these in an evolutionary context. The EVE model estimates four parameters from genome regulatory data:

- The evolutionary "optimum" of a phenotype (referred to as *theta*, θ). Shifts in this parameter during evolution can be interpreted as a change in the fitness landscape resulting in selection to express a new optimum phenotype.
- The ratio of population (within species) to evolutionary (between species) variance in regulatory phenotype (referred to as *beta*, β). A gene having a lower beta estimate compared to the genome-wide background level indicates divergent selection between species. Whereas genes with unusually high beta estimates have high expression diversity within species. This can, for

example, be due to environmentally sensitive gene regulation or loss of purifying selection pressure due to redundancy after WGD.

- Quantification of how much a regulatory phenotype is pulled towards the θ value during evolution (referred to as alpha, α). This parameter can be interpreted as the strength of selection pressure.
- Quantification of how fast the phenotype changes during evolution (referred to as sigma, σ^2). This parameter can be interpreted as the strength of neutral drift.

The regulatory phenotype data from several species are modelled as a multivariate normal distribution with means, variances, and covariances parametrized by θ, β, α, σ^2, and the phylogeny.

In Fig. 2, we have visualized the impact of a change in phenotypic optimum (θ) and selection strength (α) on the expected evolutionary changes over time. As you can see, a regulatory phenotype under weak purifying selection ($\alpha = 0.001$) fluctuates more over time than a phenotype under stronger selection ($\alpha = 0.1$) (Fig. 2a), and changes in the fitness landscape that favors evolution of a different θ will result in novel regulatory phenotypes (Fig. 2b).

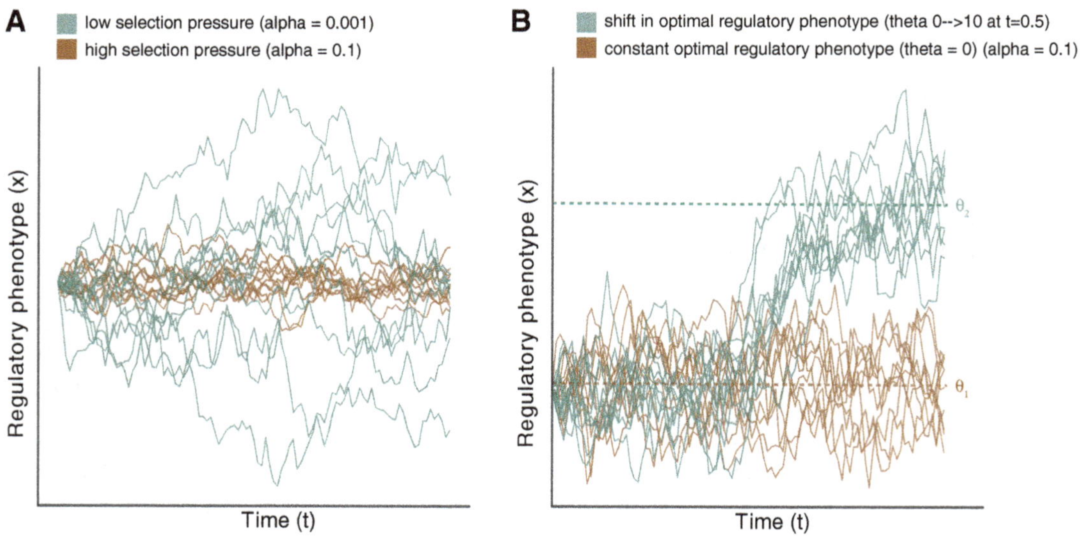

Fig. 2 Simulations of OU processes showing the change in regulatory phenotype over time. The brown lines represent 10 simulated OU processes over 100 time points with parameters: $\theta = 0$, $\alpha = 0.1$, and $\sigma^2 = 1$. The green lines represent 10 simulated OU processes where one parameter has been changed. (**a**) The effect of relaxing the selection pressure from $\alpha = 0.1$ (brown) to $\alpha = 0.001$ (green). (**b**) The effect of a constant optimum expression level of $\theta = 0$ (brown) and changing the optimum expression level halfway through the simulation from $\theta = 0$ to $\theta = 10$ (green)

5 Testing of Evolutionary Hypotheses Using the *evemodel*

Two main types of tests for genome regulatory evolution are implemented in *evemodel* and exemplified: (i) test for shifts in θ at specific times (i.e., gene tree branches) during the evolutionary history of a gene (Fig. 3a) and (ii) test for deviation in beta compared to the genome-wide background level (Fig. 3b). The results from these analyses are in essence ranked lists of genes that likely have experienced selection on regulatory phenotypes.

It is also possible to use the EVE model to compare model parameter estimates for selection pressure and phenotype optimum evolution directly. However, confidence intervals of these parameter estimates are large as long as there are relatively few species in the dataset [12]. Hence, here we focus on the use of the *evemodel* R-package as a toolbox to compare competing evolutionary hypotheses using a likelihood ratio test (LRT) framework.

5.1 Testing for WGD-Associated Theta Shift in Regulatory Phenotype Theta

WGD can result in the adaptive evolution of genome regulation in different ways. One possibility is that increased functional redundancy can release selective constraints and offer the potential for swift adaptive divergence among gene duplicates [13]. However, if WGDs lead to negative fitness effects due to dysregulation of essential cellular processes [7], we will also expect selection for new regulatory mutations that restores fitness following WGD. In both these scenarios, a prediction is that the "optimal" regulatory phenotype (θ) will shift in one or both gene duplicates following WGD.

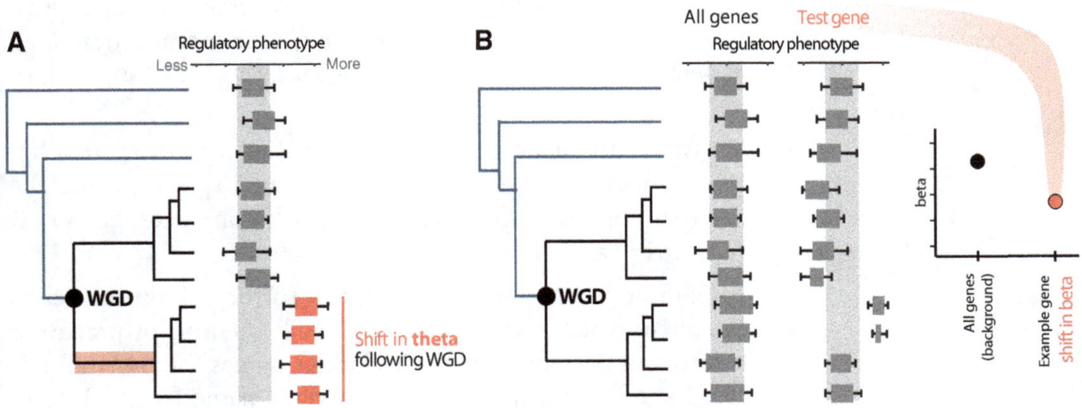

Fig. 3 Two main signatures of selection on regulatory phenotypes. (**a**) Shift in the level of a regulatory phenotype can happen if selection favors changes in total transcript dosage. (**b**) Left panel shows how novel selection pressure on gene regulatory phenotypes following WGD can change the ratio of within-to-between species variance. Right panel shows the expectation for the model parameter beta for a gene ("test gene") that has evolved novel regulatory variance compared to the background of all genes

To test this prediction across a set of genes, we first formulate two competing evolutionary hypotheses specified as two OU models, one with and one without a WGD-associated theta shift. We then fit our data (regulatory phenotypes from extant species) to the competing hypotheses of one shared theta (one theta for the entire gene tree, the gene tree describing the evolutionary relationship between genes in one ortholog group) and two thetas (θ_2 for the duplicated clade and θ_1 for the rest of the tree, Fig. 3a) using the *evemodel*. The result from these tests are as follows:

- One theta fit: a table with parameter estimates for θ, σ^2, α, and β as well as the computed log-likelihood value (LL1), one row for each ortholog group
- Two theta fit: a table with parameter estimates for θ_1, θ_2, σ^2, α, and β as well as the computed log-likelihood value (LL2), one row for each ortholog group
- Final test: a vector with likelihood ratio test (LRT) scores ($2 \cdot (LL2 - LL1)$), one for each ortholog group.

To carry out this test we need to go through the following steps:

1. Build the regulatory phenotype table (`gene.data`): When analyzing WGD-associated shifts, we tested one duplicated clade for a shift in regulatory phenotype compared to the other duplicated clade *and* the non-duplicated species. Hence, we build a combined table with ortholog groups as rows (containing one gene for each unduplicated species and two genes for the duplicated species), samples as columns (several replicates per species, with samples from duplicated species appearing twice, once for each of the duplicated clades), and the regulatory phenotype (e.g., log-transformed expression) as values (normalized using, e.g., the method described earlier). In the case of missing orthologs, the value can be set to "NA."
2. Construct the gene phylogeny (`tree`): Specify the gene phylogeny, with two clades for the duplicated species, using the Analysis of Phylogenetics and Evolution package in R (`library(ape)`).
3. Link the tip names in the phylogeny to the column names in the phenotype table (`colSpecies`): This is done by preparing a vector of tip names (typically species names for the unduplicated species and species names + clade name [e.g., "1" and "2"] for the duplicated species) with length equal to the number of columns in the phenotype table.
4. Specify the evolutionary hypothesis that you want to test (`isTheta2edge`): When in evolutionary history did the mean regulatory phenotype evolve a new optimum? This is

done by preparing a logical vector that has value TRUE for tree tips with a shift (i.e., corresponding to the duplicated clade with θ_2) and FALSE for the rest of the tree (with θ_1).

5. Run *evemodel*: `evemodel::twoThetaTest(tree, gene.data, colSpecies, isTheta2edge)`. The results are described above.
6. Compute p-values: The LRT scores are (in the asymptotic case, i.e., with very many species) chi-square distribution with one degree of freedom and p-values can thus be computed as `pchisq(LRT, df = 1, lower.tail = F)`. However, in practice, with relatively small phylogenies, we recommend obtaining empirical significance thresholds using simulations.

To illustrate the steps of the theta shift test, we simulated gene expression data using a phylogeny with six salmonid species with a recent WGD and seven unduplicated outgroup teleost species (Fig. 4a). Five genes were simulated to have experienced a shift in expression in one of the duplicates following the WGD (alternative hypothesis: $\theta_2 > \theta_1, \theta_2 = 60\text{–}100$ and $\theta_1 = 50$) while the remaining genes (985 genes) experienced no shift ($\theta_1 = \theta_2 = 50$) (Fig. 4b). Intuitively, θ_2 can be thought of as the new optimal expression value. The time it takes for expression levels to approach θ_2 depends on the level of drift (σ^2) and the strength of pull towards the optimum (α). Notice, however, that the difference between θ_1 and θ_2 also affects the strength of pull towards θ_2; the bigger the difference, the stronger the pull. As can be seen in Fig. 4b, not

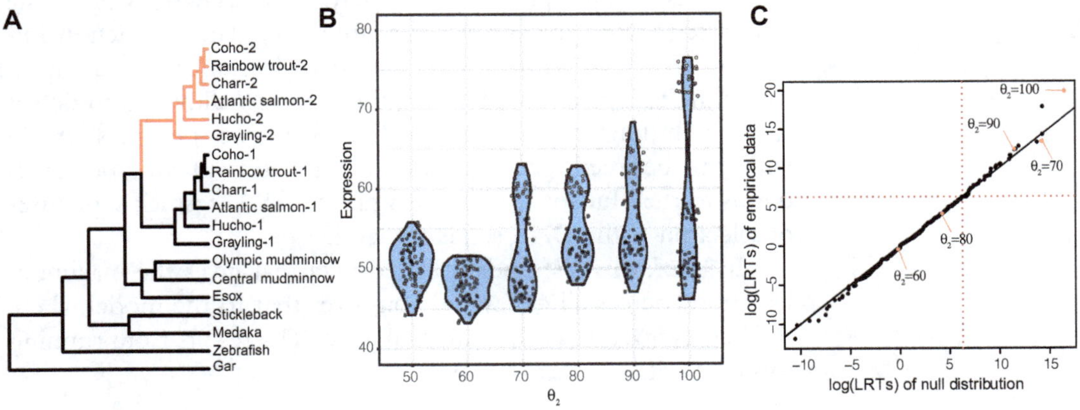

Fig. 4 The theta shift test. (**a**) The teleost phylogeny with one of the duplicated salmonid clades is highlighted in red. (**b**) Gene expression data was simulated using parameters: $\theta_1 = 50$, $\alpha = 0.005$, $\sigma^2 = 0.1$, and $\beta_{\text{shared}} = 0.1$. The violin plots show the expression values of six genes (four replicate per species) with a theta shift in the highlighted clade in A increasing from $\theta_2 = 50$ (i.e., no shift) to $\theta_2 = 100$ (i.e., doubling). The colors of the dots match the colors in the phylogeny in A. (**c**) A Q-Q plot comparing the LRTs of the dataset with five shifted genes (with the five shifted genes corresponding to the red dots) to the LRTs of an equal number of genes simulated under the null model ($\theta = 50$). The dotted red lines mark the empirically determined significance threshold to attain a false-positive rate of 0.05

enough time has passed in this simulation for the genes to reach their optimum values (i.e., θ_2): a θ_2 of 100 resulted in expression values of around 75 (increase of around 25), θs in the range pf 70–90 resulted in values around 60, and a θ_2 of 60 resulted in no shift. Running the *evemodel* theta shift test demonstrates the power to detect three of the four genes with the largest shifts in expression, with an empirically determined false-positive rate below 0.05 (p-value < 0.05) (Fig. 4c). While the gene with the clearest shift ($\theta_2 = 100$) also has the highest observed LTR score among all genes, we would have to accept five false positives to detect the two other genes, indicating that a shift of around 20% (shift from 50 to 60) is at the edge of what can be detected in this phylogeny with the set of parameters used in this simulation. The EVE model detection power will be treated in more detail later in this chapter. Since power is sensitive to various model parameters, we recommend that researchers use the provided power analysis script to perform their own customized power analysis.

Data and code to run the test, including shifts on different branches in the phylogeny, can be found in our git repository (https://gitlab.com/sandve-lab/evemodel/tutorials).

5.2 Beta Shift Following WGD

If selection pressure on a gene regulatory phenotype changes following WGD, this could result in the evolution of gene regulatory diversity (i.e., variance between species). If a gene becomes functionally redundant and starts to evolve under relaxed purifying selection pressure, cis-regulatory mutations will accumulate at a higher rate, leading to (most likely) higher diversity of gene expression levels within species. Conversely, in a scenario where one duplicated gene evolves a novel adaptive regulatory function following WGD, we expect a reduction of intraspecific regulatory variation. The beta shared test in the *evemodel* allows us to detect such evolutionary events by testing for gene-specific shifts in within-to-between species variance in regulatory phenotypes, analogous to the Hudson-Kreitman-Aguadé (HKA) test for signatures of selection at the DNA sequence level.

In brief, this test finds a shared beta that gives the maximum likelihood across all genes and compares that to the model where the beta is fitted to each individual gene. The results from running this on a data set are as follows:

- *Shared beta fit*: a table with parameter estimates for θ, σ^2, α, and β as well as the computed log-likelihood value (LL1), one row for each ortholog group
- *Individual beta fit*: a table with parameter estimates for θ, σ^2, α, and β as well as the computed log-likelihood value (LL2), one row for each ortholog group

- *Final test*: a vector with likelihood ratio test (LRT) scores ($2 \cdot (LL2 - LL1)$), one for each ortholog group

To carry out this test we need to go through steps very similar to those for the theta shift test (described in detail earlier). Briefly:

1. Prepare the regulatory phenotype table (`gene.data`).
2. Prepare the phylogeny (`tree`).
3. Link the tip names in the phylogeny to the column names in the phenotype table (`colSpecies`).
4. Run evemodel: `evemodel::betaSharedTest(tree, gene.data, colSpecies)`.
5. Compute *p*-values.

To illustrate the steps of the beta shift test, we again simulated gene expression data using the phylogeny in Fig. 4a. Five genes were simulated to have experienced an alternative beta (alternative hypothesis: $\beta_{\text{shared}} \neq \beta_{\text{alt}}$, $\beta_{\text{alt}} = 0.01$, 0.5, 2, 10, and 100 and $\beta_{\text{shared}} = 0.1$) while the remaining genes (985 genes) were simulated using the shared beta. Running the evemodel beta shared test demonstrates the power to detect genes with the largest deviation from the beta shared, with an empirically determined false-positive rate below 0.05 ($p < 0.05$) (Fig. 5). Data and code to run the test can be found in our git repository (https://gitlab.com/sandve-lab/evemodel/tutorials).

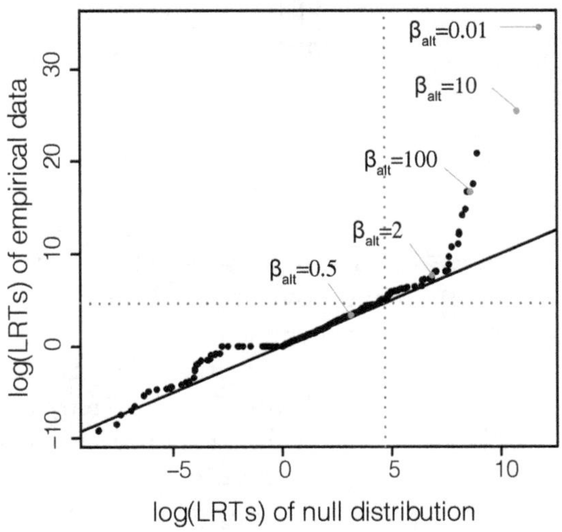

Fig. 5 The beta shift test. Expression data were simulated using parameters: $\theta = 50$, $\alpha = 0.005$, $\sigma^2 = 0.1$, and $\beta_{\text{shared}} = 0.1$. Five genes were simulated under an alternative beta: $\beta_{\text{alt}} = 0.01, 0.5, 2, 10, 100$. A Q-Q plot comparing the LRTs of the dataset where five genes have an alternative beta (with the five genes corresponding to the red dots) to the LRTs of an equal number of genes simulated under the null model ($\beta_{\text{alt}} = \beta_{\text{shared}}$). The dotted red lines mark the empirically determined false-positive rate cutoff of 0.05 (i.e., $p < 0.05$)

6 Power Analyses for Shift in Expression Variance or Level

Interpretation of *evemodel* test results is most clear when informed by the power of a specific analysis. Under the null distribution, likelihood ratio test (LRT) statistics follow a chi-square (1) distribution in the asymptotic case, i.e., when there are large numbers of species with informatively structured phylogenies and many samples per species. However, most practical analyses fall short of the expansive data required for asymptotic performance. In these realistic cases, the null distribution will take some other shape, depending on both the amount of data and the evolutionary parameters. If empirical likelihood ratio test statistics are interpreted assuming a chi-square (1) null distribution to establish significance (i.e., to compute p-values), the analysis may have an uncontrolled false-positive rate or depressed power. However, it should be noted that even for such an analysis, examining outliers compared to the genomic distribution of LRTs will still draw attention to empirically extreme gene expression levels.

A solution to the challenge outlined above is to use simulations to estimate the null and alternative LRT distributions for comparative expression data parallel to the empirical dataset. Such power analyses can be performed as part of experimental design to ensure the development of a dataset that will have high power to detect expression adaptation. Even if the data is already established, power analyses can inform the interpretation of results to avoid uncontrolled false-positive rates, as well as to take negative results from an underpowered study with a grain of salt.

A power analysis requires three major steps: (1) simulate expression data under both the null distribution and alternative distributions, (2) test evolutionary hypotheses of interest on the simulated data, and (3) analyze the resulting LRT distributions. In order to simulate expression data, EVE parameter values (i.e., θ, σ^2, α, and β) must be established. The parameter values chosen should reflect the empirical data. For power analyses preceding data collection, parameter values should result in an informative expression covariance matrix. That is, where the expression covariances are aligned with the phylogeny in such a way that the relationship is stronger for pairs of species that are closely related than for species that are more distantly related. For example, for a single theta model, if a disproportionately high alpha value is used, then expression levels may not vary at all. Once expression data has been generated, EVE can be used to find maximum likelihood (ML) parameter estimates for the empirical data. Those ML parameter estimates can then be used to simulate realistic data for power analysis.

Once the data are simulated, EVE can be used to test the evolutionary hypotheses of interest. Results for data simulated under the null hypothesis will establish the null distribution, while the results for data simulated under the alternative distribution can be used to investigate power. The null distribution can be used to gauge the significance of an LRT value. For example, the LRT value at the 95th percentile of the null distribution represents the threshold for significance for $\alpha = 0.05$. By comparing that threshold to the alternative LRT distribution, power is established. For example, if 70% of the alternative LRT distribution is greater than the threshold, then the power is 0.70. Simulating expression data under increasingly extreme deviation from the null distribution will allow researchers to estimate the power curve of their analysis.

The code for the power analyses described below can be found here: https://gitlab.com/sandve-lab/evemodel/tutorials.

6.1 Shift in Expression Level (Theta)

The following example demonstrates how to conduct a power analysis for the twoThetaTest on the previously described phylogeny with 13 species and four replicates per species. We tested two major shift points along the phylogeny (Figs. 6a and 7a). Each shift point represents the separation of the clade following the point

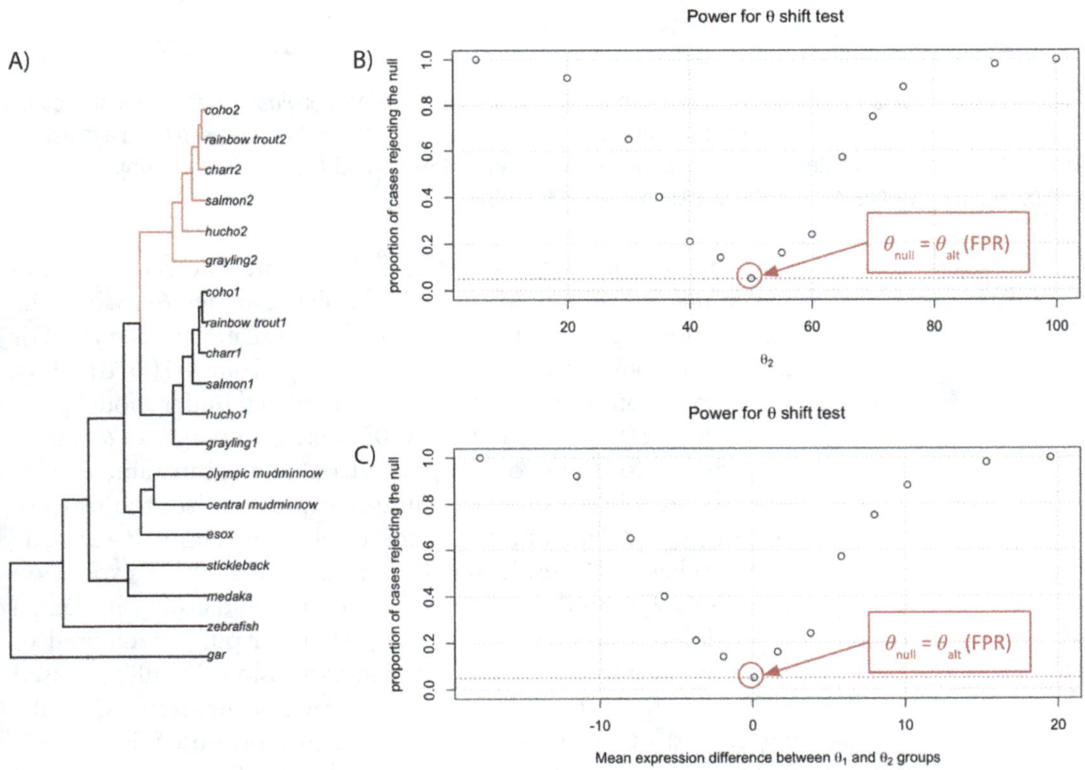

Fig. 6 Power analysis for theta shift test—salmonid clade. (**a**) The teleost phylogeny with the tested clade highlighted in red. (**b**) Power (the proportion of cases correctly rejecting the null hypothesis) plotted against θ_2. The dotted red line indicates an empirical false-positive rate (FPR) of 0.05 (i.e., p-value <0.05). (**c**) Power plotted against average observed expression change from θ_1

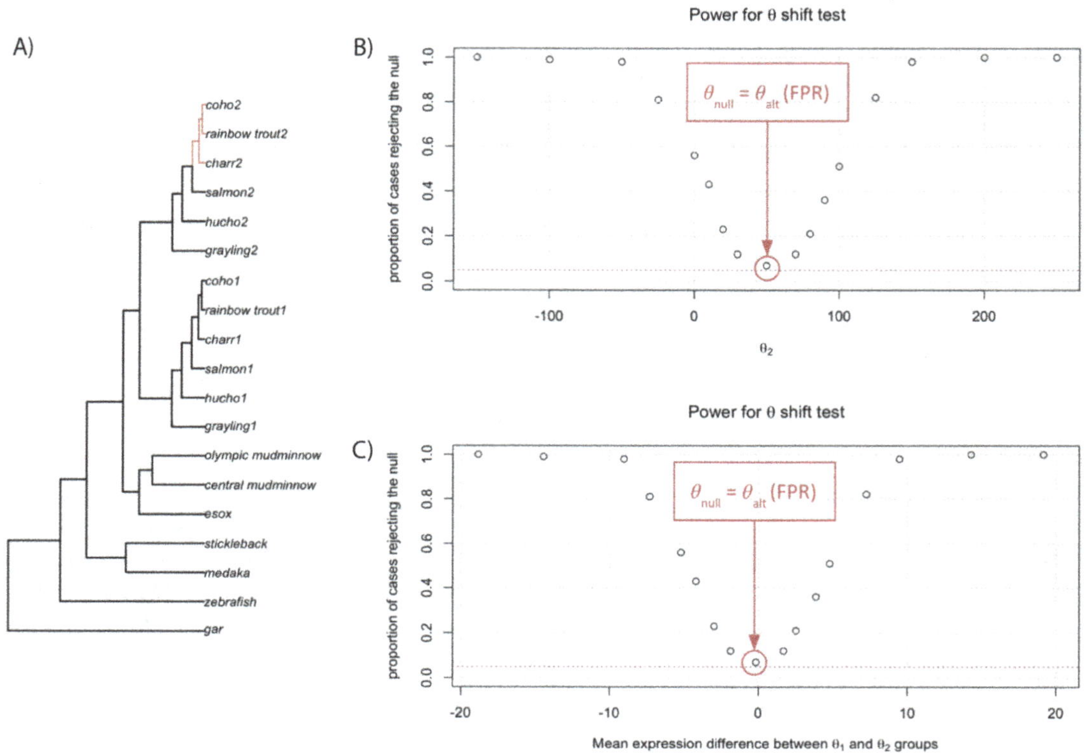

Fig. 7 Power analysis for theta shift test—salmonid subclade. (**a**) The teleost phylogeny with the tested clade highlighted in red. (**b**) Power (the proportion of cases correctly rejecting the null hypothesis) plotted against θ_2. The dotted red line indicates an empirical false-positive rate (FPR) of 0.05 (i.e., *p*-value <0.05). (**c**) Power plotted against average observed expression change from θ_1

from the set of species preceding the point. The set of species preceding the shift point will be simulated under θ_1, while the clade following the shift point will be simulated under θ_2. For each shift point and for each θ_2 (varying from -100 to 250), gene expression data for 100 genes is simulated under model parameters $\theta_1 = 50$, $\sigma^2 = 2$, and $\alpha = 0.05$. Different ranges of θ_2 values were chosen for the two shift points to obtain comparable magnitudes for the observed average shifts in expression (see the *x*-axis of Fig. 6b, c and Fig. 7b, c). As the duplicated salmonid subclade (Fig. 7a) has had considerably shorter time to shift the expression from θ_1 to θ_2 than the entire duplicated salmonid clade (Fig. 6a), a larger difference between θ_1 and θ_2 (stronger pull) is required to obtain comparable observed shifts in expression. Finally, we used EVE to perform the thetaTwoTest on each of the datasets, resulting in a list of LRTs for each one. The proportion of these LRTs that are significant ($p < 0.05$) per θ_2 and per shift point is calculated as the power.

Two main observations can be made from this power analysis. Firstly, for the given phylogeny and parameters, a shift up or down of approximately 20% (absolute shift of around 10 from $\theta_1 = 50$)

can be detected with reasonable power (power > 0.8) (Figs. 6c and 7c). Secondly, the power to detect more evolutionarily recent shifts is higher. This indicates that a given shift in expression is more detectable when happening on a shorter time frame.

6.2 Shift in Expression Variance (Beta)

The following example demonstrates how to conduct a power analysis for the betaSharedTest. Gene expression data for 100 genes is simulated under model parameters $\theta = 50$, $\sigma^2 = 2$, and $\alpha = 0.05$, with 99 of the genes under the null distribution parameter $\beta_{\text{shared}} = 1$ and the last gene under varying alternative betas ($\beta_{\text{alt}} = 0.01, 0.05, 0.1, 0.3, 0.5, 1, 5, 10, 25, 50, 75, 100$). One hundred datasets were simulated in this way for each alternative beta. Finally, we use EVE to perform the betaSharedTest on each of the datasets, resulting in a list of LRTs for each one. The LRT associated with the last gene under the alternative distribution is extracted, and the proportion of these LRTs that are significant ($p < 0.05$) per alternative beta is calculated as the power.

For this power analysis, we see an almost perfect power for the tail ends of the alternative beta range, in particular when $\beta = 0.01$ and $\beta \geq 5$ (Fig. 8). We notice a dip in power as beta approaches the null distribution $\beta_{\text{shared}} = 1$. This is expected and suggests greater power of detection for larger deviations away from the null distribution and thus greater ability to identify candidates for divergence.

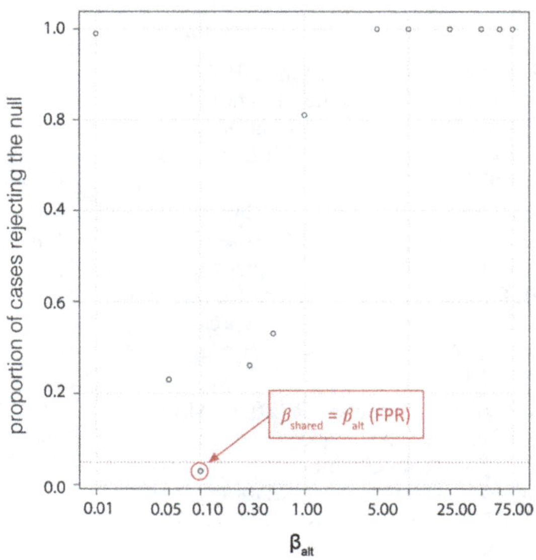

Fig. 8 Power analysis for beta shared test. Power (the proportion of cases correctly rejecting the null hypothesis) plotted against β_{alt}. The dotted red line indicates an empirical false-positive rate (FPR) of 0.05 (i.e., p-value <0.05)

7 Concluding Remarks

Sound analysis of regulatory phenotype data across species requires formal statistical frameworks that take phylogenetic relationships into account. The complexity of such frameworks may seem daunting to many researchers, which could explain why simple pairwise analysis still dominates the comparative genomics literature. Here we provide step-by-step workflows and RMarkdown scripts to help researchers conduct cross-species analysis of regulatory phenotype data using the R package *evemodel*. We provide the practical details of how to run such an analysis, describe the parameters of the framework in terms of evolutionary processes, and explain how to interpret the output in an accessible nonspecialized language. Although we use regulatory evolution after whole-genome duplication as an example, this chapter should be general enough to help researchers run other cross-species analysis as well.

Acknowledgements

The study was funded by the Norwegian Research Council through the projects The role of whole genome duplication in vertebrate adaptation (project number 274669) and Transposable elements as agents of genome evolution and adaptation following a recent whole genome duplication (project number 275310).

References

1. Sandve SR, Rohlfs RV, Hvidsten TR (2018) Subfunctionalization versus neofunctionalization after whole-genome duplication. Nat Genet 50:908–909. https://doi.org/10.1038/s41588-018-0162-4
2. Chen J, Swofford R, Johnson J, Cummings BB, Rogel N, Lindblad-Toh K, Haerty W, Palma F di, Regev A (2019) A quantitative framework for characterizing the evolutionary history of mammalian gene expression. Genome Res 29:53–63. https://doi.org/10.1101/gr.237636.118
3. Rohlfs RV, Harrigan P, Nielsen R (2014) Modeling gene expression evolution with an extended Ornstein-Uhlenbeck process accounting for within-species variation. Mol Biol Evol 31:201–211. https://doi.org/10.1093/molbev/mst190
4. Cooper N, Thomas GH, Venditti C, Meade A, Freckleton RP (2016) A cautionary note on the use of Ornstein Uhlenbeck models in macroevolutionary studies. Biol J Linn Soc Lond 118:64–77. https://doi.org/10.1111/bij.12701
5. Catalán A, Briscoe AD, Höhna S (2019) Drift and directional selection are the evolutionary forces driving gene expression divergence in eye and brain tissue of heliconius butterflies. Genetics 213:581–594. https://doi.org/10.1534/genetics.119.302493
6. Fukushima K, Pollock DD (2020) Amalgamated cross-species transcriptomes reveal organ-specific propensity in gene expression evolution. Nat Commun 11:4459. https://doi.org/10.1038/s41467-020-18090-8
7. Gillard GB, Grønvold L, Røsæg LL, Holen MM, Monsen Ø, Koop BF, Rondeau EB, Gundappa MK, Mendoza J, Macqueen DJ, Rohlfs RV, Sandve SR, Hvidsten TR (2021) Comparative regulomics supports pervasive selection on gene dosage following whole genome duplication. Genome Biol 22:103. https://doi.org/10.1186/s13059-021-02323-0
8. Emms DM, Kelly S (2019) OrthoFinder: phylogenetic orthology inference for comparative genomics. Genome Biol 20:238 https://doi.org/10.1186/s13059-019-1832-y

9. Derelle R, Philippe H, Colbourne JK (2020) Broccoli: CombiningPhylogenetic and Network Analyses for Orthology Assignment, Molecular Biology and Evolution, 37(11): 3389–3396. https://doi.org/10.1093/molbev/msaa159
10. Dillies M-A, Rau A, Aubert J, Hennequet-Antier C, Jeanmougin M, Servant N, Keime C, Marot G, Castel D, Estelle J, Guernec G, Jagla B, Jouneau L, Laloë D, Le Gall C, Schaëffer B, Le Crom S, Guedj M, Jaffrézic F, French StatOmique Consortium (2013) A comprehensive evaluation of normalization methods for Illumina high-throughput RNA sequencing data analysis. Brief Bioinform 14:671–683. https://doi.org/10.1093/bib/bbs046
11. Robinson MD, McCarthy DJ, Smyth GK (2010) edgeR: a Bioconductor package for differential expression analysis of digital gene expression data. Bioinformatics 26:139–140. https://doi.org/10.1093/bioinformatics/btp616
12. Rohlfs RV, Nielsen R (2015) Phylogenetic ANOVA: the expression variance and evolution model for quantitative trait evolution. Syst Biol 64:695–708. https://doi.org/10.1093/sysbio/syv042
13. Van de Peer Y, Mizrachi E, Marchal K (2017) The evolutionary significance of polyploidy. Nat Rev Genet 18:411–424. https://doi.org/10.1038/nrg.2017.26

Chapter 12

Beyond Transcript Concentrations: Quantifying Polyploid Expression Responses per Biomass, per Genome, and per Cell with RNA-Seq

Jeremy E. Coate

Abstract

RNA-seq has been used extensively to study expression responses to polyploidy. Most current methods for normalizing RNA-seq data yield estimates of transcript concentrations (transcripts per transcriptome). The implicit assumption of these normalization methods is that transcriptome size is equivalent between the samples being compared such that transcript concentrations are equivalent to transcripts per cell. In recent years, however, evidence has mounted that transcriptome size can vary dramatically in response to a range of factors including polyploidy and that such variation is ubiquitous. Where such variation exists, transcript concentration is often a poor or even misleading proxy for expression responses at other biologically relevant scales (e.g., expression per cell). Thus, it is important that transcriptomic studies of polyploids move beyond simply comparing transcript concentrations if we are to gain a complete understanding of how genome multiplication affects gene expression. I discuss this issue in more detail and summarize a suite of approaches that can leverage RNA-seq to quantify expression responses per genome, per cell, and per unit of biomass.

Key words Polyploidy, RNA-seq, Gene expression, Transcriptome size, Transcript concentration, Transcripts per cell, Transcripts per genome, Transcripts per biomass

1 Introduction

A major focus of polyploid research has been to quantify the effects of genome multiplication on gene expression. Early studies of gene expression, by necessity, examined only up to a handful of genes at a time using low-throughput techniques like Northern blots [1], semiquantitative RT-PCR (e.g., [2]), and quantitative RT-PCR (e.g., [3]). In the 1990s and early 2000s, higher-throughput methods, including microarrays, came online that made it possible to assay thousands of genes at a time. These methods were applied widely to investigate polyploid expression phenotypes (e.g., [4–6]),

Yves Van de Peer (ed.), *Polyploidy: Methods and Protocols*,
Methods in Molecular Biology, vol. 2545, https://doi.org/10.1007/978-1-0716-2561-3_12,

presenting the first whole genome-level view of polyploid transcription (transcriptomics).

In the late 2000s, next-generation sequencing enabled a fundamental advance in high-throughput transcriptomic studies, namely, RNA-sequencing or RNA-seq [7]. In comparison to microarrays, RNA-seq offers more quantitative and more complete coverage of the transcriptome, requires no prior knowledge of the underlying genome or gene space, and additionally gives sequence information. This makes RNA-seq a powerful tool, not only for analysis of gene expression, but also for phylogenetics, annotating genomes, and characterizing alternative splicing, to name a few. Consequently, over the last decade RNA-seq has largely supplanted microarrays in transcriptomic studies, including studies of polyploid transcriptomics (Fig. 1a).

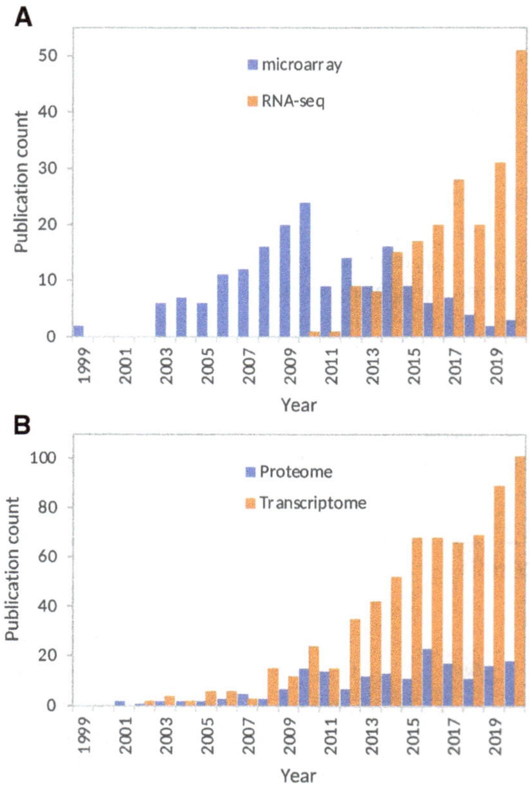

Fig. 1 Recent trends in transcriptomic and proteomic studies of polyploidy. (**a**) Counts of microarray-based vs. RNA-seq based transcriptomic studies of polyploids since 1999. Specifically, counts of PubMed hits (Publication count) for the search terms "microarray" + "expression" + "polyploid" (designated "microarray") and for the terms "RNA-seq" + "expression" + "polyploid" (designated "RNA-seq"). (**b**) Counts of PubMed hits (Publication count) when searching for the terms "proteome" + "polyploid" ("Proteome") or "transcriptome" + "polyploid" ("Transcriptome"). Searches conducted on 7/23/2021

Not long after microarrays were first utilized, high-throughput proteomic approaches were also first applied to polyploids [8], and a handful of proteomic studies of polyploids have subsequently been published [9–13]. Given that proteins are the functional products of gene expression, these studies provide a more direct readout of functional expression responses. However, proteomic experiments are more technically challenging, expensive, and less quantitative and generally survey a smaller fraction of gene space than RNA-seq [13, 14]. It is likely that proteomic studies will increase in frequency as these challenges are overcome, but as of now, with the ease and cost-effectiveness of RNA-seq, transcriptomic studies predominate (Fig. 1b) and are likely to continue to do so for some time.

Other "omic" technologies are also likely to see increased application in polyploid expression studies. Translatomics (polysome profiling, ribosome profiling, ribo-seq) [15, 16] lies between transcriptomics and proteomics in terms of what it measures (the abundance of RNA being translated) and, arguably, in technical difficulty. Either gradient centrifugation (polysome profiling) or nuclease-digestion of RNA not protected by ribosomes (ribo-seq) is utilized to enrich for RNA associated with ribosomes, followed by high-throughput sequencing to profile the translatome or translated fraction of the RNA. These approaches provide information on the nature and extent of translational regulation, and, therefore, yield information that is complementary to transcriptomic and proteomic approaches. Additionally, because they measure RNA, they avoid many of the technical challenges of proteomics. Translatomic methods may also see increased application in polyploid research, but to date only one study has been published [17]. Similarly, "multiomic" approaches, which enable the simultaneous assessment of transcript abundance and other data modalities, such as DNA sequence (e.g., [18]) or chromatin accessibility (ATAC-seq) [19], have not yet been used in polyploid studies, but may also see increasing application in the near future.

All of the above approaches have mostly been used to characterize bulk tissue (e.g., leaves, roots, flowers). These tissues comprise multiple diverse cell types, each with their own developmental trajectories and expression profiles. Consequently, expression estimates from bulk tissue represent average expression across diverse cells and lack the resolution to decipher biological responses that take place at the level of cell type, cell state, or even within individual cells. One of the newest frontiers in genomic technology is single-cell and single-nucleus sequencing, which enables transcriptomic and even multiomic characterization of single cells. These approaches enable comprehensive profiling of the transcriptomes (and other "omes") of thousands of cells, potentially characterizing every cell in an organ, tissue, or whole organism. Nature Methods, named "single-cell sequencing" method of the year in 2013 and "single-cell multimodal omics" method of the year in 2019 [20, 21], and single-cell technologies will undoubtedly be a major player in the future of polyploid research. At present, however, these techniques are still in their relative infancy, and only a couple studies have used single-cell approaches to explicitly examine the effects of polyploidy on gene expression [22, 23].

Consequently, though new and developing techniques are likely to expand our ability to quantify gene expression responses, bulk RNA-seq remains the major workhorse of polyploid expression studies. I outline below a general pipeline for generating and analyzing bulk RNA-seq experiments to quantify transcriptional responses to WGD. Central to this pipeline is a strategy to extend the capacity of RNA-seq from only estimating relative transcript concentrations, to also yielding estimates of transcripts per genome copy, per cell, or per biomass.

RNA-seq experiments, like all gene expression experiments, require normalization to correct for *technical* inter-sample differences such as amounts of input RNA and depth of sequencing [7, 24]. These methods for technical normalization generally operate on the assumption that global expression levels (total transcription, transcriptome size) are comparable between samples, and it is now known that this assumption is frequently violated (e.g., [3, 25–28]). The need for normalization methods that account for *biological* differences between samples (specifically, differences in total transcript abundance per cell) has until recently been largely overlooked [26, 29]. In the last few years there has been a growing documentation of, and appreciation for, this level of biological variation and its implications for transcriptomic studies, though normalizing for this variation has not yet become de rigueur for RNA-seq experiments. I argue that it should be and outline strategies to do so.

RNA-seq is highly quantitative because it provides a digital readout of expression level—specifically, the number of sequencing reads recovered from each gene. The more transcripts produced by a gene, the more templates for sequencing, so that the number of reads recovered correlates with expression level. However, though this is true in a general sense, it was recognized from the beginning that there are many confounding technical factors that cause this correlation to be less than perfect [7]. Among these are intra-sample factors, such as gene length, and inter-sample factors, foremost of which is the depth of sequencing [24]. Consequently, some form of data normalization is required to correct for such confounding effects. Mortazavi et al. [7] proposed normalizing raw RNA-seq data to account for differences in both gene length and sequencing depth by calculating reads per kilobase of exon per million mapped reads (RPKM), and this and related approaches (e.g., transcripts per million; TPM) remain widely used today.

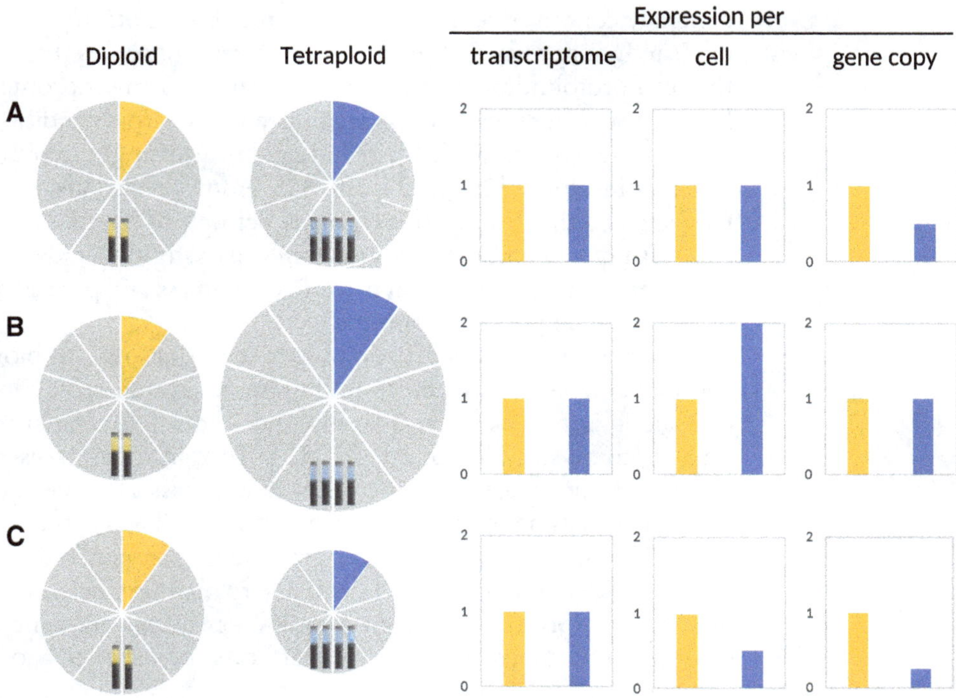

Fig. 2 Effect of transcriptome size on measures of gene expression between a diploid and a tetraploid. Gray circles indicate transcriptome size (number of transcripts per cell) in diploid and tetraploid cells. Colored wedges represent the fraction of the total transcriptome produced by a gene of interest (GOI). Black bars with colored segments represent copy number per cell for the genome and GOI, respectively. Bar plots to the right indicate relative expression per transcriptome, per cell, or per gene copy for the GOI. (**a**) Transcriptome size and the proportion of the transcriptome contributed by the GOI are equal in the diploid and the tetraploid. Consequently, the diploid and the tetraploid have equal expression per transcriptome and per cell, but the tetraploid has half the expression per gene copy. (**b**) Transcriptome size is twice as large in the tetraploid as in the diploid, and GOI expression scales with transcriptome size. Consequently, the diploid and the tetraploid have equal expression per transcriptome and per genome copy, but the tetraploid has higher expression per cell. (**c**) Transcriptome size is smaller in the tetraploid than in the diploid, and GOI expression scales with transcriptome size. Consequently, the diploid and the tetraploid have equal expression per transcriptome, but tetraploid expression is lower per cell and per genome copy. (Figure adapted from [14])

For each individual gene, this normalization approach yields, in effect, that gene's fraction of the total transcriptome, or what has been referred to as transcript concentration [27] or transcriptome-normalized expression [3] (Fig. 2). As a result, when comparing among samples, differences in normalized expression estimates (e.g., TPM) reflect differences in transcript concentration. As has now been pointed out by several research groups, transcript concentration does not necessarily reflect differences in absolute transcript abundance per cell because global expression level (overall transcriptome size) can vary substantially between samples (Fig. 2).

Consequently, though all RNA-seq studies employ data normalization at the front end of data analysis [24], normalization is generally conceived as a process to correct for *technical* biases and tends to ignore a key *biological* feature of transcriptomes (total size) that can profoundly affect data interpretation. Transcriptome size has now been documented to vary substantially across conditions in a wide variety of studies [26], including, unsurprisingly, in response to shifts in ploidy [3, 22, 27, 30–32]. Additionally, Visger et al. [27] showed that expression responses per unit of biomass can also vary with ploidy and that *Tolmiea menziesii* autopolyploids exhibit significant shifts in expression per unit of biomass and per cell more frequently than per transcriptome.

To quantify shifts in expression per cell or per unit of biomass and move beyond only detecting differences in transcript concentration, it is necessary to quantify and control for variation in transcriptome size [3, 26, 28, 29, 31]. This, in turn, necessitates methods to normalize for differences in the mass of tissue and/or number of cells per sample from which the RNA was extracted. In the following section, I outline procedures for generating and analyzing RNA-seq data from bulk tissue to quantify these different metrics of expression between samples, specifically designed for cross-ploidy comparisons, but that can be applied to any comparison.

2 Overview of Procedures and Potential Branch Points

Figure 3 summarizes the major steps in standard, transcriptome-normalized RNA-seq experiments (blue), as well as the additional or alternative steps necessary to obtain estimates of gene expression per unit of biomass (green) or per cell (two alternative strategies in light and dark orange). The basic transcriptome-normalized RNA--seq pipeline involves extraction of RNA from bulk tissue, library construction, sequencing, mapping and counting reads by feature (e.g., gene or exon), normalizing raw counts to account for differences in sequencing depth, and differential expression (DE) analysis. To quantify expression per biomass or per cell, additional steps are layered onto this basic pipeline (Fig. 3), including estimation of the biomass and/or number of cells used as input, adding exogenous controls (spike-ins) to the sample RNA, and using read counts from these spike-ins to normalize gene expression to the relevant denominator. The following sections will discuss each of these steps in more detail, highlighting key considerations and decision points and offering recommendations for best practices. For reference, Table 1 summarizes the key parameters that need to be measured and/or calculated and illustrates how these values are determined using hypothetical examples for a diploid and tetraploid.

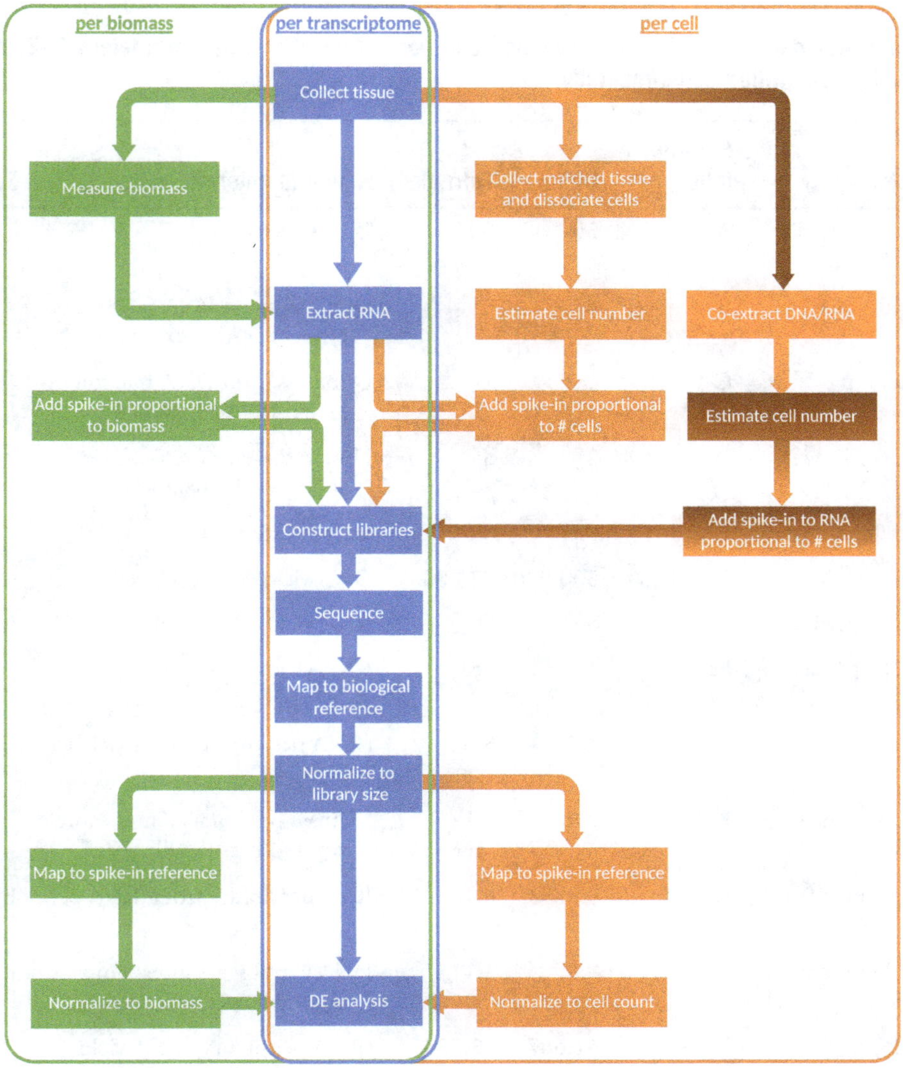

Fig. 3 Flowchart of the major steps in transcriptome-normalized RNA-seq (blue) and the additional/alternative steps involved in biomass-normalized RNA-seq (green) and cell-normalized RNA-seq (orange)

3 Detailed Procedures and Recommendations

3.1 Tissue Collection

For standard, transcriptome-normalized expression studies, no special considerations are required for the amount of tissue to use as input beyond ensuring that enough is collected to yield sufficient RNA for library construction but not so much that it exceeds the capacity of the extraction method. However, to enable normalization to biomass [27] (green boxes in Fig. 3), it is necessary to collect defined (ideally equal) amounts of tissue from diploids and polyploids. This can be achieved by weighing equal amounts of tissue or, for planar tissues like leaves, collecting equivalent surface

Table 1
Definition of key parameters for biomass- and cell-normalized RNA-seq. Parameters A–E are user-determined or quantified experimentally

Parameter	Parameter label	Diploid	Tetraploid	How it is determined	Formula
Biomass per sample (mg)	A	50	50	User selected	
RNA yield (μg)	B	6	5	Qubit assay on RNA fraction of DNA/RNA co-extraction	
DNA yield (μg)	C	1	1	Qubit assay on DNA fraction of DNA/RNA co-extraction	
DNA per cell (pg)	D	2	3	Flow cytometry (genome size × endopolyploidy index)	
RNA per library (μg)	E	0.5	0.5	User selected	
Cells per sample[a]	F	500,000	333,333	DNA yield/DNA per cell	C/D
Total RNA per DNA	G	6	5	RNA yield/DNA yield	B/C
Total RNA per cell (pg)	H	12	15	RNA yield/cells per sample	B/F
Biomass per library (mg)	I	4.2	5.0	(RNA per library/RNA yield) × biomass	(E/B) × A
Relative biomass per library	J	0.9	1.1	Biomass per library/inter-sample avg. biomass per library	I/I_{avg}
DNA per library (μg)	K	0.08	0.10	RNA per library/total RNA per DNA	E/G
Relative DNA per library	L	0.89	1.11	DNA per library/inter-sample avg. DNA per library	K/K_{avg}
Cells per library	M	41,667	33,333	(RNA per library/RNA yield) × cells per sample	(E/B) × F
Relative cells per library	N	1.1	0.9	Cells per library/inter-sample avg. cells per library	M/M_{avg}
ERCC spike-in volume per biomass (μL)	O	0.9	1.1	Relative biomass per library × 1 μL	J
ERCC spike-in volume per DNA (μL)	P	0.89	1.11	Relative DNA per library × 1 μL	L
ERCC spike-in volume per cell (μL)	Q	1.1	0.9	Relative cells per library × 1 μL	N

Parameters F–N are calculated based on other parameters. Parameters O–Q are the amounts of ERCC spike-in to add per sample for the specified normalization strategy. To illustrate the calculations, hypothetical values are given for parameters A–E

[a]Cells per sample estimated based on the DNA/RNA co-extraction method (can also be estimated via cell dissociation and counting)

[b]Volumes shown are for a 1:100 dilution of ERCC mix (ERCC guidelines are to use 1 μL of a 1:100 dilution of ERCC mix per 0.5 μg of RNA).

areas using a hole punch or cork borer [27]. Carefully measuring the amount of tissue collected is also required for the per-cell normalization strategy using cell dissociation as described below (light orange boxes in Fig. 3).

It should be noted that either way you define biomass (equal weights or equal surface areas of planar tissues) potentially introduces biases. Collecting equal masses of tissue may entail differential sampling of cell types, whereas collecting equal surface areas may entail sampling of different amounts of total biomass. These issues are illustrated in the case of leaf sampling where leaf thickness is greater in the polyploid, as has been observed [33]. If equal surface areas are collected, this will result in greater biomass in the polyploid sample than in the diploid sample. Additionally, if increased thickness is driven by cell type-specific size increases—for example, if, relative to the diploid, polyploid leaves have larger epidermal cells but equivalently sized mesophyll cells as observed in Arabidopsis [34]—then the ratio of mesophyll to epidermal cells will be higher in the polyploid than in the diploid sample. To the extent that transcriptional responses to ploidy are cell-type specific, this altered ratio will influence the overall expression patterns regardless of whether equal biomasses or equal surface areas are collected.

Expression per unit of tissue can still be measured using an appropriate spike-in strategy, but the definition of "biomass" should be explicitly defined (surface area, mass, etc.), and as with any bulk tissue expression study, it should be recognized that the values obtained will represent averages of potentially distinct cell type- or tissue layer-specific responses. Ideally, prior to collection, tissue thickness, as well as the relative contribution of different cell types to tissue size, can be assessed by microscopy. This will enable a more informed interpretation of the data, enabling, for example, expression levels per unit of surface area to be converted to expression per biomass and vice versa.

3.2 Estimating Number of Cells per Sample

To quantify expression per cell, it is necessary to estimate the number of cells from which you isolate RNA, which is one of the key challenges of solid tissue experiments. Two primary approaches have been described for doing this: (1) dissociating and counting cells from an equivalent sample and (2) co-extracting DNA and RNA from the tissue [1, 3, 26, 27, 35].

Dissociating and counting cells [36] uses acid digestion to release cells into solution where they can then be counted. This gives a direct estimate of the cell number per sample. Co-extraction, on the other hand, uses the yield of DNA, combined with estimates of genome size and ploidy, to infer cell counts. Thus, dissociation and counting is preferable, or even required, in cases where genome size and/or ploidy distribution is unknown (even if genome size is known, levels of endopolyploidy must also be accounted for; see below).

An advantage of the co-extraction approach is that cell counts can be estimated from the same tissue used for RNA-seq. This should minimize artifacts associated with biological variation in cell density and is beneficial in cases where tissue is limited. A further advantage of co-extraction is that differences in total RNA transcriptome size per cell can be estimated by comparing the ratios of RNA to DNA. Dissociation and counting gives the number of cells per unit of tissue but provides no direct information about transcriptome size.

Visger et al. [27] demonstrated that the two approaches yield similar results in terms of estimating cell count per sample. Nonetheless, in our hands and in other studies [1, 3, 27, 35], the estimates from both methods are associated with considerable variance, so it is recommended that multiple measurements be made, ideally using both methods, to zero in on the true value.

Cell dissociation and counting: Coate et al. [35] and Visger et al. [27] both used the chromic acid method first described by Brown and Rickless [36]. A tissue of defined size (e.g., a hole punch of known diameter or sample of known mass) is digested in chromic acid to dissociate cells. The concentration of cells (cells per μL) in the slurry is estimated by counting cells in a defined volume on a hemocytometer or with an automated cell counter. Based on the surface area of the original sample and the volume of chromic acid used for digestion, the number of cells per μL can be translated to cell density (cells per unit of tissue) in the intact tissue with one of the following equations:

$$\frac{\text{cell concentration (cells/}\mu\text{L)} \times \text{volume } (\mu\text{L})}{\text{tissue surface area (mm}^2\text{)}} = \text{cell density} \left(\text{cells/mm}^2\right)$$

$$\frac{\text{cell concentration (cells/}\mu\text{L)} \times \text{volume } (\mu\text{L})}{\text{tissue mass (mg)}} = \text{cell density (cells/mg)}$$

Additionally, the dissociated cells retain their cell walls and, consequently, their original shape, which can, in some cases, be utilized to distinguish and count different cell types. For example, Coate et al. [35] took advantage of cell shape differences to count palisade cells and spongy mesophyll cells separately in leaf tissue of *Glycine* subg. *Glycine* polyploids.

It should be noted that the efficiency of dissociation and recovery of individual cells is likely to be less than 100%, and consequently, the estimates of cell number per unit of tissue given will tend to underestimate the true value. Consequently, caution should be used when drawing conclusions about absolute cell numbers.

On the assumption that dissociation is equally efficient across samples (diploids and polyploids), the estimates should, however, give reasonable estimates of relative cell density.

RNA/DNA co-extraction: Numerous approaches are available to isolate both DNA and RNA from the same biological sample. Several column-based commercial kits are available for this purpose, including the AllPrep DNA/RNA kits (Qiagen; AllPrep Universal used by Song et al. 2020), *Quick*-DNA/RNA and Direct-zol DNA/RNA kits (Zymo Research), and the E.Z.N.A. DNA/RNA Isolation kit (Omega Bio-Tek). Column-based kits are simple to use and offer the benefit of collecting RNA and DNA into separate fractions as opposed to many column-free protocols that yield a single total nucleic acid fraction. Total nucleic acid must then be split into two fractions and treated with either DNase or RNase to yield pure RNA or DNA, respectively. In addition to halving the yields of each type of nucleic acid, the DNase treatment, often involving incubation and inactivation at elevated temperatures followed by a second round of purification, has the potential to cause RNA degradation and introduce variable losses of RNA that add technical noise into the estimates of RNA/DNA.

Conversely, despite the limitations mentioned above, non-column-based protocols tend to be less expensive and generally give higher total yields. These include CTAB-based approaches, as used by Visger et al. [27], and the Dr. P kit (BioChain) as used in Coate and Doyle [3].

Unfortunately, there is probably not a one-size-fits-all protocol that is optimal for all species and tissue types. For example, I developed a custom, column-based protocol that performs well with flash-frozen leaf tissue from Arabidopsis [37] and maize (unpublished), but not with maize anthers preserved in RNAlater. Consequently, it is likely that several methods will need to be tested and potentially modified to determine the optimal approach for a given sample.

In theory, if extraction efficiency is 100%, the number of cells in the original sample can be calculated by dividing the mass of recovered DNA by DNA content per cell (Table 1, parameter F, "Cells per sample"). In reality, recovery efficiency is less than 100%, meaning that this approach will underestimate the absolute cell number. Consequently, as with cell dissociation, the assumption is that the efficiency of DNA recovery is comparable between samples such that their yields can be used to estimate relative cell densities (e.g., cells per unit of surface area in polyploids relative to diploids).

As noted above, the co-extraction method relies on accurate estimates of genome size. Most natural polyploids, and even some first-generation polyploids, exhibit progressive genome downsizing, so it is not sufficient to assume that, for example, tetraploids have twice the DNA per cell as diploids. Furthermore, many plant tissues undergo somatic polyploidy (endoreduplication), and this process is partially dampened by whole plant polyploidy [37, 38], so the relationship between diploid and polyploid genome size per cell is complex. Consequently, it is necessary to obtain accurate estimates of unreduplicated genome size, as well as the extent of endopolyploidy to estimate the cell number from the DNA yield. Both measures (genome size and endopolyploidy index) are typically obtained via flow cytometry of nuclei stained with a DNA intercalator such as propidium iodide or DAPI and run with an internal size standard [32, 37–39]. The mean DNA content per cell can then be calculated as the base genome size multiplied by the endopolyploid index (the average number of endocycles per cell). With estimates of DNA content per cell from flow cytometry and DNA yield from tissue co-extractions, the number of cells in the original tissue sample can then be estimated by dividing the total mass of DNA recovered by the mean DNA content per cell (Table 1, parameter F, "Cells per sample"). Additionally, co-extractions facilitate estimation of RNA transcriptome size (RNA content per cell) by simply dividing RNA yield by the inferred number of cells in the sample (Table 1, parameter H, "Total RNA per cell"). Alternatively, transcriptome size can be estimated per unit of DNA, which indicates the relative transcriptional activity per haploid genome (Table 1, parameter G, "Total RNA per DNA"). This measure of transcriptome size can be calculated even in the absence of data on endopolyploidy.

Prior to quantifying DNA and RNA yields, it is important to estimate sample purity (in particular, checking for the absence of RNA contamination in the DNA fraction and vice versa) by UV spectroscopy and/or gel electrophoresis. This is particularly important for methods in which you isolate all nucleic acids together followed by nuclease treatments to eliminate DNA or RNA. If significant cross-contamination is present, due, for example, to incomplete nuclease digestion, this should be visible on a gel and indicated by the 260 nm/280 nm absorbance ratio (~1.8 for pure DNA, >1.9 for pure RNA). For pure samples, UV absorbance can be used to quantify concentrations. Ideally, however, concentrations will be measured using fluorescence assays, such as Qubit DNA and RNA assays. These assays are more accurate than UV absorbance measurements if the sample contains UV-absorbing impurities. Additionally, the fluorescent dyes used preferentially bind to their target species (dsDNA or RNA), making them robust to low levels of cross-contamination.

As with estimations of cell number from DNA content, estimates of transcriptome size per cell should be considered relative. It is likely that the efficiency of recovery differs for RNA and DNA in most co-extraction protocols, such that the ratio of recovered RNA to recovered DNA is not the same as the true ratio of RNA to DNA in the cells. But if the differences in DNA and RNA recovery efficiencies are comparable across samples, then comparing ratios should give reasonable estimates of relative transcriptome size.

Finally, regarding transcriptome size, it is important to be explicit about which circumscription of the transcriptome you have measured [14, 26]. The term "transcriptome" is often used to refer specifically to messenger RNA (mRNA), yet the bulk of cellular RNA is ribosomal RNA (rRNA), and several additional species of RNA (long noncoding RNA, small RNAs, etc.) are present as well. Most co-extraction procedures will recover rRNA, but not all methods recover all types of RNA; in particular, many methods will fail to recover small RNAs less than 50 nt in length. Consequently, transcriptome size, as calculated here (Table 1, parameter H, "Total RNA per cell," or parameter G, "Total RNA per DNA") will likely include rRNA and may or may not include small RNAs, and you should be explicit about which RNA fractions have been recovered.

3.3 Applying Exogenous Spike-Ins

Having quantified the biomass or cellularity of tissue from which the RNA was derived, exogenous spike-in controls are employed to transform the estimates of expression from standard RNA-seq approaches (transcriptome-normalized expression) into estimates of expression per unit of biomass, per unit of DNA, or per cell. Exogenous spike-in controls are transcripts of known sequence and concentration that are readily distinguishable from an experimental RNA pool and that are added to the RNA prior to library construction. These spike-ins are subsequently sequenced in parallel with the sample RNA. The resulting read counts from the spike-ins can be used to convert raw read counts to estimates of expression per targeted normalization parameter (e.g., biomass, cell number).

The External RNA Control Consortium (ERCC) [40] produced a set of synthetic polyadenylated RNAs with properties (e.g., length and GC content) consistent with typical eukaryotic mRNAs and premixed in defined quantities for use in validating transcriptomic studies [40, 41]. These mixes are available commercially (Spike-In Mix and ExFold Spike-In Mixes, ThermoFisher Inc.), are widely used as exogenous spike-ins for RNA-seq [27–29, 32], and are well-suited for the per-cell and per-biomass normalization strategies described here.

Measuring transcript concentrations (transcriptome-normalized expression; blue boxes in Fig. 3) does not require the use of spike-in controls, though these can be used to evaluate other elements of RNA-seq experimental performance such as the limit of detection [41]. In contrast, spike-ins are critical for converting transcriptome-normalized expression to estimates of expression per biomass or expression per cell. To do so, the ERCC (or alternative) spike-ins are added to the experimental RNA samples prior to RNA-seq library construction (Fig. 3, Table 1, parameters O-Q).

For normalizing expression to biomass, spike-ins are added proportionally to the amount of biomass from which input RNA was isolated, similar to as described by Visger et al. [27]. For per-cell normalization, spike-ins are added to each sample proportionally to the number of input cells from which the RNA was derived [32, 37]. Table 1 illustrates the underlying calculations used for determining the amount of spike-in to add per sample for a hypothetical example comparing a diploid and tetraploid.

3.4 Library Construction and Sequencing

Purified RNA containing external spike-ins can be used to construct RNA-seq libraries using any standard RNA-seq library kit. Several sequencing platforms exist (e.g., Illumina, PacBio, Oxford Nanopore), giving different combinations of read length, depth, error rate, and cost. Because Illumina offers the highest read counts, relatively low error rates, and low costs per base, it is generally the platform of choice for transcriptomic studies focused on quantifying gene expression.

Because research questions are typically focused on the expression of protein coding genes, some form of mRNA enrichment is usually employed. This serves to deplete the more abundant rRNA (>80% of total RNA) from the sample prior to sequencing so that the bulk of sequencing effort focuses on the RNA species of interest. The most common approach to enrich mRNA is polyA capture, which utilizes biotinylated oligo-d(T) beads that hybridize to and capture the polyadenylated mRNA, enabling rRNA to be washed away. Alternatively, oligos complementary to rRNA can be used to capture the rRNA fraction (e.g., TruSeq Stranded Total RNA with RiboZero kits, Illumina), or rRNA can be depleted utilizing depletion kinetics (e.g., Zymo-Seq RiboFree Total RNA Library Kit, Zymo Inc.). Non-rRNA can then be isolated for use in library construction.

PolyA capture is considerably cheaper than hybridization-based rRNA depletion and is, therefore, the most widely used method. However, polyA capture will deplete all non-polyadenylated RNAs, including lncRNA, siRNA, miRNA, and organellar RNA. If the sole focus of the experiment is on quantifying the nuclear, protein-coding fraction, polyA capture is probably the method of choice. If, however, you are interested in organelle transcription, or the expression of regulatory RNAs, rRNA depletion should be employed.

After enrichment for the target RNA species, the next significant decision with respect to library construction is what insert size to generate. Insert size refers to the mean length of RNA molecule incorporated into the sequencing library. For Illumina-based sequencing, this typically ranges from 200 to 800 nt. The optimal insert size will be dictated by the type of sequencing (e.g., single or paired end) you intend to perform. This, in turn, will depend on the specifics of your biological system (e.g., is there a high-quality reference genome?) and the nature of your questions (e.g., are you interested exclusively in transcript abundance, or do you want to obtain additional layers of biological information such as identifying splice variants?).

For quantifying expression in organisms with relatively simple (not highly repetitive) genomes and high-quality reference genomes/transcriptomes, single-end sequencing and short (e.g., 50–100 bp) read lengths are adequate and most cost-effective (longer reads are more expensive without significantly improving expression estimates). In this case, you will typically fragment the RNA sample to an average of 200 nt. In other cases, however, paired-end sequencing and/or longer read lengths (e.g., 150–250 bp) may be necessary. For example, if the study organism has a highly duplicated genome with high levels of sequence similarity among homologues (e.g., some neopolyploids), then long paired-end reads may be necessary to effectively discriminate between duplicates and map reads with confidence to the correct locus. If you lack a reference genome/transcriptome and will be performing de novo transcriptome assembly, long paired-end reads will yield better assemblies. Similarly, if you want to identify or quantify cases of alternative splicing, paired-end reads are preferable to single-end reads [42].

Additional factors to consider in library preparation include whether, and to what extent, you will be multiplexing samples prior to sequencing, whether to use single- or dual-index barcodes, and whether to include unique molecular identifiers (UMIs). Multiplexing reduces costs by enabling multiple libraries to be sequenced on a single lane, but also raises some risk of cross-contamination due to "index hopping," though this can be mitigated through the use of dual indices [43, 44]. The level of multiplexing must also be balanced with ensuring you have sufficient read count per sample to get quantitative expression estimates. What constitutes a sufficient read count, however, is open to debate [45]. Blencowe et al. [46] estimated that 700 million reads would be required to reliably estimate the expression of >95% of transcript in a mammalian-sized genome. Illumina recommends 5–25 million reads per sample "for a quick snapshot of highly expressed genes," 30–60 million reads for "a more global view of gene expression, and some information on alternative splicing," and 100–200 million reads "to get an in-depth view of the transcriptome, or to assemble new transcripts" (https://support.illumina.com/bulletins/2017/04/considerations-for-rna-seq-read-length-and-coverage-.html). In contrast, Liu et al. [47] found that increasing read count beyond 10 million per sample (in human cell line MCF7) provided little increase in power to detect differentially expressed genes. Instead, increasing biological replication yielded significant increases in statistical power independent of sequencing depth, and Liu et al. [47] recommended prioritizing maximizing replication over sequencing depth per replicate. I generally sequence libraries to a depth of 10–25 million reads, including in studies of synthetic Arabidopsis autotetraploids [32] and *Glycine* neoallopolyploids [39]. Finally, the use of UMIs makes it possible to identify and collapse PCR duplicates (for libraries that use PCR), which can make read counts more quantitative by correcting for PCR-induced biases [48]. However, Sena et al. [49] have demonstrated that "PCR stutter" can result in failure to collapse PCR duplicates using UMIs. UMIs also increase overall costs and reduce the fraction of sequence that will be derived from the sample RNA.

3.5 Sequence Data Analysis

Once sequencing is complete, you will typically receive the raw data in the form of FASTQ files. At this point, the basic steps are as follows: (1) data quality assessment and quality filtering, (2) read mapping, (3) counting reads per feature, (4) data normalization, and (5) differential expression (DE) analysis. There is a dizzying array of options for carrying out each of these steps and no clear consensus on which are optimal. To illustrate this, Table 2 shows which tools were used at each of these steps in four recent transcriptomic studies of polyploids.

In short, many combinations of tools have been used successfully in the analysis of RNA-seq data, there is no one "correct" data analysis pipeline, and normalization using spike-ins can be achieved using multiple approaches. For reference, I outline a pipeline for per-cell data normalization below, starting with 100-bp single-end Illumina sequences generated from libraries that were spiked with ERCC for per-cell normalization as described above. Refer to Fig. 3 and Table 1 for illustrations of the logic underlying these analysis steps as well as the data collection and analysis steps that precede them. The examples given below are for the hypothetical scenario in which the raw data are FASTQ files for three replicates of diploid *Arabidopsis thaliana* (with file names "diploid1.fq," "diploid2.fq,"

Table 2
Bioinformatics approaches used in recent polyploid RNA-seq studies

Study	QC	Alignment	Read counting	DE analysis	Normalization
Zorilla-Fontanesi et al., 2016 [50]	Illumina purity filter	TopHat, Bowtie	HTSeq	EdgeR	Rel. log expression (RLE)
Xiang et al., 2019 [51]	FASTX-Toolkit	STAR	STAR	DESeq2	Not specified
Braynen et al., 2021 [52]	Not specified	TopHat2	Cufflinks	DESeq	FPKM
Song et al., 2020 [32]	Trimmomatic	HISAT2	HTSeq	DEseq2	Per genome (ERCC spike-ins)

and "diploid3.fq") and three replicates of a synthetic tetraploid ("tetraploid1.fq," "tetraploid2.fq," and "tetraploid3.fq"). I also refer the reader to Visger et al. [27], who provide details of a similar approach in the form of a Jupyter Notebook as supplemental material.

In the present pipeline, quality assessment, quality filtering, mapping, and read counting are performed using tools (FastQC, Trimmomatic, HISAT2, and HTseq, respectively) that run from the command line. Data normalization and DE analysis are performed using DESeq2 in R. For the purposes of illustration, I provide examples of the relevant command line or R syntax to run each tool.

Sequence Data Preprocessing: Assess raw data quality using FastQC (https://www.bioinformatics.babraham.ac.uk/projects/fastqc/), which provides a comprehensive assessment, including the distribution of quality scores by read position and contamination by technical sequences (e.g., Illumina adapters and barcodes). From the command line, FastQC can be run using the following:

```
fastqc diploid1.fq
```

(repeat for five remaining FASTQ files)

Invariably, there is some level of technical sequence contamination, as well as some drop-off in sequence quality, typically at the 3' ends of reads, which is addressed by quality filtering using Trimmomatic [53] to output filtered fastq files (with "_trimmed" appended to the input file name):

```
java -jar trimmomatic-0.35.jar SE -phred33 diploid1.fq di-
poid1_trimmed.fq ILLUMINACLIP:TruSeq3-SE.fa:2:30:10 LEADING:3
TRAILING:3 SLIDINGWINDOW:4:15 MINLEN:36
```

(repeat for five remaining FASTQ files)

Then run FastQC on trimmed files to confirm that adapter contamination has been removed and median quality scores for remaining reads remain above Q20 across the full length of reads.

Read Mapping and Counting: Once reads have been cleaned and filtered (and deduplicated in the case of libraries with UMIs), the next step is to map the reads to a biological reference sequence (genome or transcriptome of the study organism) as well as to the ERCC reference (available from Thermo Fisher at: https://assets.thermofisher.com/TFS-Assets/LSG/manuals/cms_095047.txt). For organisms that lack a reference genome/transcriptome, one must be generated via de novo assembly of the reads. Various tools are available for generating de novo assemblies, including but not limited to Trinity [54, 55], SOAPdenovo-Trans [56], Trans-ABySS [57], and Oases [58]. Additionally, there are studies that evaluate strategies to optimize assembly when working with duplicated genomes [59, 60]. The details of these steps are beyond the scope of this chapter, and I refer the reader to those resources should you need to assemble your own reference.

Subsequent steps depend on whether the goal is to quantify expression levels of individual homoeologues in allopolyploids, or, more simply, to compare the combined expression of duplicated loci in a polyploid (either auto- or allopolyploid) to that of the unduplicated locus in the diploid progenitor(s). Several tools have been developed to analyze homoeologues separately, as reviewed by Voshall and Moriyama [61]. Here, however, I focus on the simpler task of quantifying combined gene expression of the two subgenomes in an autopolyploid or allopolyploid relative to the unduplicated genome of its diploid progenitor(s).

For this example, I use the reference genome for Arabidopsis [62], including the genome sequence in FASTA format (Athaliana_447_TAIR10.fa) and annotation file in GFF3 format (Athaliana_447_Araport11.gene_exons.gff3) downloaded from Phytozome (https://phytozome-next.jgi.doe.gov/), as well as the reference sequence and annotation file for the ERCC spike-ins (ERCC92.fa and ERCC92.gtf) available from ThermoFisher (https://assets.thermofisher.com/TFS-Assets/LSG/manuals/ERCC92.zip).

Quality-filtered/trimmed reads are subsequently mapped to the reference genome. Several read mapping tools exist that are tailored to RNA-seq experiments (e.g., Bowtie2, TopHat2, HISAT2, STAR, BWA), and many of these mappers have been used effectively in polyploid transcriptomic studies. Consequently, the choice of mapper is flexible and largely a matter of personal preference. Here, I utilize HISAT2 [63]. Because the reference genome is well-annotated (i.e., has a GTF/GFF annotation file), I take advantage of the HISAT2 Python script (extract_splice_sites.py) to extract known splice junctions, which it then uses to aid in mapping spliced reads.

To do this, first convert the Arabidopsis annotation file from GFF3 format to GTF format using the "gffread" utility of Cufflinks [64]:

```
gffread Athaliana_447_Araport11.gene_exons.gff3 -T -o Atha-
liana_447_Araport11.gene_exons.gtf
```

Then generate a text file of known splice junctions in Arabidopsis using the annotation file:

```
python extract_splice_sites.py Athaliana_447_Araport11.gen-
e_exons.gtf > Athaliana_447_Araport11_spliceSites.txt
```

To map reads, index the Arabidopsis and ERCC references:

```
hisat2-build Athaliana_447_TAIR10.fa
hisat2-build ERCC92.fa
```

After indexing, reads are mapped to the Arabidopsis genome with HISAT2, using the defined spliced sites:

```
hisat2 -p 8 -k 10 -x Athaliana_447_TAIR10 -U diploid1_trimmed.
fq --known-splicesite-infile Athaliana_447_Araport11_splice-
Sites.txt --no-unal -t -S diploid1_Athaliana.sam
```

(repeat for five remaining FASTQ files)

Reads that map equally well (same alignment score) to multiple locations in the genome are assigned a mapping score (MAPQ) of 1 by HISAT2. Because the true origin of these reads is ambiguous, exclude these reads from downstream analyses using the samtools function with "-q 2."

The trimmed FASTQ files are then remapped to the ERCC reference FASTA:

```
hisat2 -p 8 -k 10 -x ERCC92 -U diploid1_trimmed.fq --no-unal
-t -S diploid1_ERCC.sam
```

Then count reads per Arabidopsis gene and reads per ERCC transcript using HTSeq [65]. Counts per Arabidopsis gene are generated with the following:

```
python -m HTSeq.scripts.count -m intersection-nonempty -s no
-t gene -i ID diploid1_Athaliana.sam Athaliana_447_Araport11.
gene_exons.gff3 > diploid1_Athaliana_HTSeq.txt
```

Counts per ERCC transcript are generated with the following:

```
python -m HTSeq.scripts.count -m intersection-nonempty -s no
-t exon -i gene_id diploid1_ERCC.sam ERCC92.gtf > diploi-
d1_ERCC_HTSeq.txt
```

(repeat for five remaining SAM files)

Normalization and Differential Expression Analysis: As with most of the earlier steps, numerous tools are available for data normalization and statistical testing for DEGs [24]. Here, I use DESeq2 [66], which is implemented in R. For measuring fold change (tetraploid/diploid) in standard, transcriptome-normalized expression, import the Arabidopsis count files from HTSeq, generate a DESeq2 dataframe, and run the default DESeq2 analysis to normalize data, calculate fold changes, and test for significant shifts.

From R, read in the count files (e.g., "diploid1_Athaliana_HTSeq.txt"):

```
variable <- 'Athaliana'
files <- grep(variable,list.files(directory),value=TRUE)
```

Create a DESeqDataSet object:

```
cond <- c('diploid', 'diploid', 'diploid', 'tetraploid',
'tetraploid', 'tetraploid')
Attable <- data.frame(sampleName=files, fileName=files,con-
dition=cond)
At <- DESeqDataSetFromHTSeqCount(sampleTable=Attable, direc-
tory=directory, design=~condition)
```

Change "levels" to match the variables in 'cond':

```
colData(At)$condition <- factor(colData(At)$condition, le-
vels=c('diploid','tetraploid'))
```

Finally, run the standard DESeq analysis, which includes normalization for library size, calculation of fold change (tetraploid/diploid), and statistical tests for significant fold changes:

```
dds <- DESeq(At)
```

The output can then be saved to a text file:

```
Atres <- results(dds)
write.csv(as.data.frame(Atres),file= '4x_vs_2x_per_tran-
scriptome.csv', sep = "")
```

Next, use the ERCC counts to normalize expression estimates on a per-cell basis. To do this, import the ERCC read counts output by HTSeq (e.g., "diploid1_ERCC_HTSeq.txt"), generate a dataframe, and use ERCC counts to estimate per-cell size factors. First, read in the count files:

```
variable <- 'ERCC'
files <- grep(variable,list.files(directory),value=TRUE)
```

Then create a DESeqDataSet object:

```
ERCCcond <- c('diploid', 'diploid', 'diploid', 'tetraploid',
'tetraploid', 'tetraploid')
ERCCtable <- data.frame(sampleName=files, fileName=files,con-
dition=cond)
ERCC <- DESeqDataSetFromHTSeqCount(sampleTable=ERCCtable,
directory=directory, design=~condition)
```

Change "levels" to match the variables in 'cond':

```
colData(ERCC)$condition <- factor(colData(ERCC)$condition,
levels=c('diploid','tetraploid'))
```

Then estimate, view, and save size factors from ERCC counts:

```
dds <- estimateSizeFactors(ddsHTSeq)
sizeFactors(dds)
ERCCsizeFac <- sizeFactors(dds)
```

Finally, use the ERCC-based size factors to normalize Arabidopsis read counts, calculate fold changes (now on a per-cell basis), and perform statistical tests.

Specify sizeFactors from ERCC:

```
sizeFactors(At) <- ERCCsizeFac
```

Estimate dispersions, calculate fold changes, and perform the Wald tests:

```
dds <- estimateDispersions(At)
deseqtest <- nbinomWaldTest(dds)
```

Save the output as a text file:

```
ERCCres <- results(dds)
write.csv(as.data.frame(ERCCres), file= '4x_vs_2x_per_cell.
csv', sep = ""))
```

These analyses yield two text files with fold changes (tetraploid/diploid), *p* values, and *p* values adjusted to control the FDR. The first file (4x_vs_2x_per_transcriptome.csv) gives transcriptome-normalized expression (fold changes in transcript concentration in the tetraploids vs. the diploids). The second file (4x_vs_2x_per_cell.csv) gives per-cell normalization (fold changes in transcripts per cell in tetraploids vs. diploids). For mRNA-enriched libraries, relative mRNA transcriptome size (tetraploid/diploid) can then be estimated from each expressed gene by dividing fold change per cell by fold change per transcriptome [26, 32, 37]. Note that the process illustrated here uses per-cell normalization, but it can be applied equally well if you spiked in ERCC on a per-biomass or per-genome basis (Table 1).

4 Summary

As it is typically implemented today, RNA-seq yields relative estimates of transcript concentrations. It is now known that global shifts in expression per cell are common and make transcript concentrations poor, or even misleading proxies for expression responses per gene copy, per cell, or per biomass. Meanwhile, quantifying expression on these other biologically relevant scales has yielded insights into a wide range of biological phenomena—including polyploidy—that would have otherwise been missed. I have outlined here a suite of approaches that enable RNA-seq to capture expression responses on these other scales. Though these procedures require additional steps beyond what are required for standard, transcriptome-normalized RNA-seq, the increase in effort is relatively modest and yield a more comprehensive and informed understanding of transcriptional responses.

Acknowledgements

This work was supported by the National Science Foundation (grant 1257522) and eXtreme Science and Engineering Discovery Environment (XSEDE) allocation TG-BIO170018.

References

1. Guo M, Davis D, Birchler JA (1996) Dosage effects on gene expression in a maize ploidy series. Genetics 142:1349–1355
2. Adams KL, Cronn R, Percifield R et al (2003) Genes duplicated by polyploidy show unequal contributions to the transcriptome and organ-specific reciprocal silencing. PNAS 100:4649–4654
3. Coate JE, Doyle JJ (2010) Quantifying whole transcriptome size, a prerequisite for understanding transcriptome evolution across species: an example from a plant Allopolyploid. Genome Biol Evol 2:534–546
4. Hegarty MJ, Barker GL, Brennan AC et al (2008) Changes to gene expression associated with hybrid speciation in plants: further insights from transcriptomic studies in Senecio. Philos Trans R Soc Lond B Biol Sci 363:3055–3069
5. Hegarty MJ, Barker GL, Wilson ID et al (2006) Transcriptome shock after interspecific hybridization in Senecio is ameliorated by genome duplication. Curr Biol 16:1652–1659
6. Wang J, Tian L, Lee H-S et al (2006) Genome-wide nonadditive gene regulation in Arabidopsis Allotetraploids. Genetics 172:507–517
7. Mortazavi A, Williams BA, McCue K et al (2008) Mapping and quantifying mammalian transcriptomes by RNA-Seq. Nat Methods 5: 621–628
8. Albertin W, Brabant P, Catrice O et al (2005) Autopolyploidy in cabbage (Brassica oleracea L.) does not alter significantly the proteomes of green tissues. Proteomics 5:2131–2139
9. Carpentier SC, Panis B, Renaut J et al (2011) The use of 2D-electrophoresis and de novo sequencing to characterize inter- and intra-cultivar protein polymorphisms in an allopolyploid crop. Phytochemistry 72:1243–1250
10. Yao H, Kato A, Mooney B et al (2011) Phenotypic and gene expression analyses of a ploidy series of maize inbred Oh43. Plant Mol Biol 75:237–251
11. Jin K, Sixue C, Ning Z et al (2012) Comparative proteomics of the recently and recurrently formed natural allopolyploid Tragopogon mirus (Asteraceae) and its parents. New Phytol 196:292–305
12. Ng DW-K, Zhang C, Miller M et al (2012) Proteomic divergence in *Arabidopsis* autopolyploids and allopolyploids and their progenitors. Heredity 108:419–430

13. Soltis DE, Misra BB, Shan S et al (2016) Polyploidy and the proteome. Biochimica et Biophysica Acta (BBA) - Proteins and Proteomics 1864:896–907
14. Doyle JJ, Coate JE (2019) Polyploidy, the nucleotype, and novelty: the impact of genome doubling on the biology of the cell. Int J Plant Sci 180:1–52
15. Ingolia NT, Ghaemmaghami S, Newman JRS et al (2009) Genome-wide analysis in vivo of translation with nucleotide resolution using ribosome profiling. Science 324:218–223
16. Piccirillo CA, Bjur E, Topisirovic I et al (2014) Translational control of immune responses: from transcripts to translatomes. Nat Immunol 15:503–511
17. Coate JE, Bar H, Doyle JJ (2014) Extensive translational regulation of gene expression in an allopolyploid (Glycine dolichocarpa). Plant Cell 26:136–150
18. Reuter JA, Spacek DV, Pai RK et al (2016) Simul-seq: combined DNA and RNA sequencing for whole-genome and transcriptome profiling. Nat Methods 13:953–958
19. Hendrickson DG, Soifer I, Wranik BJ et al (2018) Simultaneous profiling of DNA accessibility and gene expression dynamics with ATAC-Seq and RNA-Seq. Methods Mol Biol 1819:317–333
20. (2014) Method of the year 2013. Nat Methods 11:1. https://doi.org/10.1038/nmeth.2801
21. (2020) Method of the year 2019: single-cell multimodal omics. Nat Methods 17:1–1
22. Song Q, Ando A, Jiang N et al (2020) Single-cell RNA-seq analysis reveals ploidy-dependent and cell-specific transcriptome changes in Arabidopsis female gametophytes. Genome Biol 21:178
23. Coate JE, Farmer AD, Schiefelbein JW et al (2020) Expression partitioning of duplicate genes at single cell resolution in Arabidopsis roots. Front Genet 11:1363
24. Evans C, Hardin J, Stoebel DM (2017) Selecting between-sample RNA-Seq normalization methods from the perspective of their assumptions. Brief Bioinform 19:776–792
25. Aanes H, Winata C, Moen LF et al (2014) Normalization of RNA-sequencing data from samples with varying mRNA levels. PLoS One 9:e89158
26. Coate JE, Doyle JJ (2015) Variation in transcriptome size: are we getting the message? Chromosoma 124:27–43
27. Visger CJ, Wong GK-S, Zhang Y et al (2019) Divergent gene expression levels between diploid and autotetraploid Tolmiea relative to the total transcriptome, the cell, and biomass. Am J Bot 106:280–291
28. Lovén J, Orlando DA, Sigova AA et al (2012) Revisiting global gene expression analysis. Cell 151:476–482
29. Chen K, Hu Z, Xia Z et al (2016) The overlooked fact: fundamental need for spike-in control for virtually all genome-wide analyses. Mol Cell Biol 36:662–667
30. Bourdon M, Pirrello J, Cheniclet C et al (2012) Evidence for karyoplasmic homeostasis during endoreduplication and a ploidy-dependent increase in gene transcription during tomato fruit growth. Development 139: 3817–3826
31. Pirrello J, Deluche C, Frangne N et al Transcriptome profiling of sorted endoreduplicated nuclei from tomato fruits: how the global shift in expression ascribed to DNA ploidy influences RNA-Seq data normalization and interpretation. Plant J 93:387–398
32. Song MJ, Potter BI, Doyle JJ et al (2020) Gene balance predicts transcriptional responses immediately following ploidy change in Arabidopsis thaliana. Plant Cell 32:1434–1448
33. Chen Y, Xu H, He T et al (2021) Comparative analysis of morphology, photosynthetic physiology, and transcriptome between diploid and tetraploid barley derived from microspore culture. Front Plant Sci 12:283
34. Tsukaya H (2019) Re-examination of the role of endoreduplication on cell-size control in leaves. J Plant Res 132:571–580
35. Coate JE, Luciano AK, Seralathan V et al (2012) Anatomical, biochemical, and photosynthetic responses to recent allopolyploidy in Glycine dolichocarpa (Fabaceae). Am J Bot 99: 55–67
36. Brown R, Rickless P (1949) A new method for the study of cell division and cell extension with some preliminary observations on the effect of temperature and of nutrients. Proc Royal Soc London Series B, Biol Sci 136:110–125
37. Robinson DO, Coate JE, Singh A et al (2018) Ploidy and size at multiple scales in the Arabidopsis Sepal. Plant Cell 30:2308–2329
38. Pacey EK, Maherali H, Husband BC (2020) The influence of experimentally induced polyploidy on the relationships between endopolyploidy and plant function in Arabidopsis thaliana. Ecol Evol 10:198–216
39. Coate JE, Song MJ, Bombarely A et al (2016) Expression-level support for gene dosage sensitivity in three Glycine subgenus Glycine polyploids and their diploid progenitors. New Phytol 212:1083–1093
40. External RNA Controls Consortium (2005) Proposed methods for testing and selecting the ERCC external RNA controls. BMC Genomics 6:150

41. Munro SA, Lund SP, Pine PS et al (2014) Assessing technical performance in differential gene expression experiments with external spike-in RNA control ratio mixtures. Nat Commun 5:5125
42. Rossell D, Stephan-Otto Attolini C, Kroiss M et al (2014) Quantifying alternative splicing from paired-end RNA-seq data. Ann Appl Stat 8:309–330
43. Costello M, Fleharty M, Abreu J et al (2018) Characterization and remediation of sample index swaps by non-redundant dual indexing on massively parallel sequencing platforms. BMC Genomics 19:332
44. MacConaill LE, Burns RT, Nag A et al (2018) Unique, dual-indexed sequencing adapters with UMIs effectively eliminate index cross-talk and significantly improve sensitivity of massively parallel sequencing. BMC Genomics 19:30
45. Tarazona S, García-Alcalde F, Dopazo J et al (2011) Differential expression in RNA-seq: A matter of depth. Genome Res 21:2213–2223
46. Blencowe BJ, Ahmad S, Lee LJ (2009) Current-generation high-throughput sequencing: deepening insights into mammalian transcriptomes. Genes Dev 23:1379–1386
47. Liu Y, Zhou J, White KP (2014) RNA-seq differential expression studies: more sequence or more replication? Bioinformatics 30:301–304
48. Kivioja T, Vähärautio A, Karlsson K et al (2012) Counting absolute numbers of molecules using unique molecular identifiers. Nat Methods 9:72–74
49. Sena JA, Galotto G, Devitt NP et al (2018) Unique Molecular Identifiers reveal a novel sequencing artefact with implications for RNA-Seq based gene expression analysis. Sci Rep 8:13121
50. Zorrilla-Fontanesi Y, Rouard M, Cenci A et al (2016) Differential root transcriptomics in a polyploid non-model crop: the importance of respiration during osmotic stress. Sci Rep 6: 22583
51. Xiang D, Quilichini TD, Liu Z et al (2019) The transcriptional landscape of Polyploid wheats and their diploid ancestors during embryogenesis and grain development. Plant Cell 31: 2888–2911
52. Braynen J, Yang Y, Yuan J et al (2021) Comparative transcriptome analysis revealed differential gene expression in multiple signaling pathways at flowering in polyploid Brassica rapa. Cell Biosci 11:17
53. Bolger AM, Lohse M, Usadel B (2014) Trimmomatic: a flexible trimmer for Illumina sequence data. Bioinformatics 30:2114–2120
54. Grabherr MG, Haas BJ, Yassour M et al (2011) Full-length transcriptome assembly from RNA-Seq data without a reference genome., Trinity: reconstructing a full-length transcriptome without a genome from RNA-Seq data. Nat Biotechnol 29, 29:644, 644–652
55. Haas BJ, Papanicolaou A, Yassour M et al (2013) De novo transcript sequence reconstruction from RNA-Seq: reference generation and analysis with Trinity. Nat Protoc 8:1494–1512
56. Xie Y, Wu G, Tang J et al (2014) SOAPdenovo-Trans: de novo transcriptome assembly with short RNA-Seq reads. Bioinformatics 30:1660–1666
57. Robertson G, Schein J, Chiu R et al (2010) De novo assembly and analysis of RNA-seq data. Nat Methods 7:909–912
58. Schulz MH, Zerbino DR, Vingron M et al (2012) Oases: robust de novo RNA-seq assembly across the dynamic range of expression levels. Bioinformatics 28:1086–1092
59. Gutierrez-Gonzalez JJ, Garvin DF (2017) De Novo transcriptome assembly in Polyploid species. In: Gasparis S (ed) Oat: methods and protocols. Springer, New York, pp 209–221
60. Chopra R, Burow G, Farmer A et al (2014) Comparisons of De Novo transcriptome assemblers in diploid and polyploid species using peanut (Arachis spp.) RNA-Seq data. PLoS One:9:e115055
61. Voshall A, Moriyama EN (2020) Next-generation transcriptome assembly and analysis: Impact of ploidy. Methods 176:14–24
62. Cheng C-Y, Krishnakumar V, Chan AP et al (2017) Araport11: a complete reannotation of the Arabidopsis thaliana reference genome. Plant J 89:789–804
63. Kim D, Paggi JM, Park C et al (2019) Graph-based genome alignment and genotyping with HISAT2 and HISAT-genotype. Nat Biotechnol 37:907–915
64. Trapnell C, Roberts A, Goff L et al (2012) Differential gene and transcript expression analysis of RNA-seq experiments with TopHat and Cufflinks. Nat Protoc 7:562–578
65. Anders S, Pyl PT, Huber W (2015) HTSeq—a Python framework to work with high-throughput sequencing data. Bioinformatics 31:166–169
66. Love M, Huber W, Anders S (2014) Moderated estimation of fold change and dispersion for RNA-seq data with DESeq2. Genome Biol 15:550

Chapter 13

A Robust Methodology for Assessing Homoeolog-Specific Expression

J. Lucas Boatwright

Abstract

Angiosperm evolution is marked by numerous, recurring polyploidization events. While hybridization and polyploidization have greatly increased the degree of genetic and phenotypic diversity in plants, the mechanisms underlying changes in the genotype-to-phenotype relationships remain unclear. As the field of natural sciences continues to expand during the post-genomic era, large datasets are becoming increasingly common. However, the development of tools and workflows available to robustly assess these changes have lagged behind data production. A robust homoeolog-specific expression analysis strongly depends upon proper homoeolog calling, the ability to account for reference sequence biases, flexible and accurate methods for dealing with residual bias, and a reproducible workflow. To that end, this chapter aims to provide a detailed description of the potential pitfalls encountered while estimating homoeolog-specific expression as well as provide a workflow that allows for robust inferences based on precise estimates of expression changes.

Key words Polyploid, Homoeolog, Reference bias, Homoeolog-specific expression, Bayesian inference

1 Introduction

Differential expression analysis may take numerous forms such as differential transcript expression (DTE), differential gene expression (DGE), allele-specific expression (ASE), or homoeolog-specific expression (HSE) [1–4]. In both DTE and DGE, a common reference genome is typically used across all samples to examine expression changes between samples at either the transcript or gene level, respectively [5]. However, ASE and HSE focus on differences in expression due to reference sequence differences between homologous or homoeologous chromosomes, respectively [6]. As such, they may utilize multiple references or a genotype-specific reference genome, where a reference genome has either been de novo generated for each genotype or an existing reference genome is modified using variant data specific to each

Yves Van de Peer (ed.), *Polyploidy: Methods and Protocols*,
Methods in Molecular Biology, vol. 2545, https://doi.org/10.1007/978-1-0716-2561-3_13,

genotype [7]. ASE and HSE further differentiate from each other where ASE is an analysis of expression differences for loci between homologous chromosomes in homoploid hybrids, a cross between species or varieties resulting in an offspring with the same chromosome count as the parents [8], whereas HSE examines expression differences for loci between homoeologous chromosomes in polyploid hybrids, a cross between species or varieties resulting in an offspring containing more than two sets of chromosomes [6]. Due to the similarities between ASE and HSE, both techniques share similar sources of bias and complications but may be analyzed using similar approaches [7, 9].

2 Materials

RNA-Seq data is required for the allopolyploid under study and the corresponding parents—though methods may be performed pairwise for subgenomes in higher polyploids as well. Either a single genome reference with single-nucleotide polymorphisms called for each parent—preferably called using the GATK [10], individualized reference genomes—or gene-level transcriptome assemblies for each parent, as might be obtained using Trinity [11] and Corset [12], are also required. Access to a Linux or Unix-based operating system with adequate storage, processing power, and permissions to install required software will be necessary. For working examples of the code described here, see the GitHub repository at https://github.com/jlboat/Robust-HSE-Examples, and install the required dependencies.

3 Methods

3.1 Ortholog/Homoeolog Identification

A robust analysis of HSE should begin with an annotated polyploid reference. However, in the absence of an existing annotated polyploid reference, parental genome references may be used if the polyploid is recently formed or has not significantly diverged from the parental sequences. If a well-annotated polyploid reference is available, then homoeologs may be identified between subgenomes directly [13]. Like orthologs, homoeologs are pairs of genes originating from speciation, but these gene pairs were then brought into the same nucleus via polyploidization [14]. For this discussion, we will assume that a polyploid reference is not available as this is more common due to the difficulties that arise in assembling polyploid reference genomes. Polyploid reference assemblies are often more fractionated and contain more chimeric sequences due to similarities between the subgenomes [9, 15]. Instead of assembling a polyploid reference and identifying homoeologs directly, homoeologs may be inferred using orthologs, which are genes that diverged due to a speciation event [14]. Orthologs identified between the parental references are useful since the union of the nonreduced parental genomes is equivalent to the polyploid genome [13]. To identify orthologs, it is common to use a reciprocal-best-hit approach [16], which exhibits low false-positive rates and follows a simple implementation [17]. There are numerous programs capable of quickly identifying orthologs using this approach [18–20]. Here, we use Last [21] due to its ease of use, speed, and ability to train the sequence alignment step to maximize the fidelity of ortholog calling.

For reproducibility and ease of installation, we highly recommend using containers such as Docker [22] or Singularity [23]. To obtain a local installation of Last using Singularity, this one-liner will do:

```
singularity pull docker://biocontainers/last-align:v963-2-
deb_cv1
```

We also recommend providing scripts and files after publication at sites such as BitBucket (https://bitbucket.org/) or GitHub (https://github.com) for reproducibility and to serve as supplemental documentation. After getting a local installation of Last, index the FASTA files containing the gene sequences for each parent:

```
singularity run last-align_v963-2-deb_cv1.sif \
lastdb refA_db refA.fasta
singularity run last-align_v963-2-deb_cv1.sif \
lastdb refB_db refB.fasta
```

Reciprocally align the reference sequences. Here, we use an extended tabular BLAST output format for ease of parsing. Parameters include -P 4, which assigns four processors to this alignment run, and -E0.00005, which filters alignments based on the maximum expected alignments per square gigabases. A similar method may be applied to other local-alignment software, and each comes with advantages and disadvantages—most notably in speed and implementation style. Interestingly, reciprocal best hits are largely consistent across programs [17]:

```
singularity run last-align_v963-2-deb_cv1.sif \
lastal -f BlastTab+ -P 4 -E0.00005 refA_db refB.fasta >
refB_To_refA.maf
singularity run last-align_v963-2-deb_cv1.sif \
lastal -f BlastTab+ -P 4 -E0.00005 refB_db refA.fasta >
refA_To_refB.maf
```

We can get more accurate alignments between the two references by first training on the sequences:

```
lastdb refA_db refA.fasta
last-train refA_db refB.fasta > refB_To_refA.train
lastal -p refB_To_refA.train refA_db refB.fasta > refB_To_r-
efA.trained.maf
```

The *last-train* command determines the rate of deletions, insertions, and each substitution type between the aligned sequences and writes the training results to the file "refB_To_refA.train." The *p* option in *lastal* then uses the training file to improve alignment accuracy. Using a custom python script, we then identify gene pairs that hit each other both best and reciprocally:

```
python LAST_COREs.py -a refA_To_refB.maf -b refB_To_refA.maf
```

In addition to a list of reciprocal best hits, two BED files are generated that contain the coordinates of the local alignments between orthologous pairs. These subsequences, which we refer to as Common Orthologous REgions (COREs), represent the regions of the genes with similar sequence identity that are best suited for inferring differences in HSE. Regions that lie outside the local alignments are divergent and more likely to affect the ability to map reads, and by limiting inference to COREs, differences in the lengths of orthologs, which can convolute an accurate comparison of HSE, are limited [9, 20]. Setting a proper stringency threshold during alignments also helps reduce the impact of spurious mismatches often caused by paralogs. The resulting orthologous pairs may then be used to represent comparable references for either parental reads, or they can represent the homoeologs in the polyploid.

3.2 Reference Sequence Biases

It is vital that sequence differences between homoeologous regions are accounted for to strengthen the conclusions of the analyses [15, 24]. Otherwise, these sequence differences may contribute to read-mapping bias. There are numerous sources of error for read alignment, which can affect expression quantification in a variety of ways whether using a single reference [25] or multiple references [26], and there are numerous steps required before HSE can be determined. As such, reproducibility may be difficult, and expertise is often required to ensure accurate results. Sequence bias in particular may result in reads mapping to multiple areas, especially among duplicated subsequences (i.e., duplications, repetitive regions, gene families), thereby reducing their mapping quality score and effectively eliminating them from subsequent analysis.

Reads may also map better to an incorrect location due to differences between the generated reference genome and the true reference genome [7]. Similarly, DNA duplications, loss, or divergence that occurred after the parents diverged will affect proper quantification in the polyploid and may also affect ortholog/homoeolog calling. These issues can be reduced by using genotype-specific references, which have the effect of lowering type I error and improving the estimation of expression imbalance [1, 27–30]. While genotype-specific references improve analysis quality, residual bias will likely remain due to uncorrected differences or mismatches between the sequenced genotype and the reference [7]. In such circumstances, either real [2, 28] or simulated DNA sequence [25, 26, 29] can serve as a control to estimate the degree of mapping bias present within a reference set since DNA sequences should be approximately equally distributed across the reference [7].

3.3 Bayesian Inference of Homoeolog-Specific Expression

Standard read processing and quality control (QC) assessment are vital aspects to any analysis. For adapter-trimming and read quality filtering, we recommend fastp [31] due to its fast runtime and detailed reporting. The example provided assumes illumine single-end sequences:

```
fastp -i in.reads.fq.gz -o out.reads.fq.gz -j reads.json
```

Reports from individual samples can be synthesized into a single report using MultiQC [32] for easier cross-sample comparison:

```
multiqc ./
```

We recommend using the individualized BED files containing CORE coordinates for each ortholog set to filter alignment files or allowing BayesASE, described below, to handle the filtering directly. As mentioned, this process will control for any major sequence differences between homoeologs. After filtering and assessing the read data for quality, the rest of the analysis may be completed using BayesASE [7].

Bayesian models provide the flexibility to designate prior expectations and will produce a posterior distribution representing a compromise between the prior expectations and the observed data [33]. This posterior distribution represents the range and probability of the true unobserved parameters rather than simple point estimates so that inferences drawn from the results may be tempered according to the degree of confidence [33]. To facilitate a Bayesian approach to HSE, we promote the use of BayesASE—a program designed to estimate ASE [7] or workflows based on previous iterations of this program [9, 34]. The required inputs for BayesASE are the quality-filtered FASTQ files, a reference genome, genotype-specific VCF files, and design files containing various details about the input files. Individual loci can be included by providing BED files—such as those generated from the reciprocal alignment phase.

BayesASE utilizes a negative binomial model [7, 29] that can use a fixed prior to identify and correct for mapping biases to the corresponding references. The true proportion of reads from the paternal subgenome can be estimated for each species or variety after mapping to the corresponding references using three separate estimations where the prior is set at 0.4, 0.5, and 0.6, to account for potential bias toward the paternal reference, no bias, or bias toward the maternal parent, respectively. Each locus is considered to have homoeologous expression bias if the credible interval does not overlap 0.5 for all posteriors arising from each prior condition [9, 29].

This approach has been shown to conservatively control type I error [30]. BayesASE also allows for priors to be independently specified between conditions. If both parental and polyploid reads are available, each set of parental reads may be mapped to both references to determine if loci exhibit innate bias. The expectation is that each genotype-specific reference should map reads derived from that individual best. However, reference biases may result in reads mapping better from the "wrong" parent [9]. Thus, performing a preliminary mapping of each genotype to both references will facilitate the filtering of loci that exhibit strong read mapping bias, and then subsequent inferences of HSE using polyploid reads should be limited to unbiased loci. However, it is important that reads always be mapped to the full references since removal of sequences will alter read mapping biases [9]. In BayesASE, alignments are compared and categorized according to several metrics including the references to which they successfully map, quality of mapping based on edit distance, and whether they mapped uniquely or to multiple locations [7, 29, 30]. Unlike prior methods, BayesASE comes with Nextflow scripts [35], which enable scalable and reproducible scientific workflows using software containers, to automatically execute the pipeline (https://github.com/McIntyre-Lab/BayesASE) and may be installed using Anaconda, a distribution of the Python and R programming languages for scientific computing, or the CPython package installer, pip. For full details, on running BayesASE, visit the GitHub repository.

References

1. Turro E, Su S-Y, Gonçalves Â et al (2011) Haplotype and isoform specific expression estimation using multi-mapping RNA-seq reads. Genome Biol 12:R13
2. Wittkopp PJ, Haerum BK, Clark AG (2004) Evolutionary changes in cis and trans gene regulation. Nature 430:85–88
3. Wolf JBW (2013) Principles of transcriptome analysis and gene expression quantification: an RNA-seq tutorial. Mol Ecol Resour 13: 559–572
4. Love MI, Huber W, Anders S (2014) Moderated estimation of fold change and dispersion for RNA-seq data with DESeq2. Genome Biol 15:550
5. Pertea M, Kim D, Pertea GM et al (2016) Transcript-level expression analysis of RNA-seq experiments with HISAT, StringTie and Ballgown. Nat Protoc 11:1650–1667
6. Hu G, Wendel JF (2019) Cis-trans controls and regulatory novelty accompanying allopolyploidization. New Phytol 221:1691–1700
7. Miller BR, Morse AM, Borgert JE et al (2021) Testcrosses are an efficient strategy for identifying cis-regulatory variation: Bayesian analysis of allele-specific expression (BayesASE) G3 11. https://doi.org/10.1093/g3journal/jkab096
8. Nieto Feliner G, Álvarez I, Fuertes-Aguilar J et al (2017) Is homoploid hybrid speciation that rare? An empiricist's view. Heredity 118: 513–516
9. Boatwright JL, McIntyre LM, Morse AM et al (2018) A robust methodology for assessing differential homeolog contributions to the transcriptomes of allopolyploids. Genetics 210:883–894
10. McKenna A, Hanna M, Banks E et al (2010) The genome analysis toolkit: a mapreduce framework for analyzing next-generation DNA sequencing data. Genome Res 20: 1297–1303
11. Haas BJ, Papanicolaou A, Yassour M et al (2013) De novo transcript sequence reconstruction from RNA-seq using the Trinity platform for reference generation and analysis. Nat Protoc 8:1494–1512
12. Manado Manado Press (2016) Corset: enabling differential gene expression analysis for de Novo assembled transcriptomes. CreateSpace Independent Publishing Platform
13. Soltis DE, Gitzendanner MA, Stull G et al (2013) The potential of genomics in plant systematics. Taxon 62:886–898
14. Glover NM, Redestig H, Dessimoz C (2016) Homoeologs: What are they and how do we infer them? Trends Plant Sci 21:609–621
15. Krasileva KV, Buffalo V, Bailey P et al (2013) Separating homeologs by phasing in the tetraploid wheat transcriptome. Genome Biol 14: R66
16. Moreno-Hagelsieb G, Latimer K (2008) Choosing BLAST options for better detection of orthologs as reciprocal best hits. Bioinformatics 24:319–324
17. Hernández-Salmerón JE, Moreno-Hagelsieb G (2020) Progress in quickly finding orthologs as reciprocal best hits: comparing blast, last, diamond and MMseqs2. BMC Genomics 21: 741
18. Remm M, Storm CE, Sonnhammer EL (2001) Automatic clustering of orthologs and in-paralogs from pairwise species comparisons. J Mol Biol 314:1041–1052
19. Emms DM, Kelly S (2019) OrthoFinder: phylogenetic orthology inference for comparative genomics. Genome Biol 20:238
20. Li L, Stoeckert CJ Jr, Roos DS (2003) OrthoMCL: identification of ortholog groups for eukaryotic genomes. Genome Res 13: 2178–2189
21. Kiełbasa SM, Wan R, Sato K et al (2011) Adaptive seeds tame genomic sequence comparison. Genome Res 21:487–493
22. Merkel D et al (2014) Docker: lightweight linux containers for consistent development and deployment. Linux J 2014:2
23. Kurtzer GM, Sochat V, Bauer MW (2017) Singularity: Scientific containers for mobility of compute. PLoS One 12:e0177459
24. Kuo TCY, Hatakeyama M, Tameshige T et al (2020) Homeolog expression quantification methods for allopolyploids. Brief Bioinform 21:395–407
25. Degner JF, Marioni JC, Pai AA et al (2009) Effect of read-mapping biases on detecting allele-specific expression from RNA-sequencing data. Bioinformatics 25: 3207–3212
26. Stevenson KR, Coolon JD, Wittkopp PJ (2013) Sources of bias in measures of allele-specific expression derived from RNA-seq data aligned to a single reference genome. BMC Genomics 14:536
27. Skelly DA, Johansson M, Madeoy J et al (2011) A powerful and flexible statistical framework for testing hypotheses of allele-specific gene

expression from RNA-seq data. Genome Res 21:1728–1737

28. Graze RM, Novelo LL, Amin V et al (2012) Allelic imbalance in Drosophila hybrid heads: exons, isoforms, and evolution. Mol Biol Evol 29:1521–1532

29. León-Novelo LG, McIntyre LM, Fear JM, Graze RM (2014) A flexible Bayesian method for detecting allelic imbalance in RNA-seq data. BMC Genomics 15:920

30. Fear JM, León-Novelo LG, Morse AM et al (2016) Buffering of genetic regulatory networks in Drosophila melanogaster. Genetics 203:1177–1190

31. Chen S, Zhou Y, Chen Y, Gu J (2018) fastp: an ultra-fast all-in-one FASTQ preprocessor. Bioinformatics 34:i884–i890

32. Ewels P, Magnusson M, Lundin S, Käller M (2016) MultiQC: summarize analysis results for multiple tools and samples in a single report. Bioinformatics 32:3047–3048

33. Kruschke JK (2010) Bayesian data analysis. WIREs. Cogn Sci 1:658–676

34. Shan S, Boatwright JL, Liu X et al (2020) Transcriptome dynamics of the inflorescence in reciprocally formed allopolyploid Tragopogon miscellus (Asteraceae). Front Genet 11: 888

35. Di Tommaso P, Chatzou M, Floden EW et al (2017) Nextflow enables reproducible computational workflows. Nat Biotechnol 35: 316–319

Part IV

Population Genomics Approaches to Study Polyploidy

Chapter 14

Analyzing Autopolyploid Genetic Data Using GenoDive

Patrick G. Meirmans

Abstract

Analyzing autopolyploid genetic data still presents numerous challenges due to, e.g., missing dosage information of genotypes and the presence of multiple ploidy levels within species or populations, but also because the choice of software is limited when compared to what is available for diploid data. However, over the last years, the number of software programs that can deal with polyploid data is slowly increasing. The software GENODIVE is one of the most widely used programs for the analysis of polyploid genetic data, presenting a wide array of different methods. In this chapter, I outline several frequently used types of population genetic analyses and explain how these apply to polyploid data, including possible pitfalls and biases. I then explain how GENODIVE approaches these analyses and whether and how it can overcome possible biases. Specifically, I focus on analyses of genetic diversity, Hardy-Weinberg equilibrium, quantifying population differentiation, clustering, and calculation of genetic distances. GENODIVE can be downloaded freely from http://www.patrickmeirmans.com/software.

Key words AMOVA, K-means, Population differentiation, Genetic diversity, Genetic distances, Polyploidy

1 Introduction

Despite the existence of a great number of polyploid species, the analysis of polyploid genetic data still presents a number of challenges for researchers, as should be clear from the other chapters in this book. For example, for many polyploid species complexes, the evolutionary origin of the polyploidy is unknown, and the species may be either allopolyploid or autopolyploid or something in between [1, 2], which can bias the results of the population genetic analysis [3]. Furthermore, the ploidy level of individuals may be unknown, populations may consist of multiple ploidy levels, there may be double reduction at various parts of the genome [4], or the dosage of alleles may be unknown. All in all, these problems are serious enough to lead to a general avoidance of polyploid species as study systems among researchers, except of course for those interested in polyploid evolution.

Yves Van de Peer (ed.), *Polyploidy: Methods and Protocols*,
Methods in Molecular Biology, vol. 2545, https://doi.org/10.1007/978-1-0716-2561-3_14,

Arguably the most important of these challenges is the difficulty of estimating the dosage of alleles in the different types of heterozygotes [5]. In diploids, obtaining such dosage is straightforward: if two different alleles are observed for an individual at a given locus, we know that this individual has genotype *AB*, carrying one copy of allele *A* and one copy of allele *B*. For a tetraploid, this is less straightforward as an individual with alleles *A* and *B* may be any of the three possible heterozygotes: *AAAB*, *AABB*, or *ABBB*. With increasing ploidy level and increasing number of alleles, this problem gets more complicated: at a locus with seven alleles, octoploids ($2n = 8$) have 3003 possible unique genotypes, of which only 7 are homozygotes; for diploids there are only 28 genotypes, again with 7 homozygotes. In next-generation sequencing studies, a high coverage may allow estimation of allele dosage for polyploid heterozygotes [6], but here the required coverage increases rapidly with ploidy level. The absence of dosage information can lead to appreciable biases in estimates of allele frequencies and many other statistical analyses (reviewed in [7]).

When dosage information is available, autopolyploids—which are the main focus of this chapter—are actually not so different from diploids from a statistical point of view. This means that several standard population genetic methods can quite simply be scaled from diploids to autopolyploids by allowing for the higher number of allele copies within individuals. For example, a coalescent model for autotetraploids can be obtained quite accurately by a scaling of the effective population size to $4N_e$ [8]. Also, calculation of summary statistics like F_{ST}, F_{IS}, H_E, and H_O and standard analyses like PCA are relatively straightforward when genotypes with full dosage are available. Nevertheless, their interpretation can sometimes differ from their diploid counterparts [7]. Furthermore, even when full dosage information is available, the value of these statistics may be affected by double reduction [9].

Most researchers do not calculate the population genetic summary statistics by hand, but instead rely on software for such calculations. A large body of software is available for the statistical analysis of the population genetic data, either implemented in standalone software [10–13] or in Python or R-libraries [14–16]. Unfortunately, the great majority of these programs are geared towards diploids, even if the analyses they present are equally valid for polyploids. However, a number of programs do allow for data from polyploids and this number is still growing [12, 17–19]. Whereas the first programs allowing polyploid data were often geared to basic analyses, a number of programs now also allow for more specialized types of inferences, such as linkage disequilibrium [20], population assignment [21], and clustering [18]. On the other hand, some types of analyses, most importantly detection of selection and ecological associations, are still underrepresented. Sometimes it is possible to subsample polyploid data to diploidy, to match the input requirements of software that only take diploid input data [22], by randomly drawing two alleles from the complete polyploid genotype. Unfortunately, it is not always known if this leads to bias in the results, especially when missing dosage information is not accounted for. For example, signals of selection in polyploid genomes are different than in diploid genomes, which may affect the output of methods for detecting such signals [23].

I am proud to say that my own program GenoDive [12] is one of the most widely used programs for the analysis of polyploid genetic data. GenoDive started its life as a single-purpose tool for estimating clonal diversity in populations of asexually reproducing species [24]. I wrote that tool mainly for my own use, and since I was working with apomictic triploid dandelions, it made sense to allow for polyploid input data. Over the years, GenoDive then transformed into general-purpose software with many types of inferences and options. Its most important feature is probably its fully developed graphical user interface which allows for direct manipulation of the data (*see* Fig. 1); e.g., the interface allows the user to make subsets of individual populations or loci, to create groups of populations, and to set the ploidy levels of individuals. The user interface also allows easily changing the parameters of the

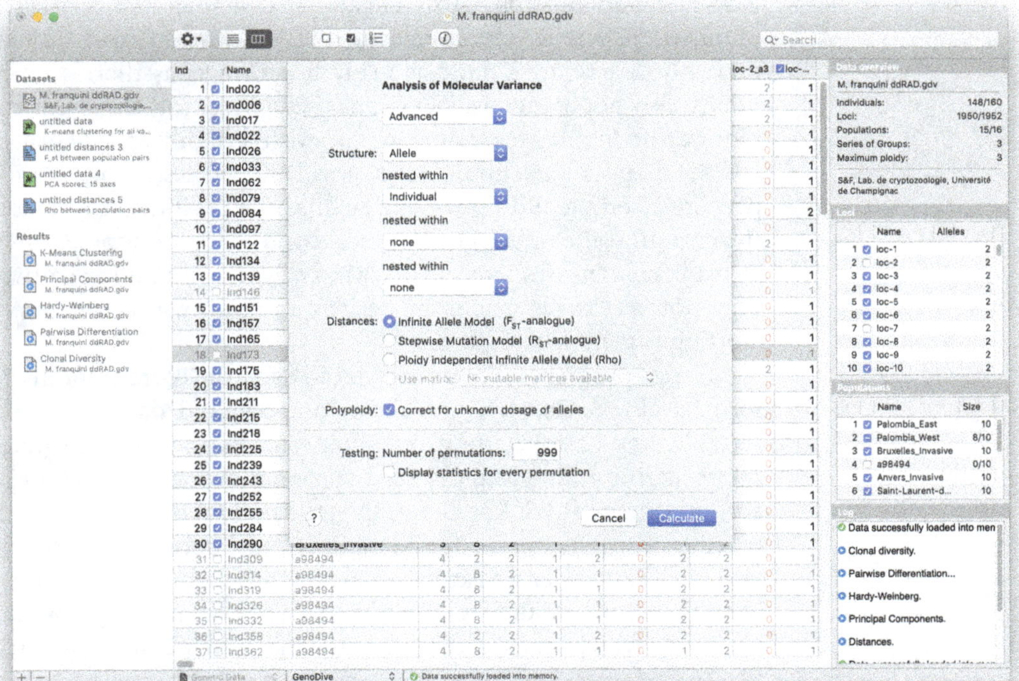

Fig. 1 The main GenoDive window, showing the options for performing an analysis of molecular variance

analyses without having to dive into command line arguments. Despite the simplicity of the interface, GenoDive is very powerful and very fast, as it is able to use all processing cores in a computer's CPU. The size of data sets supported by GenoDive depends on the user's computer, but files with 10,000 loci and 400 individuals have been run successfully. For such large data sets, the user interface becomes somewhat slow, but the calculations are faster than for comparable software. GenoDive is only available for Mac computers due to the difficulty of porting a native user interface to different operating systems. GenoDive can be downloaded from http://www.patrickmeirmans.com/software.

In this chapter I present an overview of different aspects of the analysis of polyploid data in general and how these are implemented into GenoDive specifically. I will point out polyploidy-specific problems, their effect on the results of several types of analysis, the way GenoDive works around these challenges, and the limitations of the chosen approach. I hope that this overview will help polyploidy researchers in making informed decisions on how to approach the analysis of population genetic data and will help them to avoid any biases.

2 Getting the Data in

The first and possibly most challenging step in the analysis of population genetic data is getting the software to read the data correctly. There are almost as many file formats as there are programs for population genetic analyses. Most of these formats require the file to be a plain text file with a highly specific way to encode populations, individuals, loci, and alleles; some formats permit including additional data such as the position of loci along the chromosome or the geographical coordinates of populations or individuals. Unfortunately, the output of genotype calling software does not necessarily match the desired input format so some reformatting is generally required.

As few programs support polyploid data, the corresponding file formats also usually do not allow for polyploid data. The most suitable way of formatting polyploid data is therefore to follow the formatting for polyploid-ready programs such as SPAGeDi [19] and Structure [13]. For GenoDive, the preferred way is to use the specific GenoDive input file, though it can (among many others) also read SPAGeDi and Structure files. However, it is important to note that Structure files differ in the way that missing dosage is encoded, which may lead to a bias when such a file is read by GenoDive. For Structure files, genotypes with missing dosage should be filled up to their full ploidy level with an arbitrarily chosen allele. When reading these files into GenoDive, the program will interpret these genotypes as having fully known dosage, as GenoDive expects missing dosage to be indicated as missing data.

Most analyses in GENODIVE work only properly if the ploidy of the included individuals is known. It is therefore advisable to perform flow cytometry on all sampled individuals in addition to genotyping. However, when the genotyping is done using multiallelic markers such as microsatellites, it may be possible to set the ploidy level of individuals based on the maximum number of unique alleles found at any of the loci. When reading a data file, GENODIVE automatically sets the ploidy level of an individual to the maximum number of alleles (not necessarily unique ones) found across all loci for that individual. The ploidy of each individual is then displayed in the data table of the GENODIVE interface and can be manually corrected by simply double-clicking the value and typing a new one. In addition, there is a special "Define Ploidy Levels..." option where users can set the ploidy levels of individuals from either the individual names or the population names. For example, any individual for which the given name contains either the string "pentaploid," "pen," or "5×" will be set to be pentaploid. It is also possible to set all individuals in the data file to the same ploidy level. However, the assigned ploidy levels cannot be in conflict with the genetic data itself; for example, an individual cannot be assigned to be diploid if it has four alleles at a particular locus.

3 Genetic Diversity and Hardy-Weinberg Equilibrium

The estimation of allele frequencies is at the heart of any population genetic study as many types of inferences crucially depend on them. Estimating allele frequencies in autopolyploids is actually straightforward when the full dosage of the genotypes is known. In such cases, allele frequency estimates for a sample of polyploids are even more precise than those for a similarly sized sample of diploids, because of the higher number of chromosome copies included in the polyploid sample.

When allele dosage is not known, estimates of allele frequencies will be biased, as the frequencies of low-frequency alleles will be overestimated and those of high-frequency alleles will be underestimated [5, 25]. To correct for this bias, GENODIVE uses the maximum likelihood method of De Silva et al. [26]. However, GENODIVE does not implement De Silva's method of simultaneously estimating the selfing rate for a locus, but works on the assumption of random mating within populations. GENODIVE can do this correction for any ploidy level up to octaploid and also for mixtures of ploidy levels. Higher ploidy levels than octoploid are not included as the number of possible genotypes gets very large and calculating their likelihoods gets very cumbersome. The optimization method used in GENODIVE does involve some randomness, and therefore, there may be some slight deviations if the method is run several times on the same data. Tests with simulated data sets show that this method generally performs very well when the assumption of random mating is met [7].

Given equal population sizes, polyploids are expected to show a higher level of genetic diversity than diploids, because polyploid population contain a higher total number of chromosome copies and therefore experience more mutation events and weaker genetic drift than diploid populations [27]. This is one of several reasons why it is of interest to study the level of genetic diversity in polyploid complexes. The most widely used metric for quantifying genetic diversity is the expected heterozygosity H_S (also abbreviated as H_E). For diploids, this is calculated as the expected frequency of heterozygous individuals in a population under Hardy-Weinberg equilibrium (HWE), given the estimated allele frequencies [28].

For polyploids, the situation is more complicated as these have multiple types of heterozygotes, each with a different expected frequency under HWE. Therefore, it is common to use the same equation as for diploids to measure genetic diversity, since this makes comparisons among ploidy levels much easier [7]. It is however important to note that despite being called "expected heterozygosity," this metric then does not necessarily say anything about heterozygotes in a polyploid population. It would be less confusing to use the alternative term "gene diversity" as originally proposed by Nei [28], though that would obscure the comparisons between expected and observed heterozygosity.

For the expected heterozygosity, GenoDive uses the nearly unbiased estimators as outlined in Nei and Chesser [29], which includes correction for sample size. As the original formulation by Nei and Chesser was for diploids, this has been adapted for polyploids and mixed-ploidy data sets by replacing all "$2n$"-terms (where n is the sample size) with the total number of allele copies sampled for that locus. GenoDive also includes an option to correct for unknown dosage in polyploids, in which case the allele frequencies are estimated using the method of De Silva et al. [26] described above. In addition to the expected heterozygosity, GenoDive also gives the number of alleles (allelic richness) and the effective number of alleles as measures of genetic diversity. For each statistic, the standard deviation is calculated by jackknifing over loci, and the upper and lower bounds of the 95% confidence interval are calculated by bootstrapping over loci. Both jackknifing and bootstrapping only take place when more than five loci are available in the data set.

Another important aspect of genetic diversity is the observed heterozygosity H_O. Again, this is more complicated in polyploids by the presence of different types of heterozygotes, some of which (e.g., genotype *ABCD*) can be considered more heterozygous than others (genotype *AAAB*). To weigh these heterozygotes by their level of heterozygosity and to enable comparisons among ploidy levels, Moody et al. [30] proposed to use the "gametic heterozygosity," which for a genotype is the frequency of heterozygotes among diploid gametes that are randomly drawn, without replacement, from the polyploid genotype. So the genotype *ABCD* has a gametic heterozygosity of 1, while *AAAB* has a gametic heterozygosity of 0.5. The concept of drawing diploid gametes is simply a theoretical construct to enable comparisons among ploidy levels and can therefore also be applied to polyploids that do not actually produce any diploid gametes. The calculation of H_O is strongly contingent on the availability of complete genotypes, without any missing dosage. However, GenoDive does not give a warning when a user tries to calculate H_O values for data with missing dosage, so it is the responsibility of the researcher to decide whether these values can be interpreted.

Testing for conformance to Hardy-Weinberg proportions in autopolyploids also requires data with known dosage. Deviation from Hardy-Weinberg proportions can be expected in polyploids even when there is random mating within populations, due to the presence of double reduction [4]. The rate of double reduction can also vary among loci along a chromosome, so there can be a wide variation in HWE deviation within a single genetic data set. GenoDive implements two types of tests for HWE, both using F_{IS} as a test statistic. The first uses the classical definition where F_{IS} is calculated as the complement of the ratio between the observed and the expected heterozygosity [31]. The second option calculates F_{IS} from the within-individual and within-population variance components of an analysis of molecular variance (AMOVA [32]). The results of these two different tests are generally highly comparable, though the test based on AMOVA is in most cases faster to execute than the heterozygosity-based test. For both approaches, the significance of F_{IS} is tested using Monte Carlo permutations, by permuting alleles among individuals within populations. For polyploids, the use of permutations is preferable to other methods like chi-square tests or log-likelihood tests. Chi-square and log-likelihood tests do not work well when there are genotypes for which the expected frequency is very low, which is often the case for polyploids, especially for higher ploidy levels or for multiallelic markers.

4 Quantifying Population Structure

Analysis of population structure in polyploid complexes can be a highly informative tool to make inferences about the (possibly multiple) origins of polyploidy [33], relationships between different ploidy levels [6], and the influence of historical events such as the Pleistocene ice ages [34]. In addition, the analysis of population structure forms an essential aspect of modern genome scans to detect loci that are putatively under selection [6]. There are two main ways of studying population structure. The first way is to quantify the degree of differentiation among a priori defined populations and groups of populations and test the significance of relevant statistical hypotheses. The second way is to cluster either individuals or populations into groups to detect the main structures present in a genetic data set. Even though such confirmatory and exploratory inferences represent two quite different, and possibly epistemologically incompatible, research strategies, they are typically both applied to a single data set.

Quantifying the strength of an a priori defined population structure is typically done using Wright's F_{ST} [35] or similar summary statistics. Expanding this framework to polyploid and mixed-ploidy data is mathematically relatively straightforward, but the interpretation remains challenging. One problem is that for a given population size, migration rate, and mutation rate, F_{ST} is expected to be lower for polyploid populations than for diploid populations [7, 9], making comparisons among ploidy levels difficult. Realizing this problem, Ronfort et al. [9] developed the ρ-statistic, which is unaffected by ploidy level, but also by the rate of double reduction and inbreeding. The ρ-statistic should therefore be used to compare the degree of differentiation across independent population samples of varying ploidy levels.

Unfortunately, ρ is less useful when the populations of the different ploidy levels are strongly intertwined, e.g., due to high levels of gene flow or (multiple) recent origins of the polyploids from diploid ancestors. In such a case, the ploidy correction implicit in ρ will cause the degree of divergence in the polyploids to be overestimated; when diploids and polyploids have the exact same allele frequencies, the ρ-value for the polyploids will be higher than that for the diploids. So both F_{ST} and ρ have interpretation issues, especially for mixed-ploidy data, but under different conditions. Besides F_{ST} and ρ, there are a number of other summary statistics for estimating divergence, including D_{est} [36], F'_{ST} [37], and F''_{ST} [38], which all differ in their applicability to polyploid complexes (*see* [25] for an overview).

There are various statistical frameworks for estimating the value of F_{ST} and associated statistics, though not all estimation methods can be applied to all statistics, and the naming of statistics may change depending on the estimation method (*see* Table 1). GenoDive implements the two most widely used frameworks, one based on heterozygosities [31] and one based on least squares via an analysis of molecular variance [25, 32, 39]. These are the same two frameworks that were discussed above as methods for estimating the deviation from HWE using the F_{IS}-statistics, since F_{IS} and F_{ST} are conceptually closely related. For both frameworks, Monte Carlo permutations can be used to test whether the observed F_{ST}

Table 1
Overview of summary statistics for population differentiation that can be calculated using GenoDive

Statistic	Reference	Dosage correction	Remarks
F_{ST}	[25, 39]	Imputation	Equivalent to Weir and Cockerham's [52] θ, estimated using the AMOVA framework
F'_{ST}	[37, 38]	Imputation	Estimator for Hedrick's standardized measure, using AMOVA framework
R_{ST}	[32, 53]	Imputation	Useful for markers following the stepwise mutation model
ρ	[7, 9]	Imputation	Ploidy-independent and independent of the rate of double reduction
G_{ST}	[31]	De Silva et al.	Includes correction for small sample sizes and number of populations
G''_{ST}	[37, 38]	De Silva et al.	Estimator for Hedrick's standardized measure, using heterozygosities
D_{est}	[36]	De Silva et al.	Uses different concepts of population divergence

Included is also the method that can be used to correct for missing dosage information: either estimating corrected allele frequencies under the assumption of random mating, using the method of De Silva et al. [26], or random imputation of the missing parts of the genotypes based on estimated corrected allele frequencies. All of the included summary statistics, except R_{ST}, can also be used to calculate matrices of pairwise differentiation

values are significantly different from zero. Standard deviations and confidence intervals are obtained using jackknifing and bootstrapping, respectively.

In GenoDive, the heterozygosity-based statistics can be calculated using the "*G*-statistics" command, named after the series of F_{ST} estimators developed by Masatoshi Nei [31]. This command performs a calculation of multiple related statistics, namely, G_{ST} (with and without sample size correction), G'_{ST}, G''_{ST}, and D_{est} (*see* Table 1). These statistics all have in common that they are calculated via H_S and H_T, which GenoDive estimates from the allele frequencies using the method of Nei and Chesser [29], with some modification to account for the higher number of allele copies sampled in polyploids. When dosage information is missing there is again the option to calculate the allele frequencies using the De Silva method [26], under the assumption of HWE.

In contrast with the heterozygosity-based statistics, the AMOVA-based estimation implemented in GenoDive allows the inclusion of multiple hierarchical levels in the population structure (Fig. 1). GenoDive is quite flexible in this regard and different population clusters can be defined for a single data set, allowing testing which of several evolutionary scenarios best fit the data. The AMOVA command in GenoDive allows estimation of F_{ST}, F'_{ST}, R_{ST}, and ρ (*see* Table 1), but also allows the use of a custom distance matrix, which may lead to additional, possibly undefined, F_{ST}-like statistics. For polyploids, the AMOVA variance components are estimated using the approach of Meirmans and Liu [25]. Following this approach, the ρ-statistics is then calculated using a matrix of squared Euclidean distances for all pairs of individuals, based on the within-individual allele frequencies. Again, there is an option to correct for missing dosage, but this is slightly different than above, as AMOVA is not based on allele frequencies, but directly works from the genotypes. Therefore, missing dosage is corrected by randomly filling in the missing parts of the incomplete genotype, based on the relative likelihood of the different possible underlying genotypes given the estimated, corrected, allele frequencies [26]. Since this imputation of the missing dosage data is done randomly, a slightly different result is obtained each time the algorithm is run.

5 Detecting Population Structure

Principal component analysis (PCA) is one of the oldest and most important tools for detecting the main population structure in a genetic data set. In a mixed-ploidy data set, the presence of missing dosage information in the polyploids can lead to a bias if this is not corrected for [7]. The underlying reason for this bias is that the missing dosage leads to errors in the estimation of the allele frequencies, which can lead to spurious separation between ploidy levels, even when they form a single gene pool. Even when dosage information is present, or corrected for, polyploids are expected to show different patterns in a PCA than diploids; in a PCA on mixed-ploidy data, polyploid individuals will occupy more central positions than diploid individuals. This is because in polyploids there will be less fully homozygous individuals and therefore fewer within-population allele frequencies of 0 and 1, resulting in less variance in allele frequencies among individuals.

The PCA implemented in GenoDive can be performed at both the individual and the population level, using either a matrix of within-individual or within-population allele frequencies. For polyploids, the way that missing dosage is handled depends on the level of analysis. At the individual level, dosage is corrected for by imputation of the incomplete genotypes, as for the AMOVA discussed above. At the population level, the De Silva method is used to estimate the allele frequencies. The analysis at the population level also allows a permutation test that enables testing in which axes are significant and thus worthy of interpretation. In this permutation test, individuals are randomized over populations and the PCA is recalculated; the percentage of variance explained by the axis is then used as the test statistic, where the value for the actual data is compared to the distribution of values for the permuted data. Besides a principal component analysis, GENODIVE also implements a principal coordinates analysis (PCoA). A PCoA is less suitable for mixed-ploidy data, or polyploid data with missing dosage, because it is based on a distance matrix, and there are no suitable distance metrics for such data sets (*see* below).

K-means clustering is a very general clustering algorithm that has applications ranging from community ecology to image analysis. In population genetics, its use is especially of interest as it is based on sums of squares and can therefore be implemented in the AMOVA framework and thus used to calculate *F*-statistics [40]. This means that when individuals are clustered into a given number of groups, the optimal solution is the one with the maximum possible value of F_{ST} (or ρ). For polyploids, *K*-means clustering was shown to have a good performance in recovering the true population structure in simulated data sets [22]. However, when missing dosage information was not corrected for, there was a bias where individuals were spuriously clustered by ploidy when population structure was weak.

GENODIVE offers AMOVA-based *K*-means clustering, where the ρ-statistics is used as the optimality criterion, for its efficiency and for its easy implementation with polyploid data. In addition, GENODIVE offers clustering based on the allele frequencies, which is slightly faster, but less informative than the AMOVA-based option. Both methods can not only be applied at the individual level, but also at the population level [40], which sets it apart from other clustering techniques aimed at genetic data. Two different optimization algorithms are implemented: the classic algorithm from MacQueen [41] and a customized simulated annealing algorithm [40, 42]. The former is slightly faster than the latter, but the latter presents a more thorough searching of the solution space. To estimate the optimal number of clusters, GENODIVE calculates both the pseudo-*F*-statistics [43] (highest value is best) and the Bayesian information criterion [44] (lowest value is best). One major advantage of *K*-means clustering above other clustering methods is that it is very fast: clustering of a simulated data set of 50 individuals and 100,000 loci using the simulated annealing algorithm, with 20 independent runs of the simulated annealing chain and from $k = 2$ to $k = 15$ clusters, took only 9 s on my own desktop computer. One drawback of *K*-means clustering is that it only provided so-called "hard" clustering, which means that each individual can only belong to a single cluster. Therefore, it cannot be used to detect admixed individuals.

The most widely used software for performing genetic clustering is STRUCTURE, which combines assignment tests with clustering in a Bayesian framework. In a test of multiple clustering methods for polyploid data, Stift et al. [22] found that STRUCTURE had the best overall performance. STRUCTURE can handle polyploid data and even has explicit support for missing dosage information. GENODIVE does not directly implement the STRUCTURE algorithm, but allows running the STRUCTURE program as an add-in. This means that GENODIVE will convert the input file correctly to the STRUCTURE format, run the STRUCTURE program (which should be installed separately on the user's hard drive), and read all the produced output. GENODIVE then determines the optimal number of clusters using the delta-K statistics [45] and presents a clear overview of the results. Similarly, GENODIVE allows running the clustering program INSTRUCT [46] as an add-in. INSTRUCT has support for tetraploid data and can infer the inbreeding rate in combination with the population structure and can therefore be used for selfing or otherwise inbreeding species. However, INSTRUCT was shown to have lower performance than STRUCTURE or K-means clustering for polyploid data.

6 Distances

Calculating genetic distances among populations or individuals can be an important step for further analysis. Most importantly, a matrix of pairwise distances among populations can be used to test for isolation by distance, by correlating it to a matrix of geographical distances in a Mantel test. However, it is important to stress that the strength of the correlation in a Mantel test, measured using Mantel's r statistics, does not necessarily say anything about the strength of isolation by distance, but instead indicates the strength of spatial autocorrelation in the allele frequencies [47]. In a spatial setting, lower levels of gene flow generally lead to weaker, less significant, correlations. This is because under very low, spatially restricted, gene flow, all populations will be generally very diverged (high F_{ST}), independently of their geographical distances. In contrast, high levels of gene flow (low F_{ST}) give long gradients in allele frequencies across space and thus to strong autocorrelation. For polyploid complexes, this means that differences in the strength of the autocorrelation can be expected between ploidy levels even when the migration rates are similar. As polyploids show weaker divergence (lower F_{ST}) than diploids for similar rates of gene flow, it follows that Mantel tests, based on pairwise F_{ST}, should show stronger autocorrelation in polyploids than in diploids. It remains to be tested whether the use of the ploidy-independent ρ-statistic also alleviates this difference.

For distances among populations, GENODIVE offers several options, most importantly F_{ST} and the associated statistics G_{ST}, G'_{ST}, D_{est}, and ρ. This means that all the restrictions and caveats mentioned above about these statistics also apply here, and therefore, one should be careful in applying these statistics as distance metrics with mixed-ploidy data. For polyploid data with missing dosage, GenoDive also offers dosage-corrected versions of G_{ST}, G''_{ST}, and ρ. In addition, GenoDive offers a number of other genetic distance metrics: Nei's D [48], Rogers [49], Euclidean, Manhattan, Chord [50], and chi-square. All of these additional metrics can be calculated for polyploid and mixed-ploidy data, but have not yet been tested in this regard. However, they are likely to be similarly biased as the other statistics, especially with missing dosage information.

Genetic distances between individuals can also be used for further analyses, including tests of spatial autocorrelation, principal coordinates analysis, or the creation of neighbor-joining networks. Unfortunately, mixed-ploidy data is very difficult to analyze using interindividual distances, as there is no known distance metric that does not have a bias in such analyses. This is also the case for distance metrics that were specifically designed for polyploids. Nevertheless, GenoDive offers several interindividual distance metrics: clonal, Smouse and Peakall, Bruvo, Euclidean, Manhattan, Chord, and chi-square (*see* Table 2). However, it is very important to note that their use should be limited to single-ploidy data sets.

Table 2
Overview of pairwise individual resemblance metrics that can be calculated by GenoDive

Statistic	Reference	Biased with mixed-ploidy data	Biased with missing dosage
Bruvo	[54]	Yes	Yes
Smouse and Peakall	[55]	Yes	Yes
Euclidean	[25]	Yes, but can be used in AMOVA to calculate ρ	Yes
Manhattan	–	Not tested, but likely	Not tested, but likely
Chord	[50]	Yes	Yes
Chi-square	–	Not tested, but likely	Not tested, but likely
Clonal	[24]	No, but not applicable to mixed-ploidy data as it assumes clonality	Yes
Kinship	[51]	No	No

All metrics can be calculated for polyploid and mixed-ploidy data, most of them having biases with both mixed-ploidy data and data with missing dosage information. The kinship coefficient [51] does not show clear ploidy- or dosage-related biases, but has more limited applications

In addition to distance metrics, GenoDive offers the kinship coefficient of Loiselle et al. [51]. This metric does not have a bias with mixed-ploidy data [7], but its biological interpretation is different than for distance metrics as it is an estimator for the degree of relatedness between individuals. Its use should thus be restricted to cases where actual close kinship is expected to occur at least to some degree in a sample of individuals. The kinship coefficient cannot be used for some downstream analyses such as PCoA, multidimensional scaling, AMOVA, or building trees or networks. One way in which the kinship coefficient can be used is in analyses of spatial autocorrelation as the expectation is that individuals growing closely together are more likely to be close kin [51].

7 Conclusions

Overall, GenoDive presents a wide palette of different types of analyses for population genetic data, for both diploids and polyploids. For polyploidy researchers, GenoDive stands out because it is the only general-purpose software for which almost all of the analyses support polyploid data in some way. In this chapter, I described several of the most important and most-used methods, but GenoDive offers several more, including Mantel tests, hybrid index calculation, and population assignment. A full overview of the implemented methods is given in Table 3, including a short description of how data from polyploids is supported.

One of my objectives for developing GenoDive was to make powerful and fast software that is also easy to use. I hope this makes genetic analyses more approachable to novice users and non-geneticists and helps to further increase the use of genetic methods in other fields of research such as ecology and conservation biology. I also hope that the availability of GenoDive lowers the barrier for researchers to do population genetic studies on polyploid species. However, it is important that all users are aware of the pitfalls inherent in the analysis of polyploid and especially mixed-ploidy data. Researchers should therefore always keep an eye open for possible biases and when in doubt try to test the presence of biases using simulations.

Table 3
Overview of analyses that can be performed by GenoDive, with a short description of how GenoDive handles polyploid data for that analysis

Analysis	Reference	Adaptation to polyploid data
Allele frequencies	–	Dosage correction
Genetic diversity	[31]	Dosage correction (not for H_O), gametic heterozygosity
Hardy-Weinberg equilibrium	[30]	Uses gametic heterozygosity, requires dosage information
Clone assignment and clonal diversity	[24]	None, but can be used with polyploid data
Genetic distances	–	*See* Tables 1 and 2
Spatial autocorrelation	[56, 57]	None, but can be used with polyploid data
Mantel test	[58, 59]	None, but depends on the choice of distance metric
Hybrid index	[60]	Calculates genotype likelihoods for polyploids, no support for missing dosage or double reduction
Population assignment	[21, 61]	Calculates genotype likelihoods for polyploids, no support for missing dosage or double reduction
Structure (as add-in)	[13]	Explicit support for missing dosage information
InStruct (as add-in)	[46]	Only works with diploid or tetraploid data, requires dosage information
Redundancy analysis (as add-in)	[47, 62]	Dosage correction, calculation of ρ
LFMM (as add-in)	[63]	None, only works with diploid data
K-means clustering	[40]	Dosage correction, calculation of ρ
Principal components analysis	–	Dosage correction
Principal coordinates analysis	–	None, dependence on distance matrix makes it difficult with mixed-ploidy data or missing dosage information
Compare groups	–	Dosage correction
G-statistics	[31]	Dosage correction (*see* Table 1)
AMOVA	[25, 32]	Dosage correction, calculation of ρ (*see* Table 1)

The analyses are given in the order at which they are presented in the interface

References

1. Stift M, Berenos C, Kuperus P, Van Tienderen PH (2008) Segregation models for disomic, tetrasomic and intermediate inheritance in tetraploids: a general procedure applied to *Rorippa* (yellow cress) microsatellite data. Genetics 179:2113–2123
2. Chester M, Gallagher JP, Symonds VV et al (2012) Extensive chromosomal variation in a recently formed natural allopolyploid species, *Tragopogon miscellus* (Asteraceae). Proc Nat Acad Sci 109:1176–1181

3. Meirmans PG, Van Tienderen PH (2013) The effects of inheritance in tetraploids on genetic diversity and population divergence. Heredity 110:131–137
4. Bever JD, Felber F (1992) The theoretical population genetics of autopolyploidy. Oxford Surv Evol Biol 8:185–217
5. Dufresne F, Stift M, Vergilino R, Mable BK (2014) Recent progress and challenges in population genetics of polyploid organisms: an overview of current state-of-the-art molecular and statistical tools. Mol Ecol 23:40–69
6. Monnahan P, Kolář F, Baduel P et al (2019) Pervasive population genomic consequences of genome duplication in *Arabidopsis arenosa*. Nat Ecol Evol:1–15
7. Meirmans PG, Liu S, Van Tienderen PH (2018) The analysis of polyploid genetic data. J Hered:1–36
8. Arnold BJ, Bomblies K, Wakeley J (2012) Extending coalescent theory to autotetraploids. Genetics 192:195–204
9. Ronfort J, Jenczewski E, Bataillon T, Rousset F (1998) Analysis of population structure in autotetraploid species. Genetics 150:921–930
10. Excoffier L, Lischer HEL (2010) Arlequin suite ver 3.5: a new series of programs to perform population genetics analyses under Linux and Windows. Mol Ecol Resour 10:564–567
11. Goudet J (1995) FSTAT (Version 1.2): a computer program to calculate F-statistics. J Hered 86:485–486
12. Meirmans PG (2020) GenoDive version 3.0: easy-to-use software for the analysis of genetic data of diploids and polyploids. Mol Ecol Resour 20:1126–1131
13. Pritchard J, Stephens M, Donnelly P (2000) Inference of population structure using multilocus genotype data. Genetics 155:945–959
14. Raj A, Stephens M, Pritchard JK (2014) fastSTRUCTURE: Variational inference of population structure in large SNP data sets. Genetics 197:573–U207
15. Goudet J (2005) Hierfstat, a package for r to compute and test hierarchical F-statistics. Mol Ecol Notes 5:184–186
16. Kamvar ZN, Tabima JF, Grünwald NJ (2014) Poppr: an R package for genetic analysis of populations with clonal, partially clonal, and/or sexual reproduction. PeerJ 2:e281
17. Clark LV, Jasieniuk M (2011) POLYSAT: an R package for polyploid microsatellite analysis. Mol Ecol Resour 11:562–566
18. Shastry V, Adams PE, Lindtke D et al (2021) Model-based genotype and ancestry estimation for potential hybrids with mixed-ploidy. Mol Ecol Resour 21:1434–1451
19. Hardy O, Vekemans X (2002) SPAGEDi: a versatile computer program to analyse spatial genetic structure at the individual or population levels. Mol Ecol Notes 2:618–620
20. Gerard D (2021) Pairwise linkage disequilibrium estimation for polyploids. Mol Ecol Resour 21:1230–1242
21. Field DL, Broadhurst LM, Elliott CP, Young AG (2017) Population assignment in autopolyploids. Heredity 119:389–401
22. Stift M, Kolář F, Meirmans PG (2019) Is more robust than other clustering methods in simulated mixed-ploidy populations. Heredity 123:429–441
23. Monnahan P, Brandvain Y (2020) The effect of autopolyploidy on population genetic signals of hard sweeps. Biol Lett 16:20190796
24. Meirmans PG, Van Tienderen PH (2004) GenoType and GenoDive: two programs for the analysis of genetic diversity of asexual organisms. Mol Ecol Notes 4:792–794
25. Meirmans PG, Liu S (2018) Analysis of Molecular Variance (AMOVA) for autopolyploids. Front Ecol Evol 6:217
26. De Silva HN, Hall AJ, Rikkerink E et al (2005) Estimation of allele frequencies in polyploids under certain patterns of inheritance. Heredity 95:327–334
27. Luttikhuizen PC, Stift M, Kuperus P, Van Tienderen PH (2007) Genetic diversity in diploid vs. tetraploid Rorippa amphibia (Brassicaceae). Mol Ecol 16:3544–3553
28. Nei M (1973) Analysis of gene diversity in subdivided populations. J Hered 70: 3321–3323
29. Nei M, Chesser R (1983) Estimation of fixation indexes and gene diversities. Ann Hum Genet 47:253–259
30. Moody M, Mueller L, Soltis DE (1993) Genetic-variation and random drift in autotetraploid populations. Genetics 134:649–657
31. Nei M (1987) Molecular evolutionary genetics. Columbia University Press, New York
32. Michalakis Y, Excoffier L (1996) A generic estimation of population subdivision using distances between alleles with special reference for microsatellite loci. Genetics 142:1061–1064
33. Hedrén M, Nordström S, Ståhlberg D (2008) Polyploid evolution and plastid DNA variation in the *Dactylorhiza incarnata/maculata* complex (Orchidaceae) in Scandinavia. Mol Ecol 17:5075–5091

34. Buono D, Khan G, von Hagen KB et al (2020) Comparative phylogeography of *Veronica spicata* and *V. longifolia* (Plantaginaceae) across Europe: integrating hybridization and polyploidy in phylogeography. Front. Plant Sci 11: 588354
35. Wright S (1965) The interpretation of population structure by *F*-statistics with special regard to systems of mating. Evolution 19:395–420
36. Jost L (2008) G_{ST} and its relatives do not measure differentiation. Mol Ecol 17:4015–4026
37. Hedrick P (2005) A standardized genetic differentiation measure. Evolution 59: 1633–1638
38. Meirmans PG, Hedrick P (2011) Assessing population structure: FST and related measures. Mol Ecol Resour 11:5–18
39. Excoffier L, Smouse P, Quattro J (1992) Analysis of molecular variance inferred from metric distances among DNA haplotypes: application to human mitochondrial-DNA restriction data. Genetics 131:479–491
40. Meirmans PG (2012) AMOVA-based clustering of population genetic data. J Hered 103: 744–750
41. MacQueen J (1967) Some methods for classification and analysis of multivariate observations. In: Proceedings of the fifth Berkeley symposium on mathematics, statistics, and probability, vol. 1. University of California Press, Berkeley, pp. 281–297
42. Kirkpatrick S, Gelatt C, Vecchi M (1983) Optimization by simulated annealing. Science 220: 671–680
43. Calinski T, Harabasz J (1974) A dendrite method for cluster analysis. Commun Stat 3: 1–27
44. Schwarz G (1978) Estimating the dimension of a model. Ann Stat 6:461–464
45. Evanno G, Regnaut S, Goudet J (2005) Detecting the number of clusters of individuals using the software STRUCTURE: a simulation study. Mol Ecol 14:2611–2620
46. Gao H, Williamson S, Bustamante C (2007) A Markov Chain Monte Carlo approach for joint inference of population structure and inbreeding rates. Genetics 176:1635–1651
47. Meirmans PG (2015) Seven common mistakes in population genetics and how to avoid them. Mol Ecol 24:3223–3231
48. Nei M (1978) Estimation of average heterozygosity and genetic distance from a small number of individuals. Genetics 89:583
49. Rogers JS (1972) Measures of genetic similarity and genetic distance. Studies in genetics VII. University of Texas Publication 7213, Austin, pp 145–153
50. Cavalli-Sforza LL, Edwards A (1967) Phylogenetic analysis. Models and estimation procedures. Am J Hum Genet 19:233–257
51. Loiselle B, Sork V, Nason J, Graham C (1995) Spatial genetic-structure of a tropical understory shrub, *Psychotria officinalis* (Rubiaceae). Am J Bot 82:1420–1425
52. Weir BS, Cockerham C (1984) Estimating *F*-statistics for the analysis of population structure. Evolution 38:1358–1370
53. Slatkin M (1995) A measure of population subdivision based on microsatellite allele frequencies. Genetics 139:457–462
54. Bruvo R, Michiels N, D'Souza T, Schulenburg H (2004) A simple method for the calculation of microsatellite genotype distances irrespective of ploidy level. Mol Ecol 13:2101–2106
55. Smouse P, Peakall R (1999) Spatial autocorrelation analysis of individual multiallele and multilocus genetic structure. Heredity 82: 561–573
56. Moran P (1950) Notes on continuous stochastic phenomena. Biometrika 37:17–23
57. Smouse P, Long JC, Sokal RR (1986) Multiple regression and correlation extensions of the Mantel test of matrix correspondence. Syst Zool 35:627–632
58. Mantel N (1967) The detection of disease clustering and a generalized regression approach. Cancer Res 27:209–220
59. Diniz-Filho JAF, Soares TN, Lima JS et al (2013) Mantel test in population genetics. Genet Mol Biol 36:475–485
60. Buerkle C (2005) Maximum-likelihood estimation of a hybrid index based on molecular markers. Mol Ecol Notes 5:684–687
61. Paetkau D, Slade R, Burden M, Estoup A (2004) Genetic assignment methods for the direct, real-time estimation of migration rate: a simulation-based exploration of accuracy and power. Mol Ecol 13:55–65
62. Orsini L, Mergeay J, Vanoverbeke J, De Meester L (2012) The role of selection in driving landscape genomic structure of the waterflea *Daphnia magna*. Mol Ecol 22:583–601
63. Frichot E, Schoville SD, Bouchard G, Francois O (2013) Testing for associations between loci and environmental gradients using Latent Factor Mixed Models. Mol Biol Evol 30: 1687–1699

Chapter 15

Inference of Polyploid Origin and Inheritance Mode from Population Genomic Data

Alison Dawn Scott, Jozefien D. Van de Velde, and Polina Yu Novikova

Abstract

Whole-genome duplications yield varied chromosomal pairing patterns, ranging from strictly bivalent to multivalent, resulting in disomic and polysomic inheritance modes. In the bivalent case, homeologous chromosomes form pairs, where in a multivalent pattern all copies are homologous and are therefore free to pair and recombine. As sufficient sequencing data is more readily available than high-quality cytological assessments of meiotic behavior or population genetic assessment of allelic segregation, especially for non-model organisms, bioinformatics approaches to infer origins and inheritance modes of polyploids using short-read sequencing data are attractive. Here we describe two such approaches, where the first is based on distributions of allelic read depth at heterozygous sites within an individual, as the expectations of such distributions are different for disomic and polysomic inheritance modes. The second approach is more laborious and based on a phylogenetic assessment of partially phased haplotypes of a polyploid in comparison to the closest diploid relatives. We discuss the sources of deviations from expected inheritance patterns, advantages and pitfalls of both methods, effects of mating types on the performance of the methods, and possible future developments.

Key words Polyploidy, Disomic, tetrasomic, polysomic, and mixed inheritance modes, Rediploidization, Homeologous exchanges, selfing, outcrossing

1 Introduction

Polyploid organisms have more than two sets of chromosomes as a result of whole-genome duplication (WGD) and are typically described as belonging to one of two categories: autopolyploids being formed by WGD within a single lineage and allopolyploids in the case of hybridization between divergent lineages. Though observations of chromosome behavior during meiosis were initially the only information available for assigning polyploid "types" [1], we now have to distinguish between this original cytogenetic (genetic according to Li et al.'s [2] classification of auto- or allo-polyploidy) and the now common phylogenetic classification (taxonomic according to Li et al., we prefer the term phylogenetic as

Yves Van de Peer (ed.), *Polyploidy: Methods and Protocols*,
Methods in Molecular Biology, vol. 2545, https://doi.org/10.1007/978-1-0716-2561-3_15,

evolutionary relationships are not always reflected by taxonomy). Furthermore, there is a distinction between inheritance mode (allelic segregation) in polyploids and their classification by either cytogenetic or phylogenetic means.

1.1 Polyploid "Types" and Inference of Inheritance Patterns

1.1.1 Cytogenetic Inferences

Depending on the amount of divergence and structural differences between the duplicated chromosomes, they may or may not pair and recombine during meiosis. Therefore, the origin of polyploidy is traditionally associated with the manner of chromosome segregation. In allopolyploids with sufficient divergence between the subgenomes, only true homologous chromosomes (i.e., homologous within the same subgenome) would pair and recombine, whereas homeologous chromosomes (i.e., homologous chromosomes brought together via hybridization among species) would not. This results in disomic inheritance mode [3, 4] and practically diploid-like behavior. In autopolyploids, where all chromosomal sets are homologous, chromosomes show a random assortment during meiosis and can form bivalents or multivalents [5, 6]. In, for instance, a tetraploid with four identical subgenomes (both a cyto- and phylogenetic autopolyploid), pairing can occur among any of the four homologous chromosomes during meiosis, via either formation of multivalents or random bivalents. Therefore, gametes formed by this autopolyploid can contain any combination of its four alleles, as preferential pairing is absent. Multivalent formation also presents the opportunity for segments of sister chromatids to segregate together in the same gamete (double reduction), resulting in homozygous gametes [7, 8]. In a cytogenetic allopolyploid, bivalents are formed among homologous chromosomes, while pairing of homeologous chromosomes is prevented (e.g., pairing only happens *within* subgenomes, not between or among). Thus, the types of gametes formed by an allopolyploid are disomic, and allelic combinations are limited compared to those formed by an autopolyploid.

As is often the case in biology, real-life examples deviate from binary extremes and can lie anywhere along a spectrum from bivalent pairing with disomic inheritance to fully multivalent pairing and polysomic inheritance. For example, autopolyploids initially form multivalents following their origin, but eventual diploidization will result in strictly bivalent pairings. Similarly, allopolyploids exhibit primarily bivalent pairing due to divergence among parental genomes; however, even homeologous chromosomes may pair, yielding multivalent formation.

1.1.2 Phylogenetic Inferences

With the increased availability of sequence data, the origins of countless polyploid organisms have been classified (e.g., auto, allo, or some combination thereof) via phylogenetic inference [9]. While cytogenetic classification is based on chromosome pairing behavior (e.g., formation of either bivalents or multivalents), phylogenetic classification of polyploids is based on the relatedness of a polyploid's constituent subgenomes in a tree. That is, if homeologous gene copies within the polyploid taxon are more closely related to each other than to gene copies from another taxon, it is commonly classified as an autopolyploid. Conversely, the homeologous gene copies of a phylogenetic allopolyploid would not form a clade; rather one or more homeologs would be more closely related to another taxon than to other homeologs.

Generally, phylogenetic inferences of polyploids are focused on identifying their origin, from either a past hybridization event or a duplication within a lineage. Unlike the cytogenetic state and inheritance mode, the phylogenetic context in which a polyploid emerged cannot change. However, it may become harder to infer the phylogenetic origin with the passage of time and accompanying genome evolution and lineage extinction.

1.1.3 Segregation Patterns in Offspring and Gametes

To study inheritance patterns, the gold standard is to observe genotypes in parents and their offspring. Observed inheritance patterns in parents and their F_1 hybrids can then be compared to the different possible inheritance modes, e.g., disomic or tetrasomic inheritance [10]. However, such an approach is not always feasible for a number of reasons, including the time and expense involved in breeding studies. One possible "shortcut" is single-cell sequencing of gametes [11], which permits high-throughput genotyping on many thousand haplotypes and can easily be expanded to polyploid taxa [12]. While phylogenetic inference of polyploids is typically used to understand something that happened in the past, namely, the origin of a polyploid, understanding the current process of inheritance allows insights into the ongoing evolution of polyploid taxa.

1.1.4 Conflict Between Inferences

When making inferences about the origin of a given polyploid, it is possible to observe, e.g., a cytogenetic allopolyploid that is a phylogenetic autopolyploid and other seemingly disparate combinations. For example, autotetraploid *Arabidopsis arenosa* has a tetrasomic inheritance mode despite forming bivalents during meiosis [13]. The possibility of different observations from (a) the phylogenetic classification of polyploid origin, (b) cytogenetic classification of polyploid meiosis, and (c) polyploid inheritance patterns highlights the importance of integrating multiple inference methods to understand polyploid evolution. The origin of polyploids often dictates the way we analyze genome sequencing data. Sequencing data from allopolyploids with sufficient divergence between subgenomes can often be split either by mapping to the combined reference of the parental genomes with subsequent filtering for primary alignment in proper pairs [14, 15] or by using SNP markers specific to each parent [16, 17]. Divided subgenomes are then analyzed as if they were separate taxa. Autopolyploid data are usually mapped to a single reference genome and the number of alleles called at each site is equal to ploidy [18, 19], similar to pooled sequencing experiments, where the number of alleles corresponds to the number of haploid individuals (*see* also the chapter by Bohutinska et al.). The inheritance mode of polyploids impacts the evolutionary inferences we draw from analyzing such data and deviations from the assumed mode can strongly bias the estimates of genetic diversity and population divergence [20].

1.2 Sources of Mixed Inheritance Patterns

1.2.1 Homeologous Exchanges

Auto- and allopolyploid origins do not have a strict correspondence in chromosomal pairing behavior. Although rare, the homeologous chromosomes from the different subgenomes in allopolyploids can sometimes pair and recombine [21–23]. Homeologous exchanges are common in recent allopolyploids, for example, in *Tragopogon miscellus* [24], cultivated peanut [25], and resynthesized allotetraploid rice [26]. In more established allopolyploids, homeologous exchanges are rare and appear to be under genetic control. For example, in allohexaploid wheat, *Ph1* locus [27–29] carries the major crossover gene *ZIP4* and prevents homeologous crossovers. *PrBn* locus in oilseed rape (*Brassica napus*) has a similar effect [30]. Preferential homeologous pairing in *A. suecica* also has a genetic basis [31] (*BOY NAMED SUE* locus) but also appears to be regulated on the transcriptional level of meiotic genes [14]. This suggests that homeologous exchanges are mostly deleterious, and allopolyploids do not only rely on the divergence levels between the subgenomes to prevent these generally undesirable genome reshufflings. Despite stricter regulation of preferential pairing between homologs, homeologous exchanges do happen, and we expect such regions to show apparent mixed inheritance patterns.

1.2.2 Rediploidization

Rediploidization is the process by which a polyploid gradually behaves more and more like a diploid such that, among others, (a) multivalent formation is replaced by strict bivalent pairing and (b) polysomic inheritance is no longer observed [4, 32, 33]. In general, the term rediploidization is used when discussing the evolution of autopolyploids, as allopolyploids often behave largely as diploids (cytogenetically speaking). Unlike a whole-genome duplication event, which happens quickly and is followed by a polyploid establishment process, rediploidization takes place over longer periods of time—anywhere from 20–30 generations [34] to tens of millions of years [35]. Furthermore, as rediploidization is an ongoing process, it is likely that different lineages are experiencing the process in different ways [35], such that a handful of observations may not be representative of the evolutionary dynamics at play. For example, when bivalent and multivalent pairing are both possible, different chromosomes and lines may exhibit an inconsistent pattern [36]. It is possible that parts of the genome show disomic inheritance, while other regions do not, the latter associated with, for instance, interchromosomal translocations and inversions. This drawn-out rediploidization process can therefore create mixed inheritance patterns when we observe polyploid taxa caught "in limbo" between an autopolyploid and a fully diploidized state.

1.2.3 Interspecific Introgression

Interspecific introgression can become another source of mixed inheritance patterns. Recent studies in *Arabidopsis* plants [37, 38] and *Neobatrachus* frogs [39] suggested that despite the apparent isolation of the diploid species, tetraploid lineages in those groups can hybridize. This suggests that polyploidy can mediate interspecific gene flow [40]. One explanation of this phenomenon in seed plants is based on the hypothesis that polyploidy permits the breakdown of endosperm-based incompatibility, thus lifting the postzygotic barrier and allowing hybridization among species [41, 42]. Another explanation could be the more effective masking of recessive hybrid incompatibility loci in polyploids. Hybridization can happen (a) between tetraploids with different ancestors and (b) between a tetraploid and an unreduced gamete of a non-ancestral diploid species. If genomes of both species are sufficiently diverged to prevent homologous pairing, both cases would result in mixed inheritance patterns.

Except for lab-generated polyploids, we are not privy to the polyploidization events of the taxa we study. Furthermore, researchers rarely have access to cytogenetic, genomic, and breeding resources to allow direct observation of chromosome pairing, phylogenetic context, and inheritance patterns of their study systems. The methods we describe here are meant to facilitate the study of polyploids and the consequences of their whole-genome duplications, using bioinformatic approaches to infer origin and inheritance modes from population genomic data. The first method uses allelic depth distributions within an individual and classifies their fit to expected distributions for disomic and tetrasomic inheritance modes. The second method leverages haplotype distances along the genome to infer origin of polyploidy and deviations from expected inheritance modes.

2 Materials

The computational inference of polyploid inheritance mode requires sequenced datasets of the polyploid and related diploid lineages. We demonstrate both approaches using a reference genome and whole-genome resequencing data with short paired reads; however, these methods can also work with reduced representation (e.g., RADseq) or exome-capture sequencing data

[39]. Data from several individuals per lineage (we use at least four individuals in our examples) will account for within-lineage variation (but not deep coalescence) and ensure the robustness of the approach. Analysis of the whole-genome sequencing data usually requires a high-performance computer; however, the computational load depends on the genome size and sequencing depth. To illustrate the approaches we describe in this chapter, we used previously published [15, 43–45] raw reads from *A. suecica* (SRR2084158, SRR3123767, SRR3123760, SRR2084157), *A. arenosa* (SRR2040811, SRR2040813, SRR2040820, SRR2040824, SRR3111444, SRR3111445, SRR2040800, SRR2040802), and *A. thaliana* (SRR519491, SRR519660, SRR519519, SRR519685, SRR519656, SRR519575, SRR519494, SRR519484, SRR519647, SRR519714, SRR519613, SRR519641, SRR519684, SRR519505, SRR519489, SRR519687).

3 Methods

The first approach to infer inheritance modes of sequenced samples is based on the differences in allelic read depth expectations. In a diploid sample, the distribution of allelic depth at biallelic heterozygous (Aa) sites is expected to be centered around 0.5, where 50% of the reads belong to the first (A) and another 50% to the second (a) allele. In tetraploid samples, biallelic sites can form three heterozygous genotypic states: AAAa, AAaa, and Aaaa. The expectations of the relative frequency of each genotype in tetraploids are different for autotetraploids (with tetrasomic inheritance) and allotetraploids (with disomic inheritance) [13, 46] (*see* **Note 1**). In the case of an allotetraploid origin or long-term evolution under disomic inheritance in an original autotetraploid, we expect to see an excess of AAaa genotypes due to subgenome divergence. Similarly, on a population level, we expect to see an excess of SNPs with intermediate allele frequencies. The amount of such excess depends on the evolutionary history of the polyploid and its diploid relatives and can be compared to simulated data or computationally combined diploid data into tetraploids of different origins. In order to illustrate the method, we show examples of allelic depth distributions from both simulated and real sequencing data (Fig. 1).

We first applied our method to simulated diploid and polyploid individuals (Fig. 1a). Briefly, we simulated a tree, genome, and sequences with Zombi [47]. These sequences were then used to simulate Illumina HiSeq reads using InSilicoSeq [48]. Based on the tree topology, we selected eight diploid individuals and one reference accession. The diploid genomes were combined to form four autotetraploids and four allotetraploids, mimicking the immediate outcome following polyploidization (that is, without

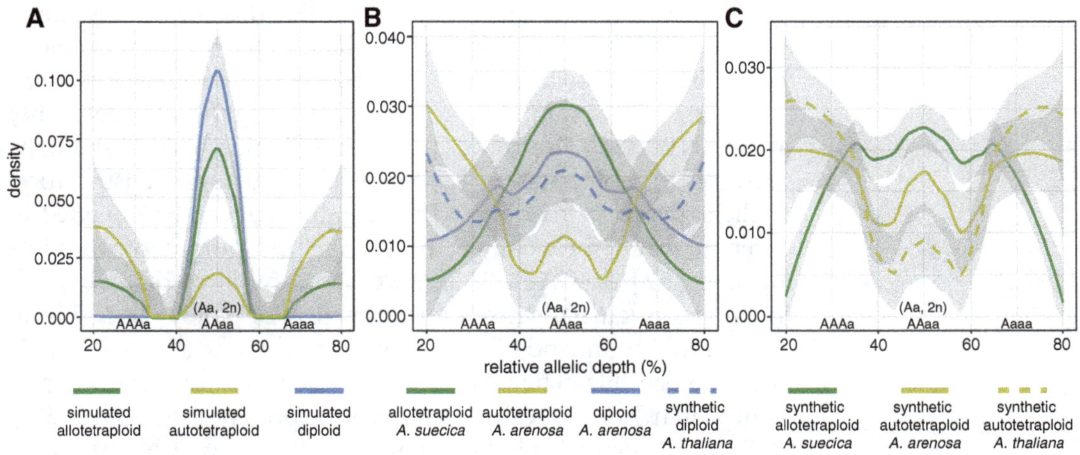

Fig. 1 Inference of inheritance modes of polyploids using allelic read depth distributions. (**a**) Simulated diploid (blue line) and allotetraploid (green line) samples show excess of intermediate ratios of allelic depth: Aa or AAaa genotypes, respectively. This contrasts with simulated autotetraploid (yellow line), where the intermediate allelic ratio does not have such a pronounced peak. (**b**) Real data examples showing the distributions of allelic read depth for diploid *A. arenosa* (blue line), autotetraploid *A. arenosa* (yellow line), and allotetraploid *A. suecica*. Because *A. thaliana* is highly inbred due to selfing, we combined sequencing data of two different accessions to show expected distribution of an "outcrossing" diploid *A. thaliana* (dashed blue line). (**c**) Synthetic controls for the real data examples. Green line shows computationally combined sequencing data of two haploid *A. thaliana* accessions and one diploid *A. arenosa* making an "outcrossing" allotetraploid with a similar genetic composition to *A. suecica*. Yellow lines show synthetic autotetraploids: dashed, four combined accessions of *A. thaliana*; solid, two combined accessions of diploid *A. arenosa*

rediploidization). These simulated accessions were then analyzed as described for the real data below. The obtained distributions of allelic ratios for the simulated data follow the expected pattern: diploid and allotetraploid samples have an excess of intermediate ratios (Aa and AAaa), when genotypes of the simulated autotetraploids are more evenly distributed.

As a real data example, we chose *Arabidopsis* species with well-described evolutionary histories and known inheritance modes [13–15, 43, 44, 49]: *A. thaliana*, a diploid; *A. arenosa*, both diploid and autotetraploid lineages; and *A. suecica*, an allotetraploid hybrid between *A. thaliana* and *A. arenosa*. This example also demonstrates the importance of taking a mating type of the species into account when estimating inheritance modes (*see* **Note 2**), as this method is based on heterozygous sites only—selfing species will produce an excess of homozygotes compared to outcrossers. Both *A. thaliana* and *A. suecica* are self-compatible and mostly selfing plants, while both diploid and tetraploid lineages of *A. arenosa* are obligate outcrossers. For our example, we mapped 12 different *A. thaliana*, 4 diploid and 4 tetraploid *A. arenosa* lineages and 4 *A. suecica* accessions to the *Arabidopsis lyrata* reference genome [50]. For mapping, we chose the bwa mem algorithm [51], which allows split reads and is therefore suitable for cross-species reference mapping with greater divergence times. For this comparative analysis between different species, we chose only the genic regions which are more conserved and usually better covered compared to the intergenic regions. Apart from increasing the average quality of the SNPs, such an approach of region filtering also substantially decreases the computational time for further analysis. Distribution of allelic depth at biallelic SNPs per individual can be obtained in different ways (for example, by parsing the output of samtools mpileup [52] or GATK [53]); we chose nQuire software [54], which was originally written for estimation of ploidy levels. Using the "create" command of nQuire, we first generate a binary .bin file from the bam file. Then we apply the "denoise" algorithm implemented in nQuire to scale down the noise present from low-quality SNPs and potential misalignments. Finally, we visualize the base frequencies for each individual using the command "histo" from nQuire, which creates a histogram in ASCII format, easily parsed in R [55].

The real data example of the distribution of allelic depth follows the expectations: the obligatory outcrossing *A. arenosa* diploid lineage (Fig. 1b, blue line) has the highest peak at the 50% allelic ratio (Aa), while in the autotetraploid *A. arenosa* (Fig. 1b, yellow line), AAAa and Aaaa genotypes seem to be more common. The small bumps in the diploid *A. arenosa* (solid blue line) and *A. thaliana* (dashed blue line) samples may be explained by paralogous mismapping to the *A. lyrata* reference genome. In the case of *A. thaliana* species, we need to consider that it is mostly a selfing plant, resulting in a dearth of heterozygous sites and therefore no peak (at genotype Aa). To show the expected patterns based on *A. thaliana* sequences, we combine reads from two different *A. thaliana* accessions to obtain a "synthetic" diploid *A. thaliana* (Fig. 1b, dashed blue line), which follow a similar pattern to a diploid outcrossing *A. arenosa*, except for the tails of the distribution. This difference in the tails, where the synthetic diploid *A. thaliana* more closely resembles true tetraploid *A. arenosa*, may be explained by rare heterozygous sites within *A. thaliana* accessions. *A. suecica* (Fig. 1b, green) is also a selfer and an allotetraploid with an extremely high divergence (about 5 My or 9%) [43, 56] between the subgenomes: the excess of intermediate alleles (AAaa genotype) is prominent.

Although in these examples (Fig. 1b) the differences in the distributions are obvious, in order to make the inferences we still prefer to compare the real data examples to the synthetic combinations of the real data forming synthetic auto- and allotetraploids from the exact diploid relatives. We formed synthetic autotetraploid *A. arenosa* by combining reads from two different outcrossing diploid *A. arenosa* individuals (Fig. 1c, solid yellow line) and formed a synthetic autotetraploid *A. thaliana* by combining four selfing *A. thaliana* accessions (Fig. 1c, dashed yellow line). Both of those distributions lack an excess of intermediate allelic ratio genotypes (AAaa) and resemble real autotetraploid *A. arenosa* data (Fig. 1b, yellow line). We have also combined two "haploid" selfing *A. thaliana* and one diploid outcrossing *A. arenosa* to simulate an outcrossing allotetraploid (Fig. 1c, green). This distribution shows an excess of intermediate frequencies, but also differs substantially from the real *A. suecica* data, because *A. suecica* is selfing and all the biallelic sites are distributed around a 50% allelic ratio, representing the differences between the subgenomes (AAaa, where AA = *A. thaliana* and aa = *A. arenosa* variant, for example). In cases where the differences are not as obvious, one could use, for example, the Mann–Whitney *U* test to compare the distributions of ratios between intermediate (40–60%) and rare (<30%) allele frequencies in real data samples and synthetically combined controls.

The main drawback of the approach based on allelic frequencies is that any deviation from the inheritance mode at the origin of the polyploid (for example, due to ongoing introgression with other species, rediploidization, or homeologous exchanges) will lead to an averaged mixed signal on the relative allelic depth distributions. Inspecting inheritance mode along the genome, in this case, would provide a more reliable result and pinpoint the regions leading to the mixed signals. Next, we describe an approach that allows investigating the inheritance mode along the genome using the same type of input data.

The second approach uses distance-based phylogenetic relatedness between haplotypes. Ideally, if one had a full high-quality assembly of all four chromosomal copies of a tetraploid and closely related diploid species, the identification of the origin, evolutionary history, and inheritance mode of such tetraploid would be a matter of inferring phylogenetic relationships between each tetraploid allele and the diploids. While this sounds straightforward, phylogenetic inference over many thousands of loci is complicated by, for instance, poor phylogenetic resolution within regions and considerable conflict among regions (*see* **Note 3**). Despite substantial advances in long-read sequencing, it is still most common to obtain a long-read-based reference and several short-read-based resequenced genomes. Short reads do not allow a complete chromosome-scale haplotype-phased resolution; however, it is possible to distinguish between the two most divergent haplotypes within relatively short and well-covered loci using paired-end information. For example, the phasing algorithm implemented in the samtools phase [52] splits a .bam file resulting from the mapping of short reads to a reference genome into the two most divergent haplotypes (which makes the method suitable only for tetraploids, but not higher ploidy-level samples, *see***Note 4**). Here, one should be aware that regions with low coverage or a low SNP density would interrupt continuous phasing and randomize haplotypes in the two output files, so the phasing of the haplotypes is typically only reliable within relatively short loci. Nevertheless, this approach allows estimation of phylogeny within, e.g., genic regions and has been applied to infer the origins of, for example, allotetraploid *Medicago* species [9].

Using the same *Arabidopsis* samples as in the first approach (*A. thaliana* – $2n = 2\times$, selfing; *A. arenosa* – $2n = 2\times$ and auto $4n = 4\times$, outcrossing; and *A. suecica* – allo $4n = 4\times$, selfing), we demonstrate the framework to infer inheritance modes using partially (*see***Note 5**) phased haplotypes. Again, we use the synthetic controls for auto- and allotetraploids (or tetrasomic and disomic inheritance modes in this case): computationally combined *A. thaliana* and *A. arenosa* samples within and between the species, respectively (Fig. 2a). We describe the procedure starting from the mapped short reads of all the samples to the same reference genome (in our case, the *A. lyrata* reference genome [50]):

1. We split the .bam files of the tetraploids using the samtools phase [52] into two .bam files—this step includes both real data polyploids and synthetic controls (*see***Note 5**, Fig. 2c).
2. Then, we call variants from all the diploid samples and the two phases of the polyploids separately, using the GATK Haplotype Caller, with setting sample ploidy to 2 [53]. To speed up the process we only call the variants on the genic regions, which we assume would also be better covered and therefore better phased. For an estimate of appropriate coverage for SNP calling, Song et al. [57] recommend $15\times$.
3. Then we combine the variants and perform joint genotyping using two subsequent steps of CombineGVCF and GenotypeGCVF from GATK. For further analysis, we filter for biallelic single-nucleotide polymorphisms (SNPs), which are present in all the diploid samples.
4. We transform the variant table to numeric values, by summing up the amount of nonreference alleles and dividing the result by the number of alleles. So, the diploid samples and split tetraploid samples with reference allele would have 0 (AA), nonreference homozygous allele 1 (aa), and heterozygous allele 0.5 (Aa).
5. Then, using R [55], we identified biallelic SNPs present in all 16 diploid samples and encoded them on a scale from 0 (homozygous reference) to 1 (homozygous alternate) in a separate table for each gene. To determine genetic distance to a real or synthetic polyploid, we added one phase from the polyploid to each gene containing at least five SNPs. We then calculated the absolute pairwise distance between all pairs of samples in each gene variant table that contained all diploids and one phase of one polyploid.

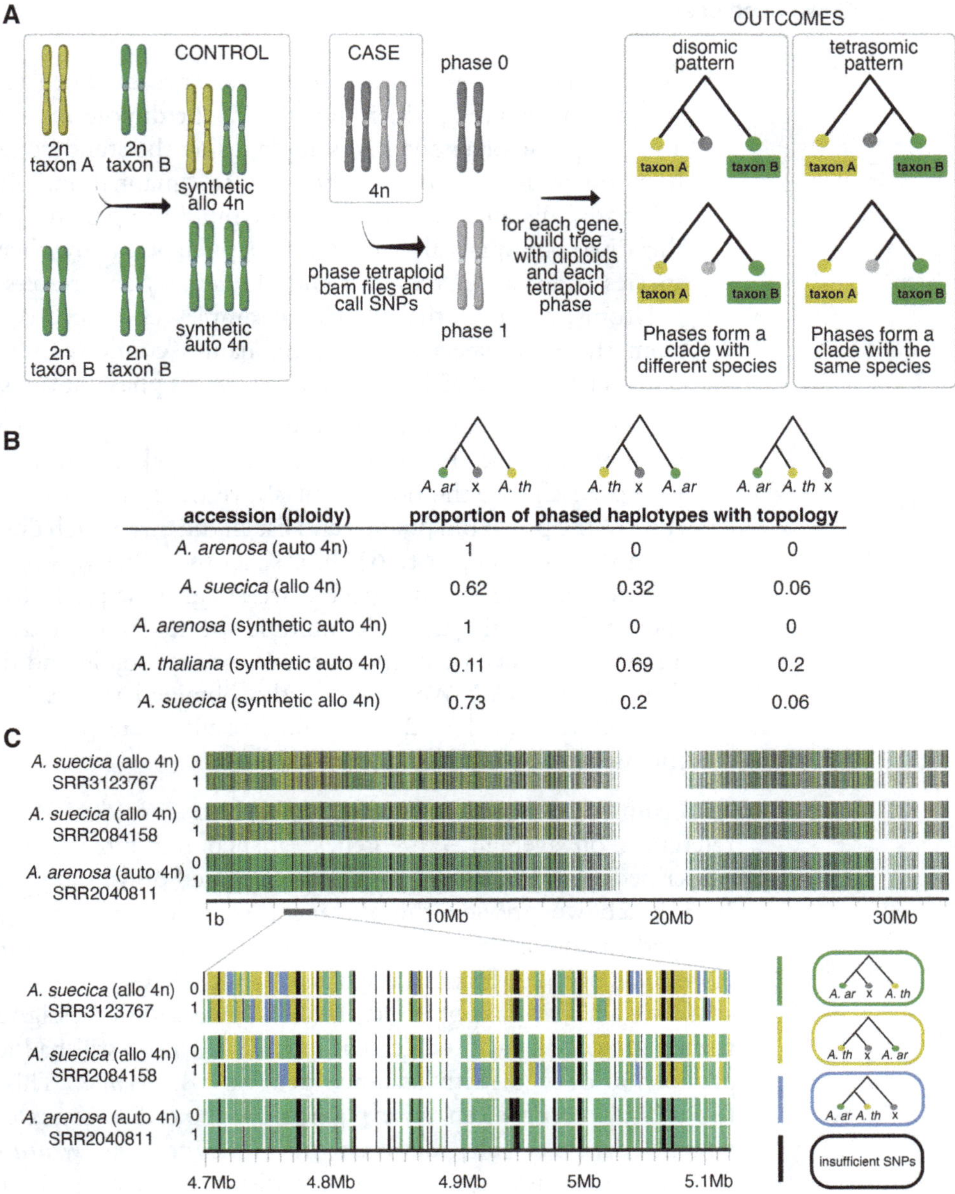

Fig. 2 Inference of origin and inheritance modes of polyploids using haplotype phylogenies. (**a**) Schematic representation of the computational experiment; created with BioRender.com. Synthetic controls are based on combinations of real sequencing data of the related diploid species (same as in the first approach, Fig. 1c). To obtain a synthetic autotetraploid, we combine two real diploid samples within the same species; to obtain a synthetic allotetraploid, we combine diploid samples of different species. Diploid samples, synthetic controls, and the real data from polyploids are mapped to the same reference genome and then real and synthetic polyploids are split into two phases. Assuming that phasing is correct within the genic regions, we investigate the phylogenetic placement of each polyploid phase within each gene/locus: the two phases from the allopolyploid samples would cluster with different diploid species, when the two phases of autopolyploid species would cluster with the same species. (**b**) Results obtained from the data of natural *A. suecica* and *A. arenosa* polyploids, as well as synthetic controls. The schematic table represents the average proportions of the genic regions assigned to *A. arenosa*, *A. thaliana*, or "NA"; topologies illustrating the assignment procedure are shown above each column. (**c**) Assignment of each gene in each phase for two natural allotetraploid *A. suecica* and one natural autotetraploid *A. arenosa* individuals, with scaffold distribution of tree topologies. The zoomed in region shows the homeologous exchange pattern on one of the *A. suecica* accessions (SRR3123767)

6. We calculate the absolute pairwise distance between all the samples in the table, which includes all the diploid individuals and one phase of a polyploid sample, using the standard "dist" function in R with the method set to "manhattan." If the calculated distance between the polyploid sample and any of the diploids equals 0, which can happen with low-diversity species (such as *A. thaliana* and *A. suecica*) and results in a polytomy, we check that the clonal samples are indeed coming from the same species (list of tip names contains only one species name) and assign this species to this phase of the gene.
7. In the rest and majority of the cases, we then calculate a neighbor-joining tree using the "ape" package in R [58], where we choose the node right above the sample of interest (one of the polyploid phases) and list all the tips, which descend from this node (*see* **Note 6**). In case all the other tips (diploid samples) belong to one species, we assign that phase to that species, in case they include multiple species (both *A. arenosa* and *A. thaliana*), we cannot assign a closest diploid and therefore call it "NA." We illustrate the obtained results for the example data on Fig. 2b. The final results were plotted with the use of karyoploteR [59].

From 5226 genes annotated on the first scaffold of *A. lyrata* reference, on average 3763 genes satisfied our filtering criteria (described in step 5 above) for the investigated polyploid samples. In Fig. 2b we provide the average proportion of these genes assigned to one or the other lineage for the natural *A. suecica* and *A. arenosa* polyploids and the synthetic controls. Genes from both synthetic and natural polyploid *A. arenosa* were always assigned to the *A. arenosa* lineage, while synthetic *A. thaliana* still had a low proportion of incorrectly assigned genes to *A. arenosa*. This bias towards *A. arenosa* can be explained by about five times closer genetic distance [56] between the reference *A. lyrata* and *A. arenosa* species compared to the distance between *A. lyrata* and *A. thaliana*. As expected for an allotetraploid, the genes from synthetic and natural *A. suecica* were more evenly assigned to both of the parental (*A. thaliana* and *A. arenosa*) lineages, however, with a certain bias towards *A. arenosa*.

The second method also allows investigation of deviations from expected patterns. These deviations along the genome can be due to reasons mentioned in the introduction (e.g., ongoing rediploidization, interspecific introgression, and homeologous exchanges between subgenomes of an allotetraploid). While these post-polyploid evolutionary processes impact allelic segregation, deep coalescence (pre-polyploidization), which also results in a conflicting signature of inheritance mode scattered across the genome, does not impact allelic segregation. This is clearly demonstrated in the case of *A. suecica*, a known allopolyploid (hybridization between *A. arenosa* and *A. thaliana*) which we expect to exhibit preferential (bivalent) chromosome pairing and thus disomic inheritance. Figure 2c shows a stretch of chromosome 1 with a different inferred inheritance mode than the rest of the genome; this is a previously described [14] homeologous exchange between the first chromosome of *A. suecica* (*A. thaliana* subgenome) and the sixth chromosome of *A. suecica* (*A. arenosa* subgenome) in SRR3123767 (AS530) accession. This region of almost 16 Mb of the *A. arenosa* subgenome has been exchanged for the *A. thaliana* copy. Despite this apparent deviation, we would still infer disomic inheritance as the norm for this accession, as this homeologous exchange is not expected to break the previously established preferential pairing among chromosomes. However, this example merits some attention as over time, allopolyploids can accumulate extensive homeologous exchanges, resulting in "candy cane" chromosomes (where the closest relative of each stretch of chromosome switches between progenitors). Furthermore, when diploidization follows significant homeologous exchange, the resulting genome will contain a mixture of genomic regions from both parental species. This case highlights the importance of evolutionary context for the correct interpretation of the observed phylogenetic patterns of the haplotypes along a polyploid genome. In contrast, the first method we describe is more flexible, as it can be applied even when information about polyploid origin is unclear, e.g., when parental genome donors are extinct, unidentified, or otherwise unavailable for sequencing.

4 Outlook

With currently available long-read sequencing, assembly of chromosome scale (or at least substantially longer) haplotypes can replace the need for genomic phasing. This, of course, does not nullify the phylogenetic approach to inheritance mode inference, but rather makes it applicable to systems with higher ploidy levels. Long-read sequencing may eventually be universally available, but in the meantime short-read-based approaches are still useful. Because long-read technology requires high-quality, high-molecular-weight input DNA, this method is inaccessible in some cases, for example, ancient or degraded samples, or those stored in formalin. At the same time, the affordability of sequencing already allows the assessment of hundreds of offspring samples or even gametes to directly measure inheritance patterns, avoiding the estimations altogether. Limitations of this approach again include ancient samples and extinct species, as well as species which are unfeasible to cross (e.g., due to generation time, difficulty of cultivation) and which are hard to obtain in the field, such as protected species.

5 Notes

1. In the method based on allelic depth distributions, we assume that polyploid samples are mapped to a single ancestral (parental) genome, even in the case of allotetraploids, where one of the subgenomes could be closer to the reference genome in this case. In the case of mapping an allotetraploid to a reference, containing both of the subgenomes, we expect to see a diploid-like allelic depth distribution.
2. When mating type is unclear, it is useful to start with estimations of overall heterozygosity. Additionally, one should be aware that an outcrossing diploid and a selfing allotetraploid can show a similar distribution of allelic depth at the heterozygous genotypes with a single peak at 50%. In order to distinguish between those types (outcrossing diploid and selfing allotetraploid), we recommend including a genome size estimation into the analysis, which can also be done bioinformatically, analyzing k-mer frequencies from the same type of input—whole-genome short-read sequencing data [60–62]. Also note that k-mer-based genome size estimation may be unreliable in selfing autopolyploids due to homozygosity; one would need to measure the genome size directly using, for example, flow cytometry.
3. Regarding our second method, it is important to consider the expected variation in topologies among gene trees—different loci will tell different stories, so care must be taken to include many gene trees. Furthermore, the methods described herein do not account for uncertainty in gene tree estimation, nor do they attribute discordance along a chromosome to deep coalescence.
4. The phasing algorithm [63] implemented in the samtools phase [52] is tailored to assemble only the two "best" haplotypes, which makes it applicable for diploid and tetraploid samples, but probably not higher ploidies. Even with the tetraploid samples, one should be aware that such phasing is only partial. That is why we describe the synthetic controls to use for the correct interpretation of the results.
5. One should be aware that the samtools phase will split one bam into two separate files, which do not represent two completely phased chromosomes, but rather short stretches of phased data mixed between the chromosomes (this is also demonstrated in our results, Fig. 2c). The length of the phased stretches will depend on the length of the short reads, the coverage depth, the mapping quality in that region. and the amount of polymorphisms, which allow the assignment of haplotypes into two separate phases. The phasing will be interrupted with the gap between the markers longer than a fragment length.

6. Here we employ distance-based phylogenetics to estimate neighbor-joining trees, which is a fast process that scales well to genome-wide analyses. Other phylogenetic inference methods (based on maximum likelihood or Bayesian inference) are also suitable, but will be more computationally intensive.
7. In the case of method 2, branch lengths are ignored but could be used for an estimate of divergence time between subgenomes. This would be best performed using tree estimation by a different method (maximum likelihood, Bayesian inference) and therefore would also require base call information to incorporate a substitution model.

Acknowledgments

We thank Robin Burns, Filip Kolář, Matt Johnson, and Arthur Zwaenepoel for their insights and advice during the preparation of this chapter.

References

1. Stebbins GL Jr (1947) Types of polyploids; their classification and significance. Adv Genet 1:403–429
2. Li Z, McKibben MTW, Finch GS et al (2021) Patterns and processes of diploidization in land plants. Annu Rev Plant Biol 72:387–410
3. Soltis DE, Soltis PS, Rieseberg LH (1993) Molecular data and the dynamic nature of polyploidy. CRC Crit Rev Plant Sci 12:243–273
4. Ramsey J, Schemske DW (2002) Neopolyploidy in flowering plants. Annu Rev Ecol Syst 33:589–639
5. Muller HJ (1914) A new mode of segregation in Gregory's tetraploid primulas. Am Nat 48: 508–512
6. Little TM (1945) Gene segregation in autotetraploids. Bot Rev 11:60
7. Darlington CD (1929) Chromosome behaviour and structural hybridity in theTradescantiae. J Genet 21:207–286
8. Butruille DV, Boiteux LS (2000) Selection-mutation balance in polysomic tetraploids: impact of double reduction and gametophytic selection on the frequency and subchromosomal localization of deleterious mutations. Proc Natl Acad Sci U S A 97(12):6608–6613
9. Eriksson JS, de Sousa F, Bertrand YJK et al (2018) Allele phasing is critical to revealing a shared allopolyploid origin of Medicago arborea and M. strasseri (Fabaceae). BMC Evol Biol 18:9
10. Stift M, Berenos C, Kuperus P et al (2008) Segregation models for disomic, tetrasomic and intermediate inheritance in tetraploids: a general procedure applied to Rorippa (yellow cress) microsatellite data. Genetics 179:2113–2123
11. Campoy JA, Sun H, Goel M et al (2020) Gamete binning: chromosome-level and haplotype-resolved genome assembly enabled by high-throughput single-cell sequencing of gamete genomes. Genome Biol 21:306
12. Sun H, Jiao W-B, Krause K et al (2021) Chromosome-scale and haplotype-resolved genome assembly of a tetraploid potato cultivar. https://www.biorxiv.org/content/10.1101/2021.05.15.444292v1
13. Hollister JD, Arnold BJ, Svedin E et al (2012) Genetic adaptation associated with genome-doubling in autotetraploid Arabidopsis arenosa. PLoS Genet 8:e1003093
14. Burns R, Mandáková T, Jagoda J et al (2020) Gradual evolution of allopolyploidy in Arabidopsis suecica. https://www.biorxiv.org/content/10.1101/2020.08.24.264432v1
15. Novikova PY, Tsuchimatsu T, Simon S et al (2017) Genome sequencing reveals the origin of the allotetraploid Arabidopsis suecica. Mol Biol Evol 34:957–968
16. Akama S, Shimizu-Inatsugi R, Shimizu KK et al (2014) Genome-wide quantification of homeolog expression ratio revealed nonstochastic

gene regulation in synthetic allopolyploid Arabidopsis. Nucleic Acids Res 42:e46

17. Paape T, Briskine RV, Halstead-Nussloch G et al (2018) Patterns of polymorphism and selection in the subgenomes of the allopolyploid Arabidopsis kamchatica. Nat Commun 9:3909
18. Yant L, Hollister JD, Wright KM et al (2013) Meiotic adaptation to genome duplication in Arabidopsis arenosa. Curr Biol 23:2151–2156
19. Monnahan P, Kolář F, Baduel P et al (2019) Pervasive population genomic consequences of genome duplication in Arabidopsis arenosa. Nat Ecol Evol 3:457–468
20. Meirmans PG, Van Tienderen PH (2013) The effects of inheritance in tetraploids on genetic diversity and population divergence. Heredity 110:131–137
21. Gaeta RT, Chris Pires J (2010) Homoeologous recombination in allopolyploids: the polyploid ratchet. New Phytol 186:18–28
22. Zhang Z, Gou X, Xun H et al (2020) Homoeologous exchanges occur through intragenic recombination generating novel transcripts and proteins in wheat and other polyploids. Proc Natl Acad Sci U S A 117:14561–14571
23. Edger PP, McKain MR, Bird KA et al (2018) Subgenome assignment in allopolyploids: challenges and future directions. Curr Opin Plant Biol 42:76–80
24. Chester M, Gallagher JP, Symonds VV et al (2012) Extensive chromosomal variation in a recently formed natural allopolyploid species, Tragopogon miscellus (Asteraceae). Proc Natl Acad Sci U S A 109:1176–1181
25. Bertioli DJ, Jenkins J, Clevenger J et al (2019) The genome sequence of segmental allotetraploid peanut Arachis hypogaea. Nat Genet 51: 877–884
26. Wu Y, Lin F, Zhou Y et al (2020) Genomic mosaicism due to homoeologous exchange generates extensive phenotypic diversity in nascent allopolyploids. Natl Sci Rev 8(5): nwaa277
27. Riley R, Chapman V (1958) Genetic control of the cytologically diploid behaviour of hexaploid wheat. Nature 182:713–715
28. Sears ER (1958) Intergenomic chromosome relationships in hexaploid wheat. Proc Int Congr Genet 2:258–259
29. Sears ER (1977) Genetics society of Canada award of excellence lecture an induced mutant with homoeologous pairing in common wheat. Can J Genet Cytol 19:585–593
30. Jenczewski E, Eber F, Grimaud A et al (2003) PrBn, a major gene controlling homeologous pairing in oilseed rape (Brassica napus) haploids. Genetics 164:645–653
31. Henry IM, Dilkes BP, Tyagi A et al (2014) The BOY NAMED SUE quantitative trait locus confers increased meiotic stability to an adapted natural allopolyploid of Arabidopsis. Plant Cell 26:181–194
32. Wolfe KH (2001) Yesterday's polyploids and the mystery of diploidization. Nat Rev Genet 2:333–341
33. Parisod C, Holderegger R, Brochmann C (2010) Evolutionary consequences of autopolyploidy. New Phytol 186:5–17
34. Weiss H, Maluszynska J (2000) Chromosomal rearrangement in autotetraploid plants of Arabidopsis thaliana. Hereditas 133:255–261
35. Robertson FM, Gundappa MK, Grammes F et al (2017) Lineage-specific rediploidization is a mechanism to explain time-lags between genome duplication and evolutionary diversification. Genome Biol 18:111
36. Santos JL, Alfaro D, Sanchez-Moran E et al (2003) Partial diploidization of meiosis in autotetraploid Arabidopsis thaliana. Genetics 165:1533–1540
37. Marburger S, Monnahan P, Seear PJ et al (2019) Interspecific introgression mediates adaptation to whole genome duplication. Nat Commun 10:5218
38. Schmickl R, Koch MA (2011) Arabidopsis hybrid speciation processes. Proc Natl Acad Sci U S A 108:14192–14197
39. Novikova PY, Brennan IG, Booker W et al (2020) Polyploidy breaks speciation barriers in Australian burrowing frogs Neobatrachus. PLoS Genet 16:e1008769
40. Schmickl R, Yant L (2021) Adaptive introgression: how polyploidy reshapes gene flow landscapes. New Phytol 230(2):457–461
41. Lafon-Placette C, Hatorangan MR, Steige KA et al (2018) Paternally expressed imprinted genes associate with hybridization barriers in Capsella. Nat Plants 4:352–357
42. Johnston SA, Hanneman RE Jr (1982) Manipulations of endosperm balance number overcome crossing barriers between diploid solanum species. Science 217:446–448
43. Novikova PY, Hohmann N, Nizhynska V et al (2016) Sequencing of the genus Arabidopsis identifies a complex history of nonbifurcating speciation and abundant trans-specific polymorphism. Nat Genet 48:1077–1082
44. The 1001 Genomes Consortium (2016) 1,135 Genomes reveal the global pattern of polymorphism in Arabidopsis thaliana. Cell 166:481–491

45. Long Q, Rabanal FA, Meng D et al (2013) Massive genomic variation and strong selection in Arabidopsis thaliana lines from Sweden. Nat Genet 45:884–890
46. Lloyd A, Bomblies K (2016) Meiosis in autopolyploid and allopolyploid Arabidopsis. Curr Opin Plant Biol 30:116–122
47. Davín AA, Tricou T, Tannier E et al (2020) Zombi: a phylogenetic simulator of trees, genomes and sequences that accounts for dead linages. Bioinformatics 36:1286–1288
48. Gourlé H, Karlsson-Lindsjö O, Hayer J et al (2019) Simulating Illumina metagenomic data with InSilicoSeq. Bioinformatics 35:521–522
49. Novikova PY, Hohmann N, Van de Peer Y (2018) Polyploid Arabidopsis species originated around recent glaciation maxima. Curr Opin Plant Biol 42:8–15
50. Hu TT, Pattyn P, Bakker EG et al (2011) The Arabidopsis lyrata genome sequence and the basis of rapid genome size change. Nat Genet 43:476–481
51. Li H, Durbin R (2009) Fast and accurate short read alignment with Burrows-Wheeler transform. Bioinformatics 25:1754–1760
52. Li H, Handsaker B, Wysoker A et al (2009) The sequence alignment/map format and SAMtools. Bioinformatics 25:2078–2079
53. McKenna A, Hanna M, Banks E et al (2010) The Genome Analysis Toolkit: a MapReduce framework for analyzing next-generation DNA sequencing data. Genome Res 20:1297–1303
54. Weiss CL, Pais M, Cano LM et al (2018) nQuire: a statistical framework for ploidy estimation using next generation sequencing. BMC Bioinform 19:122
55. R Core Team (2020) R: a language and environment for statistical computing
56. Hohmann N, Wolf EM, Lysak MA et al (2015) A time-calibrated road map of Brassicaceae species radiation and evolutionary history. Plant Cell 27:2770–2784
57. Song K, Li L, Zhang G (2016) Coverage recommendation for genotyping analysis of highly heterologous species using next-generation sequencing technology. Sci Rep 6:35736
58. Paradis E, Schliep K (2019) ape 5.0: an environment for modern phylogenetics and evolutionary analyses in R. Bioinformatics 35:526–528
59. Gel B, Serra E (2017) karyoploteR: an R/Bioconductor package to plot customizable genomes displaying arbitrary data. Bioinformatics 33:3088–3090
60. Vurture GW, Sedlazeck FJ, Nattestad M et al (2017) GenomeScope: fast reference-free genome profiling from short reads. Bioinformatics 33:2202–2204
61. Sun H, Ding J, Piednoël M et al (2018) findGSE: estimating genome size variation within human and Arabidopsis using k-mer frequencies. Bioinformatics 34:550–557
62. Becher H, Brown MR, Powell G et al (2020) Maintenance of species differences in closely related tetraploid parasitic Euphrasia (Orobanchaceae) on an Isolated Island. Plant Commun 1:100105
63. He D, Choi A, Pipatsrisawat K et al (2010) Optimal algorithms for haplotype assembly from whole-genome sequence data. Bioinformatics 26:i183–i190

Chapter 16

Population Genomic Analysis of Diploid-Autopolyploid Species

Magdalena Bohutínská, Jakub Vlček, Patrick Monnahan, and Filip Kolář

Abstract

This chapter outlines an empirical analysis of genome-wide single-nucleotide polymorphism (SNP) variation and its underlying drivers among multiple natural populations within a diploid-autopolyploid species. The aim is to reconstruct the genetic structure among natural populations of varying ploidy and infer footprints of selection in these populations, framed around specific questions that are typically encountered when analyzing a mixed-ploidy data set,e.g., addressing the relevance of natural whole-genome duplication for speciation and adaptation. We briefly review the options for the analysis of polyploid population genomic data involving variant calling, population structure, demographic history inference, and selection scanning approaches. Further, we provide suggestions for methods and associated software, possible caveats, and examples of their application to mixed-ploidy and autopolyploid data sets.

Key words *Arabidopsis*, Autopolyploidy, Genetic variation, Population differentiation, Selection scans

1 Introduction

In this chapter, we guide the reader through an analysis of genome-wide single-nucleotide polymorphism (SNP) variation in natural populations within a diploid-autopolyploid species. Fundamental assumptions of this chapter are (i) a priori knowledge of ploidy of all individuals (e.g., identified by flow cytometry or karyology), (ii) polysomic inheritance of the polyploid populations (involving random pairing among all homologous chromosomes) and thus autopolyploid origin, and (iii) availability of a suitable reference genome either for the focal species or for a closely related species. For specific analyses tailored towards allopolyploidy and mixed-inheritance systems, we refer to the chapters by Scott et al. and Roux et al. in this volume. Instead of a complete review of available models, population genetic approaches, and methodologies that

Yves Van de Peer (ed.), *Polyploidy: Methods and Protocols*,
Methods in Molecular Biology, vol. 2545, https://doi.org/10.1007/978-1-0716-2561-3_16,

are well discussed elsewhere [1, 2], we rather directly guide the reader through a set of procedures applicable to autopolyploid data. This chapter is thus primarily directed to empirical evolutionary biologists facing the challenge of analyzing genome-wide DNA polymorphism data in autotetraploid species (although most methods are directly extendable to higher ploidy levels). Although our primary focus is on wild populations, the analyses presented here are applicable to other autopolyploid systems such as autopolyploid crops (e.g., potato, alfalfa, watermelon, pak-choi), crop-wild species complexes (e.g., interploidy crop-to-wild admixture), and evolving autopolyploid cancer cells.

Specifically, we provide an example set of analyses applicable on large (tested for up to 600 resequenced genomes) genome-wide SNP data sets of multiple diploid and autotetraploid populations. When appropriate, we refer to the results of such analyses published elsewhere [3–6]. Throughout the chapter, we illustrate our examples primarily using ScanTools, a toolset specifically designed to work with autopolyploid and mixed-ploidy data. It contains a set of Python 3 scripts for window-based and single variant-based analyses of population differentiation, taking as an input a set of diploid and/or autopolyploid VCF (variant call format) files.

In each section, we will illustrate the application of the discussed analysis on a smaller example data set of two diploid and two autotetraploid populations of *Arabidopsis arenosa* (L.) Lawalrée (Brassicaceae). We also prepared a tutorial which can guide you through the practical aspects of analyses of the example data set, available at https://github.com/mbohutinska/PolyplChapter. *A. arenosa* is an obligate outcrosser [7] widely distributed across much of Europe, and its established diploid (2x) and autotetraploid (4x) cytotypes naturally meet at several contact zones across its range [8, 9]. The evolutionary history of the species and its intraspecific lineages are well-known, encompassing five distinct diploid lineages, one of which gave rise to a tetraploid cytotype approximately 11,000–30,000 generations ago [3, 10, 11]. From this area of origin, the tetraploid lineage spread and came into secondary contact with the earlier diverging diploid lineages in three regions, leaving distinct traces of interploidy (2x to 4x) gene flow in tetraploid populations [3]. Moreover, the species, predominantly dwelling in foothill rocky habitats, exhibits remarkable expansion to distinct, highly selective habitats such as toxic soils, railway tracks, or alpine areas, and those habitats have been colonized mainly by tetraploids [7]. Notably, alpine habitats have been independently colonized by both diploid and tetraploid populations in several areas, forming a distinct alpine ecotype bearing footprints of environmental adaptation [4, 12] (Fig. 1).

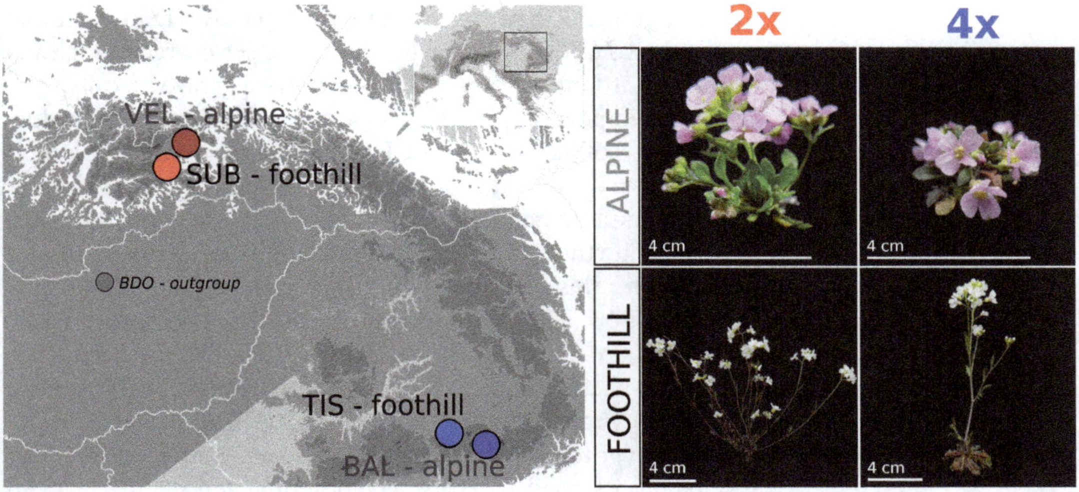

Fig. 1 Distribution of populations used in the example data set and typical phenotypes of alpine and foothill populations of *Arabidopsis arenosa*, following cultivation over two generations in a common garden (diploid = red and tetraploid = blue; dark shade = alpine, light shade = foothill). Note that while there are no striking phenotypic differences between ploidies (a frequent phenomenon in diploid-autopolyploid systems), there are conspicuous heritable differences between foothill and alpine ecotypes within each ploidy

2 Variant (SNP) Calling and Filtering

Regardless of ploidy, accurate and reliable variant calling is the foundational step for successful population genomic analysis. In our example, we will describe a relatively standardized way using one of the most popular variant calling tools, the Genome Analysis Toolkit (GATK); some alternative approaches are shortly summarized in the next section. Our genotyping is based on typical short-read sequence data (150-bp paired-end Illumina reads), and the pipeline consists of three main steps: mapping to a reference genome, variant calling, and variant filtering. These steps are nearly identical for both diploids and tetraploids; the few deviations are explained below. Mapping to the reference genome is unbiased by ploidy level because all four chromosomes are homologous (unlike allopolyploids where "composite" references of both diploid subgenomes are typically used). Individual physical reads are assigned to a matching position on the reference genome based on nucleotide sequence, e.g., using the program BWA [13]. Afterwards, we use GATK that allows joint variant calling and genotyping of diploid and polyploid individuals [14]. In this two-step process, genotypes are first inferred separately for each individual based on the number of reads supporting each allele and sequencing error rate using a designated likelihood model (*HaplotypeCaller* GATK tool) and then refined for the entire set of samples (callset) (*GenotypeGVCFs* tool). There are no ploidy-specific considerations for the *GenotypeGVCFs* step in which a mixture of diploids and tetraploids is analyzed together. On the other hand, the --ploidy flag in the initial per-individual *HaplotypeCaller* tool must be set correctly for each individual in order to produce the appropriate genotype calls.

Another important ploidy-specific consideration emerges during the process of filtering the VCF file produced by GATK. For an equivalent level of coverage, polyploid genotype calls tend to be less reliable than diploid genotypes, especially when the depth of coverage is low as further discussed in the chapters by Scott et al. and Meirmans in this volume. In diploids, sequencing errors (as evidenced by uneven coverage of an allele) are more easily detected as there are only three genotypes of a biallelic locus (*aa, Aa, AA*, with alleles *a* and *A*). In tetraploids, however, there are five possible genotypes (*aaaa, aaaA, aaAA, aAAA, AAAA*) and especially heterozygotes with uneven allele dosage (*aaaA, aAAA*) can be mistakenly assigned as errors. The potential genotyping bias can be limited by removing sites with low coverage (e.g., removing sites supported by less than *N* reads per individual). Additional filters on potentially repetitive or paralogous regions can be applied, based on criteria such as excessive heterozygosity and coverage, similarly as in diploids (e.g., [3]). However, we caution the use of relative thresholds of genotype quality (GQ annotation in the VCF file) for filtering mixed-ploidy data set. The difference between the best and second-best genotype likelihoods and thus the value of GQ are typically lower for polyploid genotypes (distinction between five genotype classes) as compared to diploids (distinction between three genotype classes) so that default filtering strategies for diploids based on relative GQ thresholds do not readily generalize to higher ploidy levels. Filtering can be performed in any tool capable of handling VCF files with polyploid genotypes, e.g., GATK (*VariantFiltration* and *SelectVariant* tools) or BCFTOOLS. For later compatibility with SCANTOOLS we recommend using GATK.

3 Allele Frequency Estimation and Inference of Genetic Diversity

Inference of allele frequencies lies at the core of population genetic analysis as it represents the very first step in the calculation of population diversity and differentiation among populations. Optimally, complete genotypes are reconstructed for each sampled individual from molecular markers (as in our GATK example explained above), allowing a straightforward calculation of observed allele frequencies. Population genomic analysis of polyploids or mixed-ploidy systems, however, has been plagued by several challenges not present in diploids [1, 2]. Firstly, the information on allele dosage—the number of copies of each allele present within an individual—may be challenging to infer (*see*, for instance, the chapter by Meirmans in this volume). As a consequence, the resulting incomplete genotypes are handled as missing data, in some cases leading to biases. To overcome them, specific methods allowing for genotype imputation or allele frequency estimation from genotype likelihoods can be applied (*see* the next paragraph). Second, certain regions of the autopolyploid genome are more susceptible to double reduction, which leads to nonstandard segregation of alleles into gametes and increased homozygosity [15, 16]. Finally, incomplete tetrasomy (or heterosomy, *see* Roux et al., this volume) may arise either by polyploid formation from partially divergent genomes of distinct lineages or by ongoing homolog divergence of initially "pure" autopolyploids. In such a case, preferential pairing of hom(e)ologous chromosomes may lead to segregation distortion in particular loci [1], which is manifested, e.g., by increased frequency of intermediate heterozygotes (*aaAA*).

Alternative approaches in allele frequency estimation of autopolyploids are thus emerging and represent novel promising avenues in polyploid research. One direction is leveraging genotype likelihoods directly for the inference of allele frequencies and other population genomic parameters. The power of such an approach is that it bypasses the need to reconstruct complete genotypes that may be challenging in autopolyploids. With simulations, Blischak et al. [17] demonstrated that by using genotype likelihoods, one can overcome genotype uncertainty resulting from both low-coverage sequencing data, allelic dosage uncertainty, and nonstandard inheritance patterns. Such an approach for estimating population genetic parameters from genotype likelihoods is implemented in the programs EBG [17] and ENTROPY [18]. Another approach is modelling the particular biases associated with polyploid data such as increased variation in sequencing error, additional variation among samples beyond a simple binomial model (overdispersion), and systematic allelic bias (e.g., in UPDOG [19] and FITPOLY [20]).

Genetic diversity is usually the first indicator inferred from allele frequency data informing about past and present evolutionary processes experienced by the population. Theory predicts that, at equilibrium, the amount of nucleotide diversity in an autotetraploid population is likely to be higher than in a similarly sized diploid population [2, 21]. This is due to the higher number of chromosome copies, which may accumulate a higher number of mutations under the same mutation rate. Owing to more chromosomes, autopolyploid populations also experience less severe effects of genetic drift than diploid populations of the same size [22]. The basic measure of genomic variation within a population, both genome-wide and locally, is nucleotide diversity, i.e., the proportion of pairwise differences among two randomly drawn chromosomes from a population (designated as π or Θ_π) [23]. π is usually calculated for a given category of variants (e.g., non-synonymous SNPs), and for a proper estimation of the proportions, invariant positions shall be also called and filtered in a similar way as SNPs.

Table 1
Nucleotide diversity (π or Θ_π) and Tajima's D calculated over ~200 k putatively neutral fourfold degenerate sites genome-wide in the four *A. arenosa* example populations (subsampled to seven individuals per population at each site)

Population	Ploidy	Type	Nucleotide diversity	Tajima's *D*
VEL	2x	Alpine	0.049	0.10
SUB	2x	Foothill	0.055	−0.01
TIS	4x	Foothill	0.059	0.09
BAL	4x	Alpine	0.054	0.44

Population codes correspond to Fig. 1

Given full-dosage genotypes reconstructed or allele frequencies estimated directly using the designated models explained above, this is a straightforward measure that has been implemented in multiple tools, including ScanTools, GenoDive (*see* Meirmans, this volume), and adegenet R package (Table 1).

A complementary way of estimating diversity is leveraging the information on allele frequencies within individuals, estimating observed (H_O) and expected heterozygosity (indicated as either H_E or H_S). While such quantification is very straightforward in diploids, it is more challenging in autopolyploids due to the presence of several classes of heterozygous individuals (e.g., *aaaA, aaAA,* and *aAAA* at an autotetraploid locus with two alleles, *a* and *A*). A proper analysis is discussed elsewhere [2] and is implemented, e.g., in GenoDive [24]. As heterozygosity estimation is based on the reliable reconstruction of complete genotypes, it is particularly challenging in polyploids, especially in genome-wide studies where per-individual genome-wide coverage is usually achieved at a cost of lower sequencing depth.

Utilizing metrics of genetic diversity to reconstruct evolutionary history is greatly aided by information on past demographic events each population has experienced. Tajima's *D* [25] calculated on a genome-wide subset of putatively selectively neutral sites serves as a useful summary statistics (Table 1), indicating the presence of past population contractions (positive genome-wide *D* values) or expansions (negative values), barring confounding effects due to selection at linked loci. In *A. arenosa,* Monnahan et al. [3] discovered on average lower Tajima's *D* values in autotetraploid populations as compared to their diploid counterparts suggesting tetraploid expansion, in correspondence with their recent origin and large geographical area occupied. As Tajima's *D* values are strongly influenced by frequencies of rare alleles, proper inference of allele frequency and robust calling and filtering even of the rarest variants (including singletons) is crucial. If the ability to detect and genotype rare alleles differs across ploidies, this will introduce systematic bias in *D* across ploidies. A novel method for calculating Tajima's *D* in autopolyploid populations from high-throughput data, accounting for individual genotype uncertainty, has been recently proposed [26].

4 Analysis of Population Genetic Structure

Detecting population structure is the first step in a thorough understanding of evolutionary forces shaping diversity of natural populations, like natural selection and hybridization. It may also provide valuable insights into quantitative genetic inquiries in cultivars and laboratory strains. Specifically, population structure defines a "neutral background," against which non-neutral processes may be highlighted and modelled accordingly. Additionally, for ploidy-variable species, determining structure within and among ploidy lineages can inform the process of polyploid origin (single vs. multiple diploid lineages involved in a polyploid's origin), spatiotemporal context of genome duplication (time, place, and number of genome doubling events), and interploidy gene flow (direction and potential asymmetry in the intensity of migration).

While inference of population genetic structure has a long tradition in population genetics and later also genomics of diploid organisms [27], the range of available tools is still very limited in polyploids and best practices need to be established. Here, we highlight several options provided by recent developments allowing for the analysis of medium-sized data sets (up to ~100 k SNPs, typically included in one VCF file comprising all scaffolds and all individuals) using a combination of ordination analyses, Bayesian clustering, and network-reconstruction algorithms which are already implemented in various programs (e.g., ADEGENET and StAMPP R packages, Structure, Treemix, and GenoDive). Where appropriate, we also highlight possible extensions towards custom-based tools designed to handle massive parallel analyses of large genome-wide sets ($\gg$100 k, up to 10–20 M SNPs over hundreds of individuals; e.g., ScanTools). Instead of a complete overview of available population genetic methods (reviewed elsewhere; [1, 2]), we present a set of approaches applicable on a large genome-wide SNP data set based on complete genotypes (full dosage) and moderate (~3–10%) levels of missing data in the diploid-autotetraploid species *A. arenosa*. Such a data set can be the expected outcome of a short-read resequencing project in a non-model species with a draft reference genome available (*see*, e.g., *Cardamine amara* [28], *Cochlearia officinalis* [29] for recent applications to non-model species). Similar to the analyses of diploid population genomic data, we start by calculating basic population differentiation statistics, followed by an exploratory analysis of the complete data set of all individuals. After outlining these descriptive approaches, we shall turn to model-based approaches for testing specific evolutionary hypotheses using coalescent theory.

4.1 Population Differentiation

Proper and unbiased estimation of population differentiation and divergence is crucial for understanding the relationships between studied populations, their evolutionary history, and migration dynamics. They are also invaluable for designing further analyses focused on specific processes affecting genetic diversity such as gene flow and selection. The most widely used metrics to estimate genetic differentiation is the fixation coefficient, F_{ST}, summarizing genetic differentiation between populations inversely scaled by the diversity within populations [30]. If autopolyploids have on average higher genetic diversity than diploids, F_{ST} values between tetraploid populations should on average be lower than between diploids [2]. This may prove a problem if the study aims at direct comparison of the differentiation between pairs of diploid vs. pairs of tetraploid populations. For these purposes, one may use an alternative metric, *rho* (ρ), which is designed to be comparable between ploidy levels [22]. It expresses the average differentiation between individuals from compared populations and should be the statistics of choice for studies of population structure in polyploids since its expected value is independent of ploidy, frequency of polysomic inheritance, rate of double reduction, and mating system [2]. On the other hand, *rho* is less suitable for comparisons across ploidy levels in a single data set as the degree of divergence between ploidy levels quantified by *rho* typically gets inflated [2] (*see* Table 2). Thus, presenting both F_{ST} and *rho* is advisable for a general overview about the differentiation in the entire mixed-ploidy data set. There are multiple programs designed to simultaneously calculate F_{ST} and *rho*, e.g., GenoDive (smaller data sets,

Table 2
Population differentiation metrics calculated over ~200 k fourfold degenerate sites genome-wide in a pairwise manner between the four *A. arenosa* example populations

Pop. pair	Ploidy contrast	*Rho*	F_{ST}	Fixed differences[a]
SUB-VEL	2x–2x	0.17	0.12	63
SUB-BAL	2x–4x	0.25	0.15	82
SUB-TIS	2x–4x	0.23	0.13	14
VEL-BAL	2x–4x	0.26	0.15	33
VEL-TIS	2x–4x	0.26	0.15	25
TIS-BAL	4x–4x	0.22	0.09	3

[a]Total number of fixed differences between the corresponding populations

also including incomplete dosage information) and ScanTools (larger data sets, complete dosage) and additional options are listed elsewhere [1, 2].

Finally, analysis of molecular variance (AMOVA) offers a formalized quantification and test of hierarchical differentiation between individuals within a population, between populations, and optionally also groups of populations or lineages. While originally developed for diploids, its extensions to autopolyploids with both complete and incomplete dosage have been recently developed and implemented in GenoDive [31] and polygene [32].

4.2 Principal Component Analysis

The unconstrained ordination method such as principal component analysis (PCA) provides perhaps the most frequently used visualization of genome-wide differentiation among individuals. The advantages are fast analysis of large multivariate data sets (consisting of thousands of SNPs) and straightforward visualization of the major orthogonal trends in variation. As input, a set of allele frequencies within individuals (0, 0.25, 0.5, 0.75, or 1 for a non-reference allele of a biallelic locus of an autotetraploid) or populations of all loci are taken. Such analyses are very straightforward when full genotypes are resolved and missing data (typically replaced by average allele frequencies for that locus prior to the analysis) are not pervasive. However, large amounts of missing data (the presence of individuals with >10–20% missing genotypes) greatly hamper the resolution of these analyses, and such problematic individuals, more typically tetraploids in case of low coverage and/or incomplete dosage inference, tend to be dragged towards the center of the PCA plot. Note that even without missing data, tetraploids may be positioned more centrally than diploids due to the fewer occurrences of the extreme genotype classes, i.e., the full homozygotes. This results in less 0 and 1 frequencies and therefore a lower variance in allele frequencies in tetraploids compared to diploids [2]. Such an effect might be hard to separate from real demographic effects causing lower differentiation among tetraploids such as recent and potentially recurrent origin, slower divergence, and intense gene flow. Additionally, as PCA is designed to capture major orthogonal trends in the variation, it is particularly sensitive to sampling (e.g., few highly diverged samples in an otherwise homogeneous data set may strongly affect the outcome). It is thus advisable to present multiple PC axes to fully capture the population structure and to complement PCA with additional approaches, listed below.

4.3 Clustering Approaches

A vital complement to simple plotting of individual relatedness is the statistical inference of each individual's ancestry to an a priori defined number of groups. *K*-means clustering, which is not based on an explicit statistical model, offers the simplest (and quickest) approach directly applicable to any ploidy level. However, this approach is unable to reconstruct the mixed ancestry of individuals and is associated with ploidy-related biases when dosage information is incomplete [33].

More elaborate, model-based approaches have been developed, allowing for inference of mixed ancestry from a hypothetical ancestral population and thus inference of potential hybrids and/or individuals exhibiting shared polymorphism. STRUCTURE [34] has been the most popular tool in this respect for the last 20 years. However, it has been primarily developed for the analysis of Sanger-sequence, microsatellite, and restriction-fragment data sets, which significantly limits its application on massive parallel sequencing data. Thus, several alternative clustering methods based on similar principles have been developed for large population genomic data sets of diploids, e.g., ADMIXTURE [35] or FASTSTRUCTURE [36]. While the original STRUCTURE program also allowed for the analysis of polyploid data sets (and with certain encoding drawbacks such as mixed ploidy [33]), until very recently there was no faster alternative for large sequencing data in polyploids. Thus, the few available studies dealing with the diploid-autopolyploid population structure so far relied on workarounds such as random subsampling of two alleles for analysis using programs designed to analyze diploids, e.g., ADMIXTURE or FASTSTRUCTURE [3, 37]. Encouragingly, a recent simulation-based study demonstrated that such approaches lead to only minor biases in the analysis of codominant data with fully resolved genotypes (such as genome-wide SNPs) [33]. Recently, Shastry et al. [18] developed a clustering software specifically designed to infer the population structure and ancestry in large genome-wide mixed-ploidy data sets. ENTROPY represents the very first program allowing such analysis in populations and species of varying ploidy using both high- and low-depth sequencing data, using genotype likelihoods as a direct input. If established in the community, ENTROPY may become a standard tool for postulating evolutionary hypotheses on polyploid origin and population structure.

4.4 Tree and Network Reconstruction Algorithms

Standard phylogenetic tools, such as maximum likelihood reconstruction of concatenated sequences or reconstruction under multispecies coalescent, are also generally applicable for diploid-autopolyploid systems as they typically operate with per-individual consensus sequences. However, the effect of collapsing polyploid genotypes to consensus remains to be evaluated. Applicability of phylogenetic methods is limited in systems of recent divergence, such as many polyploid complexes, where much of the variation is still segregating, ancestral polymorphism is frequent, and gene flow is still ubiquitous. For example, while likelihood-based tree reconstruction approaches provided robust phylogeny of diploid lineages in *A. arenosa*, they failed to reconstruct relationships among the recently expanding autotetraploid lineages [3].

Thus, alternative approaches based on allele frequencies may be more informative. Perhaps the simplest (and quickest) method is the calculation of genetic distances from genome-wide allele frequency data followed by their visualization in a network, e.g., using neighbor-joining networks. As the conversion of genome-wide polymorphisms into a single distance index means dramatic simplification, selection of a suitable distance index is the crucial step in this analysis. There are multiple options for polyploid or mixed-ploidy data sets, varying in the fidelity of representation of within vs. interploidy relationships and potential biases associated with missing data including incomplete genotypes, discussed elsewhere [1, 2].

Allele frequency covariance graphs, implemented in TREEMIX [38], represent an exciting alternative, which is now frequently used in large genomic data sets in diploids and has recently also been applied to polyploid species [3, 28, 39, 40]. Unlike standard phylogenetic tools assuming bifurcation events, TREEMIX also allows for the visualization of additional residual variation, unexplained by the backbone tree; they are considered as migration edges, which could be interpreted as additional pulses of gene flow. However, certain interpretation issues may arise when the levels of gene flow are high enough to dominate the allele frequency covariance patterns, and complex scenarios involving multiple polyploid origins combined with recurrent interploidy introgression may occur (such as in *Neobatrachus* frogs [40]). Thus, TREEMIX serves as a hypothesis-generating method, not a formal test. To infer a realistic number of migration events, TREEMIX can be run in replicates assuming different numbers of migration events, and a partition when these models reach likelihood saturation can be identified (*see* [40]).

4.5 Interpretation and Example Application in A. arenosa

In *Arabidopsis arenosa*, applying the processes outlined above allowed us to unravel the complex evolutionary history of this diploid-autotetraploid complex [3, 11]. First, we revealed high overall nucleotide diversity, in line with large outcrossing populations typical for this species, which have likely survived in multiple glacial refugia. Surprisingly, nucleotide diversity of diploid and tetraploid populations was nearly the same. On the other hand, across the range, the differentiation was higher among diploid populations, as quantified by both F_{ST} and *rho*, corresponding with relatively long divergence among diploid lineages and a recent tetraploid origin. Such a finding was further corroborated by ordination analysis (PCA), maximum likelihood phylogenies, and Bayesian clustering analyses that demonstrated five divergent diploid lineages, one of which was closely related with a single yet still very diverse autotetraploid cluster. On top of that, further genetic substructuring within the tetraploid cluster including the tendency for clustering of spatially close diploid and tetraploid populations suggested additional locally restricted interploidy gene flow. This hypothesis was further addressed by coalescent simulations (see the next section).

A similar trend is recapitulated also in our simplified example data set presented below (populations corresponding to Fig. 1). The populations were clustered primarily according to region (and thus ploidy) as indicated by separation along PC1 (Fig. 2a), distance-based neighbor-joining tree (Fig. 2b), TREEMIX graph (Fig. 2c, d), and clustering in ENTROPY (Fig. 2e). Differentiation appeared to be higher between diploid populations than between tetraploids as identified by the degree of separation along the second PC axis (PC2; Fig. 2a) and the higher F_{ST} values (Table 2). However, the opposite is found when *rho* values are compared, which are more appropriate for such comparisons, showing that the PC2 differentiation results are biased by ploidy itself, in line with simulation studies [2]. There was only slightly higher diversity in tetraploid populations (Table 1).

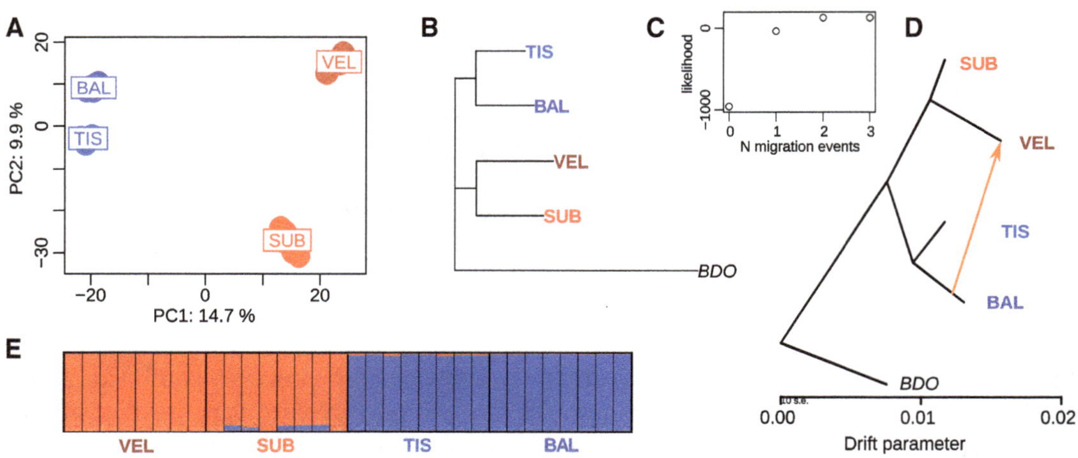

Fig. 2 Relationship between diploid and autotetraploid populations of *Arabidopsis arenosa* from two regions and of two elevational ecotypes reconstructed using ~200 k putatively neutral fourfold degenerate SNPs. (**a**) Principal component analysis of the individuals, colored according to ploidy (corresponding with the region) and alpine/foothill ecotype (red, diploid; blue, tetraploid; foothill light, alpine dark). (**b**) Neighbor-joining tree based on Nei's genetic distances between populations calculated from allele frequencies. (**c**, **d**) Allele frequency covariance graph of populations, inferred by TREEMIX, and the likelihood of models differing by the number of migration events. (**c**) One migration event has been selected based on the highest improvement in likelihood values. (**d**) Distances along the x-axis show the drift parameter estimates, corresponding to the number of generations separating the two populations and their effective population sizes ($t/2N_e$). The arrow indicates the most likely migration event. The outgroup (BDO) is a diploid population from the Pannonian Basin, previously inferred as the earliest-diverging lineage of the entire species. (**e**) Posterior probability of assignment of each sequenced individual to the two groups as inferred by ENTROPY based on ~7 k unlinked SNPs (subsampled to a minimum distance of 1 kb to avoid linked sites)

4.6 Best Practices

- Based on genome annotation information, extract putatively selectively neutral sites such as intergenic (if data quality permits) and/or fourfold-degenerate genic sites (GATK or BCFTOOLS) *[input: .vcf file].*
- Load the VCF file into R (using VCFR) and calculate PCA (ADEGENET) and inspect for outlier individuals (erroneous species, contaminants, hybrids/allopolyploids?) and biases stemming from low-coverage individuals with excessive amounts of missing data, positioned towards the center of the plot *[input: .vcf file].*
- Follow-up with quality checks: calculate genetic distances among individuals (STAMPP), plot them in a neighbor-joining network (SPLITSTREE), and check for additional spurious clustering in the network, e.g., individuals not clustering with other members of their populations (contaminants, labelling errors?) or connected with multiple alternative splits of similar weight (hybrids/allopolyploids?). Remove such individuals, clearly document the criteria used for such a decision, and check the PCA and neighbor-joining network once again *[input: .vcf for R packages and distance matrix .dst file for SPLITSTREE].*
- For the clean data set, calculate diversity statistics and genome-wide Tajima's *D* for each population (SCANTOOLS or GENODIVE), investigate potential sources of diversity variation among populations (ploidy, spatial distribution, etc.) to postulate hypotheses about past population size changes *[input: SCANTOOLS– .vcf file + population key, GENODIVE- special format or several population genetic formats,* e.g., *structure-formatted .str file].*
- Calculate F_{ST} and *rho* differentiation among populations (SCANTOOLS or GENODIVE); in comparisons focused on the relative differentiation among 2x–2x vs. 4x–4x population pairs, use *rho*. Optionally, the matrix of interpopulation distances may also be plotted as a network using SPLITSTREE.
- Run clustering analysis (STRUCTURE, ENTROPY, and/or *K*-means clustering in ADEGENET or GENODIVE) following general recommendations for mixed-ploidy data [33] and infer major genetic groups/lineages and potential admixed individuals/populations *[input: structure-formatted .str file].*
- Infer the relationships among individuals (e.g., neighbor-joining networks in SPLITSTREE) and populations (e.g., TREEMIX) in a network-like manner *[input: .vcf file].*
- Integrate the information from population differentiation, PCA, clustering, and network-based analyses, and formulate hypotheses on population relationships, the extent of intra- and inter-ploidy gene flow, and the number and geographic locations of autopolyploid origins.

5 Inference of Population Demographic History

Coalescent simulations followed by a model-selection approach are becoming a standard tool for selecting among competing scenarios of population demographic history as well as for the inference of important demographic parameters such as past and present effective population size (*Ne*), divergence time, and past gene flow (migration) (*see* also Roux et al., this volume). Comparing coalescent-based simulations of population genetic data for different models of demographic history allows statistical testing of hypotheses formulated a priori (e.g., based on ploidy differentiation, spatial distribution, ecological preference, or addressing controversial scenarios resulting from the exploratory approaches). For autotetraploids with tetrasomic inheritance, Arnold et al. [41] demonstrated that at a locus without recombination, the ancestral genetic process in a large population may be approximated using Kingman's standard coalescent, with a coalescent effective population size $4N$. Thus, coalescent simulation programs developed for diploids can be adapted to study the population history in autotetraploids by interpreting the timescale in units of $4N$ generations.

There is an expanding set of tools for coalescent simulations available; here we will focus on FASTSIMCOAL2, a versatile tool that has been used most frequently in the few so far available inquiries of autotetraploid and mixed-ploidy systems [3, 10, 12, 39, 42]. A first step when analyzing empirical data is retrieving allele frequency spectra (AFS, also referred to as site frequency spectra, SFS) that could be directly calculated from existing allele frequency data stored in a VCF file (e.g., using SCANTOOLS). A specialized approach focused on the estimation of joint AFS in autotetraploids, additionally taking into consideration potential deviations from random mating, has been recently developed yet not implemented in software [26].

Coalescent simulations are a demanding multistep process that is strongly conditional on a priori knowledge on the system studied, not only the results of population structure reconstruction, but also overall biology of the species, mode of polyploid origin (auto- vs. allopolyploidy), local geological history, etc. Thus, it requires careful construction of realistic yet not overly complicated models, as runtimes increase drastically as the number of populations and thus possible population iterations and parameters to be estimated increase. Briefly, expected AFS are simulated under a given set of parameters of each modelled scenario and their composite likelihood is computed. An efficient optimization algorithm is used to find optimal parameter values from which simulations best match the observed AFS, thus maximizing the likelihood (for more details, *see* dedicated studies [43]). Finally, likelihood values inferred for different scenarios are compared, e.g., using the Akaike

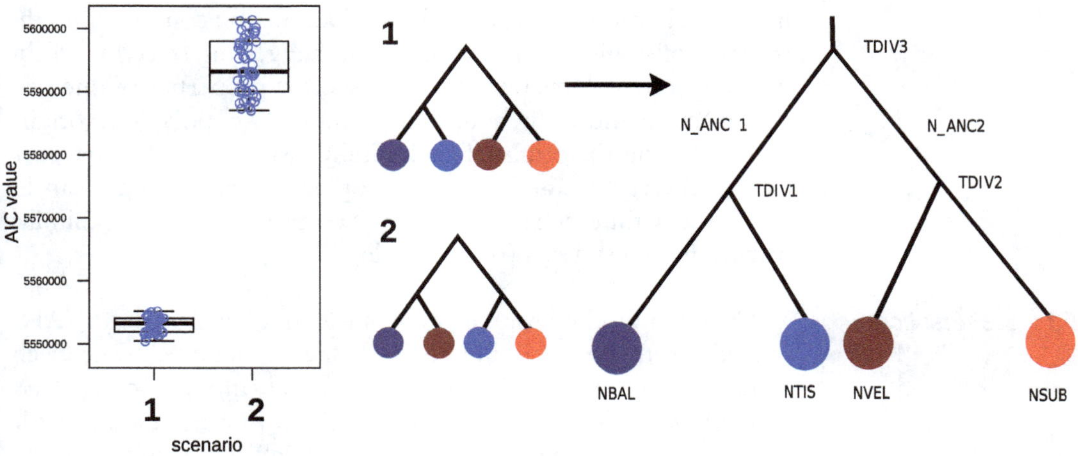

Fig. 3 Comparison of evolutionary scenarios simulated in a coalescent framework in FASTIMCOAL2. Although population structure analyses (Fig. 2) strongly indicate regional (ploidy) clustering, formal rejection of an alternative scenario (single origin of alpine ecotype followed by independent genome duplication within each region) can be done by a model comparison of distinct scenarios. We simulated a four-population model assuming either regional = ploidy (scenario 1) or ecotype (scenario 2) sister relationships. Alpine (dark shade) vs. foothill (light shade) ecotype, diploid (red) and tetraploid (blue) cytotype. Boxplot summarizes AIC values over 50 FASTSIMCOAL2 optimization runs for each scenario, demonstrating that the scenario of parallel origin of the alpine ecotype consistently reaches lower AIC and thus is preferred. Scenario 1 is further depicted with all parameters used in the simulations, namely, effective population sizes (codes starting with "N") and divergence times ("TDIV")

information criterion (AIC). Alternatively, approximate Bayesian computation (ABC) methods may be used to perform simulation-based inference and model selection in a Bayesian framework (*see* Roux et al., this volume).

For mixed-ploidy systems, a scenario of single vs. multiple autopolyploid origin can be compared in a computationally efficient way via a population-quartet design, if the source diploid populations are sampled [3]. Two general topologies can be contrasted: (i) sister position of populations of each ploidy (scenario 1 in Fig. 3) and (ii) sister position of a diploid and tetraploid population paired by, e.g., sympatry or similar morphology (scenario 2 in Fig. 3). Such topologies may or may not be accompanied by additional inter- or intraploidy gene flow (e.g., within regions) and population size changes (e.g., initial bottleneck followed by population growth in expanding polyploid). If more populations per each ploidy/lineage are sampled, it is advisable to run the analysis repeatedly, iterating different (meaningful) population quartets to leverage such natural replicates while keeping the models manageably simple. For example, when comparing scenarios of repeated polyploid origin vs. single origin followed by regional interploidy gene flow, Monnahan et al. [3] used such quartets and for each scenario iterated different populations of each ploidy and lineage. They found that regardless of population combination, the scenario of single origin was preferred, suggesting selection of the winning scenario was not driven by specific properties of one particular population. In addition to modelling polyploid origin, hypotheses on the origin of ecologically distinct populations, population divergence, and historical population size changes can be addressed within each ploidy or in mixed-ploidy systems in a similar way to diploid systems [4, 5, 12] (Fig. 3).

5.1 Best Practices

- Create multidimensional joint allele frequency spectra (AFS, SFS) from the well-filtered VCF file of putatively selectively neutral sites of a particular population set (module *.generateFSC2input* in ScanTools). Filtering should be done carefully in order to estimate proper frequencies of all classes of mutations, including the rarest categories (e.g., do not apply minor allele frequency filters, do not remove singletons). Missing data can be handled by downsampling to a smaller number of individuals for which complete genotype information is available *[input: .vcf file]*.
- Carefully consider the scenarios to be compared and encode them into designated fastsimcoal files (*.est* and *.tpl* files). Construct as simple models as possible, in order to avoid exploring an unrealistic broad parameter space and/or comparing too many scenarios. To achieve generality, consider running a similar set of scenarios over multiple combinations of natural populations (natural replicates). Set realistic, sufficiently broad ranges for initial parameter values.
- Specifically to polyploidy, it is good to consider specific possibilities such as limited/asymmetrical gene flow between ploidy levels, independent polyploid origins, large and increasing population sizes for expanding polyploid lineages, or, in contrast, a bottleneck if the recent polyploid lineage had not yet restored diversity by recurrent origins or gene flow.
- Run the simulations following recommendations in the fastsimcoal2 manual (e.g., at least 50 replicates per scenario and population combination), collect the likelihoods, calculate AIC, and perform model comparison for each particular population combination. It is also worth exploring the distribution of estimated parameter values to check whether the replicates are converging to similar (and realistic) optimal values under each scenario. Likely, various combinations of initial parameter settings and scenarios have to be explored prior to reaching final conclusions *[input: empirical joint allele frequency spectra in .obs files, parameter .est and .tpl files]*.

- For the selected scenario, bootstrap the SNP matrix used for creating the empirical AFS (bootstrapping option is available within the module *.generateFSC2input* in ScanTools), rerun the analysis, and calculate confidence intervals for your parameters of interest.

6 Population Genomic Inference of Selection

Identification of genomic regions differentiated by positive selection is a way to gain insights into the genomic basis of adaptive evolution. Specifically to polyploid systems, such analyses in mixed-ploidy species may be particularly relevant for understanding the genomic basis of adaptation towards organismal- and cellular-level challenges imposed by whole-genome duplication [6, 44] and significance of (adaptive) interploidy introgression [45, 46] as well as mechanisms of selection and (local) adaptation in polyploid populations [4, 42].

Numerous statistical methods have been developed to detect genomic regions bearing footprints of positive selection in diploid data sets (called selection scans hereafter). Applications of these methods range from the study of recent speciation (scanning for microevolutionary changes between subdivided populations) to selection scans for macroevolutionary changes between species [47]. While the broad diversity of selection scan methods covers various timescales of divergence between compared populations and species (Fig. 4), they mostly assume (and were tested for) diploid outcrossing populations without sharp demography changes [48]. This makes it difficult to directly infer their applicability for the analysis of polyploid data sets and simulation-based validations or tests are generally lacking. Nevertheless, here we present several complementary approaches which we have applied to analyze the footprints of selection in genome resequencing data sets representing both mixed-ploidy and purely autopolyploid populations [4, 5] (Table 4).

Very roughly, selection scans can be divided into four main categories: (1) substitution rate-based selection scans, suitable for between-species or deeper divergences; (2) population differentiation-based scans, primarily identifying signals of rapid allele frequency changes (hard sweeps) between populations; (3) linkage disequilibrium-based scans, searching for recent modest allele frequency changes (soft sweeps) within populations; and (4) environmental association analyses, which attempt to link environmental variables with associated genetic variation, providing more direct evidence for the drivers of local adaptation. For more details, *see* dedicated reviews [47, 49–51].

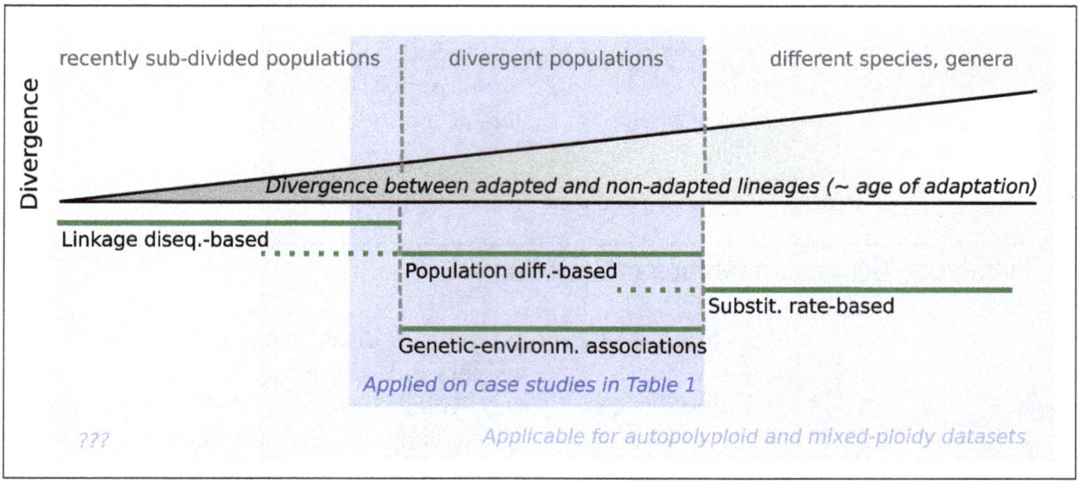

Fig. 4 Overview of selection scan methods to detect signatures of positive selection and their applicability for autopolyploid and mixed-ploidy data sets. The question marks denote so far unexplored approaches of haplotype-based selection scans

In mixed-ploidy and autopolyploid data sets, we applied three of these approaches, each suitable for different divergence between the compared units: substitution rate-based and population differentiation-based scans and environmental association analyses (Table 4, Fig. 4). All three approaches are based on genotype/allele frequencies, whose estimation is feasible in autopolyploids (conditioned on detection of full genotypes, reliable imputation of partially resolved genotypes, and/or likelihood-based allele frequency estimation; for details, *see* the Subheading 3 of this chapter). The use of linkage disequilibrium-based methods is perhaps the most challenging as it requires proper inference of genomic linkage information from short-read data which is very difficult in autopolyploids [52]. Although this was recently achieved in highly heterozygous autotetraploid potato [53], we will introduce here only the three approaches based on genotype/allele frequencies. On the bright side, there is empirical evidence of positively selected standing alleles (swept by soft sweeps, [54]) identified in autotetraploid *A. arenosa* using allele frequency-based methods [5, 6], suggesting that soft sweeps can be at least partially captured by the population-differentiation approaches.

Selection scan approaches are typically based on a specific design, comparing sets of populations from contrasting environments (called here "background" and focal "adapted" populations). Such design may, for example, involve the contrast of a plant population growing on toxic soil ("adapted"), with its close relative occupying nontoxic soil ("background"—note that even a background population is expected to be adapted to its local environment). Specifically to mixed-ploidy and autopolyploid data sets, there are three possible selection scan designs (Table 4). First, one may wish to scan for selection between diploids and tetraploids (contrast 2x–4x), for example, to ask which genes are involved in adaptation to whole-genome duplication. Such a comparison in *A. arenosa* revealed a set of genes involved in meiosis likely underlying the adaptation to the polyploid state [44]. Second, the study design may include multiple adapted and background population pairs, each pair comprising populations of the same ploidy but the pairs differing in ploidy (i.e., multiple 2x–2x and 4x–4x pairs). This setup tests for the presence of parallel adaptation in both ploidies. Examples include adaptation to harsh alpine environments, repeatedly occurring in diploid *Arabidopsis halleri* as well as diploid and autotetraploid *A. arenosa* [4]. Third, the data set might contrast various autopolyploid populations only, e.g., following adaptive divergence of established autopolyploid lineages. An example is adaptation of an autotetraploid *A. arenosa* to toxic serpentine soils [5]. While in the first example, polyploidization represents the selection agent of interest, in the second and third case it acts rather as a confounding factor. Nevertheless, the selection scan methods listed below are applicable to all three of these designs. Some of the candidate genes which we identified using these approaches have also been followed further experimentally. This follow-up research provided multiple hints that the adapted alleles code a new function in comparison to the background alleles [5, 55] (Table 4).

Each of the selection scan approaches is implemented in various methods and programs. First, for simple population differentiation scans, any software which enables calculation of F_{ST}, D_{XY}, *rho*, or related estimators of population differentiation and divergence from allele frequency data in autopolyploids should be applicable (we used ScanTools). Calculation of population differentiation for each window/SNP should be then followed by identification of the most outlier regions, either by dedicated programs (e.g., ScanTools) or by custom scripts. Differentiation scans are intuitive and straightforward, and they may provide a convenient selection detection tool. However, their application requires good understanding of the evolutionary history of studied populations (*see* Roux et al., this volume) and correct setting of contrasts of interest (i.e., carefully selecting pairs of "adapted" and phylogenetically close "background" populations). Further, one may wish to apply scans for differentiation that are model-based, explicitly accounting for the covariance structure among the population allele frequencies reflecting common history of the populations under study. This may improve the robustness of differentiation scans by considering the population structure while estimating differentiation (*see* dedicated papers, i.e., [57, 58]). With this motivation, we applied BayPass [57], which requires reference and alternative allele counts as an input, enabling us to include autopolyploid data, and identified a set of candidate loci potentially involved in alpine adaptation. By overlapping such candidate lists across multiple foothill-alpine

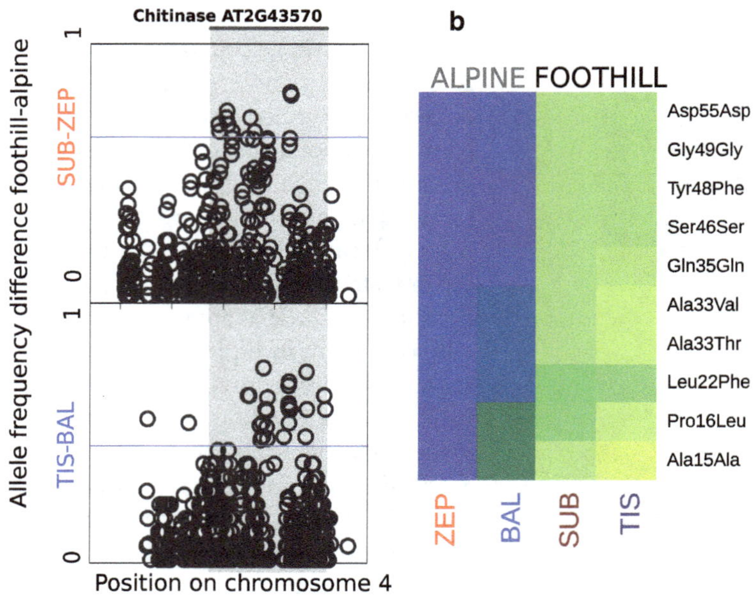

Fig. 5 Example of a candidate locus for alpine adaptation identified by overlapping simple divergence scanning approaches while leveraging detailed knowledge of the evolutionary history of the system. (**a**) Pattern of allele frequency differentiation between foothill and alpine population in a selected locus that has been revealed as a top candidate in both independent population contrasts (mountain ranges) investigated. (**b**) Heatmap showing allele frequency difference of this candidate in the complete data set, the frequency of derived allele scales with the depth of blue (yellow: AF = 0, green: AF = 0.5, blue: AF = 1)

population contrasts, we also identified a subset of such candidates which probably evolved in parallel [4] (*see* Fig. 5 for an example). Finally, to approximate a likely functional consequence of candidate regions, one might combine population differentiation measures with estimation of the impact of corresponding amino acid substitutions on the protein function. This approach is implemented in the method FineMAV [59], originally developed for diploid human data, but has been modified also for non-model autopolyploid data sets [6].

Second, for tests for selection based on the distribution of substitutions between populations, like the McDonald-Kreitman test [60], one requires knowledge of the number of synonymous and nonsynonymous substitutions and polymorphisms. Apart from identifying specific candidate genes for positive selection, this type of procedure can also establish the genome-wide adaptive proportion of genetic variation, which may inform the debate about possible higher adaptability of autopolyploids. However, estimates of adaptive evolution from the McDonald-Kreitman test may be biased downwards by the presence of slightly deleterious mutations

[61]. Thus, filtering out low-frequency polymorphisms and other workarounds were suggested by simulations [61] and applied for diploids [62, 63]. Although we miss similar simulation-based recommendations for autopolyploids, their capacity to harbor higher recessive variability than diploids [21] may make them even more prone to these biases, calling for applying similar recommendations as those developed for diploids.

Third, to identify alleles associated with population-specific covariates (e.g., environmental variables, phenotypic traits) in diploids and tetraploids, we used BayPass [57] and latent factor mixed models (LFMM; [64]). As mentioned above, BayPass is a Bayesian method to identify genomic regions subjected to adaptive divergence across populations. On top of this framework, it also involves models enabling the assessment of association of allele frequency differences with population-specific information about environmental conditions or phenotype [57]. LFMM uses similar logic, testing the association between genetic markers and variables of interest, yet here using linear mixed models with a discrete number of ancestral population groups modelled as latent factors. LFMM is, however, much faster and computationally less demanding. Both methods, BayPass and LFMM, require population allele counts as an input and their application in mixed-ploidy and autopolyploid data sets is thus straightforward.

6.1 Assumptions and Limitations

To account for possible biases connected to the analysis of autopolyploid data, one should first understand the population genomic consequences of autopolyploidy to the pattern of allele frequency changes left by positive selection [21]. In short, simulations showed that peaks of genetic differentiation formed by the action of positive selection are narrower in autopolyploids, likely as a result of elevated recombination rates and longer times to fixation. This may lead to better localization of the specific gene or region targeted by positive selection. On the other hand, differentiation tends to be more subtle in autopolyploids, especially when the selected allele is (partially) dominant. That leads to overall slower allele frequency shifts under positive selection. For a comprehensive overview of the effect of autopolyploidy on the patterns of selection sweeps, *see* [21]; here we focus on possible practical consequences of these effects.

First, all approaches introduced above strongly depend on the assumption that analyzed autopolyploids exhibit random segregation of homologous chromosomes (general assumption of the entire chapter). Deviations from this rule may lead to segregation distortions and bias our estimation of genetic differentiation [2, 65]. Second, if the divergence between background and adapted population is low and/or adaptation proceeds via recessive alleles, the peak of differentiation may be too low to be detected under stringent outlier/significance thresholds [21]. This again highlights the need for good understanding of the evolutionary history of the studied system. Still, the overlap of candidate genes discovered using a higher number of complementary selection scanning approaches with lower thresholds (*see*, e.g., [4, 44]) may partially help to overcome these issues, at the cost of allowing potentially more false positives in the discovery process. Third, while the better localization of selection sweeps in autopolyploids in principle provides an advantage [21], more sweeps might remain undetected when using reduced representation sequencing techniques like RADseq as the linkage is weaker in autopolyploids. Finally, although genetic differentiation estimators may be biased when comparing values inferred for population pairs of different ploidy (e.g., F_{ST} in case of 2x–2x vs. 4x–4x comparisons [2]), they are still helpful in outlier selection scans based on relative comparison to a genome-wide distribution for that particular pair, e.g., 5% of the top-differentiated regions. While keeping in mind drawbacks of using relative thresholds (e.g., allowing an unknown amount of false positives; [49]), we recommend against the usage of absolute value of genetic differentiation measures as outlier threshold over multiple population pairs differing by ploidy, as the absolute values may be affected by ploidy.

In conclusion, selection scans may provide interesting insights into the genomics of adaptive evolution in both diploids and autopolyploids [7]. However, we stress that similar to the diploid case, and likely even more, selection scans should serve as hypothesis-generating tools to stimulate further research, not an approach to confidently identify the adaptive genes. Asserting an adaptive role of specific SNPs will require more rigorous functional or in silico validation of candidate selected alleles, i.e., modelling the impact of candidate allele on a protein structure or introduction of adaptive allele into a common background and testing for its effect in a common garden experiment.

6.2 Best Practices

- Carefully design population sampling based on a good knowledge of the evolutionary and ecological background of the study system (known ploidy and population structure, number of polyploidization events—*see* the Subheading 4) and additional information on the adaptive events of interest (information on selective factor in natural populations, e.g., abiotic and biotic parameters; optimally complemented by transplant or common garden experiments determining heritable basis of the adaptive traits and overall fitness response).
- Consider the divergence between adapted and background populations (*see* Subheading 5) and up-to-date knowledge about available selection scan methods to design the optimal selection scanning procedure. Ideally, combine multiple complementary selection scan approaches:

Table 3
Reference to programs noted in the "Best practices" examples

Software/ method	Purpose	Link
GATK	Variant calling, genotyping, and filtering	https://gatk.broadinstitute.org/hc/en-us
GATK-wrapper	Variant calling and genotyping pipeline for parallelized analysis on a cluster	https://github.com/vlkofly/Fastq-to-vcf
EBG	Polyploid genotyping	https://github.com/pblischak/polyploid-genotyping
BCFTOOLS	Filtering and manipulation of VCF files	http://samtools.github.io/bcftools/bcftools.html
SCANTOOLS	Population genomic analysis using large genomic data sets	https://github.com/mbohutinska/ScanTools_ProtEvol
GENODIVE	Population genetic analysis of diploids and polyploids; data conversion	https://www.bentleydrummer.nl/software/software/GenoDive.html
ENTROPY	Inferring ancestry proportions of individuals to genetic groups, tailored to mixed-ploidy systems and large SNP data	https://anaconda.org/bioconda/popgen-entropy
STRUCTURE	Inferring ancestry proportions of individuals to genetic groups	https://web.stanford.edu/group/pritchardlab/structure.html
STRUCTURE-input	Data conversion VCF file to structure format	https://github.com/MarekSlenker/snipStrup
TREEMIX	Inferring population relationships	https://bitbucket.org/nygcresearch/treemix/wiki/Home
FASTSIMCOAL2	Coalescent simulations	http://cmpg.unibe.ch/software/fastsimcoal2/
McDonald-Kreitman test	Selection test	http://benhaller.com/messerlab/asymptoticMK.html

– Substitution-based methods, calculating the number of substitutions, i.e., fixed nonreference alleles and polymorphisms for each category and applying the McDonald-Kreitman test. Use SNPEFF [66] to classify synonymous and nonsynonymous variants and then use either custom scripts or a dedicated model-based web tool (Table 3). Be aware that this test is not applicable to recent selection events, in which both adaptive and background alleles are still segregating within populations [*input: .vcf file (SNPEFF)*].

Table 4
Three different selection scan designs, varying in the combination of diploid and polyploid populations and divergence between them

Selective factor	Ploidy contrast	Divergence	Focus	Methods	Validation	Reference
Whole-genome duplication	2x–4x	~20,000	Protein evolution	Divergence scan on amino acid substitutions, FINEMAV	Functional assays of two candidates	[6, 28, 44]
Alpine environment	2x–2x, 4x–4x	~10,000	Gene evolution	BayPass, divergence scan	Modelling* (signal confirmed in 151 out of 190 candidates)	[4]
Serpentine soil	4x–4x	~5000	Gene evolution	Environmental association analysis, divergence scan	Modelling* (signal confirmed in 19 out of 61 candidates)	[5, 42]

Note: Ploidy is given for the focal contrast(s). Divergence between adapted and background populations is shown, in generations. *Distinguishing among modes of convergent adaptation (DMC, [56]) has been used to formally test selection signal against a neutral scenario for each candidate

  - Differentiation-based methods (F_{ST}, D_{XY}, *rho* – ScanTools, BayPass, FineMAV + SnpEff) [*input: .vcf file (ScanTools, SnpEff), allele count tables (BayPass)*].
  - Environmental association analyses (BayPass, LFMM) [*input: allele count tables*].
- Visualize the differentiation metrics at the candidate loci, use visual feedback to fine-tune your selection scan design, yet anticipate that visual manifestations of selection are influenced by ploidy: ScanTools, scripts. Consider using different method settings and thresholds; explore how changing thresholds alter the results while avoiding cherry-picking.
- Collapse the resulting candidate regions (sites/windows) to genes, and interpret them in a functional context (GO enrichment, KEGG pathway analysis, STRING protein interaction networks, functional descriptions from databases like Uniprot).
- Consider follow-up predictive or functional validation of the top candidate loci.

Acknowledgments

We are grateful to Arthur Zwaenepoel, Patrick Meirmans, Josselin Clo, Nélida Padilla García, and Polina Novikova for very useful comments on an earlier version of this chapter. We thank Veronika Konečná, Nélida Padilla García, and Gabriela Šrámková for help with running particular analyses of the example data set and Doubravka Požárová for photos of alpine plants. This work was supported by the Czech Science Foundation (project 20-22783S to FK) and the long-term research development project no. RVO 67985939 of the Czech Academy of Sciences. Access to computing and storage facilities has been provided by the National Grid Infrastructure MetaCentrum provided under the program "Projects of Large Research, Development, and Innovations Infrastructures" (CESNET LM2015042).

References

1. Dufresne F, Stift M, Vergilino R, Mable BK (2014) Recent progress and challenges in population genetics of polyploid organisms: an overview of current state-of-the-art molecular and statistical tools. Mol Ecol 23:40–69. https://doi.org/10.1111/mec.12581
2. Meirmans PG, Liu S, van Tienderen PH (2018) The analysis of polyploid genetic data. J Hered 109:283–296. https://doi.org/10.1093/jhered/esy006
3. Monnahan P, Kolář F, Baduel P et al (2019) Pervasive population genomic consequences of genome duplication in Arabidopsis arenosa. Nat Ecol Evol 3:457. https://doi.org/10.1038/s41559-019-0807-4
4. Bohutínská M, Vlček J, Yair S et al (2021) Genomic basis of parallel adaptation varies with divergence in Arabidopsis and its relatives. PNAS 118. https://doi.org/10.1073/pnas.2022713118
5. Konečná V, Bray S, Vlček J et al (2021) Parallel adaptation in autopolyploid Arabidopsis arenosa is dominated by repeated recruitment of shared alleles. bioRxiv 2021.01.15.426785. https://doi.org/10.1101/2021.01.15.426785
6. Bohutínská M, Handrick V, Yant L et al (2021) De novo mutation and rapid protein (co-) evolution during meiotic adaptation in Arabidopsis arenosa. Mol Biol Evol 38:1980–1994. https://doi.org/10.1093/molbev/msab001
7. Yant L, Bomblies K (2017) Genomic studies of adaptive evolution in outcrossing Arabidopsis species. Curr Opin Plant Biol 36:9–14. https://doi.org/10.1016/j.pbi.2016.11.018
8. Kolář F, Lučanová M, Záveská E et al (2016) Ecological segregation does not drive the intricate parapatric distribution of diploid and tetraploid cytotypes of the Arabidopsis arenosa group (Brassicaceae). Biol J Linn Soc Lond 119:673–688. https://doi.org/10.1111/bij.12479
9. Morgan EJ, Čertner M, Lučanová M et al (2020) Niche similarity in diploid-autotetraploid contact zones of Arabidopsis arenosa across spatial scales. Am J Bot 107: 1375–1388. https://doi.org/10.1002/ajb2.1534
10. Arnold B, Kim S-T, Bomblies K (2015) Single geographic origin of a widespread autotetraploid Arabidopsis arenosa lineage followed by interploidy admixture. Mol Biol Evol 32: 1382–1395
11. Kolář F, Fuxová G, Záveská E et al (2016) Northern glacial refugia and altitudinal niche divergence shape genome-wide differentiation in the emerging plant model Arabidopsis arenosa. Mol Ecol 25:3929–3949. https://doi.org/10.1111/mec.13721
12. Knotek A, Konečná V, Wos G et al (2020) Parallel Alpine Differentiation in Arabidopsis arenosa. Front Plant Sci 11. https://doi.org/10.3389/fpls.2020.561526
13. Li H, Durbin R (2009) Fast and accurate short read alignment with Burrows–Wheeler transform. Bioinformatics 25:1754–1760. https://doi.org/10.1093/bioinformatics/btp324
14. McKenna A, Hanna M, Banks E et al (2010) The Genome Analysis Toolkit: a MapReduce framework for analyzing next-generation DNA

sequencing data. Genome Res 20:1297–1303. https://doi.org/10.1101/gr.107524.110

15. Parisod C, Holderegger R, Brochmann C (2010) Evolutionary consequences of autopolyploidy. New Phytol 186:5–17. https://doi.org/10.1111/j.1469-8137.2009.03142.x
16. Butruille DV, Boiteux LS (2000) Selection--mutation balance in polysomic tetraploids: impact of double reduction and gametophytic selection on the frequency and subchromosomal localization of deleterious mutations. PNAS 97:6608–6613
17. Blischak PD, Kubatko LS, Wolfe AD (2018) SNP genotyping and parameter estimation in polyploids using low-coverage sequencing data. Bioinformatics 34:407–415. https://doi.org/10.1093/bioinformatics/btx587
18. Shastry V, Adams PE, Lindtke D et al (2021) Model-based genotype and ancestry estimation for potential hybrids with mixed-ploidy. Mol Ecol Resour 21:1434–1451. https://doi.org/10.1111/1755-0998.13330
19. Gerard D, Ferrão LFV, Garcia AAF, Stephens M (2018) Genotyping polyploids from messy sequencing data. Genetics 210:789–807. https://doi.org/10.1534/genetics.118.301468
20. Voorrips RE, Gort G, Vosman B (2011) Genotype calling in tetraploid species from bi-allelic marker data using mixture models. BMC Bioinform 12:172. https://doi.org/10.1186/1471-2105-12-172
21. Monnahan P, Brandvain Y (2020) The effect of autopolyploidy on population genetic signals of hard sweeps. Biol Lett 16:20190796. https://doi.org/10.1098/rsbl.2019.0796
22. Ronfort J, Jenczewski E, Bataillon T, Rousset F (1998) Analysis of population structure in autotetraploid species. Genetics 150:921–930. https://doi.org/10.1093/genetics/150.2.921
23. Nei M, Li WH (1979) Mathematical model for studying genetic variation in terms of restriction endonucleases. PNAS 76:5269–5273
24. Meirmans PG GenoDive version 3.0: easy-to-use software for the analysis of genetic data of diploids and polyploids. Mol Ecol Resour. https://doi.org/10.1111/1755-0998.13145
25. Tajima F (1989) Statistical method for testing the neutral mutation hypothesis by DNA polymorphism. Genetics 123:585–595
26. Ferretti L, Ribeca P, Ramos-Onsins SE (2018) The site frequency/dosage spectrum of autopolyploid populations. Front Genet 9. https://doi.org/10.3389/fgene.2018.00480
27. Schraiber JG, Akey JM (2015) Methods and models for unravelling human evolutionary history. Nat Rev Genet 16:727–740. https://doi.org/10.1038/nrg4005
28. Bohutínská M, Alston M, Monnahan P et al (2021) Novelty and convergence in adaptation to whole genome duplication. Mol Biol Evol. https://doi.org/10.1093/molbev/msab096
29. Bray SM, Wolf EM, Zhou M et al (2020) Convergence and novelty in adaptation to whole genome duplication in three independent polyploids. bioRxiv 2020.03.31.017939. https://doi.org/10.1101/2020.03.31.017939
30. Hudson RR, Slatkin M, Maddison WP (1992) Estimation of levels of gene flow from DNA sequence data. Genetics 132:583–589
31. Meirmans PG, Liu S (2018) Analysis of molecular variance (AMOVA) for autopolyploids. Front Ecol Evol 6. https://doi.org/10.3389/fevo.2018.00066
32. Huang K, Wang T, Dunn DW et al (2021) A generalized framework for AMOVA with multiple hierarchies and ploidies. Integr Zool 16:33–52. https://doi.org/10.1111/1749-4877.12460
33. Stift M, Kolář F, Meirmans PG (2019) Structure is more robust than other clustering methods in simulated mixed-ploidy populations. Heredity 123:429–441. https://doi.org/10.1038/s41437-019-0247-6
34. Pritchard JK, Stephens M, Donnelly P (2000) Inference of population structure using multilocus genotype data. Genetics 155:945–959
35. Alexander DH, Novembre J, Lange K (2009) Fast model-based estimation of ancestry in unrelated individuals. Genome Res 19:1655–1664. https://doi.org/10.1101/gr.094052.109
36. Raj A, Stephens M, Pritchard JK (2014) fastSTRUCTURE: Variational inference of population structure in large SNP data sets. Genetics 197:573–589. https://doi.org/10.1534/genetics.114.164350
37. Novikova PY, Hohmann N, Nizhynska V et al (2016) Sequencing of the genus Arabidopsis identifies a complex history of nonbifurcating speciation and abundant trans-specific polymorphism. Nat Genet 48:1077–1082. https://doi.org/10.1038/ng.3617
38. Pickrell JK, Pritchard JK (2012) Inference of population splits and mixtures from genome-wide allele frequency data. PLoS Genet 8:e1002967. https://doi.org/10.1371/journal.pgen.1002967
39. Wos G, Mořkovská J, Bohutínská M et al (2019) Role of ploidy in colonization of alpine

habitats in natural populations of Arabidopsis arenosa. Ann Bot 124:255–268. https://doi.org/10.1093/aob/mcz070

40. Novikova PY, Brennan IG, Booker W et al (2020) Polyploidy breaks speciation barriers in Australian burrowing frogs Neobatrachus. PLoS Genet 16:e1008769. https://doi.org/10.1371/journal.pgen.1008769
41. Arnold B, Bomblies K, Wakeley J (2012) Extending coalescent theory to autotetraploids. Genetics 192:195–204. https://doi.org/10.1534/genetics.112.140582
42. Arnold BJ, Lahner B, DaCosta JM et al (2016) Borrowed alleles and convergence in serpentine adaptation. PNAS 113:8320–8325. https://doi.org/10.1073/pnas.1600405113
43. Excoffier L, Dupanloup I, Huerta-Sánchez E et al (2013) Robust demographic inference from genomic and SNP data. PLoS Genet 9:e1003905
44. Yant L, Hollister JD, Wright KM et al (2013) Meiotic adaptation to genome duplication in Arabidopsis arenosa. Curr Biol 23:2151–2156. https://doi.org/10.1016/j.cub.2013.08.059
45. Marburger S, Monnahan P, Seear PJ et al (2019) Interspecific introgression mediates adaptation to whole genome duplication. Nat Commun 10:1–11. https://doi.org/10.1038/s41467-019-13159-5
46. Schmickl R, Yant L Adaptive introgression: how polyploidy reshapes gene flow landscapes. New Phytol. https://doi.org/10.1111/nph.17204
47. Vitti JJ, Grossman SR, Sabeti PC (2013) Detecting natural selection in genomic data. Annu Rev Genet 47:97–120. https://doi.org/10.1146/annurev-genet-111212-133526
48. Stephan W (2019) Selective sweeps. Genetics 211:5–13. https://doi.org/10.1534/genetics.118.301319
49. Hoban S, Kelley JL, Lotterhos KE et al (2016) Finding the genomic basis of local adaptation: pitfalls, practical solutions, and future directions. Am Nat 188:379–397. https://doi.org/10.1086/688018
50. Oleksyk TK, Smith MW, O'Brien SJ (2010) Genome-wide scans for footprints of natural selection. Philos Trans R Soc B Biol Sci 365:185–205. https://doi.org/10.1098/rstb.2009.0219
51. Booker TR, Jackson BC, Keightley PD (2017) Detecting positive selection in the genome. BMC Biol 15:98. https://doi.org/10.1186/s12915-017-0434-y
52. Motazedi E, Finkers R, Maliepaard C, de Ridder D (2018) Exploiting next-generation sequencing to solve the haplotyping puzzle in polyploids: a simulation study. Brief Bioinform 19:387–403. https://doi.org/10.1093/bib/bbw126
53. Siragusa E, Haiminen N, Finkers R et al (2019) Haplotype assembly of autotetraploid potato using integer linear programing. Bioinformatics 35:3279–3286. https://doi.org/10.1093/bioinformatics/btz060
54. Hermisson J, Pennings PS (2005) Soft sweeps: molecular population genetics of adaptation from standing genetic variation. Genetics 169:2335–2352. https://doi.org/10.1534/genetics.104.036947
55. Morgan C, Zhang H, Henry CE et al (2020) Derived alleles of two axis proteins affect meiotic traits in autotetraploid Arabidopsis arenosa. PNAS. https://doi.org/10.1073/pnas.1919459117
56. Lee KM, Coop G (2017) Distinguishing among modes of convergent adaptation using population genomic data. Genetics 207:1591–1619. https://doi.org/10.1534/genetics.117.300417
57. Gautier M (2015) Genome-wide scan for adaptive divergence and association with population-specific covariates. Genetics 201:1555–1579. https://doi.org/10.1534/genetics.115.181453
58. Cheng JY, Racimo F, Nielsen R (2019) Ohana: detecting selection in multiple populations by modelling ancestral admixture components. bioRxiv 546408. https://doi.org/10.1101/546408
59. Szpak M, Mezzavilla M, Ayub Q et al (2018) FineMAV: prioritizing candidate genetic variants driving local adaptations in human populations. Genome Biol 19:5. https://doi.org/10.1186/s13059-017-1380-2
60. McDonald JH, Kreitman M (1991) Adaptive protein evolution at the Adh locus in Drosophila. Nature 351:652–654. https://doi.org/10.1038/351652a0
61. Charlesworth J, Eyre-Walker A (2008) The McDonald–Kreitman test and slightly deleterious mutations. Mol Biol Evol 25:1007–1015. https://doi.org/10.1093/molbev/msn005
62. Fay JC, Wyckoff GJ, Wu CI (2001) Positive and negative selection on the human genome. Genetics 158:1227–1234
63. Zhang L, Li W-H (2005) Human SNPs reveal no evidence of frequent positive selection. Mol Biol Evol 22:2504–2507. https://doi.org/10.1093/molbev/msi240
64. Frichot E, Schoville SD, Bouchard G, François O (2013) Testing for associations between loci and environmental gradients using latent factor

mixed models. Mol Biol Evol 30:1687–1699. https://doi.org/10.1093/molbev/mst063

65. Meirmans PG, Van Tienderen PH (2013) The effects of inheritance in tetraploids on genetic diversity and population divergence. Heredity 110:131–137. https://doi.org/10.1038/hdy.2012.80

66. Cingolani P, Platts A, Wang LL et al (2012) A program for annotating and predicting the effects of single nucleotide polymorphisms, SnpEff. Fly 6:80–92. https://doi.org/10.4161/fly.19695

Chapter 17

Inferring the Demographic History and Inheritance Mode of Tetraploid Species Using ABC

Camille Roux, Xavier Vekemans, and John Pannell

Abstract

Genomic patterns of diversity and divergence are impacted by certain life history traits, reproductive systems, and demographic history. The latter is characterized by fluctuations in population sizes over time, as well as by temporal patterns of introgression. For a given organism, identifying a demographic history that deviates from the standard neutral model allows a better understanding of its evolution but also helps to reduce the risk of false positives when screening for molecular targets of natural selection. Tetraploid organisms and beyond have demographic histories that are complicated by the mode of polyploidization, the mode of inheritance, and different scenarios of gene flow between sub-genomes and diploid parental species. Here we provide guidelines for experimenters wishing to address these issues through a flexible statistical framework: approximate Bayesian computation (ABC). The emphasis is on the general philosophy of the approach to encourage future users to exploit the enormous flexibility of ABC beyond the limitations imposed by generalist data analysis pipelines.

Key words Speciation, Demographic history, Approximate Bayesian computation, Tetraploids, Inheritance

1 Introduction

The living world is characterized by an organization of individuals into different discrete groups, partially or totally reproductively isolated from each other: species. An important challenge in evolutionary biology is to understand the historical demographic context in which species barriers evolved. Distinguishing whether lineages have diverged in sympatry or allopatry leads to different hypotheses about the origin of studied species, for instance, increasing the contribution of local adaptive processes when lineages are genetically connected by migration events [1]. However, an instantaneous mode of speciation can emerge in sympatry: polyploidization. Newly formed polyploid individuals are likely to become reproductively isolated from their parental genotypes since the first generation [2]. Mating between a newly formed tetraploid

Yves Van de Peer (ed.), *Polyploidy: Methods and Protocols*,
Methods in Molecular Biology, vol. 2545, https://doi.org/10.1007/978-1-0716-2561-3_17,

individual and a diploid parent produces hybrids that are expected to suffer because of uneven number of chromosome copies. However, triploid sterility is not a systematic consequence as unreduced gametes can result in the formation of hexaploids, allowing gene flow between different ploidy levels [3]. Studying the origin of a given polyploid lineage (autopolyploid *versus* allopolyploid), the mode of inheritance (disomic *versus* tetrasomic) as well as the putative contribution of gene flow during polyploid speciation thus improves our knowledge about the relationship between speciation and polyploidization in nature.

The democratization of genome sequencing in non-model organisms that has taken place over the last decade, as well as advances in inferential statistical methods, notably through machine learning, makes it possible to rapidly draw robust conclusions about the demographic and genomic history of a polyploid species. How and when does polyploidization take place? How have effective population sizes fluctuated over time? How important is migration in the genetic composition of the population? Although answering these questions allows for the testing of hypotheses about the biological history of species, it also avoids inferential biases when scanning the genome for natural selection when demography is neglected. This can be illustrated by current practices in population genomics. A popular way to identify genomic regions that have recently experienced spatially divergent selection is based on the "genomic scan" of the F_{ST} statistic [4], i.e., a survey of variation of this statistic across genomic regions aiming to identify regions with higher or lower values than the genomic background. The aim of these genomic scans is to identify F_{ST} values that cannot be explained by a drift-mutation-migration equilibrium alone. The usual interpretation of an apparent F_{ST} increased is the action of strong and local adaptive processes, detected by a departure from a null genomic distribution [5]. However, this null distribution depends on the demographic history of the species studied (Fig. 1 [6–9]).

Hence, the variance of F_{ST} across genomic regions can be strongly increased according to the demographic scenario experienced by the studied organism, leading to false positives when scanning the genomes to detect selection. Riverine or coastal organisms are thus not expected to share the same null envelope as pelagic organisms for the same genomic average F_{ST} [10]. Fortunately, methodological developments in population genomics since the early 2000s have greatly facilitated the exploration of sequenced data to capture the main demographic features behind the observed genetic patterns.

The underlying rationale is simple: since the genomic patterns of polymorphism and divergence obtained by population sequencing are the product of the demographic history of the sample, then it is possible to reconstruct such a history from the observed

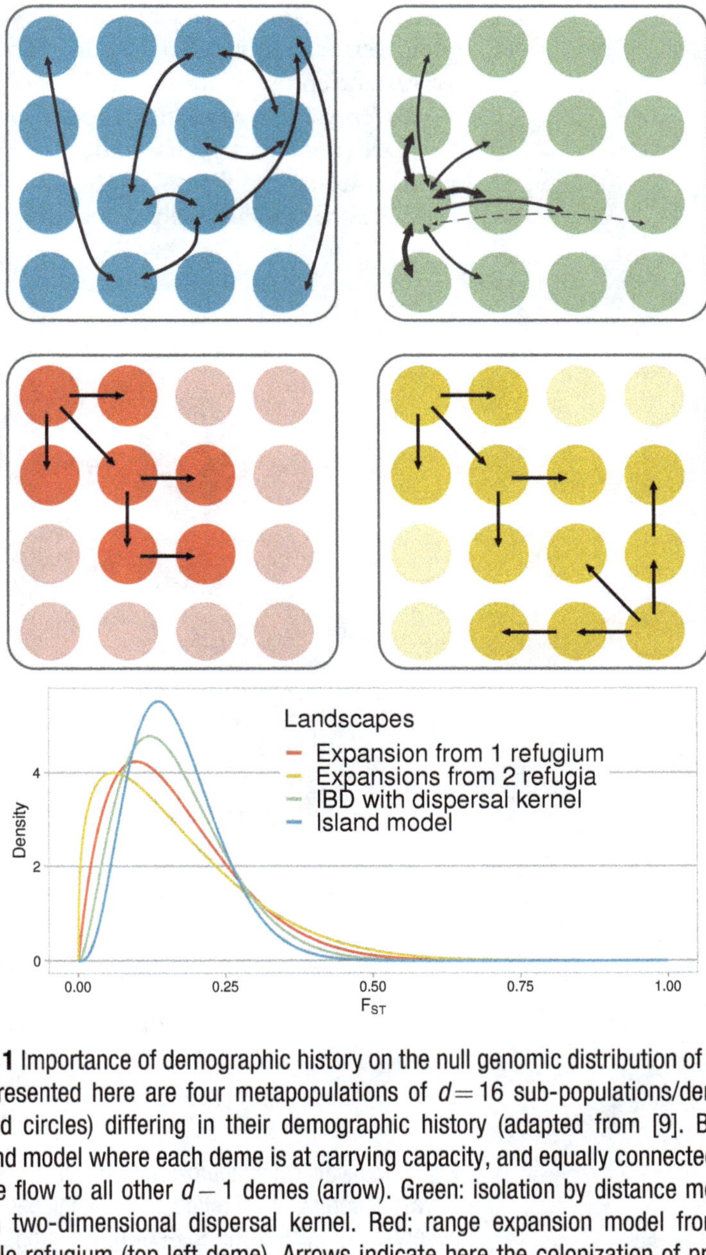

Fig. 1 Importance of demographic history on the null genomic distribution of F_{ST}. Represented here are four metapopulations of $d = 16$ sub-populations/demes (filled circles) differing in their demographic history (adapted from [9]. Blue: island model where each deme is at carrying capacity, and equally connected by gene flow to all other $d-1$ demes (arrow). Green: isolation by distance model with two-dimensional dispersal kernel. Red: range expansion model from a single refugium (top left deme). Arrows indicate here the colonization of previously free niches. Yellow: range expansion model from two refugia (top left and bottom right). Curves represent the genomic distribution of F_{ST} under each model. The four models produce different genomic F_{ST} variation for the same mean. In particular, areas beyond the blue curve represent false positive F_{ST} outlier regions if the island model is taken as the null model

genomic signatures. Reconstructing the demographic history of species consists of classifying arbitrarily proposed scenarios according to their ability to reproduce observed patterns. This involves comparing the likelihoods of alternative models in order to distinguish evolutionary processes. Not exclusively : (*i*) autopolyploid *versus* allopolyploid; (*ii*) disomic *versus* tetrasomic; (*iii*)recent *versus* ancient speciation and/or polyploidization; (*iv*)gene flow *versus* reproductive isolation models. While some measurable quantities are robustly expected under simple statistical models (Fig. 2a), models deviating from Wright and Fisher make the likelihood estimates computationally demanding (Fig. 2b).

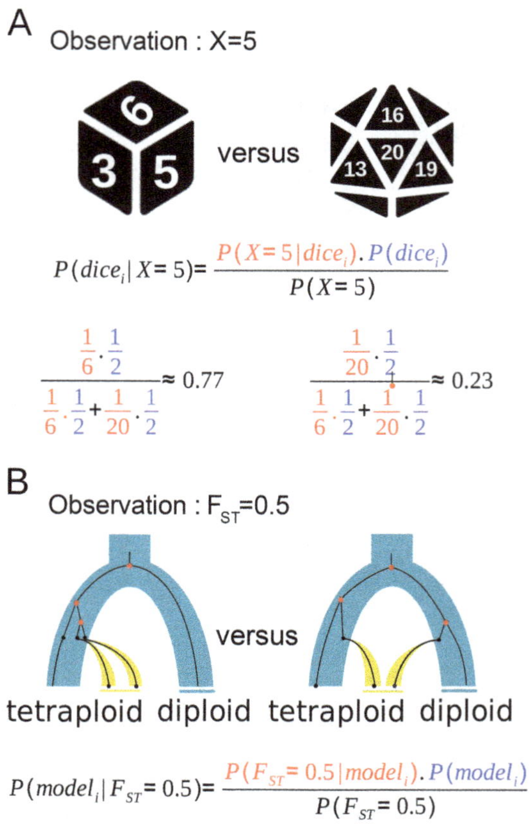

Fig. 2 Motivation for approximate Bayesian computation. The common use of Bayes' theorem is to relate the conditional probability of a model *M* given observations *X*, ($P(M|X)$), to $P(X|M)$, the probability of the observations *X* given *M*, by following the equation: $P(M|X) = \frac{P(X|M).P(M)}{P(X)}$. **a** Inferring which dice most likely produced an observation *X* is robust by applying Bayes' formula because the probability of observing *X* according to the model M_i (the $dice_i$) is known: one over the number of sides. Illustrated here is a comparison between the 6-sided dice (D6) and the 20-sided dice (D20). **b** With genomic data, the real likelihood $P(X|model_i)$ is not known for all the evolutionary scenarios to be compared. The ABC method takes its name from the approximation of the likelihood function using intensive simulations performed in a given parameter space. The autopolyploidization *versus* allopolyploidization comparison requires approximations since a closed form for the probability of observing data under a given model is not known for these two models (described in Fig. 5)

Approximate Bayesian computation (ABC) is a framework for inferential statistics that bypasses the analytical estimation of a likelihood function [11]. For models where the likelihood function is analytically untractable, ABC will use intensive simulations in a given parameter space to approximate it. These simulations empirically provide expected joint distributions of *N* summary statistics describing the genomic data under a complex evolutionary model. The final outcome of the ABC is, therefore, to return the best combination of model and parameters (under this best model) producing the combination of *N* statistics closest to the combination observed from the observed dataset. Because nowadays the whole ABC analysis can be done in four short command lines in a Linux terminal, then the aim of this chapter is not to detail one particular inferential tool among the many available to date (for more details on the current state of the art, see the review by Bourgeois and Warren [12]), but to detail the steps that lead to the exploitation of the great flexibility of ABC, in particular, to tackle the difficult issue of exploiting polyploid genomic data, and by extension, all approaches based on comparisons between simulations and observations.

2 Methods

2.1 Demography Inferred from Population Genomic Data

It is important to begin this chapter by defining what demographic inference is, particularly to avoid confusion with phylogenetic reconstruction. First, the time scale involved is not the same : phylogenetic reconstruction aims to group divergent organisms according to their relative divergence for certain molecular and/or morphological metrics and then to propose a temporal sequence leading to the established groups. In contrast, demographic reconstruction aims to describe the temporal changes in patterns of migration and in effective population sizes that occurred within a system of closely related species. In the early days of molecular population genetics, relationships between genetic patterns and demographic processes were proposed. For instance, since the per generation probability of having a coalescent event in a sample of gene copies is proportional to 1/2Ne, where *Ne* is the number of diploid individuals making the population, then the total size of the sample pedigree increases linearly with *Ne*. It follows that the number of SNPs in a population, for a given neutral locus, increases linearly with the size of the coalescent tree, and hence, with *Ne*. Thus, assuming a population whose size was constant in its recent history, it became possible to estimate this value of *Ne* from the number of segregating polymorphic sites observed at the sequenced neutral loci [13]. In addition to this relationship with the number of SNPs at a given neutral locus, *Ne* is also linearly related to the expected average number of nucleotide differences between pairs of sequences per site (π; [14, 15]). The effective number of individuals in the population, therefore, directly impacts both the number of SNPs observed in a given nucleotide sequence alignment but also the distribution of allelic frequencies. Deviations from this standard demographic model of constant population size leave different signatures in the genomes, including the effects of variations in population sizes over time [16, 17], migration between populations and/or introgression between species [18, 19], and ancient population structure [20]. Demographic inferences thus aim to control for the influences of different processes on patterns of intra-population polymorphism and on patterns of interspecific divergence by capturing the signatures they have left in the genomes. By integrating these different genomic signatures let by *Ne* variation, genetic structure, introgression, and polyploidization into an inferential framework, it becomes possible to interpret sequence data in the light of models deviating from the standard neutral model (Fig. 3a).

These more complex demographic models can be defined by sets of parameters that constrain the evolution of one or more populations/species. For instance, four parameters are sufficient to describe the simplest model of divergence of two gene pools: the divergence time T_{split} (in number of generations) and the effective sizes of the three populations (the ancestral one plus its two daughters). Under this model, the age of subdivision of the ancestral population will affect the genomic F_{ST} between the two diverging daughter populations A and B in different ways. First, F_{ST} increases monotonously with divergence from zero to one in the absence of continuous migration or secondary contact between the diverging daughter populations (Fig. 3b). Second, the variation in F_{ST} across loci, initially low in the early stages of divergence, will increase under the effect of drift and recombination, generating genomic heterogeneity in differentiation. Stochastic and progressive fixation of different alleles between the two populations at different positions (initially from standing variation, then, from derived mutations) finally tends to reduce the variation of the F_{ST}, producing somewhat homogeneously differentiated genomes (Fig. 3b).

The patterns of diversity encapsulated in the two-dimensional Site Frequency Spectrum (SFS) also evolve as a result of genetic drift over time (Fig. 3c). In the first generations immediately after the split a strong correlation in allele frequencies is expected between closely related populations (Fig. 3c for $T_{split} = 0.0001\,Ne$ generations). At this level of divergence, the majority of SNPs correspond to shared polymorphic sites between the two populations, inherited from the ancestral population. Due to the sampling effect, some SNPs may appear to have fixed an allele in one population that still segregates at high frequency in the other population. During the divergence process, the proportion of genotyped SNPs

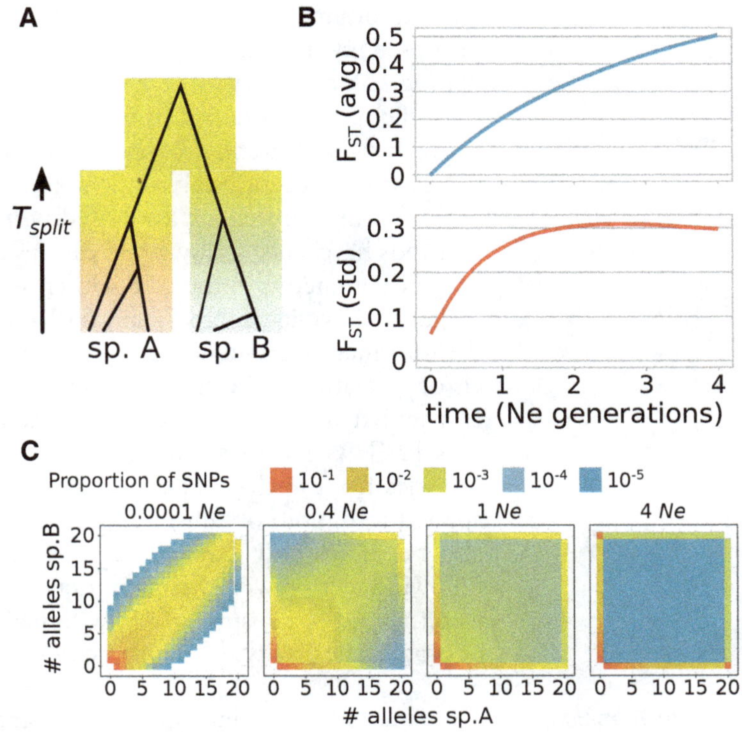

Fig. 3 Example of genomic signatures left by demographic history during the divergence of two diploid species. **a** In its simplest version, the model describes the subdivision, T_{split} generations ago, of an ancestral population with *Ne* individuals into two closely related populations giving the future species A and B. This demographic model can be extended to add (i) population structure and (ii) migration between different populations at different times. **b** Evolution of F_{ST} during divergence. The blue line indicates the evolution of the genomic mean of F_{ST}, the red line indicates its genomic variance. **c** Joint allelic frequency spectra (jSFS) during divergence. Four spectra describe here the allelic frequencies in two diploid sister species at four different divergence times expressed in *Ne* generations, where *Ne* is the effective population size of the two daughter populations and their common ancestor. From left to right: $T_{\text{split}} = 0.0001\,Ne$ generations ago, $0.4\,Ne$, $1\,Ne$ and $4\,Ne$. These spectra are formed from 20 haploid gametes sampled and sequenced from the two diploid populations/species A and B. For each SNP identified among the 40 sequences, the number of copies of the derived allele found in species A (*x*-axis) and B (*y*-axis) is counted. Colors indicate the expected proportions of genotyped SNPs falling within each bin of the jSFS. Shared polymorphisms are located in the inner square. Polymorphisms that are private to a single species are in the external contour. Derived mutations that have become fixed in a single species are in the lower right or higher left corners

exhibiting shared polymorphisms in intermediate frequencies decreases ($T_{\text{split}} = 0.4\,Ne$ generations), until it almost disappears (for $T_{\text{split}} > 1\,Ne$) to the relative benefit of polymorphisms specific to one of the two populations, and finally, to fixed differences between them [21].

Exploiting information from this two-dimensional SFS was first proposed by Li and Stephan [22] in order to reconstruct the demographic history of *Drosophila melanogaster*. They showed that the jSFS contained the information necessary to capture major historical events : (*i*) past population expansions in Africa; (*ii*) relatively recent establishment of *D. Melanogaster* in Europe; (*iii*) a strong bottleneck associated with this colonization. Other methods have since followed up on this approach by fitting demographic parameters to the jSFS, either by maximum likelihood [23–25] or by Bayesian approximation [26] procedures. More recently, the information contained in the SFS has allowed alternative models of speciation to be compared in order to highlight a secondary contact between the Mediterranean and Atlantic lineages of the sea bass [27]. While these methods of reconstructing demographic histories from nucleotide sequencing data have since been widely applied to diploid organisms, some work has shown that they can also be successfully extended to tetraploid species [28, 29]. The following section will, therefore, detail how to apply such methods to tetraploid organisms, and how to take advantage of their methodological flexibility to adapt to certain biological properties.

2.2 From Tetraploid Sequences to First Inferences

To date, demographic inferences using ABC approaches have largely been applied on diploid species, notably to: (*i*) test demographic expansions and/or contractions [30]; (*ii*) test whether two populations/species are connected by ongoing gene flow [31]); (*iii*) test different demo-genetic models, e.g., to take into account the effects of indirect selection on the genomic distribution of diversity [32, 33], or to identify molecular targets of selection against hybrid incompatibilities [34, 35]. In addition to these issues which are relevant for both diploid and polyploid species, other questions specific to polyploids arise. One is often interested in distinguishing between the establishment of a polyploid lineage following the doubling of its genome within a single population (autopolyploidization) and its establishment as a result of hybridization associated with genome doubling between two divergent populations/species (allopolyploidization). Although the distinction between auto- and allopolyploidy is conceptually important to gain insight into the origin of a given tetraploid species under study, it is often achieved indirectly by deducting the evolution mechanism from the mode of inheritance. Usually, disomic inheritance, i.e., preferential pairings between only two homologous chromosomes, is interpreted as indirect evidence of an allopolyploid origin. In contrast, polysomic inheritance (random pairings among all homologous chromosomes) is interpreted as indirect evidence of an autopolyploid origin. But tetrasomic inheritance in a recently duplicated tetraploid lineage will generally gradually evolve to disomic inheritance [36], a process called “diploidization.” This process results from the progressive divergence between pairs of homeologous chromosomes, accelerated by reduced local recombination rates (i.e., in case of large chromosomes), as well as by the subfunctionalization and neofunctionalization processes acting on homeologous genes. Because this process of diploidization can be prevented at islands of recombination between the two sub-genomes [37], differences in gene exchange between homeologous pairs across the genome can lead to mixed or intermediate inheritance models that are called "heterosomy," with some loci showing disomic and others tetrasomic inheritance. Such mixed inheritance models are somewhat neglected in the population genomic literature despite being well reported [38–40]. The evolution of disomy in autotetraploids suggests that inferring the mode of polyploidization from the mode of inheritance only may be misleading, but there is much to learn by decoupling these properties. Here we describe how this can be achieved by detailing key steps in demographic inference in tetraploids (Fig. 4):

1. Defining the models to be explored.
2. Processing of observed and simulated sequences.
3. Comparison between observation and simulations.
4. Checking the results.

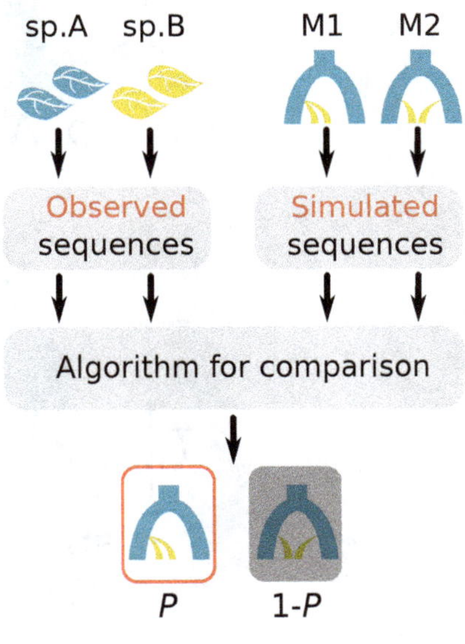

Fig. 4 Three steps in ABC for demographic inferences. The observed data is obtained by DNA/RNA sequencing of individuals from natural populations/ species A and B. Sequences are then summarized by a vector of chosen statistics to quantify the genomic patterns of polymorphism and divergence. Simulated data are exploited to obtain null joint distributions of the same summary statistics under each of the demographic scenarios to be evaluated. The last step in model classification is a statistical comparison between simulated and observed statistics in order to assign a probability *P* to the best supported model. *P* quantifies the relative abilities of the compared models to produce simulations close to the observation

2.2.1 Defining the Models to be Explored

The principle of ABC-like approaches is to first classify an arbitrary set of models according to their ability to reproduce the observed data. This requires upstream design of the alternative scenarios that an experimenter wishes to evaluate. It is, therefore, crucial to put the hypothesis to be tested at the center of the design. In the speciation genomics literature, one of many questions of interest is whether or not there is ongoing gene flow between two divergent species. So the experimenter proceeds to design alternative demographic scenarios that may or may not lead to species currently connected by gene flow [41–44]. The first question of interest concerning polyploids could, for instance, be the auto- *versus* allo-polyploid origins (Fig. 5). In such models, each sub-genome making a tetraploid taxon is considered to be a diploid population.

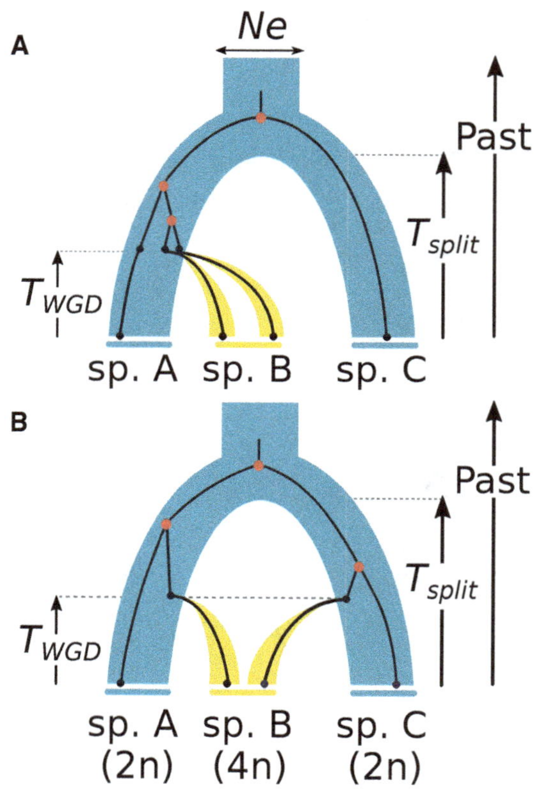

Fig. 5 Demographic speciation scenarios with polyploidization. Blue: diploid species/populations. Yellow: tetraploid species/populations, resulting from **a** autopolyploidization or **b** allopolyploidization occurring T_{WGD} generations ago. Each of the two sub-genomes forming the tetraploid taxon is simulated as two diploid populations with or without gene flow among them to mimic tetrasomic *versus* disomic inheritance, respectively. A strict bottleneck is assumed here at the origin of the tetraploid with a single founding individual, followed by exponential population growth

Autopolyploidization, therefore, involves simulating the divergence of a diploid species into three diploid lineages at a time T_{WGD} [28]: one lineage resulting in a daughter diploid species (species A in Fig. 5a), which is closely related to the other two lineages making the daughter tetraploid species (species B in Fig. 5a). Similarly, allopolyploidization is simulated by modelling two divergence events taking place simultaneously at time T_{WGD}, one between parental species A and sub-genome A of species B, and the other between parental species C and sub-genome C of the tetraploid species B (Fig. 5b).

This is an opportunity to raise a central point in model design: a demographic hypothesis test frequently involves a test between at least two different categories of scenarios that differ for the hypothesis to evaluate. In the popular example of a test for recent introgression between two species of interest, several scenarios can lead to ongoing gene flow. Not exclusively, ongoing gene flow between two species can be the result of secondary contact between them after a past allopatric divergence. But such ongoing gene flow may also have occurred continuously between these two lineages, without interruption, since their divergence from their common ancestor. Similarly, the absence of migration between two species may be the result of a reduction in gene flow which may be sudden or progressive over time. Hence, testing the hypothesis of current gene flow between two species would consist in comparing two categories of scenarios: secondary contact or continuous migration *versus* sudden interruption or progressive interruption of gene flow. Categorizing the hypotheses to be tested in this way allows possible confounding effects to be taken into account, which, when neglected, can bias the biological conclusions drawn from the inferences [33]. Applied to the elucidation of the mode of polyploidization, several sub-models can compose each of the two categories auto- and allopolyploidization. Both categories can be simulated assuming tetrasomic autopolyploids only and disomic allopolyploids only. In this case, an ABC approach will risk interpreting a functionally disomic autopolyploid as a species with allopolyploid origin. Assessing the mode of polyploidization, therefore, requires implementing the different modes of inheritance (disomic, heterosomic, tetrasomic) for both origins [29]. This is an illustration of the power of simulation-based approaches such as ABC, where confounding effects (for instance: disomic = allopolyploid) can be avoided by regrouping variants of the same scenario. This is now possible thanks to the great flexibility offered by the various simulators available [45–53]. Because the two diploid sub-genomes constituting the tetraploid species can be simulated as two diploid populations, then the process of recombination between the sub-genomes under tetrasomic inheritance is equivalent to infinite gene flow between sub-populations. In contrast, disomic inheritance is equivalent to a strict isolation between sub-populations. In between, mixed inheritance (heterosomy) corresponds to a proportion α of the genome where homeologous chromosomes are connected by gene flow, and $1-\alpha$ are isolated. Hence, if α is assumed to be equal to 1, then inheritance is strictly tetrasomic, and strictly disomic if one assumes $\alpha = 0$. Detecting whether a species is autopolyploid or allopolyploid using ABC or supervised deep learning approaches requires, at a minimum, simulating 6 different demographic scenarios, corresponding to all combinations between origin ([autopolyploidy, allopolyploidy]) and inheritance ([disomy, heterosomy, tetrasomy]) [29]. Recently, Booker et al. [54] have gone further by adding different migration sub-scenarios within these two categories and successfully applied this approach to hundreds of loci obtained using exome capture from the North American gray treefrogs Hyla. They test the same 6 models, with the addition of: (1) models with gene flow from *Hyla chrysoscelis* (diploid) to *H. versicolor* (tetraploid); (2) gene flow in both directions (but at non-necessarily symmetrical rates). Using an ABC approach, they strongly reject models with strict reproductive isolation between these two frog species suggesting introgression is effective despite differences in ploidy levels. It is important to be mindful that while the ABC or deep learning method will provide greater support to one of the proposed models (or categories of models) relative to the others, this relative support depends solely on the arbitrarily chosen set of alternative models. By analogy, the ABC will strongly support the "20-sided dice" model in a 6 *versus* 12-sided comparison, if the actual model is the untested 12-sided dice. The philosophy of this approach, in essence, is to classify models relatively to each other. Thus, the interpretation of the results must always be expressed in the light of the set of compared models rather than as definitive conclusions decoupled from the initial model design.

In our initial study [29], we proposed a set of models for which a sample mimicking that of the experimenter (in terms of number of individuals, number and length of sequenced loci) would be simulated for a number of replicates set by the user. The purpose of such simulations is to empirically produce joint distributions of N summary statistics (more details about them in the next section) under the different scenarios to be evaluated. In practice, this is done by running the script *run_ABC_polyploid_v2.py* of our *ABC_WGD* pipeline (https://github.com/popgenomics/ABC_WGD). Our intention here being to accompany the new user in the logic of the analysis, but we invite the interested reader to follow the Readme of the GitHub repository for a more technical tutorial that will be adapted to possible future updates. Our initial analysis [29] also aimed to investigate the origin of polyploidization in *Capsella Bursa-Pastoris* based on population sequencing data in

C. rubella, *C. grandiflora*, and *C. bursa-pastoris*. We thus make available in this same GitHub repository the *Capsella* molecular dataset as well as the scripts as a tutorial to learn step by step about ABC inferences in polyploid organisms. Changes in future versions of the Python and R libraries used will thus be considered in keeping the tool available online and in the associated tutorial.

2.2.2 Processing of Observed and Simulated Sequences

ABC methods are based on the comparison between simulations performed according to different alternative models and observations derived from the analysis of real data. The comparison is made through summary statistics which are used to describe the observed patterns of polymorphism and divergence genome-wide: the model receiving the strongest relative support being the one that produces combinations of summary statistics closest to the observed data. A step that becomes crucial is, therefore, the choice of relevant summary statistics. We will not develop here the issue concerning the number of statistics to choose, as major improvements brought by Random Forest algorithms now allow the use of a large number of them without loss of power [55]. Nowadays, the experimenter wishing to apply a custom ABC approach needs to focus on two important aspects of the choice of summary statistics:

1. Which statistics best capture past demographic events?
2. Which statistics can be quantified on the observed dataset (i.e., statistics requiring phased data? Is the identification of parental sub-genomes required? etc.)

A strength of ABC is its adaptability to different sampling and sequencing strategies. This flexibility circumvents problems that arise when calculating summary statistics for a polyploid. One of them is to phase the haplotypes and to attribute each sequenced copy to one of the two parental sub-genomes. We proposed in 2015 [29] a simple approach, with the aim of being transferable to the widest range of sampling and sequencing strategies (and ploidy levels), by using only summary statistics based on allelic frequencies (Figs. 6 and 7).

This approach assumes having a good confidence in the allelic count for each SNP, regardless of the ploidy level. This means an absence of sequencing bias, where all of the four copies of a gene carried by an auto- or allotetraploid individual are taking into account when assessing the genotype. Different tools propose to genotype sequencing data from polyploid organisms, by dealing more or less efficiently with sequencing errors, allelic bias and overdispersion, among them: the method of Li [56, 57], GATK [58], fitPoly [59], and updog [60]. Once the sampled genotypes are genotyped, it becomes possible without phasing to define a minimum but still efficient set of statistics suitable for demographic inference, among them: π; Watterson's θ, Tajima's D, F_{ST}, raw

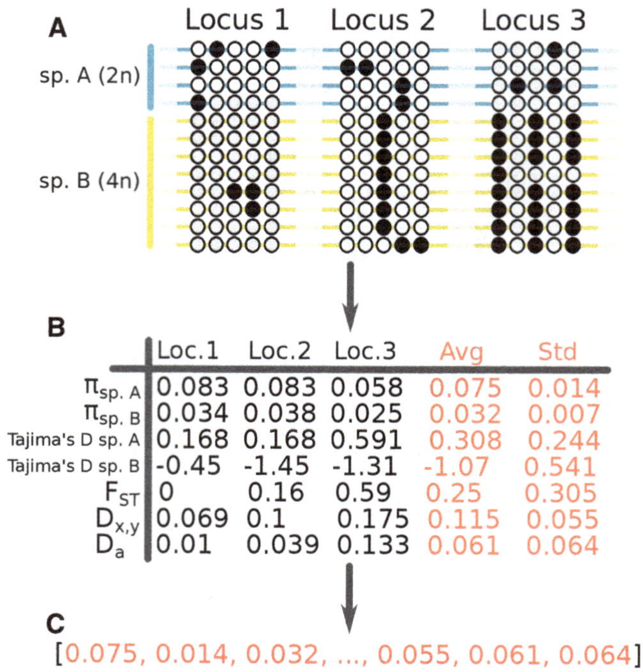

Fig. 6 Summarizing the data by using classical statistics in population genetics. Three loci were sequenced in two diploid (species A in blue) and two tetraploids (species B in yellow) individuals. The white and black circles indicate alternative alleles at the identified SNPs. A set of 7 summary statistics is calculated for each locus. The vector containing the means and standard deviations of each statistic will be used as the input for inferences

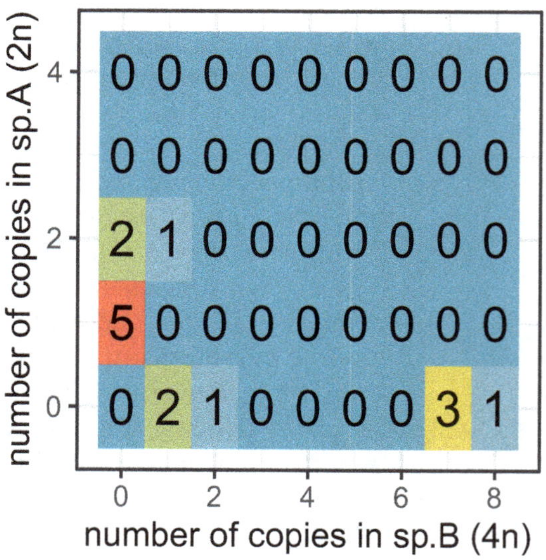

Fig. 7 Joint site frequency spectrum as summary statistics. The example found in Fig. 6a is now summarized by an SFS, where each bin corresponds to the allelic count of the derived allele (black) within the sampled and sequenced populations/species. The number of SNPs within each bin can be used as a statistic for inferences

divergence $D_{x,y}$, net divergence D_a. The purpose of these statistics is to quantify the patterns of polymorphism and divergence, in order to find the model that best reproduces them. Experimenters wishing to apply/develop an ABC approach to studying the history of their model organism must bear in mind that there is no single "magic" statistic that will resolve on its own the history of an organism: demographic signatures are essentially retrieved in the joint combination of different statistics. More precisely, demography impacts the genomic distributions of different statistics [61]. Quantiles of these genomic distributions, usually mean and standard deviation, should then be used as summary statistics, although any method describing a distribution would be relevant (Fig. 6a, b). A dataset consisting of several sequenced loci in different individuals from related species is finally summarized into a vector of statistics (Fig. 6c) from which inferences will be made, in order to propose the simulated model-parameter combination reproducing the best this vector of observations. In addition to standard population genetics statistics, SFS can also be treated as a vector of summary statistics (Fig. 7). The latter does not require phased data nor attribution to one of the parental sub-genomes, making it applicable to genomic data in tetraploids. Each bin making the SFS becomes an independent statistic summarizing the observed data.

The representation of closely related diploid and tetraploid populations in the form of a SFS highlights the effects of modes of inheritance on genomic patterns (Fig. 8). During autopolyploidization with tetrasomic inheritance, the bottleneck associated with the emergence of the tetraploid taxon rapidly purges a large amount of the polymorphism segregating in the ancestral population. As a result of such strong bottleneck, each bin of the inner square (i.e., corresponding to shared polymorphism) or on the bottom edge (i.e., polymorphisms exclusive to the tetraploid lineage) of the SFS represents about a proportion of 0.0001 of the total number of SNPs found in a diploid/tetraploid alignment 0.01 *Ne* generations after genome duplication. In this autopolyploidization scenario, the first tetraploid individual is the result of a panmictic mating within the parental diploid population. The associated bottleneck thus generates a peak of heterozygosity (representing about twenty per cent of the genotyped SNPs) in the generations immediately after genome duplication, but which will quickly fade as a result of recombination associated with unbiased Mendelian segregation. At this same stage in the divergence process, the most represented category of SNP is the polymorphism segregating only in the diploid species (Fig. 8a, left edge). Although the polymorphism within the newly established tetraploid population represents a very small part of the genome, alleles can segregate in frequencies between zero and one due to the recombination

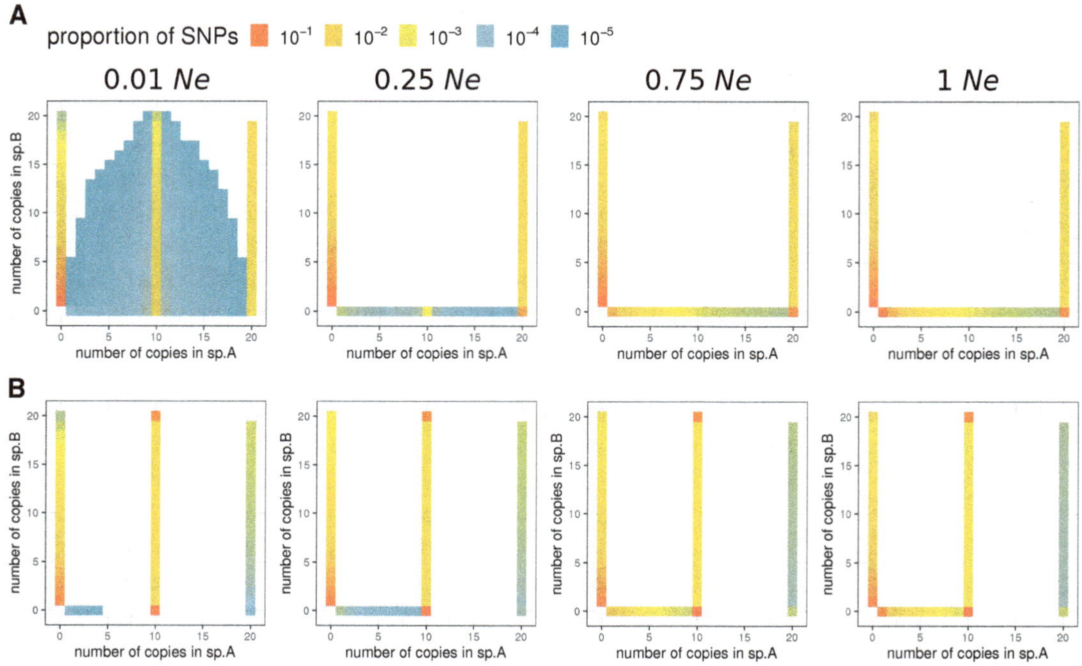

Fig. 8 Joint allelic frequency spectra (jSFS) after whole-genome duplication. Joint spectra describing the distribution of allelic frequencies for a pair of diploid (10 sampled individuals) and tetraploid species (5 sampled individuals). The genome duplication took place *Ne* generations after the separation of the ancestral population. The four jSFS represent genomic patterns after 0.01, 0.25, 0.75, and 1 *Ne* generation after the **a** autopolyploidization (tetrasomic) or **b** allopolyploidization (disomic)

between the four copies. From the early stages of divergence, genomic patterns differ between tetrasomic autopolyploid and disomic allopolyploid. In the latter, the most prevalent category of SNPs is the stabilization of alleles at frequency 0.5 in the tetraploid (Fig. 8b).

The other difference is the absence of alleles segregating at frequencies above 0.5 in the disomic allotetraploid. Hence, the process of polyploidization has a drastic immediate effect on polymorphism patterns. The time since genomic duplication will mainly contribute to increase the relative importance of segregating polymorphism in the tetraploid. These marked differences in the SFS mean that an ABC approach will be able to correctly distinguish a tetrasomic autotetraploid from a disomic allotetraploid with near 100% accuracy. If all three types of inheritance (disomy, heterosomy, tetrasomy) are taken into account for each of the two origins (auto and allo), then the rate of identification of the correct auto- *versus* alloploidization remains high but is strongly impacted by the extent of genomic tetrasomy. Thus, ABC will correctly support the (auto or allo) tetraploid origin in ~99% of cases if the species is disomic, in ~90% of cases if the species is heterosomic, and ~80% of cases [29] if the species is tetrasomic. Increasing the extent of recombination between homeologous chromosomes leads to a loss of information about the mode of polyploidization. This is a consequence of genetic drift, which increases the possibility of losing one of the two parental genomes when the new tetraploid population is still restricted to a few individuals. This biological explanation of the error rate of the ABC in classifying models puts it into perspective. Indeed, if one of the two parental genomes has been lost by drift in the early generations of a sufficiently tetrasomic allotetraploid population, then the species is genomically autotetraploid.

2.2.3 Comparison Between Observation and Simulations

In order to make statistical inferences using an ABC approach, three objects are needed:

1. A vector of summary statistics obtained from the observations (Fig. 6c).
2. A reference table obtained from the simulations, with columns corresponding to the vector of summary statistics and rows corresponding to different simulation replicates.
3. A vector of the variable to be estimated (categorical indicator variable in the case of a model comparison; continuous variable in the case of a demographic parameter under a given model), in correspondence with rows of the reference table.

The reference table is obtained thanks to the pipeline described in the previous sections in order to simulate the experimenter's sample (number of individuals, number of sequenced loci/chromosomes) under the demographic scenarios to be studied and to obtain summary statistics from the performed simulations. While the aim of this book chapter is to encourage experimenters to appropriate existing simulation tools in order to develop a pipeline for their biological questions, such a project may scare neophytes. For this reason, we have developed a turnkey version of such a pipeline, freely available on GitHub with the *Capsella* example: https://github.com/popgenomics/ABC_WGD.

The *run_ABC_polyploid_v2.py* command alone will produce the reference table needed for downstream analysis by combining two output files named *ABCstat.txt* and *ABCjsfs.txt*, for different combinations of polyploidization and inheritance modes. The third file is produced, *priorfile.txt*, which contains the values of parameters, randomly drawn from appropriate distributions according to the priors set by the experimenter, which were used to produce each simulation replicate. The command *run_ABC_polyploid_v2.py* simply takes as arguments the name of the model to be simulated (auto, allo), the inheritance mode (disomic, heterosomic, tetrasomic), the limits of the parameter values to be explored (priors), and the number of multilocus iterations. The reference table (obtained by combining *ABCstat.txt* and *ABCjsfs.txt*, if *run_ABC_polyploid_v2.py* from our pipeline is used) is composed in columns by the same summary statistics than the statistics used to describe the observed dataset (vector of summary statistics), one row per simulated iteration. There are no magic numbers concerning the summary statistics or the number of iterations per scenario to be tested. In practice, it is usual to see in the literature between 10,000 and 50,000 iterations per scenario if a random forest algorithm is used downstream (e.g., R library *abcrf* [62], and around 1 million iterations if a rejection/regression algorithm is preferred (e.g., R library *abc* [63]). It is up to the experimenter to explore around these orders of magnitude to test whether the number of iterations has an effect on the accuracy of M_0 (Model with autopolyploidization) *versus* M_1 (Model with allopolyploidization) comparison. Regarding the number of summary statistics, typically less than 50 are used, but this number can be greatly increased by including each SFS bin in the list of statistics to be used, which is a function of the sampling size. In the case of a model comparison, the third required object is a vector containing the model indicator variable. This vector, of the same length as the number of rows in the reference table (containing the statistics simulated under all models compared, such as M_0 and M_1), is intended to label each row in the reference table with the actual model (e.g., either M_0 or M_1) used to produce the row in the table. Using the *abcrf* package in R, once these three objects are produced, the comparison of M_0 *versus* M_1 models is done in two simple and rapid steps, corresponding to two short R functions to execute. The first step is a training step of the random forest (using the *abcrf()* R function, from the eponymous package), involving the reference table and the vector with the model indicator variable. The returned object is a random forest of *ntree* decision trees, trained to predict the model that best explains a given vector of summary statistics. The effect of the *ntree* value on the accuracy of the M_0 *versus* M_1 comparison can be empirically explored. To provide an indication, a value of *ntree* = 1000 appears in the literature as standard but may vary depending on the number of models compared and the number of summary statistics. The second step is a prediction step (using the *predict()* R base function) that uses the random forest trained in the previous step and the vector of observations. This step returns three values of interest:

1. The selected model (M_0 or M_1 in the case of a comparison between two models, but M_0 and M_1 can also be categories of models).
2. The number of trees in the forest that voted for each of the models compared.

3. The posterior probability of the best model (or the best category of models), approximated by $1 - \epsilon$, where ϵ is the error rate of the inference.

Once the best scenario among those proposed has been identified using the *abcrf()* and *predict()* functions; the next step is to estimate the parameter values that best explains the observed data (of the selected model). Again, using the *abcrf()* R library, this can be achieved in two short steps. The major difference with the previous step of model comparisons is that the objective here is to predict a continuous variable (the model parameter to be estimated) and not a categorical variable (the model M_0 or M_1 to be inferred). To do this, a random forest must first be trained using the same reference table as previously for the model comparison, but replacing the model indicator vector with a vector containing the value of the parameter X that one is attempting to infer. The function *regAbcrf()* from the R package *abcrf* will build a regression random forest from the simulated statistics and the used parameters. For the second step, the *predict()* function will, taking as objects the trained regression forest and the vector of observed statistics, predict the expected value of the parameter, the variance of the posterior and its quantiles. To provide an idea about the computation time required, we have here estimated the time of whole-genome duplication (T_{WGD}) for the four combinations "autopolyploidization or allopolyploidization" × "disomic or tetrasomic." For each of these four combinations, we simulated 10,000 datasets to train the regression forest and 30,000 datasets to test the trained forest (Fig. 9a).

On a 2017 laptop, with 6 dedicated threads, learning the four forests and estimating T_{WGD} for the $4 \times 30,000$ pseudo-observed datasets [65] took 13 minutes. This execution time is not impacted by the sample size in terms of number of individuals and loci. Such a re-analysis of simulated datasets (not used this time to train the forest) is necessary to assess the quality of inference along the range of explored values. In this way, we show that the age of polyploidization T_{WGD} can be accurately estimated for recent, intermediate, and ancient events, and for each of the four explored models, without making any assumptions about the relationship between polyploidization origin and inheritance (Fig. 9a). Having in mind such accuracy of ABC to estimate times of polyploidization, we have applied this approach to molecular data in *Capsella* [29]. For this purpose, 20 nuclear loci were sequenced in more than 5 *C. grandiflora* (diploid) individuals, and 13 loci in more than 5 *C. rubella* (diploid) individuals. For each pairwise ABC analysis involving one of the two diploid species with *C. bursa-pastoris* (tetraploid), we found strong statistical support for an allopolyploidization scenario, with an estimated age of $\approx$72, 000 years ago (95% confidence interval of 37–132 kya for the *C. bursa-pastoris*

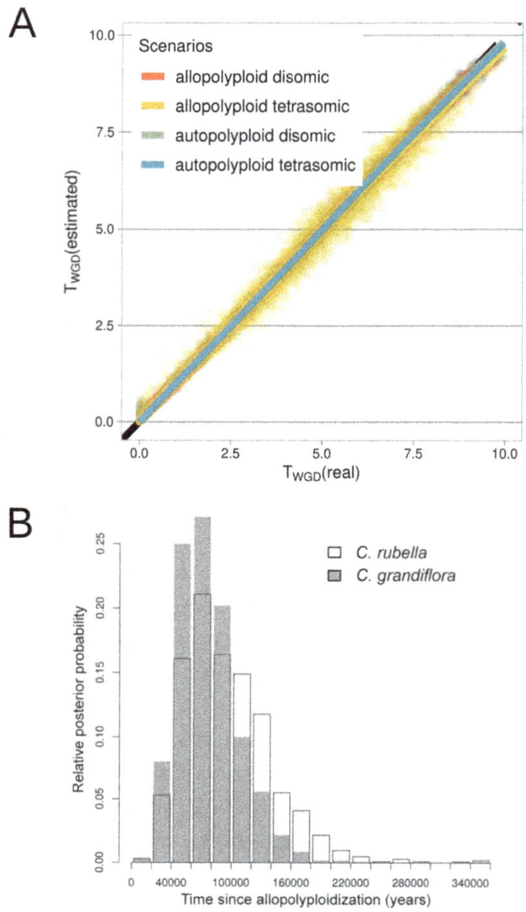

Fig. 9 Estimating time of whole-genome duplication (T_{WGD}) using ABC. **a** Estimated T_{WGD} as a function of true T_{WGD} for simulated datasets of 5000 loci of 1 kb from 20 diploid and 10 tetraploid individuals for four different models of polyploid speciation (in distinct colors). T_{WGD}, (expressed in *Ne* generations), as well as all the parameters of the model, are randomly drawn from a uniform distribution between 0 and 10. *x*-axis is the real value of T_{WGD}; *y*-axis is T_{WGD} predicted by the trained random forest. Points represent $4 \times 30,000$ simulations. **b** Estimated T_{WGD} by Roux and Pannell [29], using 13–20 nuclear loci in two ABC pairwise analysis: *C. rubella–C. bursa-pastoris* (empty bars) and *C. grandiflora–C. bursa-pastoris* (grey bars) (but *see* [64] for a convergent study)

versus C. rubella comparison; Fig. 9b from [29]). This result of an allopolyploid origin in *Capsella Bursa-Pastoris* was also published independently six months later by Douglas and colleagues [64], using different data and approaches on the same species, but with a slightly older polyploidization event, ≈128-kya (95% confidence interval of 22–177 kya), reassuring about the power of ABC to identify important events in the history of the species.

3 Conclusion

While turnkey pipelines are nowadays available in order to efficiently infer the speciation history of tetraploids from genomic data (https://github.com/popgenomics/ABC_WGD), it is still important to highlight that the strength of ABC-like approaches is their flexibility. Thus, pre-existing pipelines can be immediately useful for analyzing newly acquired data, but more importantly, they can be useful as a basis for customization. The first layer of customization may involve the demographic models to explore, either by adding genomic introgression, or by changing the population growth function of the neo-polyploid, or by increasing the ploidy level, etc. The second layer of customization can be applied to the genomic models, notably by implementing more relevant inheritance patterns, different natural selection models. The third layer of customization may concern the set of summary statistics, notably by implementing those that can be brought to light in future. And finally, the prediction algorithm itself can be modified, notably by taking advantage of the promise of convolutional neural networks in population genomics [66]. As with every inferential method, it is always important that the experimenter is not content with the results of the inferences on the observed data only but also explores the space delimited by the choice of the models to compare, as well as by the choice of the parameter range (Fig. 9). Compared to a decade ago, nowadays the execution time of an inference pipeline (model comparisons, parameter estimation) has become almost negligible compared to the preliminary steps: identification of the populations to be sampled, sampling campaign, nucleic acid extraction, sequencing, raw data processing. This time saving should benefit the reflection on the scenarios to be explored and compared and on exploring the limits of the inferential approach adopted, notably by relying on the analysis of custom-made simulated data.

Acknowledgements

C.R has greatly benefited from the support of the I-SITE ULNE foundation (Université Lille Nord-Europe) through the grant "Jeunes Chercheurs 2020" (GAFA project). The simulations were performed on the Core Cluster of the Institut Français de Bioinformatique (IFB) (ANR-11-INBS-0013).

References

1. Mayr E (1942) Systematics and the origin of species, Columbia University Press, New York, p 334
2. Linder CR, Rieseberg LH (2004) Reconstructing patterns of reticulate evolution in plants. Am J Bot 91(10):1700–1708
3. Ramsey J, Schemske DW (1998) Pathways, mechanisms, and rates of polyploid formation in flowering plants. Ann Rev Ecol Syst 29(1): 467–501
4. Wright S (1949) The genetical structure of populations. Ann Eugen 15(1):323–354
5. Lewontin RC, Krakauer J (1973) Distribution of gene frequency as a test of the theory of the selective neutrality of polymorphisms. Genetics 74(1):175–195
6. Robertson A (1975) Remarks on the Lewontin-Krakauertest. Genetics 80(2):396
7. Nei M, Maruyama T (1975) Lewontin-Krakauertest for neutral genes. Genetics 80(2):395
8. Bierne N, Roze D, Welch JJ (2013) Pervasive selection or is it...? Why are f_{ST} outliers sometimes so frequent?
9. Lotterhos KE, Whitlock MC (2014) Evaluation of demographic history and neutral parameterization on the performance of f_{ST} outlier tests. Mol Ecol 23(9):2178–2192
10. Fourcade Y, Chaput-Bardy A, Secondi J, Fleurant C, Lemaire C (2013) Is local selection so widespread in river organisms? Fractal geometry of river networks leads to high bias in outlier detection. Mol Ecol 22(8): 2065–2073
11. Csilléry K, Blum MGB, Gaggiotti OE, François O (2010) Approximate Bayesian computation (abc) in practice. Trends Ecol Evol 25(7): 410–418
12. Bourgeois YXC, Warren BH (2021) An overview of current population genomics methods for the analysis of whole-genome resequencing data in eukaryotes. Mol Ecol 30(23): 6036–6071
13. Watterson GA (1975) On the number of segregating sites in genetical models without recombination. Theor Popul Biol 7:188–193
14. Li W-H (1977) Distribution of nucleotide differences between two randomly chosen cistrons in a finite population. Genetics 85(2): 331–337
15. Tajima F (1983) Evolutionary relationship of DNA sequences in finite populations. Genetics 105(2):437–460
16. Slatkin M (1994) Linkage disequilibrium in growing and stable populations. Genetics 137(1):331–336
17. Tajima F (1989) Statistical method for testing the neutral mutation hypothesis by DNA polymorphism. Genetics 123(3):585–595
18. Navascués M, Legrand D, Campagne C, Cariou M-L, Depaulis F (2014) Distinguishing migration from isolation using genes with intragenic recombination: detecting introgression in the *Drosophila simulans* species complex. BMC Evol Biol 14(1):1–13
19. Liang M, Nielsen R (2014) The lengths of admixture tracts. Genetics 197(3):953–967
20. Wall JD (2000) Detecting ancient admixture in humans using sequence polymorphism data. Genetics 154(3):1271–1279
21. Clark AG (1997) Neutral behavior of shared polymorphism. Proc Natl Acad Sci 94(15): 7730–7734
22. Li H, Stephan W (2006) Inferring the demographic history and rate of adaptive substitution in drosophila. PLoS Genet. 2(10):e166
23. Hernandez RD, Hubisz MJ, Wheeler DA, Smith DG, Ferguson B, Rogers J, Nazareth L, Indap A, Bourquin T, McPherson J et al. (2007) Demographic histories and patterns of linkage disequilibrium in chinese and indian rhesus macaques. Science 316(5822):240–243
24. Gutenkunst RN, Hernandez RD, Williamson SH, Bustamante CD (2009) Inferring the joint demographic history of multiple populations from multidimensional SNP frequency data. PLoS Genet 5(10):e1000695
25. Nielsen R, Hubisz MJ, Hellmann I, Torgerson D, Andrés AM, Albrechtsen A, Gutenkunst R, Adams MD, Cargill M, Boyko A et al. (2009) Darwinian and demographic forces affecting human protein coding genes. Genome Res 19(5):838–849
26. Fraisse C, Popovic I, Mazoyer C, Spataro B, Delmotte S, Romiguier J, Loire E, Simon A, Galtier N, Duret L et al. (2020) Dils: Demographic inferences with linked selection by using abc. Mol Ecol Resour 21(8):2629–2644
27. Tine M, Kuhl H, Gagnaire P-A, Louro B, Desmarais E, Martins RST, Hecht J, Knaust F, Belkhir K, Klages S et al. (2014) European sea bass genome and its variation provide insights into adaptation to euryhalinity and speciation. Nat Commun 5(1):1–10
28. St. Onge KR, Foxe JP, Li J, Li H, Holm K, Corcoran P, Slotte T, Lascoux M, Wright SI (2012) Coalescent-based analysis distinguishes between allo-and autopolyploid origin in shepherd's purse (*Capsella bursa-pastoris*). Mol Biol Evol 29(7):1721–1733
29. Roux C, Pannell JR (2015) Inferring the mode of origin of polyploid species from next-

generation sequence data. Mol Ecol 24(5): 1047–1059

30. Fraimout A, Debat V, Fellous S, Hufbauer RA, Foucaud J, Pudlo P, Marin J-M, Price DK, Cattel J, Chen X et al. (2017) Deciphering the routes of invasion of *Drosophila suzukii* by means of abc random forest. Mol Biol Evol 34(4):980–996
31. Roux C, Castric V, Pauwels M, Wright SI, Saumitou-Laprade P, Vekemans X (2011) Does speciation between *Arabidopsis halleri* and *Arabidopsis lyrata* coincide with major changes in a molecular target of adaptation? PloS One 6(11):e26872
32. Charlesworth B, Nordborg M, Charlesworth D (1997) The effects of local selection, balanced polymorphism and background selection on equilibrium patterns of genetic diversity in subdivided populations. Genet Res 70(2): 155–174
33. Cruickshank TE, Hahn MW (2014) Reanalysis suggests that genomic islands of speciation are due to reduced diversity, not reduced gene flow. Mol Ecol 23(13):3133–3157
34. Roux C, Fraïsse C, Castric V, Vekemans X, Pogson GH, Bierne N (2014) Can we continue to neglect genomic variation in introgression rates when inferring the history of speciation? A case study in a *Mytilus* hybrid zone. J Evol Biol 27(8):1662–1675
35. Roux C, Tsagkogeorga G, Bierne N, Galtier N (2013) Crossing the species barrier: genomic hotspots of introgression between two highly divergent *Ciona intestinalis* species. Mol Biol Evol 30(7):1574–1587
36. Le Comber SC, Ainouche ML, Kovarik A, Leitch AR (2010) Making a functional diploid: from polysomic to disomic inheritance. New Phytol 186(1):113–122
37. Meirmans PG, Van Tienderen PH (2013) The effects of inheritance in tetraploids on genetic diversity and population divergence. Heredity 110(2):131–137
38. Stift M, Berenos C, Kuperus P, van Tienderen PH (2008) Segregation models for disomic, tetrasomic and intermediate inheritance in tetraploids: a general procedure applied to *Rorippa* (yellow cress) microsatellite data. Genetics 179(4):2113–2123
39. Koning-Boucoiran CFS, Gitonga VW, Yan Z, vd Dolstra O, Van Der Linden CG, Van Der Schoot J, Uenk GE, Verlinden K, Smulders MJM, Krens FA et al. (2012) The mode of inheritance in tetraploid cut roses. Theor Appl Genet 125(3):591–607
40. Allendorf FW, Danzmann RG (1997) Secondary tetrasomic segregation of MDH-B and preferential pairing of homeologues in rainbow trout. Genetics 145(4):1083–1092
41. Ross-Ibarra J, Tenaillon M, Gaut BS (2009) Historical divergence and gene flow in the genus *Zea*. Genetics 181(4):1399–1413
42. Roux C, Fraisse C, Romiguier J, Anciaux Y, Galtier N, Bierne N (2016) Shedding light on the grey zone of speciation along a continuum of genomic divergence. PLoS Biol 14(12): e2000234
43. Duranton M, Allal F, Fraïsse C, Bierne N, Bonhomme F, Gagnaire P-A (2018) The origin and remolding of genomic islands of differentiation in the european sea bass. Nat Commun 9(1):1–11
44. Duvaux L, Belkhir K, Boulesteix M, Boursot P (2011) Isolation and gene flow: inferring the speciation history of European house mice. Mol Ecol 20(24):5248–5264
45. Hartfield M, Wright SI, Agrawal AF (2016) Coalescent times and patterns of genetic diversity in species with facultative sex: effects of gene conversion, population structure, and heterogeneity. Genetics 202(1):297–312
46. Excoffier L, Foll M (2011) Fastsimcoal: a continuous-time coalescent simulator of genomic diversity under arbitrarily complex evolutionary scenarios. Bioinformatics 27(9): 1332–1334
47. Hudson RR (2002) Generating samples under a Wright–Fisher neutral model of genetic variation. Bioinformatics 18(2):337–338
48. Ewing G, Hermisson J (2010) MSMS: a coalescent simulation program including recombination, demographic structure and selection at a single locus. Bioinformatics 26(16): 2064–2065
49. Pavlidis P, Laurent S, Stephan W (2010) msABC: a modification of Hudson's ms to facilitate multi-locus abc analysis. Mol Ecol Resour 10(4):723–727
50. Kelleher J, Lohse K (2020) Coalescent simulation with msprime. In: Statistical population genomics. Humana, New York, NY, pp 191–230
51. Currat M, Arenas M, Quilodràn CS, Excoffier L, Ray N (2019) Splatche3: simulation of serial genetic data under spatially explicit evolutionary scenarios including long-distance dispersal. Bioinformatics 35(21): 4480–4483
52. Neuenschwander S, Michaud F, Goudet J (2019) Quantinemo 2: a Swiss knife to simulate complex demographic and genetic scenarios, forward and backward in time. Bioinformatics 35(5):886–888

53. Haller BC, Messer PW (2019) Slim 3: forward genetic simulations beyond the Wright–Fisher model. Mol Biol Evol 36(3):632–637
54. Booker WW, Gerhardt HC, Lemmon AR, Ptacek M, Hassinger ATB, Schul J, Lemmon EM (2020) The complex history of genome duplication and hybridization in north American gray treefrogs. bioRxiv
55. Marin J-M, Pudlo P, Robert CP, Cornuet J-M, Estoup A (2014) Abc model choice *via* random forests. Stat 1050:24
56. Li H (2011) A statistical framework for SNP calling, mutation discovery, association mapping and population genetical parameter estimation from sequencing data. Bioinformatics 27(21):2987–2993
57. Blischak PD, Kubatko LS, Wolfe AD (2018) SNP genotyping and parameter estimation in polyploids using low-coverage sequencing data. Bioinformatics 34(3):407–415
58. McKenna A, Hanna M, Banks E, Sivachenko A, Cibulskis K, Kernytsky A, Garimella K, Altshuler D, Gabriel S, Daly M et al. (2010) The genome analysis toolkit: a MapReduce framework for analyzing next-generation DNA sequencing data. Genome Res 20(9): 1297–1303
59. Voorrips RE, Gort G, Vosman B (2011) Genotype calling in tetraploid species from bi-allelic marker data using mixture models. BMC Bioinf 12(1):1–11
60. Gerard D, Ferrão LFV, Garcia AAF, Stephens M (2018) Genotyping polyploids from messy sequencing data. Genetics 210(3):789–807
61. Alcala N, Jensen JD, Telenti A, Vuilleumier S (2016) The genomic signature of population reconnection following isolation: from theory to HIV. G3: Genes Genomes Genetics 6(1): 107–120
62. Raynal L, Marin J-M, Pudlo P, Ribatet M, Robert CP, Estoup A (2019) Abc random forests for Bayesian parameter inference. Bioinformatics 35(10):1720–1728
63. Csilléry K, François O, Blum MGB (2012) abc: an r package for approximate Bayesian computation (abc). Meth Ecol Evol 3(3):475–479
64. Douglas GM, Gos G, Steige KA, Salcedo A, Holm K, Josephs EB, Arunkumar R, Ågren JA, Hazzouri KM, Wang W et al. (2015) Hybrid origins and the earliest stages of diploidization in the highly successful recent polyploid *Capsella bursa-pastoris.* Proc Natl Acad Sci 112(9):2806–2811
65. Cornuet J-M, Pudlo P, Veyssier J, Dehne-Garcia A, Gautier M, Leblois R, Marin J-M, Estoup A (2014) Diyabc v2. 0: a software to make approximate Bayesian computation inferences about population history using single nucleotide polymorphism, DNA sequence and microsatellite data. Bioinformatics 30(8): 1187–1189
66. Flagel L, Brandvain Y, Schrider DR (2019) The unreasonable effectiveness of convolutional neural networks in population genetic inference. Mol Biol Evol 36(2):220–238

Part V

Experimental Approaches to Study Polyploidy

Chapter 18

Studying Whole-Genome Duplication Using Experimental Evolution of *Chlamydomonas*

Quinten Bafort, Lucas Prost, Eylem Aydogdu, Antoine Van de Vloet, Griet Casteleyn, Yves Van de Peer, and Olivier De Clerck

Abstract

In this chapter, we present the use of *Chlamydomonas reinhardtii* in experiments designed to study the evolutionary impacts of whole genome duplication. We shortly introduce the algal species and depict why it is an excellent model for experimental evolution. Subsequently, we discuss the most relevant steps and methods in the design of a ploidy-related *Chlamydomonas* experiment. These steps include strain selection, ploidy determination, different methods of making diplo- and polyploid *Chlamydomonas* cells, replication, culturing conditions, preservation, and the ways to quantify phenotypic and genotypic change.

Key words *Chlamydomonas reinhardtii*, Experimental evolution, Polyploidy, Whole-genome duplication, Evolve and resequence (E&R)

1 Introduction

"A harnessed soil dwelling animal with multiple stomachs, a single small eye and a filamentous proboscis" (Fig. 1). Ehrenberg's [1] description of *Chlamydomonas* seems to come straight out of a medieval bestiary. While nineteenth-century *Chlamydomonas* was little more than a curiosity for naturalists to observe with early microscopes, these first observations set the stage for the rise of *Chlamydomonas* in the twentieth century as a model for genetics. Pascher [2] pioneered "chlamygenetics" by crossing morphologically different *Chlamydomonas* species and demonstrating Mendelian segregation in the four meiospores derived from a single zygote [3]. But the use of *Chlamydomonas* really took off with Moewus' disputed claims of relative sexuality in *Chlamydomonas* (e.g., [4]), marking the start of microbial biochemical genetics [5, 6]. He was soon followed by others such as Smith (e.g., [7]), Lewin (e.g., [8]), and Sager (e.g., [9, 10]). Within the North-American community, Smith's *Chlamydomonas reinhardtii* isolates soon became the

Yves Van de Peer (ed.), *Polyploidy: Methods and Protocols*,
Methods in Molecular Biology, vol. 2545, https://doi.org/10.1007/978-1-0716-2561-3_18,

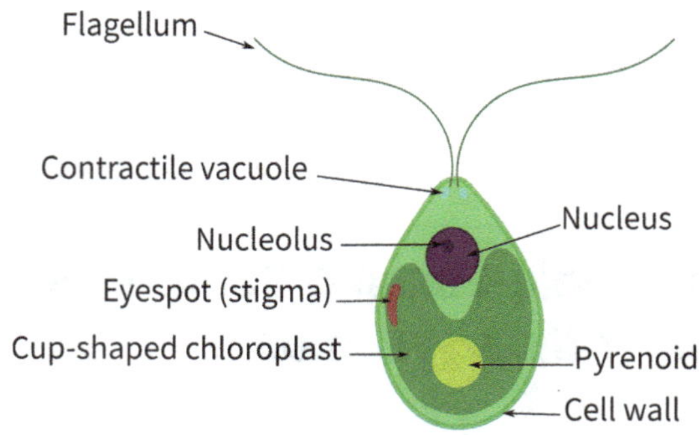

Fig. 1 Labeled diagram of a *Chlamydomonas reinhardtii* cell

preeminent model system. Factors contributing to their establishment were a high growth rate under laboratory conditions, a well-known life cycle that could be fully completed on a chemically defined medium, the ability to grow in the absence of light, and easily inducible sexual reproduction [11]. Experimental work on *Chlamydomonas* laid the foundations for our understanding of photosynthesis, the eukaryotic flagellum (cilium), basal bodies, and more [12]. Moreover, the genome of *C. reinhardtii* was one of the first algal genomes sequenced [13] and is currently one of the eight flagship JGI plant genomes[1]. Consequently, *Chlamydomonas* is probably the best studied green cell and a pivotal model organism whose importance transcends the plant kingdom and has even been dubbed as the green yeast [14].

Apart from its importance for physiology and genetics, the fast growth rate with a doubling time of 8 hours [15], the ease of culturing both sexually and asexually, and the low space requirements make *C. reinhardtii* an excellent system to study evolution. Bell and colleagues [16–21] were some of the first to realize this and, to the best of our knowledge, designed the first experimental evolution with *Chlamydomonas* [22–24]. During the past three decades, *C. reinhardtii* has been used to investigate a wide variety of evolutionary topics related to sex [25–27], population size [28], epigenetics [29], evolutionary rescue [30, 31], adaptation to different rates of environmental change [32], and the emergence of multicellularity [33, 34]. In this chapter, we present methods and considerations that will allow the use of *C. reinhardtii* to study shifts in ploidy and their importance for evolution.

Ploidy shifts, especially whole-genome duplications (WGD), are a prominent phenomenon and are often considered as a major force in evolution. WGDs are thought to alter the rate of adaptation and mutation [35], and the redundancy of the extra genome copy is believed to increase robustness and evolvability by allowing the accumulation of mutations that might eventually cause neo-

[1] https://jgi.doe.gov/our-science/science-programs/plant-genomics/plant-flagship-genomes/

and subfunctionalization [36–40]. Although these hypotheses are largely supported by observation and simulations, experimental work is crucial to test their validity. So far, the yeast species *Saccharomyces cerevisiae* has been the main model for experiments considering the evolutionary effects of polyploidization (e.g., [41–48]). Yet, although polyploidy resulting from WGD has been observed in animals and fungi, WGD has been most prevalent in plants and the so-called green lineage. Therefore, to fully understand the mechanistic underpinnings of polyploidy as well as the impact of WGD on evolution, we believe additional experimental evidence, preferably from plants (*see* Chap. 19) or plant-like organisms (this chapter) is needed.

2 Selecting the Experimental Strains

Testing effects of genome copy numbers requires organisms of different ploidy levels. *C. reinhardtii* (further referred to as *Chlamydomonas*) cells are haploid for most of their active life, although they shortly go through a diploid phase during sexual reproduction where two cells of opposite mating-type fuse to form a diploid zygote. Nevertheless, vegetative diploids can be found [49]. They are formed as a result of a defect during sexual reproduction where a small proportion of the gametic fusion products fails to form thick-walled resting cells (i.e., the zygospores) and continue to grow mitotically. *Chlamydomonas* is a heterothallic species, and as all vegetative diploids have the same mating type (*mt-* is dominant), they cannot form vegetative tetraploids or stress resistant zygospores, reducing their long-term survival chances. Luckily, there are a couple of methods that allow to create artificial diplo- and polyploids (discussed in detail in Subheadings 4 and 5). Even though some diploid strains can be ordered from culture collections such as the *Chlamydomonas* Resource Center[2] and are readily available in many labs, polyploids come in many shapes and sizes, and we would advise to put some time in creating and/or selecting strains that are well suited for your experiments.

Polyploids can be autopolyploid or allopolyploid. Traditionally, this classification was binary and based either on the origin (intraspecific for autopolyploids, interspecific for allopolyploids) or on cytological criteria (multivalent pairing in autopolyploids and bivalent pairing for allopolyploids). The existence of segmental allopolyploids, interracial autopolyploids, and a "speciation gray-zone" [50] makes these traditional categories difficult to work with [51–54] and supports the notion of a continuous spectrum starting from pure autopolyploidy with virtually identical genomes going

[2] https://www.chlamycollection.org/

over interracial autopolyploids and segmental allopolyploids toward allopolyploids with stronger genetic divergence. We here use the term "autopolyploid" for full autopolyploids that are solely affected by the effects of genome doubling and allopolyploid for organism that are subjected to the effects of heterozygosity and hybridization as well. An additional factor that should be considered when creating polyploids or even when selecting haploids is the genotype and phenotype of the strain(s) itself. Genotype-wise *Chlamydomonas* is one of the most diverse eukaryotic species. Its nucleotide diversity at synonymous sites is approximately 3% [55]. The 36 field isolates with sequenced genomes (deposited in either the *Chlamydomonas* resource center or the NIES culture collection) can be subdivided in three geographical clusters [56], making *Chlamydomonas* exceptionally well suited to study the effects of allopolyploidization. Phenotypical variation is equally great and some strains might be more suitable for experimental manipulation than others. The list of wishful features depends on the experiment, but in general, a good experimental strain does not grow in palmelloids (i.e., clumps of nonflagellated cells) that hinder phenotyping and cell counts and have an intact circadian clock mechanism (important for synchronization) and a high mating efficiency (especially when crosses are needed to create polyploids). Unfortunately, the use of standard laboratory strains does not guarantee that these prerequisites are met.

All laboratory strains trace back to a single zygote colony obtained by Gilbert Smith from soil collected on a potato field in Amherst, Massachusetts, in 1945 [11]. Molecular evidence has confirmed that all strains contain material from two haplotypes but that the contribution of both parents is not equal. It is now generally accepted that Gilbert Smith performed several backcrosses before distributing the five strains that became the ancestors of all contemporary lab strains to other laboratories [57]. Currently, some laboratory strains are available in isogenic pairs (differing only in mating type), but the most divergent strains have a maximum of 25% of their genomes coming from different ancestral haplotypes (thanks to the unequal contribution of both parents in the strains distributed by Smith). Despite their common ancestry, laboratory strains can show considerable phenotypical differences [57].

3 Determining Ploidy Level

When working with strains of different ploidy, ploidy determination is common practice. It can serve as an early indicator of strain contamination or mislabeling and is indispensable when making polyploids (see Subheadings 4 and 5).

A first method to determine ploidy takes advantage of known differences between the genome copies present. If such differences exist and are known, the ploidy can be assessed either phenotypically, for example, when the difference consists of auxotrophic mutations or by using molecular techniques such as a simple chelex DNA extraction [58] followed by mating-type PCR [59, 60] that is commonly used on natural diploids in which the mating-type locus differs between both genome copies. Although very useful, the method is not entirely foolproof, when the genome size is reduced (i.e., diploidization/haploidization), this cannot be detected as long as the mating-type loci remain present in their original copy number.

Determining DNA content through flow cytometry, a second method, is the golden standard for ploidy determination in *Chlamydomonas* and organisms alike. Because DNA content of *Chlamydomonas* cells varies during the cell cycle, it is important to use synchronized cultures or even gametes. Synchronization can be obtained by growing the cells in acetate-free medium under an alternating light-dark regime (usually 12:12, see [61]). Under these settings, cells increase in size during the light period and divide mitotically in the dark, while the number of divisions depends on the size at the end of the light phase [62]. Cells collected at the end of the dark phase (in the early G1 phase) are ideal for flow cytometry. Gametes can be produced by harvesting cultures at a cell density of 0.5 to 1 million cells ml^{-1} (by eye, the cultures are light green), resuspending them in nitrogen-free medium, phosphate buffer, or even distilled water, and incubating them agitated (shaking or bubbling) for 8 to 24 hours at 20–25 °C under 50–100 PPFD (μmol m^{-2} s^{-1}). After harvesting, the cells can be fixed in a 3:1 EtOH:Acetic acid mix for 15 min and kept at −20 °C for longer storage. Before staining, the cell should be washed twice with PBS and treated with RNAse (100 μg ml^{-1}) for 1 hour at 37 °C. PI and Sybr-green can penetrate the cell wall, but DAPI cannot. Consequently, the cultures should be treated by a cell wall resolving solution such as the commercially available enzyme subtilisin [63] or autolysin. Autolysin can be produced by bringing gametes of both mating types together and centrifugation to remove the cells when the majority of cells is quadriflagellate (30 min-1 hour). Autolysin can be stored at −80 °C and should be filter-sterilized before use [64].

A third method to determine ploidy is fluorometric-based quantification of DNA extracted from a known number of cells. DNA should be extracted using a method that produces a high yield, such as traditional cetrimonium bromide (CTAB) and phenol-chloroform extractions or genomic DNA extraction kits, while Qubit (Invitrogen, Carlsbad, CA, USA) can be used for fluorometry. This method was, for example, used by Kariyawasam et al. [65], and the quantification of the DNA from our own extractions seems to support the accuracy of this method reasonably well. Nevertheless, we would recommend treating the results with care. There might be differences in extraction efficiency between strains or ploidy levels, the efficiency could be restricted to a certain range of starting material quantities, and differences in pipetting – especially during the phase separation – will influence the results. A phase separating gel can be used to address this last problem [65].

Finally, cell size can be used as an indicator of ploidy, but since this character is highly variable both between [55, 57] and within strains depending on the growth conditions, the use of cell size as a ploidy-proxy should only be considered for strains of which the cell size and cell size plasticity are well known and should be combined with another method preferentially flow cytometry. As for flow cytometry, the problem of cell cycle heterogeneity can be overcome by using either synchronized cultures or gametes.

4 Creating Allopolyploid Strains

There are two methods that allow the creation of diplo-/polyploid cells containing divergent genomes. The first one is the laboratory version of the natural diploid formation discussed above. First, two strains of opposing mating type are selected; these strains should be fully interfertile and preferably have a high mating efficiency. If the mating type is unknown, it can be assessed via PCR of the mating-type locus or by crossing with a strain of known mating type. Once the strains are cultivated and gametogenesis is induced (see above), the gametes are mixed and left steady for 60–90 minutes to allow cell fusion. Subsequently the mating products are plated on agar plates, and the vegetative diploids are selected. The resulting agar plates will contain a mix of parental cells, vegetative diploids, and zygotes. Because vegetative diploids are formed at a low frequency, selection of diploids is often facilitated by using parental strains bearing mutations with easily assessable phenotypical effects such as auxotrophic or antibiotic resistance mutations in combination with selective medium; for instance, an arginine requiring *arg7-2* mutant can be crossed with an arginine requiring *arg2* mutant. This way the parents cannot grow on the selective agar plate (here without arginine), and because zygospore maturation, meiosis, and hatching take several days, the vegetative diploids will be the only colony forming cells present (Figs. 2 and 3). For more examples of these (double) complementation schemes, see [66–70]. When strains of interest do not have any useful marker mutations, transformation with an antibiotic resistance gene might be an option [9, 71–73]. Because mutations in these markers are not uncommon and early hatching zygotes can produce viable offspring due to recombination, ploidy should always be verified using a mating-

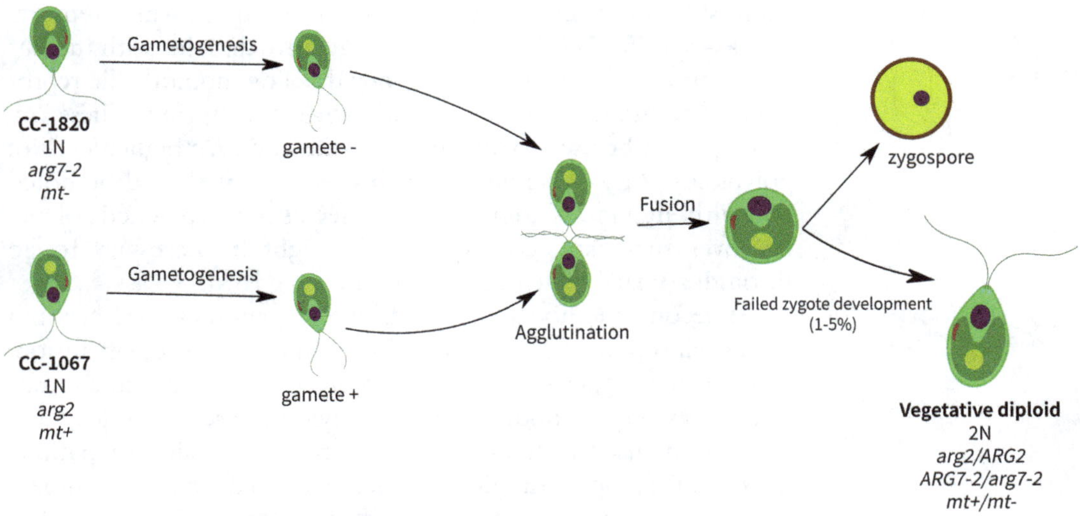

Fig. 2 Sexual cell fusion: example of double complementation scheme with two arginine-requiring strains. Haploid parents CC-1820 and CC-1067 have an auxotrophic mutation that prevents growth in absence of arginine. However, when both haploid genomes are merged in a single nucleus during mating and subsequent failed zygote development, both markers complement and the diploid can grow in a medium without arginine, allowing isolation of the vegetative diploid

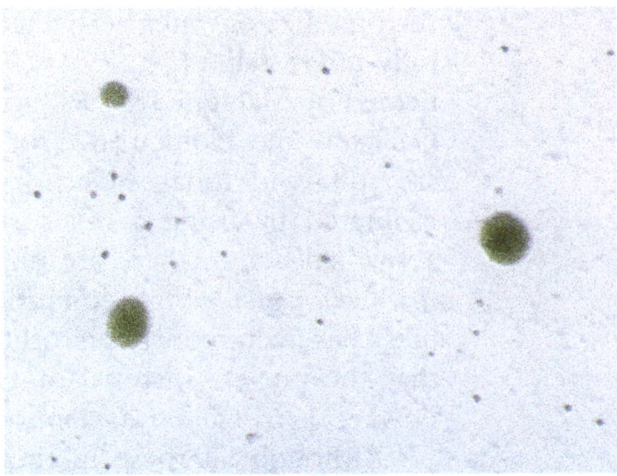

Fig. 3 Vegetative diploid colonies growing on selective medium after double complementation mating

type PCR or flow cytometry. Although the mating-type dominance makes it challenging to produce anything but diploids using this method, it is not impossible because homozygous *mt+* diploids can be generated using a different method (e.g., PEG-fusion, see below). Additionally, there is a mutation that creates a loophole in the genetic makeup of the heterothally that allows to produce diploid gametes behaving as *mt+* [65, 74]. The responsible mutation, *iso1*, discovered by Campbell et al. in 1995 [75], comes only to expression during gametogenesis in *mt-* strains and results in a mixture of regular *mt-* and pseudo *mt+* gametes. In haploids, *iso1* expression results in isoagglutination but no fusion which requires the *mt+* specific *FUS1* gene. But as vegetative diploids with the *iso1* mutation do have a *FUS1* gene, they do allow homothallic reproduction resulting in *mt+/mt+* and *mt-/mt-* offspring. These can subsequently be used to obtain tetraploids and even higher levels of polyploidy [65]. When applying this method, it should be noted that, although *iso1* is dominant, there seems to be epigenetic silencing over time, and consequently it might be necessary to go through a sexual cycle to reactivate the gene [65].

A second method to bring divergent genomes together in a single cell is polyethylene glycol (PEG)-induced protoplast fusion. As the name suggests, it is also based on cell fusion, but in contrast to the previous method, it does not rely on the sexual cycle and as such can be used to merge cells that are not sexually compatible. This opens up opportunities to generate diploids that are homozygous for one mating type and to make interspecific crosses. For instance, Fowke et al. [76] used this method to fuse *Chlamydomonas* and carrot (*Daucus carota*) cells. PEG-fusion was developed in the 1970s to allow the production of hybrid cells that could not be obtained by crosses (e.g., [77]). The exact mechanism is not entirely understood, but it is believed that the hydrophilic polymer induces agglutinations and increases cell-to-cell contact by making the presence of water between the cell membranes thermodynamically unfavorable [78, 79]. *Chlamydomonas* PEG-fusion was pioneered by Matagne [80, 81] for cell wall-deficient mutants and by Galloway and Holden [82] for regular strains, and their protocols are still valid today. Selection of fusion products can again be facilitated by using a cell sorter or using markers as described above. Although there are no reports on unstable *Chlamydomonas-Chlamydomonas* fusion products, it should be noted that instability has been reported in other organisms such as yeast [83] and that the nuclear material of *Chlamydomonas* disappeared fast in Fowkes' [76] *Chlamydomonas*-carrot fusion products.

Although all these methods allow the production of stable vegetative diploids, the stability of higher ploidy strains obtained using this method remains an open question. And as artificial polyploids in other unicellular eukaryotes such as yeasts are known to fall back to their ancestral ploidy level [44], the stability of all diploid or polyploid lines should be tested by regularly checking the ploidy of these lines (see Subheading 3).

5 Creating Autopolyploid Strains

When fully autopolyploid strains are required, one can use colchicine or some other mitotic inhibitor such as oryzalin and amiprophos-methyl (APM). Colchicine, an alkaloid that naturally occurs in the plant *Colchicum autumnale,* binds to the β-tubulin units in the microtubules, where it has a destabilizing and depolymerizing effect. During mitosis, microtubule formation is blocked, which hinders separation of the chromosome pairs, causing failure of cytokinesis and metaphase cell cycle arrest [84]. Colchicine is considered the most reliable chemical to induce chromosome doublings in plants and has been used extensively to establish higher ploidy in plant and macroalgal lines since the mid-1930s [85, 86]. Inspired by the early successes with plants, the first attempts to polyploidize *Chlamydomonas* cells followed shortly thereafter [87–89]. There are conflicting results on the efficiency of colchicine to increase ploidy in *Chlamydomonas*. Wetherel and Krauss [90] were the first to obtain stable diploid lines of *C. reinhardtii* after exposure of 8 hours to 10 g L^{-1} colchicine. However, it is unclear how long these *Chlamydomonas* lines remained polyploid. Subsequent experiments (e.g., [91–93]) are less positive and report that the cells that do recover eventually fall back to their haploid size. Nevertheless Walne [92] reported the presence of some larger cells 5 days after transfer to regular medium; consequently, it is not unthinkable that colchicine treatment can induce autopolyploidy in *Chlamydomonas*, although probably at way lower efficiency than in land plants. Based on our own experience, we recommend exposing the cultures to 2 g L^{-1} colchicine for 24 hours. Lower incubation times produce less polyploid cells that return faster to their original ploidy. When using shorter exposure times, one should verify that the exposure interval includes the short (30–40 minutes) S/M phase of the cell cycle during which the genetic material is doubled and mitosis occurs (for information on the cell cycle, see [62]).

Alternatively, the methods described above for creating strains with divergent genome copies can be adopted. PEG-fusion can be applied to fuse cells from a single strain. The absence of markers to distinguish fused from parental cells will make the selection of the polyploids strenuous, but not infeasible especially when a cell sorter can be used. The same is true for sexual fusion which allows to create autopolyploids that are only heterozygous at the mating type. The isogenic parental strains can be obtained by mating two closely related strains and performing a series of backcrossings till the offspring becomes nearly isogenic to one of the original parents (see, e.g., [94, 95]). Isogenicity should however be verified because the strains generated by Lin et al. [95] proved to be not entirely isogenic after 10 backcrosses, possibly the result of unintentional selection for mating efficiency [57].

As in the previous section, we want to emphasize that considering the potential instability of neopolyploids (especially colchipolyploids), it is key to monitor ploidy evolution in the weeks after their production.

6 Designing an Evolution Experiment with *Chlamydomonas*

Every experimental design is shaped by the hypothesis that it is supposed to test. The design of experiments aiming to quantify and qualify mutations, that is, mutation accumulation experiments (MAE) will be different from those of laboratory natural selection experiments (LNSE). Consequently, there is no standardized way to perform an evolution experiment. There are, however, several general choices that should be carefully considered when designing an experimental evolution study. We discuss them here specifically with the limitations and advantages of *Chlamydomonas* in mind.

Replication Replication and comparability are the main advantages of an evolutionary experiment that set it apart from the natural population-based and correlative methods that are otherwise used to study evolution. When brought in a new environment, adaptation can proceed along many different paths. Replication creates the unique opportunity to replay Gould's [96] "tape of life" allowing an assessment of the diversity of these adaptive solutions and the repeatability of evolution [97]. As in any other experiment, replication is also the key to statistical hypothesis testing; in this case, evolutionary hypothesis testing by comparing evolved replicates in the condition of interest with lines evolved in a control environment or even the preserved ancestral lines. More replicates will increase the statistical power, and although every replicate population increases the workload, it is strongly recommended to go for as many replicates as possible.

Growth medium *Chlamydomonas* laboratory cultures can be grown either in liquid medium or on agar plates (i.e., Petri dishes containing growth medium solidified with agar). Liquid growth medium allows the fastest growth, guarantees global interactions between the cells, offers the most practical way of subculturing, and facilitates counting and phenotyping of the cells, making it the medium of choice for most evolution experiments (especially LNSEs). On agar plates (typically 15 g agar L^{-1}), the cells are immobilized and give rise to colonies. This speeds up the detection of contamination and creates a pronounced spatial structure where interactions are localized and restricted to closely related cells. Furthermore, colony morphology can reveal features of individual genotypes. Even though the slower growth on agar is a serious disadvantage for evolution experiments, agar plates definitely do have their uses, for example, in MAE, where single cell bottlenecking is required and can even be indispensable for instance when studying adaptation in spatially heterogeneous environments.

Whether liquid or solidified, *Chlamydomonas* can grow on a wide variety of chemically defined media of which TAP [98], Bold's [99], and HSMT [100] are most commonly used. The species is vitamin independent, so there is no need for vitamin supplements, but since the laboratory strains cannot assimilate nitrate, a reduced nitrogen source is usually present [15]. The growth rate is highest under continuous light [101] and can be further increased by supplementing the medium with acetate (e.g., TAP) which is metabolized for mixo- or heterotrophic growth. Nevertheless, the higher growth rate comes at a cost as mixotrophic growth impedes synchronization even under a strict light-dark regime [61]. The optimal growth temperature for laboratory strains is 20–25 °C [15].

Culturing In liquid medium, cultures can be grown either continuously in a chemostat or by serial transfer in batch culture. Chemostats, that is, bioreactors with a continuous inflow of fresh medium and outflow of culture, provide the most efficient way to culture *Chlamydomonas*. The conditions are very stable and the growth rate is determined by the concentration of the limiting nutrient in the vessel and the rate of inflow (see [102]). But as chemostats are notoriously prone to contamination and the cost and complexity of the set up reduces the number of replicates, they are rarely used for *Chlamydomonas* experimental evolution.

Serial transferring is probably the simplest growth regime for *Chlamydomonas* lines (Fig. 4). A vessel with fresh medium is inoculated with cells, and after a few growth cycles, a fixed proportion of this culture is transferred to a new vessel with fresh medium. Models show that under such a regime there is strong selection for an increased growth rate and a reduced lag time. If the dilution

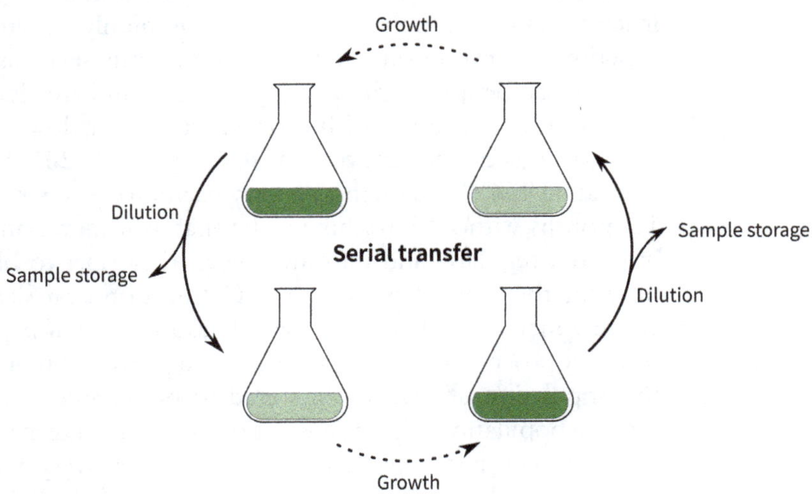

Fig. 4 Serial transfer in microbial experimental evolution

factor is large, selection will predominantly act on growth rate, whereas lag time will be affected more if the dilution factor is small [103]. If the cultures are transferred when the nutrients are limiting selection for starvation, survival might become important as well. Beside dilution factor and transfer time, population size is a third important parameter for serial transfer experiments. It is mainly determined by the amount of nutrients available or the size of the vessel when the growth medium is kept constant. Classical theory predicts higher rates of adaptation in larger populations, but population size is especially relevant to limit the effects of drift. Drift is mainly effective in small vessels and when bottlenecks are strong. Consequently, it is advised to use an effective population size of at least 5×10^4 cells [28]. The effective population size in asexual populations in batch culture is approximately the harmonic mean of the population size during the growth cycle and can be calculated as $N_e = N_{start} \times g$, where N_e is the effective population size, N_{start} is the number of cells in the inoculum, and g is the number of generations in between transfers [104]. Hubbarde and Wahl [105] calculated that a dilution factor of 5 maximizes the fixation probability of beneficial mutations and as such the rate of adaptation, but they did not take the potential effects of clonal interference into account.

Experimental lines can be grown in any laboratory vessel. Well plates are commonly used in *Chlamydomonas* experimental evolution, but they are prone to cross contamination when transferring the strains, so individually sealable vessels as Erlenmeyer flasks, culture flask, or test tubes should be considered as well. To reduce the risk on contamination, cultures should be handled in a sterile environment (i.e., a clean laminar flow) and filter tips should be used at all time.

Preservation The ability to preserve populations in a metabolically inactive state at different time points is highly advantageous in experimental evolution. Preserved strains can serve as a general backup, can be used for replay experiments, and are vital for direct comparison of ancestral with evolved lines (see below). *Chlamydomonas* strains can be kept on agar slants (typically 20 g L^{-1}) in dim light at ~18 °C. This method is very useful to keep strains for 2 to 12 months without refreshment; be that as it may, slants must be inspected regularly and refreshed when they start to bleach, turn olive, or the agar starts shrinking. Cultures on agar slants are not entirely metabolically inactive, and because the whole plating and recovery process could impose a selection pressure on or bottleneck the population, it is less well suited to preserve a heterogeneous evolved population. Cryopreservation is a better alternative for this purpose. The combined effects of a high recovery rate (>40%) when 5% methanol is used as cryoprotectant [106], and the fact that millions of cells can be preserved simultaneously in a single cryo-tube minimizes the effects of drift and potential selection during the freeze-thaw cycle. Apart from storage of heterogenous cultures, cryopreservation is the method of choice for long-term storage (i.e., years). Good protocols for cryopreservation can be found in Crutchfield et al. [106] and in the methods section on the website of the *Chlamydomonas* resource center[3].

7 Quantifying Evolution

Evolution is defined as change in the heritable properties of a population over different generations. Quantification of such change is a central theme in evolutionary experiments. As both the phenotype and genotype are subject to change, evolution can be assessed at both levels.

7.1 Phenotype

Fitness is without doubt the most relevant phenotypical trait to monitor from the perspective of natural selection, even in MAE, where it is often of interest to measure the effects of new mutations on fitness (e.g., [107]). Yet, fitness is a complex trait, even for asexual single-cell organisms that reproduce by fission such as *Chlamydomonas.* Fitness does not depend uniquely on growth rate but also on the quality of the offspring produced. This is rather nicely illustrated by "prodigal son dynamics," that is, the fact that fast-living populations in an enriched environment have to return to ancestral growth rates because the fast growth compromises the quality of the cells [108]. Even so, growth parameters are still the most commonly used proxies for microbial fitness. Furthermore, fitness depends on the environment, which is often variable, consequently in experimental evolution the environment is often held constant (or at least assumed to be constant).

Growth assays The assessment of *Chlamydomonas* growth is relatively easy; it requires the transfer of a small number of cells to fresh medium containing enough resources to support substantial growth, followed by repeated measurement to quantify the increasing population size over time.

The resulting growth curves can be used to extract growth parameters such as yield at stationary phase, yield at the transfer interval, growth rate, and lag time. When studied separately, these parameters might give contradictory results, but a more

---

[3] https://www.chlamycollection.org/methods/

comprehensive estimate of fitness can be obtained by studying them together.

The number of individuals in the population can be measured either directly by cell counts or by using a proxy such as optical density (OD). For counting, traditional methods such as hemocytometer counts and plating dilution series are an option, but high-throughput-methods using coulter counters and (imaging) flow cytometry simultaneously provide additional information on the phenotype such as cell size and shape. As mentioned above, when using these methods for phenotyping, the use of synchronized cultures or gametes is a prerequisite to ensure that all cells are at the same stage of the cell cycle. Spectrophotometric evaluation of the optical density as a proxy for cell numbers is probably the fastest method to track growth over time especially when a plate reader is used. The relationship between optical density and cell number is not linear over the entire range of OD values and is influenced to some extent by cell size (and consequently by ploidy) and pigment content. By measuring at a wavelength with low absorption by chlorophyll a and b, such as 550 or 750 nm, linearity can be maintained for cell densities between 200,000 and 10 million cells mL^{-1} [109, 110]. Even though the lower concentration seems to approximate the detection threshold of many plate readers, one should be wary that measurements at low and high densities might be inaccurate, and it remains important to make separate calibration curves for every strain and growth condition combination especially if one wants to extract cell count information such as yield at the stationary phase.

Growth parameters are commonly extracted from growth curves by fitting a growth model. We found the Baranyi model [111], fitted with the Levenberg-Marquardt algorithm [112, 113], to provide the best results but depending on the strain and the conditions other models might fit the data better.

Relative fitness In experimental evolution, researchers are often more interested in change in fitness or differences in fitness rather than in fitness itself (e.g., in the context of ploidy level difference, an important result might be differences in fitness between haploid, diploid, and tetraploid strains). Normalization of the fitness measurements depends entirely on the hypotheses and the design but often involves a comparison of the ancestral and evolved lines – $(evolvedfitness - ancestralfitness)/ancestralfitness$ – to compare adaptation between different lines [25].

Competition experiments Alternatively, relative fitness of the evolved compared to the ancestral lines can be quantified by growing both in competition starting from an equal number of cells. The competitive fitness is more inclusive than a comparison of, for example, the maximal growth rates of isogenic cultures and is closest to the standard meaning of fitness in evolutionary theory [114]. Although widely applied in other microorganisms such as yeast and bacteria, this method has as far as we know never been used in a *Chlamydomonas* evolution experiment because of the technical limitation to distinguish between the two strains in mixed cultures. Indeed, the competing strains have to be marked (e.g., by any genetic marker followed by qPCR, by phenotypical markers such as color or even by resistance genes followed by selective plating), and the marking methods are technically quite difficult, time consuming, and might come at a fitness cost. Nevertheless, due to increasing efficiency of *Chlamydomonas* transformations and precision on genome editing (CRISPR), we expect to see competing fluorescently labeled or genetically tagged *Chlamydomonas* strains soon. Another drawback of competition assays is that the fitness difference between the competing strains should be limited. If one strain is substantially less fit than the other, measuring errors might cause a considerable error on the estimate of its abundance and fitness [114].

In addition to relative fitness, competition assays can be used to quantify evolution by revealing selective sweeps. To this purpose, the competing/marked strains should be identical except for the marker, and the assay is not performed to quantify the outcome of an evolutionary experiment but is the experiment itself. The replicate populations all contain an equal number of individuals of each marker type and evolve in competition. The rate of fixation or the frequency and the amplitude of the oscillations are consequently used to make inferences about the underlying evolutionary processes. Despite the technical differences related to the use of marked strains mentioned above, this method has been applied successfully in *Chlamydomonas* [32]. Within the context of whole-genome duplication, the first experimental evolution focusing on ploidy differences utilized an related method. In this experiment, Paquin and Adams [41] followed the mutation-based dynamics of a neutral marker in the asexual population to infer the number of adaptive mutations fixed, which they incorrectly synonymized with adaptation rate [44]. Later on, Paquin and Adams method was deemed unreliable because the observed fixation rates were too high given the population size [115]. Nevertheless, their work inspired others to use competitive evolution in their comparison of haploid, diploid, and tetraploid yeast (see, e.g., [47]).

Although the environment in most experimental evolution studies is held constant, this is not always the case, and for good reason in nature, environmental change is often gradual, and the dynamics of adaptation in gradually changing environments have shown to differ from those in which gradual change is applied (e.g., [32]). In those cases, competition or comparison of the evolved and ancestral strains might have lost much of its original value, especially once the stress conditions become lethal to the ancestors. For these experiments and experiments on tolerance and resistance in general, the minimum inhibitory concentration might be a more valuable fitness proxy – examples of use of this proxy can be found in Weinreich et al., 2006 [116] and in the first experimental evolution ever conducted (at the end of the nineteenth century, the Reverend W.H. Dallinger built a bioreactor, inoculated it with flagellates, and increased the temperature over time from 15 °C to 70 °C).

Other phenotyping techniques Evolution is more than adaptation and fitness, and in the end, natural selection is only one of the major forces in evolution (next to recombination, drift, and mutation). Consequently, change in any phenotypical or genotypical (see next section) trait can be used to study evolution. However, one must be aware of the potential confounding effects of phenotypical plasticity, the evolution of which can be studied as well. Interesting phenotypic traits for *Chlamydomonas* are cell size, photosynthetic efficiency (measured, e.g., with a PAM fluorometer), respiration, pigment composition, nutrient composition, fatty acid composition, colony morphology, protein content, transcription, etc.

7.2 Genotype

Genetic changes can take many forms such as single nucleotide polymorphisms (SNPs), small and large insertions and deletions, structural variants (SVs) such as copy number variants (CNVs), variation caused by transposable element (TE) activity, inversions, translocations, and duplications. In most evolutionary experiments and population genomic studies, there is a strong focus on SNPs because these variants are relatively easy to call in next-generation sequencing data and there is a rich variety of detection techniques available. On the other hand, the adaptive role of other variants, such as CNVs, has been somewhat neglected, and there is evidence that these might be equally, or in some cases even better, fit to establish genetic differentiation and genotype-environment associations [117]. In the context of polyploidy, which is known to be linked to an increased TE-activity, large-scale deletions, and complex structural variation [118], SNPs alone might give a very incomplete picture of the genetic changes that shaped phenotypic evolution during experiments. As such, it is highly recommended to expand analyses to these other variants.

Sequencing technology One way to facilitate the detection of non-SNP variants is the use of long-read sequencing such as nanopore sequencing (e.g., by Oxford Nanopore Technologies, ONT), and single molecule real-time (SMRT) sequencing (e.g., by Pacific Biosciences, PacBio). These third-generation sequencing technologies typically provide reads with lengths over 10 kb (up to 25 kb for SMRT and even Mbs for ONT) and can be extremely useful for the detection of complex structural variants. Error rates are rather high, consequently when accuracy is important, for example, for SNP calling, they should be combined with second-generation high-accuracy sequencing technologies such as Illumina. Long reads facilitate haplotype phasing, that is, separating the different genome copies present within the same individual (see [119] for an overview of phasing techniques) which is highly relevant in the context of polyploidy. Finally, third-generation sequencing has the potential to provide information about epigenetic base modifications [120–122].

Sequencing strategy Irrespective of which variants one wants to detect, two sequencing strategies can be applied. The first, classic, strategy consists of sequencing isogenic populations derived from single cells/colonies in each of the lines. This allows establishing the links between the different variants. When the sample size per evolutionary line is large, allele frequency estimates are reasonably accurate, and low-frequency variants and mutator genotypes can be easily detected. Alternatively, the evolved lines can be sequenced as a pool of individuals, that is, pool-seq. Pool-seq produces more accurately allele frequency estimates than single sample sequencing (at least when coverage is high enough). Additionally, this method reduces library prep costs substantially, as a single library suffices for each evolved line. Nevertheless, pool-seq does have some drawbacks. Because the frequency of a variant can take every value between 0 and 1, it is difficult to distinguish low-frequency variants from sequencing errors unless there is temporal replication (e.g., [123]). Consequently, under the single time-point scenario, a new variant should reach a certain abundance within the population before it can be called with certainty. As such, the method is probably most useful to detect changes in allele frequencies in populations with a high-standing genetic variation and less for the detection of new variants in originally isogenic populations. When comparing populations of different ploidy, this problem is exacerbated by the fact that the frequency of any variant decreases with ploidy, that is, if present in 10% of the population, you will get a frequency of 0.1, 0.05, 0.025 for haploids, diploids, and tetraploids, respectively. Accordingly, when a threshold is applied to remove the noise, you are bound to lose an unequal amount of real variants for samples of different ploidy. A related problem presents itself when a variant should have a certain absolute coverage to be detected. This latter problem can be easily remediated by standardizing coverage in a ploidy dependent way. To minimize loss of data, it is advisable to try to obtain a difference in coverage of the different samples at the sequencing level. Additionally, in the absence of temporal replication, pool-seq is not the most optimal tool to establish the links between mutations, especially when short reads are used.

8 Concluding Statement

During the past 6 years, we have experimented extensively with haploid and diploid *Chlamydomonas* strains, completing a first evolution experiment and starting a second one. Many of the recommendations given above come from the experience acquired during these experiments, sometimes by trial and error. We are convinced that *Chlamydomonas reinhardtii* is an excellent system to study WGD and subsequent evolution. We hope this chapter will inspire you to follow in our footsteps, and if so, when you need any additional advice or you encounter any problems, feel free to contact us.

Acknowledgments

This work was supported by Ghent University (Methusalem funding, BOF.MET.2021.0005.01), Research Foundation Flanders (FWO Research Project – G0C0116N and a personal grant to Lucas Prost and Quinten Bafort), and the European Research Council (ERC) under the European Union's Horizon 2020 research and innovation program (grant agreement no 833522).

References

1. Ehrenberg CG (1834) Dritter Beitrag zur Erkenntniss grosser Organisation in der Richtung des kleinsten Raumes. Abh. Konigl. Akad. Wiss Berlin 1833:145–336
2. Pascher A (1916) Über die Kreuzung einzelliger, haploider Organismen. ChlamydomonasChlamydomonas. Ber Deutsch Bot Ges 34:228–242
3. Pascher A (1918) Ueber die Beziehung der Reduktionsteilung zur Mendelschen Spaltung. Ber Deutsch Bot Ges 36:163–168
4. Moewus F (1938) Vererbung des Geschlechts bei Chlamydomonas eugametos und verwandten Arten. Biol Zentralbl 58:516–536
5. Sapp J (1987) What counts as evidence or who was Franz Moewus and why was everybody saying such terrible things about him. Hist Philos Life Sci 9:277–308
6. Sapp J (1990) Where the truth lies: Franz Moewus and the origins of molecular biology. Cambridge University Press
7. Smith GM, Regnery DC (1950) Inheritance of sexuality in Chlamydomonas Reinhardi. Proc Natl Acad Sci U S A 36:246–248
8. Lewin RA (1953) The genetics of Chlamydomonas moewusii Gerloff. J Genet 51: 543–560
9. Sager R (1954) Mendelian and non-Mendelian inheritance of streptomycin resistance in Chlamydomonas reinhardi. Proc Natl Acad Sci U S A 40:356–363
10. Sager R (1955) Inheritance in the Green Alga Chlamydomonas Reinhardi. Genetics 40: 476–489
11. Harris EH (2009) The Chlamydomonas sourcebook: introduction to Chlamydomonas and its laboratory use, vol 1. Academic Press
12. Sasso S, Stibor H, Mittag M et al (2018) From molecular manipulation of domesticated Chlamydomonas reinhardtii to survival in nature. Elife 7:e39233
13. Merchant SS, Prochnik SE, Vallon O et al (2007) The Chlamydomonas genome reveals the evolution of key animal and plant functions. Science 318:245–250
14. Goodenough UW (1992) Green yeast. Cell 70:533–538

15. Harris EH (2001) Chlamydomonas as a model organism. Annu Rev Plant Physiol Plant Mol Biol 52:363–406
16. Bell G, Hill WG (1990) The ecology and genetics of fitness in Chlamydomonas. I. Genotype-by-environment interaction among pure strains. Proc Royal Soc London B Biol Sci 240:295–321
17. Bell G, Hill WG (1990) The ecology and genetics of fitness in Chlamydomonas. II. The properties of mixtures of strains. Proc Royal Soc London B Biol Sci 240: 323–350
18. Bell G (1991) The ecology and genetics of fitness in Chlamydomonas. III. genotype-by-environment interaction within strains. Evolution 45:668–679
19. Bell G (1991) The ecology and genetics of fitness in Chlamydomonas. IV. the properties of mixtures of genotypes of the same species. Evolution 45:1036–1046
20. Bell G (1992) The ecology and genetics of fitness in Chlamydomonas. V. The relationship between genetic correlation and environmental variance. Evolution 46:561–566
21. Silva JD, Bell G (1992) The ecology and genetics of fitness in Chlamydomonas. VI. Antagonism between natural selection and sexual selection. Proc R Soc Lond Ser B Biol Sci 249:227–233
22. Bell GAC (1997) Experimental evolution in Chlamydomonas . I. Short-term selection in uniform and diverse environments. Nature 78:490–497
23. Bell G, Reboud X (1997) Experimental evolution in Chlamydomonas II. Genetic variation in strongly contrasted environments. Heredity 78:498–506
24. Reboud X, Bell G (1997) Experimental evolution in Chlamydomonas . III. Evolution of specialist and generalist types in environments that vary in space and time. Heredity 78: 507–514
25. Colegrave N (2002) Sex releases the speed limit on evolution. Nature 420:664–666
26. Kaltz O, Bell G (2002) The ecology and genetics of fitness in Chlamydomonas. XII. Repeated sexual episodes increase rates of adaptation to novel environments. Evolution 56:1743–1753
27. Lachapelle J, Colegrave N (2017) The effect of sex on the repeatability of evolution in different environments. Evolution 71: 1075–1087
28. Lachapelle J, Reid J, Colegrave N (2015) Repeatability of adaptation in experimental populations of different sizes. Proc R Soc B 282:20143033
29. Kronholm I, Bassett A, Baulcombe D et al (2017) Epigenetic and genetic contributions to adaptation in Chlamydomonas. Mol Biol Evol 34:2285–2306
30. Lachapelle J, Bell G (2012) Evolutionary rescue of sexual and asexual populations in a deteriorating environment: evolutionary rescue of sexual and asexual populations. Evolution 66:3508–3518
31. Bell G (2013) Evolutionary rescue of a green alga kept in the dark. Biol Lett 9:20120823
32. Collins S, de Meaux J (2009) Adaptation to different rates of environmental change in Chlamydomonas. Evolution 63:2952–2965
33. Ratcliff WC, Herron MD, Howell K et al (2013) Experimental evolution of an alternating uni- and multicellular life cycle in Chlamydomonas reinhardtii. Nat Commun 4: 2742
34. Herron MD, Borin JM, Boswell JC et al (2019) De novo origins of multicellularity in response to predation. Sci Rep 9:2328
35. Otto SP, Whitton J (2000) Polyploid incidence and evolution. Annu Rev Genet 34: 401–437
36. Comai L (2005) The advantages and disadvantages of being polyploid. Nat Rev Genet 6: 836–846
37. Crow KD, Wagner GP (2006) What is the role of genome duplication in the evolution of complexity and diversity? Mol Biol Evol 23: 887–892
38. Van de Peer Y, Maere S, Meyer A (2009) The evolutionary significance of ancient genome duplications. Nat Rev Genet 10:725–732
39. Van de Peer Y, Mizrachi E, Marchal K (2017) The evolutionary significance of polyploidy. Nat Rev Genet 18:411–424
40. Van de Peer Y, Ashman T-L, Soltis PS et al (2021) Polyploidy: an evolutionary and ecological force in stressful times. Plant Cell 33: 11–26
41. Paquin C, Adams J (1983) Frequency of fixation of adaptive mutations is higher in evolving diploid than haploid yeast populations. Nature 302:495–500
42. Mable BK (2001) Ploidy evolution in the yeast Saccharomyces cerevisiae: a test of the nutrient limitation hypothesis. J Evol Biol 14: 157–170
43. Anderson JB, Sirjusingh C, Ricker N (2004) Haploidy, diploidy and evolution of antifungal drug resistance in Saccharomyces cerevisiae. Genetics 168:1915–1923

44. Gerstein AC, Otto SP (2009) Ploidy and the causes of genomic evolution. J Hered 100: 571–581
45. Gerstein AC, Cleathero LA, Mandegar MA et al (2011) Haploids adapt faster than diploids across a range of environments: Haploids adapt faster than diploids. J Evol Biol 24: 531–540
46. Keane OM, Toft C, Carretero-Paulet L et al (2014) Preservation of genetic and regulatory robustness in ancient gene duplicates of Saccharomyces cerevisiae. Genome Res 24: 1830–1841
47. Selmecki AM, Maruvka YE, Richmond PA et al (2015) Polyploidy can drive rapid adaptation in yeast. Nature 519:349–352
48. Lu Y-J, Swamy KBS, Leu J-Y (2016) Experimental evolution reveals interplay between Sch9 and polyploid stability in yeast. PLoS Genet 12:e1006409
49. Ebersold WT (1963) Heterozygous diploid strains of Chlamydomonas reinhardi. Genetics 48:888
50. De Queiroz K (2007) Species concepts and species delimitation. Syst Biol 56:879–886
51. Stebbins GL (1950) Variation and evolution in plants. Columbia University Press
52. Stebbins GL (1971) Chromosomal evolution in higher plants. Edward Arnold Ltd.
53. Doyle JJ, Sherman-Broyles S (2017) Double trouble: taxonomy and definitions of polyploidy. New Phytol 213:487–493
54. Spoelhof JP, Soltis PS, Soltis DE (2017) Pure polyploidy: closing the gaps in autopolyploid research: pure polyploidy. J Syst Evol 55: 340–352
55. Flowers JM, Hazzouri KM, Pham GM et al (2015) Whole-genome Resequencing reveals extensive natural variation in the model green alga Chlamydomonas reinhardtii. Plant Cell 27:2353–2369
56. Craig RJ, Böndel KB, Arakawa K et al (2019) Patterns of population structure and complex haplotype sharing among field isolates of the green alga Chlamydomonas reinhardtii. Molecul Ecol 28:3977–3993
57. Gallaher SD, Fitz-Gibbon ST, Glaesener AG et al (2015) Chlamydomonas genome resource for laboratory strains reveals a mosaic of sequence variation, identifies true strain histories, and enables strain-specific studies. Plant Cell 27:2335–2352
58. Cao M, Fu Y, Guo Y et al (2009) Chlamydomonas (Chlorophyceae) colony PCR. Protoplasma 235:107–110
59. Werner R, Mergenhagen D (1998) Mating type determination of Chlamydomonas reinhardtii by PCR. Plant Mol Biol Report 16: 295–299
60. Zamora I, Feldman JL, Marshall WF (2004) PCR-based assay for mating type and diploidy in Chlamydomonas. BioTechniques 37: 534–536
61. Surzycki S (1971) [4] Synchronously grown cultures of Chlamydomonas reinhardi. In: Methods in enzymology. Academic Press, pp 67–73
62. Cross FR, Umen JG (2015) The Chlamydomonas cell cycle. Plant J 82:370–392
63. Hwang H-J, Kim YT, Kang NS et al (2018) A simple method for removal of the Chlamydomonas reinhardtii cell wall using a commercially available Subtilisin (Alcalase). MIP 28: 169–178
64. Schlösser UG, Sachs H, Robinson DG (1976) Isolation of protoplasts by means of a "species-specific" autolysine in Chlamydomonas. Protoplasma 88:51–64
65. Kariyawasam T, Joo S, Goodenough U et al (2019) Novel approaches for generating and manipulating diploid strains of Chlamydomonas reinhardtii. Algae 34:35–43
66. Ebersold WT (1967) Chlamydomonas reinhardi: heterozygous diploid strains. Science 157:447–449
67. Loppes R (1969) A new class of arginine-requiring mutants in Chlamydomonas reinhardi. Mol Gen Genet 104:172–177
68. Matagne RF, Orbans A (1980) Somatic segregation in diploid Chlamydomonas reinhardii. Microbiology 119:71–77
69. Palombella AL, Dutcher SK (1998) Identification of the gene encoding the tryptophan synthase β-Subunit from Chlamydomonas reinhardtii. Plant Phisiol 117:455–464
70. Bellafiore S, Ferris P, Naver H et al (2002) Loss of Albino3 leads to the specific depletion of the light-harvesting system. Plant Cell 14: 2303–2314
71. Mets LJ, Bogorad L (1971) Mendelian and uniparental alterations in erythromycin binding by plastid ribosomes. Science 174: 707–709
72. Kindle KL (1990) High-frequency nuclear transformation of Chlamydomonas reinhardtii. PNAS 87:1228–1232
73. Shimogawara K, Fujiwara S, Grossman A et al (1998) High-efficiency transformation of Chlamydomonas reinhardtii by electroporation. Genetics 148:1821–1828

74. Lee J-H, Lin H, Joo S et al (2008) Early sexual origins of homeoprotein heterodimerization and evolution of the plant KNOX/BELL family. Cell 133:829–840
75. Campbell AM, Rayala HJ, Goodenough UW (1995) The iso1 gene of Chlamydomonas is involved in sex determination. MBoC 6: 87–95
76. Fowke LC, Gresshoff PM, Marchant HJ (1979) Transfer of organelles of the alga Chlamydomonas reinhardii into carrot cells by protoplast fusion. Planta 144:341–347
77. Kao KN, Michayluk MR (1974) A method for high-frequency intergeneric fusion of plant protoplasts. Planta 115:355–367
78. Lentz BR, Lee J (1999) Poly(ethylene glycol) (PEG)-mediated fusion between pure lipid bilayers: a mechanism in common with viral fusion and secretory vesicle release? (Review). Mol Membr Biol 16:279–296
79. Malinin VS, Frederik P, Lentz BR (2002) Osmotic and curvature stress affect PEG-induced fusion of lipid vesicles but not mixing of their lipids. Biophys J 82: 2090–2100
80. Matagne RF, Deltour R, Ledoux L (1979) Somatic fusion between cell wall mutants of Chlamydomonas reinhardi. Nature 278: 344–346
81. Matagne RF (1981) Transmission of chloroplast alleles in somatic fusion products obtained from vegetative cells and/or 'Gametes' of Chlamydomonas reinhardi. Curr Genet 3:31–36
82. Galloway RE, Holden LR (1984) Transmission and recombination of chloroplast genes in asexual crosses of Chlamydomonas reinhardii. Curr Genet 8:399–405
83. Molnar M, Sipiczki M (1993) Polyploidy in the haplontic yeast Schizosaccharomyces pombe: construction and analysis of strains. Curr Genet 24:45–52
84. Eng W-H, Ho W-S (2019) Polyploidization using colchicine in horticultural plants: A review. Sci Hortic 246:604–617
85. Blakeslee AF, Avery AG (1937) Methods of inducing doubling of chromosomes in plants: by treatment with colchicine. J Hered 28: 393–411
86. Dermen H (1940) Colchicine polyploidy and technique. Bot Rev 6:599–635
87. Gavaudan P, Kobozieff N (1938) Action de la colchicine sur la caryocinese et la cytodierese des Chlamydomonadinees. Compt Rend Soc Biol 127:790–793
88. Vandendries RENÉ, Gavaudan P (1939) Action de la colchicine sur quelques organismes inférieurs. C R Acad Sci 208:675–677
89. Cornman I (1942) Susceptibility of Colchicum and Chlamydomonas to Colchicine. Botan Gazzet 104:50–62
90. Wetherell DF, Krauss RW (1956) Colchicine-induced polyploidy in Chlamydomonas. Science 124:25–26
91. Buffaloe ND (1959) Some effects of colchicine on cells of Chlamydomonas eugametos Moewus. Exp Cell Res 16:221–231
92. Walne PL (1966) The effects of colchicine on cellular organization in Chlamydomonas. I. Light microscopy and cytochemistry. Am J Bot 53:908–916
93. Walne PL (1967) The effects of Colchicine on cellular organization in Chlamydomonas. II. Ultrastructure. Am J Bot 54:564–577
94. Dent RM, Haglund CM, Chin BL et al (2005) Functional genomics of Eukaryotic Photosynthesis using insertional mutagenesis of Chlamydomonas reinhardtii. Plant Physiol 137:545–556
95. Lin H, Miller ML, Granas DM et al (2013) Whole genome sequencing identifies a deletion in protein phosphatase 2A that affects its stability and localization in Chlamydomonas reinhardtii. PLoS Genet 9:e1003841
96. Gould SJ (1989) Wonderful life: The Burgess shale and the nature of history. W.W. Norton
97. Futuyma DJ, Bennett AF (2009) The importance of experimental studies In evolutionary biology. University of California Press
98. Gorman DS, Levine RP (1965) Cytochrome f and plastocyanin: their sequence in the photosynthetic electron transport chain of Chlamydomonas reinhardi. Proc Natl Acad Sci U S A 54:1665–1669
99. Bischoff HW (1963) Phycological studies. IV. Some algae from enchanted rock and related algal species. Univ Texas Publ 6318(95)
100. Sueoka N (1960) Mitotic replication of deoxyribonucleic acid In Chlamydomonas Reinhardi. Proc Natl Acad Sci U S A 46:83–91
101. Tamburic B, Zemichael FW, Maitland GC et al (2012) Effect of the light regime and phototrophic conditions on growth of the H2-producing green alga Chlamydomonas Reinhardtii. Energy Procedia 29:710–719
102. Dykhuizen DE, Hartl DL (1983) Selection in chemostats. Microbiol Rev 47:150–168

103. Lin J, Manhart M, Amir A (2020) Evolution of microbial growth traits under serial dilution. Genetics 215:767–777
104. Lenski RE, Rose MR, Simpson SC et al (1991) Long-term experimental evolution in Escherichia coli. I. Adaptation and divergence during 2,000 generations. Am Nat 138: 1315–1341
105. Hubbarde JE, Wahl LM (2008) Estimating the optimal bottleneck ratio for experimental evolution: The burst-death model. Math Biosci 213:113–118
106. Crutchfield A, Diller K, Brand J (1999) Cryopreservation of Chlamydomonas reinhardtii (Chlorophyta). Eur J Psychol 34:43–52
107. Kraemer SA, Morgan AD, Ness RW et al (2016) Fitness effects of new mutations in *Chlamydomonas reinhardtii* across two stress gradients. J Evol Biol 29:583–593
108. Collins S (2016) Growth rate evolution in improved environments under Prodigal Son dynamics. Evol Appl 9:1179–1188
109. Sager R, Granick S (1953) Nutritional studies with Chlamydomonas reinhardi. Ann N Y Acad Sci 56:831–838
110. Haire TC, Bell C, Cutshaw K et al (2018) Robust microplate-based methods for culturing and in vivo phenotypic screening of Chlamydomonas reinhardtii. Front Plant Sci 9: 235
111. Baranyi J, Roberts TA, McClure P (1993) A non-autonomous differential equation to modelbacterial growth. Food Microbiol 10: 43–59
112. Levenberg K (1944) A method for the solution of certain non-linear problems in least squares. Quat Appl Math 2:164–168
113. Marquardt DW (1963) An algorithm for least-squares estimation of nonlinear parameters. J Soc Indust Appl Math 11:431–441
114. Wiser MJ, Lenski RE (2015) A comparison of methods to measure fitness in Escherichia coli. PLoS One 10:e0126210
115. Dykhuizen DE (1990) Experimental studies of natural selection in bacteria. Ann Rev 21: 373–398
116. Weinreich DM, Delaney NF, DePristo MA et al (2006) Darwinian evolution can follow only very few mutational paths to fitter proteins. Science 312:111–114
117. Dorant Y, Cayuela H, Wellband K et al (2020) Copy number variants outperform SNPs to reveal genotype–temperature association in a marine species. Mol Ecol 29: 4765–4782
118. Parisod C, Senerchia N (2012) Responses of transposable elements to polyploidy. In: Grandbastien M-A, Casacuberta JM (eds) Plant transposable elements: impact on genome structure and function. Springer, Berlin, Heidelberg, pp 147–168
119. Zhang X, Wu R, Wang Y et al (2020) Unzipping haplotypes in diploid and polyploid genomes. Comput Struct Biotechnol J 18:66–72
120. Simpson JT, Workman RE, Zuzarte PC et al (2017) Detecting DNA cytosine methylation using nanopore sequencing. Nat Methods 14: 407–410
121. Liu Q, Fang L, Yu G et al (2019) Detection of DNA base modifications by deep recurrent neural network on Oxford Nanopore sequencing data. Nat Commun 10:2449
122. Tse OYO, Jiang P, Cheng SH et al (2021) Genome-wide detection of cytosine methylation by single molecule real-time sequencing. PNAS 118:e2019768118
123. Good BH, McDonald MJ, Barrick JE et al (2017) The dynamics of molecular evolution over 60,000 generations. Nature 551:45–50

Chapter 19

Studying Whole-Genome Duplication Using Experimental Evolution of *Spirodela polyrhiza*

Tian Wu, Annelore Natran, Lucas Prost, Eylem Aydogdu, Yves Van de Peer, and Quinten Bafort

Abstract

In this chapter, we present the use of *Spirodela polyrhiza* in experiments designed to study the evolutionary impact of whole-genome duplication (WGD). We shortly introduce this duckweed species and explain why it is a suitable model for experimental evolution. Subsequently, we discuss the most relevant steps and methods in the design of a ploidy-related duckweed experiment. These steps include strain selection, ploidy determination, different methods of making polyploid duckweeds, replication, culturing conditions, preservation, and the ways to quantify phenotypic and transcriptomic change.

Key words *Spirodela polyrhiza*, Duckweed, Experimental evolution, Polyploidy, Whole-genome duplication, Evolve and resequence (E&R)

1 Introduction

Ploidy shifts or whole-genome duplications (WGDs) are omnipresent, both within the tree of life and at all levels of biological organization [1]. With over 35% of all flowering plant species being polyploids, 15% of all spermatophyte speciation events coinciding with ploidy increase (for lycophytes and monilophytes, this amounts to even 31%) [2], and a growing number of documented paleopolyploidization events, WGD was and still is especially important for plant evolution. The exact mechanisms by which WGD fuels evolution are largely unknown, but it is widely believed that WGDs increase the adaptive potential, especially in stressful environments [3]. Most of the evidence for this hypothesis comes from observational studies relying either on the comparison of natural polyploids with the contemporary representative(s) of their assumed ancestral lineage(s) of lower ploidy or of diploids

Yves Van de Peer (ed.), *Polyploidy: Methods and Protocols*,
Methods in Molecular Biology, vol. 2545, https://doi.org/10.1007/978-1-0716-2561-3_19,

with artificial neopolyploids. Both comparisons have their flaws. The first comparison does not allow a separation of the effects of polyploidization from those of the subsequent evolution of the "ancestral" diploid(s) and the polyploid. The second comparison is limited by a restricted time frame that is generally too short to make inferences about evolution. This last problem is not imposed by the method itself but by the rather lengthy generation time of plants, and its solution lies in the use of organisms with a less constraining generation time such as the unicellular eukaryotes *Saccharomyces cerevisiae* (yeast) and *Chlamydomonas reinhardtii* (unicellular green alga) which have both been used to compare the evolution of ploidy variants (see also chapter by Bafort et al. in this issue [4]).

We here suggest the use of fast-growing duckweeds and more specifically *Spirodela polyrhiza* (from this point onward referred to as *Spirodela*), as an additional and a taxonomically more appropriate model system to study angiosperm polyploidization. Duckweeds or Lemnoideae is a clade of small aquatic macrophytes in the Araceae [5]. Their morphology is an example of extreme reduction and miniaturization. A plant typically consists of the developmental hybrid between leaf and stem, called the frond [6], and, depending on the species, none, a single, or multiple root(s). Reproduction is predominantly vegetative, and with doubling times as low as 1.34 days [7], they are the fastest growing embryophytes. Lemnoideae went through two rounds of genome duplication during the Cretaceous [8], making them an ideal model system for ploidy research. Furthermore, their aquatic lifestyle, low space requirements (they are also the smallest among the higher plants), and high clonal growth rate made duckweeds a popular plant model system in the pre-molecular era when plant research was not yet dominated by *Arabidopsis*, and they seem to be regaining popularity in recent years. Consequently, the already considerable pool of methods, resources, and knowledge relevant to duckweed research is only growing, and there is a vibrant community centered around The International Steering Committee on Duckweed Research and Applications (ISCDRA, website: http://www.ruduckweed.org/).

With a frond size of up to 15 mm, *Spirodela polyrhiza* (Fig. 1) is the largest of all duckweed species, hence its common name greater/giant duckweed. *Spirodela polyrhiza* can be distinguished from its morphologically similar sister species *Spirodela intermedia*, by its capacity to form turions, manifested as small dark-brown/green-thick rootless fronds, the prophyllum, that is, a scalelike leaflet covering the root attachment point (Fig. 1b), being penetrated by only one or rarely two roots (in *S. intermedia*, two to five roots penetrate the prophyllum), and under some growth conditions by the presence of a red dot just above the node on the upper face of the fronds [9]. *Spirodela* has a nearly cosmopolitan distribution, occurring in temperate and tropical freshwater habitats

Fig. 1 Morphology of the dorsal (**a**) and ventral (**b**) side of a *Spirodela polyrhiza* plant. Mf: mother frond; df: daughter frond; gdf: granddaughter frond; n: nerve; p: pouch; pr: prophyllum; r: root; rc: rootcap

around the world (with the notable exception of southern South America, some islands, and New Zealand) [10]. Its genome size is the smallest of all known duckweeds, that is, 138 Mb [8], making it an exemplary system for evolve and resequence experiments.

2 Strain Selection

Any species used to study the effects of WGD must come in strains of different ploidy. Ploidy determination is unfortunately not required when submitting a strain to a germplasm collection. Luckily, Urbanska-Worytkiewicz [11], Geber [12], Wang and coworkers [13], and Hoang and coworkers [14] have gone through a great deal of work to determine the ploidy of over 180 cultivated *Spirodela* strains. One of these, the Puerto Rican strain 7110 (Landolt Duckweed Collection), proved to be a tetraploid [12]. A single tetraploid strain will not suffice for most experiments, but there are methods (described below) that allow the production of neo-tetraploids.

Diploid *Spirodela* has a limited genetic diversity with a clear geographic population structure consisting of an American, a European, an Indian and a Southeast Asian cluster [15]. On the other hand, differences in relative growth rate, plastic responses to temperature, salt tolerance, turion production, etc., [7, 16–19] reveal a rich phenotypic diversity both between and within these genetic clusters. Strains of all different ecotypes are readily available and can be obtained from several duckweed collections. The Landolt collection, founded by and named after Elias Landolt, the godfather of duckweed research, is the oldest and once was the most important collection harboring over 1000 strains collected from all over the world [20]. Today, the Rutgers Duckweed Stock Cooperative (RDSC) has taken over its role as preeminent collection, although there are many other collections [21]. Alternatively, strains can easily be isolated from the wild. Because of the similarity with *S. intermedia* (see above), the morphological identification should always be verified using molecular techniques such as barcoding [22, 23] or PCR-based methods such as amplified fragment length polymorphism (AFLP) fingerprinting [23], tubulin-based polymorphism (TBP) fingerprinting [24], polymorphic NB-ARC-related genes [25], or microsatellite arrays [26]. Axenicity of fronds isolated from nature can be achieved by submerging the plants for thirty seconds to three minutes in a 10–20% bleach solution with agitation and subsequently washing the fronds twice in demi-water to remove all traces of bleach. Usually, this procedure must be repeated several times. After recovery (up to two weeks after bleaching the fronds), the presence of any remaining bacterial contaminants can be checked microscopically or by growing the plants in medium containing 25 mM glucose.

3 Determining Ploidy

Ploidy level determination, whether of newly isolated strains, of newly synthesized potential polyploids, or of possibly diploidizing experimental strains, is one of the most important procedures in ploidy-related experiments. Polyploidy in *Spirodela* causes a series of characteristic changes in morphology that can be used as a first indicator of ploidy change, including an increased frond (Fig. 2a) and stomatal guard cells size and a decrease in stomatal density (Fig. 2b). To facilitate the visibility of the stomatal guard cell, the fronds can be submerged in 70% ethanol solution, boiled for five minutes at 80 °C, washed for two minutes in fresh 70% ethanol solution, transferred to 10% bleach solution to remove any remaining pigmentation, and washed twice in demineralized water to remove all remaining ethanol and bleach. Because size and stomatal characteristics are rather plastic and depend on both strain and environment, these proxies should always be used with care and in combination with more reliable methods such as chromosome counts or flow cytometry.

Due to the rarity of flowering in *Spirodela*, chromosome counts are typically performed on metaphase spreads. Although any young tissue fixed in a 3:1 ethanol:acetic acid solution can be used [11], the efficiency is highest when using root tips treated with a mitotic inhibitor such as 2 μM of 8-hydroxyquinoline for 1 hour to cause metaphase arrest (for a full protocol, see, e.g., [14, 27]). Considering a diploid chromosome number of 40 and the size of these chromosomes, counting chromosomes can be difficult, especially when you are working with tetra- or octoploids.

Flow cytometry (Fig. 2c) offers a less cumbersome and higher-throughput alternative for ploidy determination. The nuclei can be extracted by taking a single *Spirodela* frond, washing it in water or fresh growth medium and chopping it for about a minute with a

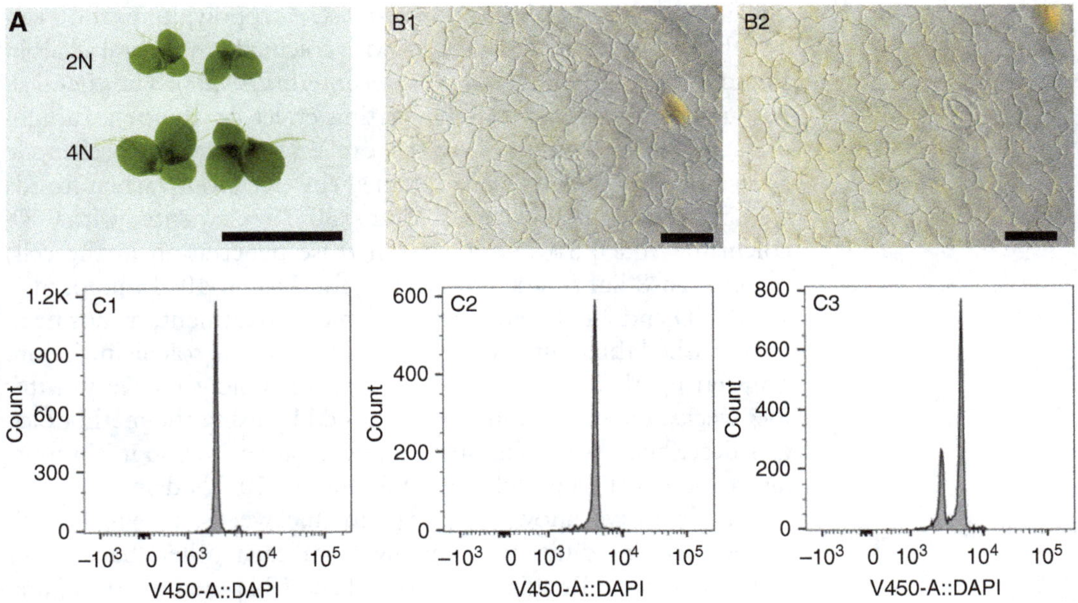

Fig. 2 Differences between diploid (2N) and tetraploid (4N) plants: (**a**) general morphology (scale bar 20 mm); (**b**) stomatal size and density of diploid (B1) and tetraploid (B2) plants (scale bar 40 μm); (**c**) flow cytometry of diploid (C1), tetraploid (C2), and mixed (C3) samples

clean scalpel or razor blade in a Petri dish containing 500 μl ice-cold Galbraith's buffer [28] with 1% PVP brought to pH 7.0 with NaOH (filter sterilized using a 0.22 μm membrane filter). After chopping, slant the Petri dish, collect and mix all material, and leave inclined for another minute before mixing the material and transferring it over a 50 μm cell strainer to a centrifuge tube. Keep the extracted nuclei on ice. Add 10 μl of 0.2 mg ml^{-1} DAPI stock solution to 500 μl of extract, mix by vortexing, and run on the flow cytometer with appropriate channel. Keep the samples protected from light as DAPI is light sensitive, and fluorescence fades quickly when exposed to light. Analyze at least 5000 nuclei. Apart from DAPI, other stains such at propidium iodide or Sybr green I can also be used; the choice of strain will depend on the lasers of the flow cytometer; DAPI has an excitation peak at 359 nm.

4 Making Polyploids

As mentioned above, the scarcity of known natural tetraploids for *Spirodela* invokes a need to create synthetical polyploids. Autopolyploidization of duckweeds is relatively easy and has been reported for *Landoltia* [29], *Lemna* (Tia-Lynn Ashman, personal communication), and *Spirodela* (unpublished work within our group). Currently, we have autotetraploids from strains belonging to each of the geographical subpopulations described by Xu et al. [15], available to all researchers upon request. Autopolyploidization can be achieved by treating the strains with colchicine, a natural alkaloid found in the autumn crocus (*Colchicum autumnale*). The chemical has an affinity for and depolymerizing effect on the beta tubulin units of the microtubules, which can cause metaphase cell cycle arrest. The best results were obtained (by our group) when fronds were incubated in Hoagland medium [30] supplemented with 0.7% colchicine and 0.5% DMSO (to increase penetration in the cell) under a 16-8 light-dark light regime for 24 h at 40–45 μmol m^{-2} s^{-1} PPFD and 24 °C. Following colchicine treatment, fronds need to be washed three times in ddH_2O to remove the colchicine before transferring them to regular medium for a slow recovery. After 3–5 weeks, the ploidy can be determined by using the methodologies described above. The success rate is rather low, so it is important to start with a population of minimum 50 fronds.

As far as we know, allopolyploid duckweeds, in which both genome copies differ substantially from each other, have never been reported. Braglia and coworkers [24] described natural hybrids between *Lemna minor* and *Lemna turionifera*, but they did not determine their karyotype, so it remains unclear whether polyploidization was involved. The difficulties to induce flowering in *Spirodela* makes natural hybridization followed by allopolyploidization unlikely, nevertheless, it should in theory be possible to create synthetic allopolyploids using somatic cell fusion. The successful creation of such a somatic hybrid depends on protoplast generation, protoplast fusion, and regeneration. Regeneration from callus is possible [31], and although protoplasts can be generated from *Spirodela* root material, the protocols can be (and are being) improved in terms of efficiency (personal communication with Anthony Bishopp, Alex Ware, and Carolin Seyfferth). The remaining step, protoplast fusion, still must be tested on *Spirodela*, but symmetrical cell fusion using electrical fields has been tried already on the related Aracean genera *Spathiphyllum* and *Anthurium* [32]. Although no hybrids were recovered and growth stopped after the production of microcalli, this was an attempted intergeneric hybridization, and we believe interspecific crosses could be brought to a more satisfactory end.

5 Designing *Spirodela* Evolution Experiments

Experimental evolution is the study of replicate populations under defined and reproducible conditions over the course of many generations [33] and is one of the most potent methods to study evolution [34]. Although natural or anthropological-driven events such as introduction events and environmental pollution can be secondarily analyzed and considered as evolutionary experiments [33], here, we restrict ourselves to a discussion of more controlled laboratory evolution experiment. Such experiments can take many forms, and the design should be neatly adjusted to the specific question(s) asked. Nevertheless, some general types of laboratory evolution experiments can be distinguished. We will shortly discuss the four that are most used before diving into the more practical and general aspects of these experiments.

5.1 General Designs

Laboratory natural selection experiments (LNS) are probably the most typical evolutionary experiments. Replicate populations are brought into a novel-controlled environment, imposing them to a new selection pressure. A random sample of the population, naturally enriched for fitter individuals, is used to initiate the next generation, that is, serial transfer. Because this is a random sample that does not necessarily include the fittest individuals as in culling and artificial selection experiments, population sizes of natural selection experiments should be larger to obtain the same rate of change. By controlling the population size, it is possible to adjust the relative contribution of drift and selection [35].

In laboratory culling experiments, the experimental populations are exposed for a short period of time to a lethal environmental stress. When only the most resistant individuals remain, these are recovered and brought into a benign environment where they are maintained to rear the next generation which will, in turn, be exposed to the same stressor and so on. In these experiments, any trait that increases survival in the lethal environment contributes to the fitness. Phenotypical heterogeneity, i.e., a form of variability is supposed to be favored under such a regime.

Artificial truncating selection experiments use the same principles that have been used for centuries by plant and animal breeders. All individuals within the population are scored for a particular (phenotypic) trait, and the obtained ranking is subsequently used to select the fittest fraction of individuals (commonly a fixed percentage of the population size) to start the next generation. Often the highest or lowest ranking individuals are selected (directional selection), but the method also allows for disruptive and stabilizing selection. By manually selecting the "fittest" individuals based on a specific trait, the experimenter creates a direct link between fitness and this trait. Although this holds to some extent for LNS, there the strong selection pressure for increased growth rate imposed by serial transfer can be counteracted if high growth rates are disadvantageous (see, e.g., Prodigal Son dynamics [36]). For this reason, some evolutionary biologists [37] do not consider selective breeding as experimental evolution.

Mutation accumulation experiments are a completely different type of evolution experiments, by applying strong population bottlenecks, for example, a single individual per generation, genetic drift is maximized and selection reduced, which allows the accumulation of beneficial, neutral, and nonlethal deleterious mutations.

The method is useful to estimate the mutation rate, and when combined with phenotyping, it can be informative on the fitness effects of those mutations.

5.2 Replication

Much of the strength of experimental evolution lies in replication. When a population faces a new environmental condition, adaptation can proceed along different paths. Depending on the path taken, the evolutionary outcome might be quite different. Replication allows an assessment of the variation in these outcomes and the paths taken to get there. As such, they might offer insights into the repeatability of evolution and can increase the power of statistical conclusions.

5.3 Growth Conditions

Duckweeds can be grown on a rich variety of growth media, but Hoagland solution [9, 30], Steinberg medium [38], N-medium [39], and Schenk-Hildebrand medium [40], or modifications thereof are most used. A more extensive overview and recipes can be found in Appenroth [41]. When fast mixotrophic growth is needed or to detect contamination, the medium can be supplemented with glucose (50 mM) or sucrose (25 mM) [41].

The culturing vessel used will depend entirely on the experimental design, especially on the population size and whether the exact number of generations must be known. Because duckweed grows vegetatively and fronds can survive for several weeks, it becomes hard to express evolution in terms of the number of generations in large populations under serial transfer. If one has to keep track of the number of generations, for example, for mutation accumulation experiments, fronds can be kept separated and individually labeled. Well plates are an ideal vessel for this purpose. When small populations are needed, for example, for phenotyping, Erlenmeyer flask, beakers, or small plastic boxes can be used. For long-term evolution experiments that require a large population to reduce the waiting time in between beneficial mutations, large open systems such as open boxes, aquariums, cattle tanks, or even ponds are more suitable. When using smaller vessels, a transparent bottom will allow bottom illumination which facilitates phenotyping using pictures taken from above, whereas for medium-scale open systems, black or dark boxes offer the best solution as they limit the exposure of the medium to light and as such suppress the growth of algae.

Optimal growing conditions might vary depending on the strain, but in general Spirodela is grown under a 16:8 light dark regime, temperatures of 23–25 °C, and 30–300 μmol m^{-2} s^{-1} PPFD. Appenroth [42] reported an increasing growth rate between 30 and 90 μmol m^{-2} s^{-1} PPFD, followed by a plateau and a decrease in growth rate from 300 μmol m^{-2} s^{-1} PPFD onward. We grow our cultures at 40–45 μmol m^{-2} s^{-1} PPFD. Open outdoor systems are commonly covered by nets to reduce direct radiation and to create some shading (e.g., [43, 44]). Within such systems, it is possible to make smaller enclosures to quantify growth parameters.

Irrespective of the conditions, when duckweeds are transferred to a new environment, they will need some time to adjust (phenotypical plasticity). As such, it is important to acclimatize your cultures sufficiently long to the new environment before quantifying growth; according to the literature, it might take 4 weeks (even 6 when coming from agar) to get a constant growth rate when fronds are transferred to a new environment [7, 9]. Landolt sees this as a result of the buds being formed long before they become visible.

5.4 Preservation

During evolution experiments, it is fundamental that ancestral strains are preserved, for example, for competition assays. Currently there are two ways to achieve this. A first method is growing the plants under a low-maintenance slow growth regime. Plants can be kept in liquid medium under low light and lower temperature. Although duckweeds can be embedded in 0.8–0.9% agar, *Spirodela* stocks are best kept in liquid medium as the plant has the tendency to grow vertically on agar which will result in overtopping and even death, when the top layer loses its connection to the agar [21]. The ISCDRA recommends keeping duckweed stock cultures at 15 °C [45], but for some (tropical) strains, higher temperatures such as 17 °C might be better [21]. A second method is preservation as turions. These starch-rich overwintering propagules can be stored safely in the dark at 4 °C for years without requiring any maintenance. Although turion-based preservation is still in its infancy, the RDSC has developed a method for turion induction that seems to work [45]. A third method, cryopreservation of duckweed fronds, is still in its infancy but recently progress has been made at Kyoto U and IPK Gatersleben (Shogo Ito and Manuela Nagel, personal communication).

6 Quantifying Phenotypic Change

6.1 Fitness

6.1.1 Growth Rate

Fitness can be assessed in different ways. A first proxy is population growth rate. Typically, exponential growth is assumed and as such growth is quantified as the exponential growth rate or relative growth rate (RGR), expressed as $RGR = (\ln P1 - \ln P0)/(t1 - t0)$, where P1 and P0 represent the population size and t1 and t0 the corresponding points in time. Often, population size is measured only twice, for example, 7 [7] or 10 [46] days apart. Landolt [9] however recommended the use of more measurements (at least three), followed by a linear regression. Although this method is more labor intensive, it will provide a more accurate estimate of growth rate and allows the detection of deviations from the assumed exponential growth.

Population size can be expressed in terms of fresh weight, dry weight, coverage (i.e., surface area), or frond number. When frond number is used, the growth rate comes closest to the rate of clonal reproduction that is often used as a proxy of microbial fitness. Nevertheless, it should be noted that this measure fails to account for changes in area and thickness of the fronds, unless these are measured separately. Additionally, counting fronds is labor intensive, especially when done manually. However, thanks to advances in image analysis software and machine learning algorithms, automated counts have become quite reliable. Such software works best if the pictures are taken from above with a light source put underneath the culture. Area measurements, dry and fresh weight, do not necessarily correspond well to the number of fronds, as the average frond size can change over time. Nevertheless, area is particularly relevant if competition for space is important. Under such a scenario, the strain that covers a large area fastest could be considered fittest, irrespective whether the higher coverage is obtained by a few large fronds or by many small ones with a high rate of reproduction. Fresh weight measurements are somewhat more inclusive as they not only give an idea of area but also of thickness and starch content. That said, fresh weight is relatively imprecise as the measurement might depend on the condition under which the samples are taken and prepared. Dry weight is more reliable as the samples are dried till the weight stabilizes and all water is gone. Nevertheless, fresh weight measurements can be done faster, and the same population can be used for the two time points, whereas extra replicates or representative subsamples are necessary for dry weight estimation, which destroys the material. As all these different measurements reflect different aspects of growth and fitness, a combination of different measures gives the most inclusive results. Irrespective of the method used, there might be considerable variation between replicate growth rate estimates. This variation is mainly caused by differences in the starting material and can be reduced by starting with a larger population of 15–20 fronds [10].

6.1.2 Competition Assays

Competition assays provide a second way to assess fitness. When talking about competition, we must distinguish between exploitative and interference competition [47], both forms of competition can be inter- and intraspecific [48]. Exploitative competition is competition for a limited resource. For duckweeds and floating aquatic macrophytes, in general, these resources can be space, light, and essential nutrients [47, 49]. The relative importance of each of these resources is however unknown. Some argue that space is less important because growth continues when the entire surface is covered (i.e., overlapping), [43] but others are convinced that light is less important [50]. Interference competition is competition in which at least one of the competitors inhibit the growth of the other (e.g., allelopathy by using a chemical) or inhibits its competitors' access to one of the resources directly (e.g., by overtopping). Although interspecific allelopathy is common among floating aquatic macrophytes [47] and the negative allelopathic potential of some duckweed species has been established [51, 52], negative allelopathic effects of *Spirodela* seem to be mainly intraspecific [52]. Overtopping on the other hand has been reported as an important factor in interspecific duckweed competition assays including *Spirodela* [52]. Because competition has so many aspects, competition assays can be designed differently according to the aspects of competition one wants to study. In the most general competition assay, the evolved strains are competed, depending on the questions asked, against each other, their ancestor, or control lines evolving under standard benign conditions, and the strain that can outcompete or obtain the highest population sizes is considered as fittest. These assays are only an option if it is possible to distinguish between the competing lines. Manually labeling the fronds, for example, with permanent marker and monitoring the cultures regularly to label the new daughter fronds, is a simple but time-consuming option that can be applied without too much difficulties as long as population densities remain low. When densities become higher and overtopping begins, special care should be taken not to disturb the fronds. Alternatively, if the competing lines are not isogenic, genetic markers such as microsatellites, for example, [26, 43, 53], inter-simple sequence repeats [54], amplified fragment length polymorphisms (AFLP), for example, [23], tubulin-based polymorphisms (TBP) fingerprinting [24], and polymorphic NB-ARC-related genes [25] can be used to distinguish the fronds at the end of the assay. When competing isogenic lines of different ploidy, bulk flow cytometry can be considered. A calibration curve using mixtures of different tetraploids and diploids can be used to account for endopolyploidy. Multiple large samples, that is, minimum 50 plants, should be used to obtain an accurate estimate of the proportion of each ploidy variant in the population. If only competition for nutrients is important, the cultures can also be grown in half-open enclosures that allow free movement of nutrients and metabolites but do not allow duckweed to move (see, e.g., [46, 55]). This setup can equally well be applied for the study of allelopathy-related questions, but it will be impossible to distinguished allelopathic effects from these of exploitative competition if some resource is limited. Alternatively, plant extracts of the competitor can be added directly to the medium, or populations can be grown in medium that was previously used to grow the potential competitor, for example, [52]. The confounding effects of exploitative competition can be removed by supplementing the nutrient levels in the used medium till they reach the levels of the control medium, for example, [56].

6.1.3 Stress Resistance

A third method consists of measuring the stress resistance directly through a tolerance assay. By exposing the evolving populations to a gradient of concentration of a stressor, it is possible to determine the half-maximal effective concentration (EC50) and to use these as a proxy of fitness. This method is particularly interesting if the stress level is increased over time and reaches levels where growth of the control population is entirely inhibited, making competition assays redundant. Alternatively, other indicators of stress can be used to detect increased stress resistance. Starch content, for example, is known to increase in duckweeds under some stressful conditions such as phosphate depleted, saline (NaCl), and heavy metal (Cd, Ni)-enriched environments [17, 57, 58] and can be easily quantified qualitatively by cleaning the fronds with 70% ethanol followed by Lugol staining [59]. Alternatively, the starch content can also be quantified colorimetrically using the method of Magel [57, 60], conversion to glucose followed by glucose quantification using an enzymatic assay [61, 62] or HPLC [63]. Other indicators of stress that can be used are photosynthetic parameters as measured by PAM fluorometry (see below) and root regrowth [64].

6.2 General Morphology

In addition to fitness, there are many other phenotypical traits that can be monitored during experimental evolution. Changes in general morphology and anatomy can be easily detected by eye, but due to the very simple morphology of *Spirodela*, the number of traits is rather limited. Frond size (area), length, width, length/width ratio, fronds per colony, amount of necrotic tissue, and general health based on false-color imprints [65] are commonly used and can be easily measured on photographs. Frond thickness can be quantified using a thickness gauge but is regularly substituted by specific leaf mass (SLM), that is, the ratio of dry weight to area of leaf/frond. SLM is the product of leaf density and thickness [66]. Because root length varies with frond and root age, it is common practice to cut the roots and measure the length of new roots within a fixed amount of time [64].

6.3 Pigments and Photosynthetic Parameters

Pigment content can be quantified colorimetrically, for example, using the method of Lichtenthaler [67] for chlorophyll a, chlorophyll b, and carotenoids and the method of Mancinelli [68] for anthocyanins. But for more specific results, a combination of high/ultra-performance liquid chromatography (HPLC/UPLC) and mass spectrometry (MS) can be used. Pulse-amplitude modulation (PAM) fluorometry is commonly used to assess parameters related to chlorophyll fluorescence. Particularly interesting parameters are the maximum quantum yield of photosystem II, that is, the Fv/Fm (measured after 30 minutes dark adaptation), non-photochemical fluorescence quenching (NPQ), that is, (Fm-Fm')/Fm', and the effective quantum yield of photosystem II or YII of illuminated samples, that is, the proportion of absorbed light that is used in photo system II photochemistry [69, 70].

7 Quantifying Transcriptomic Change

Although variable in time and extent, polyploidization can induce extensive transcriptomic changes [71]. The magnitude of change depends at least partially on the type of polyploidization. The effects of allopolyploidy are in general more pronounced than those of autopolyploidy, because the latter only involves genome doubling whereas the former includes the merging of two genomes with different regulatory mechanisms as well [72]. Surprisingly, the duplication of the genetic material is not necessarily translated into a doubling of transcription; instead the few studies in which transcriptome size was studied seem to suggest that the response can be anywhere between full dosage compensation (i.e., the tetraploid has a similar transcriptome size as its parents) and 1:1 dosage effects (i.e., transcriptome size is proportional to genome size) and that the effects are not uniform but may differ across genes [73].

Expression profiles are mainly studied using RNA-sequencing (RNA-Seq). RNA-Seq has several advantages over older transcriptomic technologies [74], and for polyploidy research in particular, this technology should be preferred for two reasons. First, RNA-Seq reveals sequence differences (e.g., SNPs), allowing identification of homoeologs (especially interesting for allopolyploids). Second, it can detect transcripts with a large range of expression levels, which is important when major expression level changes are expected, for instance, after genome merging. In the past decade, RNA-Seq has relied almost exclusively on Illumina high-throughput deep-sequencing technology [75]. Nevertheless, other (the so-called third-generation) sequencing technologies such as nanopore sequencing (e.g., by Oxford Nanopore Technologies, ONT) and single molecule real-time (SMRT) sequencing (e.g., by Pacific Biosciences, pacbio) are gaining popularity [75], and the length of these reads (complete RNA molecules converted to cDNA) might, in the near future when throughput increases and error rate decreases, become very attractive to those interested in homoeolog (duplicates created in a WGD event) discrimination. Extensive reviews on RNA-Seq technology, RNA-Seq applications, and differential expression analysis can be found elsewhere [74–77].

Although most RNA-seq experiments focus exclusively on differential expression analysis, transcriptome size should not be ignored, as it will likely differ between cells of different ploidy. Traditionally, gene expression is depth-normalized, for example, the reads per million (RPM) value. This "traditional" normalization method does not provide any information about the transcriptome size or the relative transcriptome size, and when comparing two samples, no conclusions can be drawn about expression per cell or genome. Failure to account for differences in transcriptome size is often problematic in polyploidy research, for example, potential differences in parental transcriptome size might alter the outcome of expression bias studies. Additionally, dosage response studies are needed to shed light on the effects of polyploidy on gene expression [78, 79]. Consequently, the quantification of transcriptome size or at least the relative transcriptome size of the compared organisms should always be considered for ploidy related experiments and is even a necessity when dosage response and expression bias are subjects of interest.

When the number of cells used for RNA extraction is known (e.g., single-celled organisms [4]), gene expression can be normalized per cell by adding heterologous RNA (e.g., External RNA Controls Consortium RNA) proportional to the cell number and looking at the expression of the gene of interest relative to the proportion of exogenous RNA. If the RNA is extracted from an unknown number of cells, which will be the case for *Spirodela*, the (relative) transcriptome size must be quantified. J. E. Coate and J. J. Doyle [80, 81] pointed out that the relative transcriptome size, that is, the tetraploid to diploid ratio (TDR) of transcriptome size per cell can be estimated by dividing the TDR of the expression of a specific gene per cell by the TDR of the expression of that gene relative to the transcriptome size. Whereas the denominator can be easily derived from regular RNA-seq, the numerator requires a more creative solution. First, the relative expression per genome is calculated using RNA-genomic DNA (gDNA) coextraction, followed by reverse transcription from RNA to cDNA and a qRT-PCR to quantify the RNA/gDNA ratio. Subsequently, the ploidy information is used to translate this ratio to the relative expression per cell. To increase the accuracy of the estimated relative transcriptome size, it should be calculated using several transcriptional markers all using the geometric mean of the amplification efficiency of at least three genomic markers.

If you consider such a normalizations, we highly recommend you to read Coates' excellent in-depth discussion of this topic elsewhere in this issue [82].

8 Quantifying Genetic Change

The quantification of genetic change in the context of experimental evolution and, in particular, evolve and resequence experiments has been discussed in detail elsewhere in this issue [4].

9 Concluding Statement

Although *Spirodela* and duckweeds have been used as a model system for many decades, the use of duckweed for evolution experiments is relatively novel. During the past 2 years, we have experimented extensively with diploid and tetraploid duckweeds, and many of the recommendations given above come from our own experience acquired during these experiments, sometimes by trial and error. During these experiments, we became convinced of the suitability of *Spirodela* to study WGD and subsequent evolution. We hope this chapter inspired the reader to follow in our footsteps and are always willing to provide additional advice on the use of duckweeds, in particular, *Spirodela*, in evolution experiments.

Acknowledgments

This work was supported by Ghent University (Methusalem funding, BOF.MET.2021.0005.01), Research Foundation Flanders (personal grants to Lucas Prost and Quinten Bafort), and the European Research Council (ERC) under the European Union's Horizon 2020 research and innovation program (grant agreement No 833522).

References

1. Fox DT, Soltis DE, Soltis PS, Ashman TL, Van de Peer Y (2020) Polyploidy: a biological force from cells to ecosystems. Trends Cell Biol 30: 688
2. Wood TE, Takebayashi N, Barker MS, Mayrose I, Greenspoon PB, Rieseberg LH (2009) The frequency of polyploid speciation in vascular plants. Proc Natl Acad Sci 106(33): 13,875–13,879
3. Van de Peer Y, Ashman TL, Soltis PS, Soltis DE (2020) Polyploidy: an evolutionary and ecological force in stressful times. Plant Cell 33(1):11–26. https://doi.org/10.1093/plcell/koaa015. https://academic.oup.com/plcell/article-pdf/33/1/11/36659456/koaa015.pdf
4. Bafort Q, Prost L, Aydogdu E, Van de Vloet A, Casteleyn G, Van de Peer Y, De Clerck O (2022) Studying whole-genome duplication using experimental evolution of Chlamydomonas. In: Van de Peer Y, Zwaenepoel A, Li Z (eds) Polyploidy. Methods in Molecular Biology. Humana Press
5. Group AP (2009) An update of the angiosperm phylogeny group classification for the orders and families of flowering plants: APG III. Bot J Linn Soc 161(2):105–121
6. Lemon GD, Posluszny U (2000) Comparative shoot development and evolution in the lemnaceae. Int J Plant Sci 161(5):733–748
7. Ziegler P, Adelmann K, Zimmer S, Schmidt C, Appenroth KJ (2015) Relative in vitro growth rates of duckweeds (lemnaceae)-the most rapidly growing higher plants. Plant Biol 17(33–41):19
8. Wang W, Haberer G, Gundlach H, Glaßer C, Nussbaumer T, Luo M, Lomsadze A, Borodovsky M, Kerstetter R, Shanklin J et al (2014) The spirodela polyrhiza genome reveals insights into its neotenous reduction fast growth and aquatic lifestyle. Nat Commun 5(1):1–13
9. Landolt E, Kandeler R (1987) Biosystematic investigations in the family of duckweeds (lemnaceae), vol. 4: the family of lemnaceae-a monographic study, vol. 2 (phytochemistry, physiology, application, bibliography). Veroff Geobot Inst ETH, Stiftung Rubel, Zurich (Switzerland)
10. Landolt E (1986) Biosystematic investigations in the family of duckweeds (lemnaceae)(vol. 2.) the family of lemnaceae-a monographic study. vol. 1. Veroff Geobot Inst ETH 71:1–563
11. Urbanska-Worytkiewicz K (1980) Cytological variation within the family of lemnaceae. Veroff Geobot Inst ETH 70:30–101
12. Geber G (1989) Zur Karyosystematik der Lemnaceae
13. Wang W, Kerstetter RA, Michael TP et al (2011) Evolution of genome size in duckweeds (lemnaceae). J Bot 2011:1–9
14. Hoang PT, Schubert V, Meister A, Fuchs J, Schubert I (2019) Variation in genome size, cell and nucleus volume, chromosome number

and rdna loci among duckweeds. Sci Rep 9(1): 1–13 15
15. Xu S, Stapley J, Gablenz S, Boyer J, Appenroth KJ, Sree KS, Gershenzon J, Widmer A, Huber M (2019) Low genetic variation is associated with low mutation rate in the giant duckweed. Nat Commun 10(1):1243
16. Davidson D, Simon JP (1981) Thermal adaptation and acclimation of ecotypic populations of spirodela polyrhiza (lemnaceae): thermostability and apparent activation energy of nad malate dehydrogenase. Can J Bot 59(6): 1061–1068
17. Sree KS, Adelmann K, Garcia C, Lam E, Appenroth KJ (2015) Natural variance in salt tolerance and induction of starch accumulation in duckweeds. Planta 241(6):1395–1404
18. Wozakowska-Natkaniec H (1977) Ecological differentiation of Lemna minor l. and Spirodela polyrrhiza (l.) schleiden populations. Acta Soc Bot Pol 46(2):201
19. Kuehdorf K, Jetschke G, Ballani L, Appenroth KJ (2014) The clonal dependence of turion formation in the duckweed Spirodela polyrhiza an ecogeographical approach. Physiol Plant 150(1):46–54
20. Lammler W, Bogner J (2014) Elias Landolt and the duckweeds. Aroideana 37:81–88
21. Sree KJ, Appenroth KS (2020) Worldwide Genetic Resources of Duckweed: Stock Collections. Springer International Publishing, pp 39–46
22. Borisjuk N, Chu P, Gutierrez R, Zhang H, Acosta K, Friesen N, Sree K, Garcia C, Appenroth K, Lam E (2015) Assessment, validation and deployment strategy of a two-barcode protocol for facile genotyping of duckweed species. Plant Biol 17:42–49
23. Bog M, Lautenschlager U, Landrock MF, Landolt E, Fuchs J, Sree KS, Oberprieler C, Appenroth KJ (2015) Genetic characterization and barcoding of taxa in the genera Landoltia and Spirodela (lemnaceae) by three plastidic markers and amplified fragment length polymorphism (aflp). Hydrobiologia 749(1): 169–182
24. Braglia L, Lauria M, Appenroth KJ, Bog M, Breviario D, Grasso A, Gavazzi F, Morello L (2021) Duckweed species genotyping and interspecific hybrid discovery by tubulin-based polymorphism fingerprinting. Front Plant Sci 12:270
25. Chu P, Wilson GM, Michael TP, Vaiciunas J, Honig J, Lam E (2018) Sequence-guided approach to genotyping plant clones and species using polymorphic nb-arc-related genes. Plant Mol Biol 98(3):219–231
26. Feng B, Fang Y, Xu Z, Xiang C, Zhou C, Jiang F, Wang T, Zhao H (2017) Development of a new marker system for identification of Spirodela polyrhiza and Landoltia punctata. Int J Genom 2017:5196763
27. Cao HX, Vu GTH, Wang W, Appenroth KJ, Messing J, Schubert I (2016) The map-based genome sequence of s pirodela polyrhiza aligned with its chromosomes, a reference for karyotype evolution. New Phytol 209(1): 354–363
28. Galbraith DW, Harkins KR, Maddox JM, Ayres NM, Sharma DP, Firoozabady E (1983) Rapid flow cytometric analysis of the cell cycle in intact plant tissues. Science 220(4601): 1049–1051
29. Vunsh R, Heinig U, Malitsky S, Aharoni A, Avidov A, Lerner A, Edelman M (2015) Manipulating duckweed through genome duplication. Plant Biol 17:115–119
30. Hoagland D (1920) Optimum nutrient solutions for plants. Science 52(1354):562–564
31. Yang J, Li G, Hu S, Bishopp A, Heenatigala P, Kumar S, Duan P, Yao L, Hou H (2018) A protocol for efficient callus induction and stable transformation of Spirodela polyrhiza (l.) schleiden using agrobacterium tumefaciens. Aquat Bot 151:80–86
32. Duquenne B, Eeckhaut T, Werbrouck S, Van Huylenbroeck J (2007) Effect of enzyme concentrations on protoplast isolation and protoplast culture of Spathiphyllum and Anthurium. Plant Cell Tissue Organ Cult 91(2):165–173
33. Rose MR, Garland J (2009) Darwin's other mistake. In: Experimental evolution. University of California Press, pp 3–14
34. Futuyma DJ, Bennett AF (2009) The importance of experimental studies in evolutionary biology. In: Experimental Evolution. University of California Press, pp 15–30
35. Lachapelle J, Reid J, Colegrave N (2015) Repeatability of adaptation in experimental populations of different sizes. Proc R Soc Lond B Biol Sci 282(1805):20143033
36. Collins S (2016) Growth rate evolution in improved environments under prodigal son dynamics. Evol Appl 9(9):1179–1188
37. Kawecki TJ, Lenski RE, Ebert D, Hollis B, Olivieri I, Whitlock MC (2012) Experimental evolution. Trends Ecol Evol 27(10):547–560
38. Steinberg RA (1946) Mineral requirements of Lemna minor. Plant Physiol 21(1):42
39. Appenroth KJ, Teller S, Horn M (1996) Photophysiology of turion formation and germination in spirodela polyrhiza. Biol Plant 38(1):95

40. Schenk RU, Hildebrandt A (1972) Medium and techniques for induction and growth of monocotyledonous and dicotyledonous plant cell cultures. Can J Bot 50(1):199–204
41. Appenroth KJ (2015) Useful methods 3: Media for in vitro-cultivation of duckweed. Duckweed Forum 3:180–186
42. Appenroth KJ (2002) Co-action of temperature and phosphate in inducing turion formation in spirodela polyrhiza (great duckweed). Plant Cell Environ 25(9):1079–1085
43. Hart SP, Turcotte MM, Levine JM (2019) Effects of rapid evolution on species coexistence. Proc Natl Acad Sci 116(6):2112–2117
44. Morris RS (2019) Birth order as a source of diversification in spring phenology and its potential effects on performance in greater duckweed, Spirodela polyrhiza. PhD thesis, Carleton University
45. Pasaribu B, Che YF, Lam E (2021) Optimizing a protocol for long-term storage of duckweed clones as turions. Duckweed Forum 9(2): 34–39
46. Szabo S, Scheffer M, Roijackers R, Waluto B, Braun M, Nagy PT, Borics G, Zambrano L (2010) Strong growth limitation of a floating plant (Lemna gibba) by the submerged macrophyte (Elodea nuttallii) under laboratory conditions. Freshw Biol 55(3):681–690
47. Gopal B, Goel U (1993) Competition and allelopathy in aquatic plant communities. Bot Rev 59(3):155–210
48. Holomuzki JR, Feminella JW, Power ME (2010) Biotic interactions in freshwater benthic habitats. J N Am Benthol Soc 29(1): 220–244
49. Landolt E (1957) Physiologische und okologische untersuchungen an lemnaceen. habilitationsschrift eth, 1957. published in: Berichte der schweizerischen botanischen gesellschaft, vol. 67, pp. 271-410. Berichte der Schweizerischen Botanischen Gesellschaft 67:271–410
50. Zhang LM, Jin Y, Yao SM, Lei NF, Chen JS, Zhang Q, Yu FH (2020) Growth and morphological responses of duckweed to clonal fragmentation, nutrient availability, and population density. Front Plant Sci 11:618
51. Bich TTN, Kato-Noguchi H (2012) Allelopathic potential of two aquatic plants, duckweed (Lemna minor l.) and water lettuce (Pistia Stratiotes l.), on terrestrial plant species. Aquat Bot 103:30–36
52. Wolek JA (1974) A preliminary investigation on interactions (competition, allelopathy) between some species of Lemna, Spirodela and Wolfia. Berichte des Geobotanischen Institutes der Eidg Techn Hochschule 42:140–162
53. Xu N, Hu F, Wu J, Zhang W, Wang M, Zhu M, Ke J (2018) Characterization of 19 polymorphic ssr markers in Spirodela polyrhiza (lemnaceae) and cross-amplification in Lemna perpusilla. Appl Plant Sci 6(5):e01,153
54. Xue H, Xiao Y, Jin Y, Li X, Fang Y, Zhao H, Zhao Y, Guan J (2012) Genetic diversity and geographic differentiation analysis of duckweed using inter-simple sequence repeat markers. Mol Biol Rep 39(1):547–554
55. Szabo S, Roijackers R, Scheffer M (2003) A simple method for analysing the effects of algae on the growth of Lemna and preventing algal growth in duckweed bioassays. Archiv Fur Hydrobiologie 157(4):567–575
56. Strzalek M, Kufel L, Wysokinska U (2019) How does Stratiotes aloides l. affect the growth and turion formation of spirodela polyrhiza (l.) schleiden? Aquat Bot 154:45–52
57. Appenroth KJ, Krech K, Keresztes A, Fischer W, Koloczek H (2010) Effects of nickel on the chloroplasts of the duckweeds spirodela polyrhiza and Lemna minor and their possible use in biomonitoring and phytoremediation. Chemosphere 78(3):216–223
58. Sree KS, Appenroth KJ (2014) Increase of starch accumulation in the duckweed Lemna minor under abiotic stress. Albanian J Agric Sci 13:11–17
59. Stewart JJ, Adams WW III, Escobar CM, Lopez-Pozo M, Demmig-Adams B (2020) Growth and essential carotenoid micronutrients in Lemna gibba as a function of growth light intensity. Front Plant Sci 11:480
60. Magel E (1991) Qualitative and quantitative determination of starch by a colorimetric method. Starch-Starke 43(10):384–387
61. Smith AM, Zeeman SC (2006) Quantification of starch in plant tissues. Nat Protoc 1(3): 1342–1345
62. Liu Y, Xu H, Wang Y, Tang X, He G, Wang S, Ma Y, Kong Y, Yu C, Zhou G (2020) A submerged duckweed mutant with abundant starch accumulation for bioethanol production. GCB Bioenergy 12(12):1078–1091 16
63. Xiao Y, Fang Y, Jin Y, Zhang G, Zhao H (2013) Culturing duckweed in the field for starch accumulation. Ind Crop Prod 48: 183–190
64. Park J, Lee H, Han T (2020) Comparative paraquat sensitivity of newly germinated and mature fronds of the aquatic macrophyte Spirodela polyrhiza. Am J Plant Sci 11(7): 1008–1024
65. Sree KS, Keresztes A, Mueller-Roeber B, Brandt R, Eberius M, Fischer W, Appenroth KJ (2015) Phytotoxicity of cobalt ions on the

duckweed Lemna minor-morphology, ion uptake, and starch accumulation. Chemosphere 131:149–156
66. Witkowski E, Lamont BB (1991) Leaf specific mass confounds leaf density and thickness. Oecologia 88(4):486–493
67. Lichtenthaler HK (1987) Chlorophylls and carotenoids: pigments of photosynthetic biomembranes. Methods in enzymology 148: 350–382
68. Mancinelli AL (1984) Photoregulation of anthocyanin synthesis: VIII. effect of light pretreatments. Plant Physiol 75(2):447–453
69. Klughammer C, Schreiber U (2008) Complementary ps ii quantum yields calculated from simple fluorescence parameters measured by pam fluorometry and the saturation pulse method. PAM Appl Notes 1(2):201–247
70. Murchie EH, Lawson T (2013) Chlorophyll fluorescence analysis: a guide to good practice and understanding some new applications. J Exp Bot 64(13):3983–3998
71. Adams KL, Wendel JF (2005) Novel patterns of gene expression in polyploid plants. Trends Genet 21(10):539–543
72. Parisod C, Holderegger R, Brochmann C (2010) Evolutionary consequences of autopolyploidy. New Phytol 186(1):5–17
73. Doyle JJ, Coate JE (2018) Polyploidy, the nucleotype, and novelty: The impact of genome doubling on the biology of the cell. Int J Plant Sci 180(1):000–000
74. Wang Z, Gerstein M, Snyder M (2009) RNA-seq: a revolutionary tool for transcriptomics. Nat Rev Genet 10(1):57-63 18
75. Stark R, Grzelak M, Hadfield J (2019) RNA sequencing: the teenage years. Nat Rev Genet 20(11):631–656
76. Conesa A, Madrigal P, Tarazona S, Gomez-Cabrero D, Cervera A, McPherson A, Szczesniak MW, Gaffney DJ, Elo LL, Zhang X et al (2016) A survey of best practices for rna-seq data analysis. Genome Biol 17(1):1-19 14
77. Pereira MA, Imada EL, Muniz Guedes R (2017) RNA-seq: applications and best practices. In: Applications of RNA-Seq and omics strategies: from microorganisms to human health. Intechopen, p 1
78. Jackson S, Chen ZJ (2010) Genomic and expression plasticity of polyploidy. Curr Opin Plant Biol 13(2):153–159
79. Yoo MJ, Liu X, Pires JC, Soltis PS, Soltis DE (2014) Nonadditive gene expression in polyploids. Annu Rev Genet 48:485–517
80. Coate JE, Doyle JJ (2010) Quantifying whole transcriptome size, a prerequisite for understanding transcriptome evolution across species: an example from a plant allopolyploid. Genome Biol Evol 2:534–546
81. Coate JE, Doyle JJ (2015) Variation in transcriptome size: are we getting the message? Chromosoma 124(1):27–43
82. Coate J (2022) Beyond transcript concentrations: Methods to quantify polyploid expression responses per biomass, per genome, and per cell with RNA-Seq. In: Van de Peer Y, Zwaenepoel A, Li Z (eds) Polyploidy. Methods in molecular biology. Humana Press

Chapter 20

Experimental Approaches to Generate and Isolate Human Tetraploid Cells

Sara Vanessa Bernhard, Simon Gemble, Renata Basto, and Zuzana Storchova

Abstract

Cancer cells are frequently affected by large-scale chromosome copy number changes, such as polyploidy or whole chromosome aneuploidy, and thus understanding the consequences of these changes is important for cancer research. In the past, it has been difficult to study the consequences of large-scale genomic changes, especially in pure isogenic populations. Here, we describe two methods to generate tetraploid cells induced either by cytokinesis failure or mitotic slippage. These treatments result in mixed population of diploids and tetraploids that can be analyzed directly. Alternatively, tetraploid populations can be established by single cell clone selection or by fluorescence activated cell sorting. These methods enable to analyze and compare the consequences of whole-genome doubling between the parental cell line, freshly arising tetraploid cells, and post-tetraploid aneuploid clones.

Key words Tetraploidy, Aneuploidy, Cancer, Whole-genome doubling, Cytokinesis failure, Mitotic slippage, Flow cytometry, FACS

1 Introduction

Cancer cells are frequently characterized by large-scale genome changes including polyploidy and whole chromosome aneuploidy [1]. Polyploidy is defined as the occurrence of additional whole chromosome sets, whereas aneuploidy is characterized by the presence of an unbalanced number of chromosomes. In most cases, cancer cells exhibit massively altered karyotypes, and several routes have been proposed to explain how cells can reach these complex karyotypes. One of the hypotheses states that cells become aneuploid by passing through an intermediate unstable state by doubling their genome, a condition called tetraploidy (4n) [2–4]. Recent genomic studies estimate that around 37% of solid tumors

Yves Van de Peer (ed.), *Polyploidy: Methods and Protocols*,
Methods in Molecular Biology, vol. 2545, https://doi.org/10.1007/978-1-0716-2561-3_20,

undergo at least one round of whole-genome duplication (WGD), ranging from almost no doubling events in hematopoietic cancers to 70% in cancers such as non-small lung carcinomas [1, 5]. Studies show that WGD events during tumorigenesis affect tumor evolution, progression, as well as patient prognosis [6]. WGD favors the continuous gain and loss of chromosomes, the so-called chromosomal instability (CIN) [7, 8]. Further studies addressing the role of WGD in cancer may help find new therapies and improve existing treatments.

There are several different approaches to induce tetraploidy in human cells. The methods include induction of cytokinesis failure using chemical (e.g., cytochalasin treatment) or genetic (e.g., anillin knockdown) treatments, cell to cell fusion (e.g., PEG-mediated fusion or electrofusion), or induced mitotic slippage (e.g., via nocodazole or taxol treatment) [9–14]. Cell fusion methods are usually inefficient and time-consuming, while extended mitotic arrest to attain mitotic slippage may produce DNA damage and even cell death and therefore should be combined with a treatment with spindle assembly checkpoint inhibitors. Cytochalasin treatment is a simple and effective way to induce WGD [15]. Furthermore, inducing cytokinesis failure using small molecule inhibitors can be applied to a large cell population without generating mitotic aberrations or DNA damage [10, 16]. Cytokinesis failure is usually induced via actin inhibition using different cytochalasins, with dihydrocytochalasin D or B being the most used ones [17–20]. However, induction of cytokinesis failure or mitotic slippage generates a mixed population consisting of diploid and tetraploid cells, which may hinder some analyses. One solution to achieve a pure tetraploid population is through fluorescence-activated cell sorting (FACS), where one can distinguish diploid and tetraploid cells by comparing the DNA content.

Here, we describe two methods to generate tetraploid populations from diploid cell lines. Additionally, we describe how tetraploid populations can be obtained either by isolating post-tetraploid/aneuploid single cell clones or by sorting the 4n population. Using these methods, one can obtain near-tetraploid cells with chromosome content validated via flow cytometry, fluorescence in situ hybridization (FISH), next-generation sequencing, or single nucleotide polymorphism array (SNP) analysis to study the consequences of whole-genome doubling without additional mutations [21]. Both techniques enable to study consequences of tetraploidy by comparison between the isogenic parental cell line, freshly generated tetraploid cells, and subsequently arising post-tetraploid aneuploid clones.

2 Material

2.1 Induction of Cytokinesis Failure by DCD Followed by Clone Isolation

1. Dihydrocytochalasin D (Inhibitor of actin polymerization, C8273 from Sigma-Aldrich, Taufkirchen; Germany, *see* **Note 1**).
2. DMEM growth medium (31966047 from Thermo Fisher Scientific, Waltham, USA) supplemented with 10% FBS (10270106 from Thermo Fisher Scientific, Waltham, USA) and 1% penicillin/streptomycin (15140122 from Thermo Fisher Scientific, Waltham, USA) (DMEM +/+).
3. 0.25% trypsin-EDTA (25200056 from Thermo Fisher Scientific, Waltham, USA)
4. Cloning discs 3.2 mm diameter (Z374431 from Sigma-Aldrich, Taufkirchen, Germany)
5. Fixation-permeabilization solution (*see***Note 2**).
6. Phosphate-buffered saline (PBS; 10X): 1,37 M NaCl; 27 mM KCl; 100 mM Na_2HPO_4; 18 mM KH_2PO_4 in ddH_2O (*see* **Note 3**).
7. RNase A (A2760, AppliChem, Darmstadt, Germany, *see* **Note 4**).
8. DNA stain, i.e., DAPI (6335.1 from Carl Roth, Karlsruhe, Germany).

2.2 Instruments

- Attune NxT flow cytometer (from Thermo Fisher Scientific, Waltham, USA).

2.3 Induction of Mitotic Slippage by Monastrol Followed by FACS Sorting

1. Monastrol (Eg5 inhibitor, S8439 from Selleckchem, Houston, USA).
2. MPI-0479605 (MPS1 inhibitor, S7488 from Selleckchem, Houston, USA).
3. Palbociclib (CDK4/6 inhibitor, S4482 from Selleckchem, Houston, USA).
4. Hoechst 33258 (94403 from Sigma Aldrich, Saint Louis, USA).
5. DAPI (D9542 from Sigma Aldrich, Saint Louis, USA).
6. hTERT RPE-1 cells (ATCC, Cat# CRL-4000, RRID: CVCL_4388).
7. DMEM F12 (10565018 from Thermo Fisher, Waltham, USA).
8. PBS 1x

2.4 Instruments

- FACS Sorter SH800 (from Sony, Tokyo, Japan).

3 Methods

3.1 Generation of Tetraploid Cells by Cytokinesis Failure

The following protocol describes the generation of tetraploid cells by cytokinesis failure and subsequent isolation of post-tetraploid cell lines arising from single surviving cells (Fig. 1a). The protocol presented here was established for human colorectal carcinoma cell line HCT116 using dihydrocytochalasin D, but was also used to successfully generate tetraploid populations in other cell lines, e.g., RPE1 or DLD-1.

1. Seed the cells in 10 cm dish to obtain approximately 70% confluency at the time of treatment and let cells grow for 24 h.
2. Replace DMEM with a fresh DMEM containing 0.75 μM DCD and incubate for 18 h over night at 37 °C, 5% CO_2.

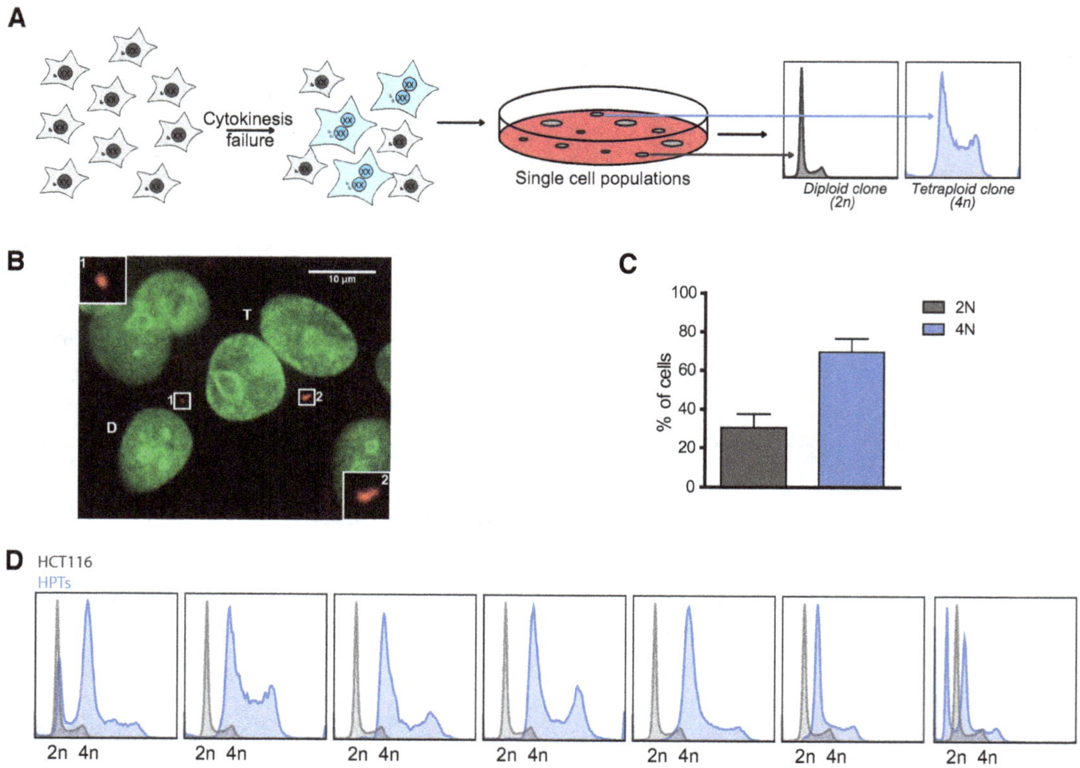

Fig. 1 (**a**) Induction of cytokinesis failure by DCD treatment creates a mixed population of 2n and 4n cells. Single cell colonies can be selected after plating a small cell number in 15 cm dishes. DNA content of the acquired single cell clones can be analyzed via flow cytometry. (**b**) Diploid (D) and tetraploid (T) cells can be distinguished by the nucleus size and numbers, by the cell size, as well as by counting the number of centrosomes per cell. DNA is in green (H2B-GFP) and centrosomes are in red (γ-tubulin-mRuby). D diploid; T tetraploid; scale bar, 10 μm. (**c**) On average, 70% of cells fail to complete cytokinesis after the DCD treatment (ranges from 50 to 80%). (**d**) DNA content profile of seven different post-tetraploid cell lines (blue) in comparison to a diploid DNA profile (gray) after staining with Vybrant™ DyeCycle™ green stain (V35004 from Thermo Fisher Scientific, Waltham, USA) acquired by flow cytometry

3. Wash out DMEM with DCD without trypsinization three times using PBS 1x and add fresh DMEM (*see* **Note 5**). Large fraction of the cells is now binucleated and carry a tetraploid chromosome set (Fig. 1b, c).

3.2 Growth of Post-Tetraploid Cells and the Analysis of DNA Content

1. After the DCD treatment, let cells grow for additional 24 h at 37 °C, 5% CO_2. The cells now contain one single nucleus, which is larger in size compared to diploids cells.
2. Seed approx. 100 single cells in one 15 cm dish.
3. Let cells grow until colonies are visible.
4. Circle colonies you want to pick on bottom of plate.
5. Pick selected single colonies using cloning discs (*see* **Note 6**).
6. Transfer single clones to 24-well plate and let grow (*see* **Note 7**).
7. As soon as you have a sufficient number of cells for analysis, trypsinize the cells, take half of the cell suspension for analysis; the second half stays in the plate for further growth.
8. Spin down at 1400 rpm for 3 min, remove supernatant, and resuspend the pellet in Fix Perm solution (*see* **Note 8**). Incubate the cell suspension for 15 min.
9. Add 1 ml PBS 1x.
10. Spin down at 1400 rpm for 3 min and remove supernatant.
11. Resuspend the pellet in 300–500 μl (depending on the pellet size) PBS 1x containing 2 μg/ml DAPI and 10 μg/ml RNase A and incubate for 5 min.
12. After incubation, the samples are ready for the analysis of the DNA content (*see* **Note 9**, Fig. 1d).

3.3 Generation of Tetraploid Cells by Inducing Mitotic Slippage and Synchronization in G1

This protocol is designed to isolate tetraploid cells generated by mitotic slippage (Fig. 2a). The protocol was established using RPE-1 cells expressing the *f*luorescent *u*biquitination-based *c*ell *c*ycle *i*ndicator (FUCCI) system (pBOB-EF1-FastFUCCI-Puro, ref 86849 from Addgene, Cambridge, USA) [22]. All procedures are carried out at room temperature unless otherwise specified.

1. At day 0, seed the wells with a sufficient number of adherent cells to achieve a confluence of about 80% on the day of the experiment (i.e., day 2).
2. At day 1, add a combination of monastrol 50 μM + MPI-0479605 1 μM to the cell culture medium and incubate during 24 h at 37 °C, 5% CO_2 to induce mitotic slippage (*see* **Note 10**) [23, 24].

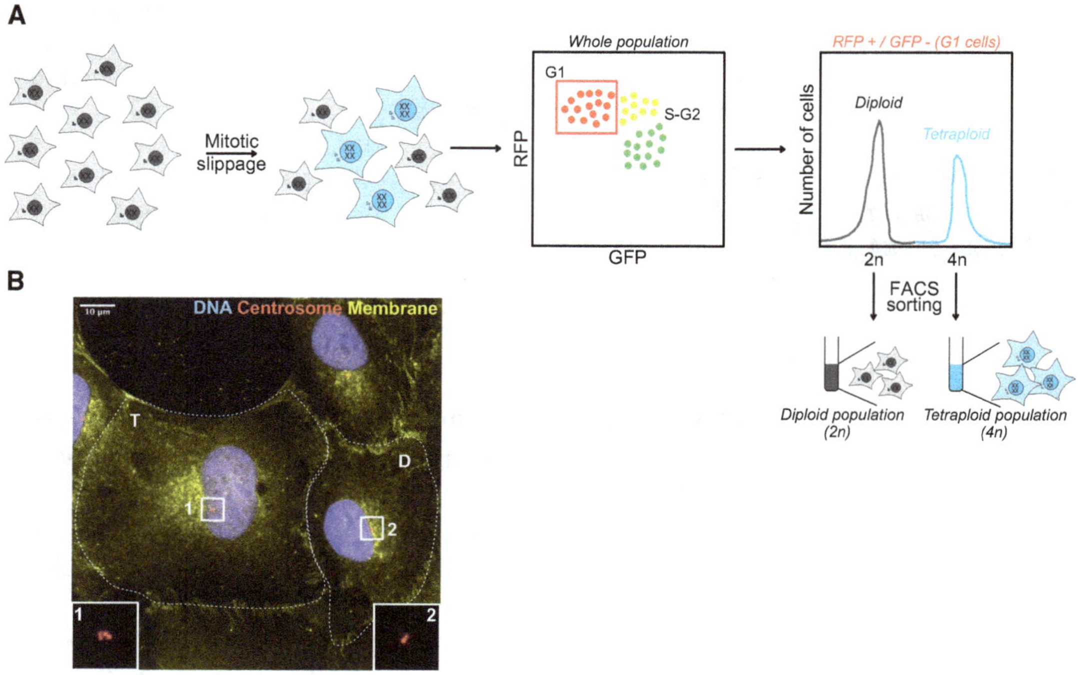

Fig. 2 (**a**) Induction of mitotic slippage by treatment with monastrol and MPI-0479605 creates a mixed population of 2n and 4n cells. G1 cells are selected as RFP-positive/GFP-negative FUCCI cells. By analyzing DNA content in this population (Hoechst channel), diploid (2n) and tetraploid (4n) G1 cells can be segregated. Diploid and tetraploid G1 cells can then be sorted into external collection tubes via FACS. (**b**) Diploid (D) and tetraploid (T) cells can be distinguished using parameters such as nuclear and cell size and the number of centrosomes. DNA is in blue (DAPI, D9542 from Sigma Aldrich, Saint Louis, USA), centrosomes are in red (CEP192 antibody, [27]) and membranes are in yellow (β-catenin antibody, C2206 from Sigma Aldrich, Saint Louis, USA). Dotted line indicates the cell membranes. D diploid; T tetraploid; scale bar, 10 µm

3. After two hours of incubation, add 1 µM palbociclib in the cell culture medium and continue the incubation at 37 °C, 5% CO_2 to block the cells in G1 [25].
4. At day 2, a mix of G1 diploid (2n) and tetraploid (4n) cells is obtained (Fig. 2b).

3.4 FACS Sorting to Isolate Diploid and Polyploid Populations

1. At day 2, add 2 µg/ml Hoechst to the cell culture medium for 1 h at 37 °C, 5% CO_2.
2. After the incubation, generate a single cell suspension.
3. Wash the cells using PBS 1x.
4. Centrifuge at 1200 rpm for 5 min.
5. Remove the supernatant and resuspend cells in cold cell culture medium at 1×10^7 cell per ml and keep them at 4 °C throughout the experiment (*see* **Note 11** and **12**).
6. Perform compensation using the appropriate negative control samples to correct for fluorescence spillover, i.e., removing the

signal of any given fluorochrome from all detectors except the one devoted to measuring the dye of interest (*see* **Note 13**).

7. Record the experimental sample to be sorted and use gating tools to select the populations of interest.
   7.1. Select G1 cells by selecting the RFP-positive/GFP-negative cells (Fig. 2a).
   7.2. By analyzing the DNA content in this population (Hoechst channel), diploid (2n) and tetraploid (4n) G1 cells can be segregated.
   7.3. Once gates have been determined, the diploid and tetraploid G1 cells can be sorted into external collection tubes until the required amount of cells has been obtained.
   7.4. Perform a post-sort analysis to determine the purity of the sorted cell populations (*see* **Note 14**).

4 Notes

1. Dissolve 1 mg DCD in 263 μl DMSO to achieve a concentration of 7.5 mM. The drug can be stored at −20 °C for up to 1 year; afterwards the efficiency drastically decreases. Alternatively, other dihydrocytochalasins can be used, as most of these small molecules have a similar effect on actin [26].
2. Mix 1 part of 4x fixation/permeabilization concentrate with 3 parts of fixation/permeabilization diluent (12736048 and 12746048 from Thermo Fisher Scientific, Waltham, USA).
3. Add 100 ml PBS 10x to a graduate 1 l cylinder, fill it up to 1 l using 900 ml ddH_20. Sterilize by autoclaving.
4. Dissolve RNase A at a concentration of 10 mg/ml in double-distilled water. Boil at 100 °C for 15 min to inactivate contamination DNases and slowly cool it down to room temperature. RNase can be stored in aliquots at −20 °C.
5. After cytokinesis inhibition with DCD, on average 60% of all cells will be binucleated [11].
6. Soak the required number of discs in a small dish filled with trypsin. Remove the medium from 15 cm dish and rinse once with PBS 1x. Place a soaked disc on colony using sterilized forceps and incubate at 37 °C, 5% CO_2 for 5 min.
7. Check the 15 cm dish under microscope to confirm the transfer of the colony.
8. The amount of fixation/permeabilization solution depends on the pellet size. For the number of cells collected from one well of a 6-well plate, 200–500 μl should be sufficient.
9. DAPI does not need to be washed out for the analysis.

10. Eg5 inhibition blocks the cells in mitosis and MPS1 inhibition forces mitotic exit; together, these drugs induce mitotic slippage, thus generating tetraploid cells with one single nucleus.
11. Control samples should be prepared similarly to the sorted sample and from the same type of cells as the ones used in the experiment.
12. Pre-coating 5 ml tubes (062978 from Dutscher, Issy-les-Moulineaux, France) overnight with medium is recommended.
13. FUCCI-negative cells and untreated cells should be used as negative control samples.
14. By analyzing nuclear size, cell size, and centrosome number, it is possible to distinguish diploids from tetraploid cells by immunofluorescence (Fig. 2a) and to quantify the purity of the sorted population.

Acknowledgments

We thank Lea Guyonnet, Anna Chipon, and Coralie Guerin from the cytometry core facility at Institut Curie for help in establishing and developing the methods to sort diploid from tetraploid cells. Work in the Storchova lab is supported by Rhineland Palatine Research Initiative BioComp and by Deutsche Forschungsgemeinschaft STO 918/7. Work in the Basto lab is supported by Institut Curie, CNRS and an ERC CoG (Chromo Number 725907). The Basto lab is a member of the LabEx Cell(n)Scale.

References

1. Zack TI, Schumacher SE, Carter SL, Cherniack AD, Saksena G, Tabak B, Lawrence MS, Zhsng CZ, Wala J, Mermel CH, Sougnez C, Gabriel SB, Hernandez B, Shen H, Laird PW, Getz G, Meyerson M, Beroukhim R (2013) Pan-cancer patterns of somatic copy number alteration. Nat Genet 45(10):1134–1140
2. Ganem NJ, Storchova Z, Pellman D (2007) Tetraploidy, aneuploidy and cancer. Curr Opin Genet Dev 17(2):157–162
3. Shackney SE, Smith CA, Miller BW, Burholt DR, Murtha K, Giles HR, Ketterer DM, Pollice AA (1989) Model for the genetic evolution of human solid tumors. Cancer Res 49(12): 3344–3354
4. Lens SMA, Medema RH (2019) Cytokinesis defects and cancer. Nat Rev Cancer 19(1): 32–45
5. Carter SL, Cibulskis K, Helman E, McKenna A, Shen H, Zack T, Laird PW, Onofrio RC, Winckler W, Weir BA, Beroukhim R, Pellman D, Levine DA, Lander ES, Meyerson M, Getz G (2012) Absolute quantification of somatic DNA alterations in human cancer. Nat Biotechnol 30(5):413–421
6. Bielski CM, Zehir A, Penson AV, Donoghue MTA, Chatila W, Armenia J, Chang MT, Schram AM, Jonsson P, Bandlamudi C, Razavi P, Iyer G, Robson ME, Stadler ZK, Schultz N, Baselga J, Solit DB, Hyman DM, Berger MF, Taylor BS (2018) Genome doubling shapes the evolution and prognosis of advanced cancers. Nat Genet 50(8): 1189–1195
7. Thomas R, Marks DH, Chin Y, Benezra R (2018) Whole chromosome loss and associated breakage-fusion-bridge cycles transform mouse tetraploid cells. EMBO J 37(2): 201–218
8. Jusino S, Fernández-Padín FM, Saavedra HI (2018) Centrosome aberrations and chromosome instability contribute to tumorigenesis

and intra-tumor heterogeneity. J Cancer Metastasis Treat 4:43

9. Zhang S, Nguyen LH, Zhou K, Tu HC, Sehgal A, Nassour I, Li L, Gopal P, Goodman J, Singal AG, Yopp A, Zhang Y, Siegwart DJ, Zhu H (2018) Knockdown of Anillin Actin binding protein blocks cytokinesis in hepatocytes and reduces liver tumor development in mice without affecting regeneration. Gastroenterology 154(5):1421–1434
10. Ganem NJ, Cornils H, Chiu SY, O'Rourke KP, Arnaud J, Yimlamai D, Thery M, Camargo FD, Pellman D (2014) Cytokinesis failure triggers hippo tumor suppressor pathway activation. Cell 158(4):833–848
11. Kuffer C, Kuznetsova AY, Storchova Z (2013) Abnormal mitosis triggers p53-dependent cell cycle arrest in human tetraploid cells. Chromosoma 122(4):305–318
12. Wong C, Stearns T (2005) Mammalian cells lack checkpoints for tetraploidy, aberrant centrosome number, and cytokinesis failure. BMC Cell Biol 6(1):6
13. Zhou X, Merchak K, Lee W, Grande JP, Cascalho M, Platt JL (2015) Cell fusion connects oncogenesis with tumor evolution. Am J Pathol 185(7):2049–2060
14. Wen B, Li R, Cheng K, Li E, Zhang S, Xiang J, Wang Y, Han J (2017) Tetraploid embryonic stem cells can contribute to the development of chimeric fetuses and chimeric extraembryonic tissues. Sci Rep 7(1):3030
15. Orth JD, Loewer A, Lahav G, Mitchison TJ (2012) Prolonged mitotic arrest triggers partial activation of apoptosis, resulting in DNA damage and p53 induction. Mol Biol Cell 23(4): 567–576
16. Panopoulos A, Pacios-Bras C, Choi J, Yenjerla M, Sussman MA, Fotedar R, Margolis RL (2014) Failure of cell cleavage induces senescence in tetraploid primary cells. Mol Biol Cell 25(20):3105–3118
17. Andreassen PR, Lohez OD, Lacroix FB, Margolis RL (2001) Tetraploid state induces p53-dependent arrest of nontransformed mammalian cells in G1. Mol Biol Cell 12(5): 1315–1328
18. Crockford A, Zalmas LP, Gronroos E, Dewhurst SM, McGranahan N, Cuomo ME, Encheva V, Snijders AP, Begum J, Purewal S, Cerveira J, Patel H, Renshaw MJ, Swanton C (2017) Cyclin D mediates tolerance of genome-doubling in cancers with functional p53. Ann Oncol 28(1):149–156
19. Krzywicka-Racka A, Sluder G (2011) Repeated cleavage failure does not establish centrosome amplification in untransformed human cells. J Cell Biol 194(2):199–207
20. Fujiwara T, Bandi M, Nitta M, Ivanova EV, Bronson RT, Pellman D (2005) Cytokinesis failure generating tetraploids promotes tumorigenesis in p53-null cells. Nature 437(7061): 1043–1047
21. Kuznetsova AY, Seget K, Moeller GK, de Pagter MS, de Roos JA, Durrbaum M, Kuffer C, Muller S, Zaman GJ, Kloosterman WP, Storchova Z (2015) Chromosomal instability, tolerance of mitotic errors and multidrug resistance are promoted by tetraploidization in human cells. Cell Cycle 14(17):2810–2820
22. Zielke N, Edgar BA (2015) FUCCI sensors: powerful new tools for analysis of cell proliferation. Wiley Interdiscip Rev Dev Biol 4(5): 469–487
23. Mayer TU, Kapoor TM, Haggarty SJ, King RW, Schreiber SL, Mitchison TJ (1999) Small molecule inhibitor of mitotic spindle bipolarity identified in a phenotype-based screen. Science 286(5441):971–974
24. Tardif KD, Rogers A, Cassiano J, Roth BL, Cimbora DM, McKinnon R, Peterson A, Douce TB, Robinson R, Dorweiler I, Davis T, Hess MA, Ostanin K, Papac DI, Baichwal V, McAlexander I, Willardsen JA, Saunders M, Christophe H, Kumar DV, Wettstein DA, Carlson RO, Williams BL (2011) Characterization of the cellular and antitumor effects of MPI-0479605, a small-molecule inhibitor of the mitotic kinase Mps1. Mol Cancer Ther 10(12):2267–2275
25. Fry DW, Harvey PJ, Keller PR, Elliott WL, Meade M, Trachet E, Albassam M, Zheng X, Leopold WR, Pryer NK, Toogood PL (2004) Specific inhibition of cyclin-dependent kinase 4/6 by PD 0332991 and associated antitumor activity in human tumor xenografts. Mol Cancer Ther 3(11):1427–1438
26. Yahara I, Harada F, Sekita S, Yoshihira K, Natori S (1982) Correlation between effects of 24 different cytochalasins on cellular structures and cellular events and those on actin in vitro. J Cell Biol 92(1):69–78
27. Vargas-Hurtado D, Brault JB, Piolot T, Leconte L, Da Silva N, Pennetier C, Baffet A, Marthiens V, Basto R (2019) Differences in Mitotic Spindle Architecture in mammalian neural stem cells influence mitotic accuracy during brain development. Curr Biol 29(18): 2993–3005 e9

Chapter 21

Measuring Cellular Ploidy In Situ by Light Microscopy

Delisa E. Clay, Benjamin M. Stormo, and Donald T. Fox

Abstract

Determining cellular DNA content is valuable in the study of numerous biological processes, including organ development and injury repair. While FACS analysis of dissociated cells is a widely used method for assaying ploidy in a tissue cell population, for many tissue samples, it is possible and convenient to measure ploidy in situ using light microscopy. Here, we present two protocols for measuring cellular ploidy in tissues. These protocols are based on our studies in *Drosophila melanogaster*, but these are applicable to other settings as well. We present example results from *Drosophila* hindgut, midgut, and wing imaginal disc as examples. The first protocol focuses on measuring DNA content from decondensed interphase nuclei, while the second protocol details the visualization of condensed chromosomes for ploidy determination, either from mitotic cells or from interphase cells with drug-induced chromosome condensation. These techniques can be completed in 1 day and require standard lab supplies as well as a fluorescence light microscope.

Key words Cellular ploidy, Polyploidy, Microscopy

1 Introduction

Polyploidy, or whole genome duplication, is a widespread tissue property throughout evolution. As an example of its prevalence, polyploidy is now appreciated to occur in at least nine out of eleven human organ systems [1]. Polyploidy often arises in specific cell types during tissue development or following tissue injury. Cellular ploidy levels can differ within an organism by over 1000-fold, and the degree of polyploidy can have important biological consequences [2, 3].

Here, we present two protocols for measuring cellular ploidy in situ. Both protocols use physical force-based tissue squashing of DAPI-stained nuclei to facilitate consistent and accurate DNA content measurements by fluorescent light microscopy. These protocols should be applicable to any tissue sample that does not require sectioning and can be flattened using physical pressure as described here. This protocol is adapted from other long-used

Yves Van de Peer (ed.), *Polyploidy: Methods and Protocols*,
Methods in Molecular Biology, vol. 2545, https://doi.org/10.1007/978-1-0716-2561-3_21,

protocols in *Drosophila* for visualizing polytene chromosomes or mitotic neuroblast metaphase spreads [4–6]. In general, we find that tissues of 50–100 um thickness can be compressed to 10–15 um thickness using this protocol. This reduction of thickness both minimizes and evens out the distance of each nucleus to the objective, which is critical when using light intensity to determine DNA content. Siliconized coverslips are used, which cause liquid to bead easily on the coverslip. This facilitates transfer of the tissue to a positively charged slide during the squash process. Subheading 3.1 is for measuring the DNA content of populations of interphase nuclei. This protocol requires a known tissue ploidy standard such as haploid sperm. Subheading 3.2 is for visualizing condensed chromosomes, either from preparations containing mitotic chromosomes or for interphase cells expressing the phosphatases PP2A and PP1, which can be inhibited by the drug Calyculin A to cause premature chromosome condensation [7, 8]. We have used both protocols to assay ploidy previously [8–12]. Both protocols can be completed in a short amount of time and rely on light microscopy to determine DNA content. These protocols should be of interest to anyone interested in measuring ploidy in situ in samples that are amenable to physical flattening using the procedures described here.

2 Materials

2.1 Siliconized Coverslips

Pour several milliliters of Sigmacote (Sigma) into a small plastic container, such as the lid of a pipette tip container. Using forceps, transfer several 18 × 18 cm coverslips into the Sigmacote solution. Leave for 5 min. Set the coverslips to dry by leaning them at a 45-degree angle against the side of an Eppendorf tube rack. Dried siliconized coverslips can be stored for several months.

2.2 Solutions

For all protocols:

0.5% Na Citrate

1× PBS (diluted in H_2O from 10× stock [13]).

95% EtOh in Coplin jar at −20 °C.

For Subheading 3.1*:*

3.7% Paraformaldehyde (Diluted from 37% Stock Using Phosphate-Buffered Saline, PBS)

For Subheading 3.2*:*

11:11:2 MeOH:Acetic Acid:H_2O.

45% acetic acid.

200 nM Calyculin A (Cell Signaling Technology, diluted in 1× PBS, for interphase chromosome visualization only).

50 ug/mL colcemid (Sigma, diluted in 1× PBS, if an enrichment in metaphase cells is desired).

2.3 Squashing Supplies

Dissection supplies as needed (e.g., forceps, dissection dish).

Platform shaker.

Dewar of liquid nitrogen.

Tongs for handling slides.

Positively charged slides (Superfrost Plus, VWR).

Standard and gel loading pipette tips.

Metal dissection probe (Fisher).

Paper towels.

Vise (Avenger Gold, Penn Tool Co).

Razor blade.

2.4 Staining Supplies

5 ug/mL DAPI.

1× PBS.

Mounting media (Vectashield, Vector Labs).

Clear nail polish.

22 × 22 coverslips.

3 Methods

3.1 Tissue Squash Protocol for Measuring Nuclear Ploidy from Decondensed Interphase Nuclei

3.1.1 Tissue Preparation

1. Dissect tissue of interest with forceps in 1× PBS. Remove any excess tissue not needed for ploidy measurements, as is possible. If the ploidy of your tissue sample is unknown, also include a known ploidy standard. A convenient standard for *Drosophila* tissue is the testis, which contains elongated haploid sperm nuclei (Fig. 1).
2. Transfer tissue to a siliconized coverslip with 50 uL of Na citrate for 10 min. This serves to swell the tissue and can impact the final size of the nuclei in your sample.
3. While viewing your tissue under a dissecting microscope, remove as much of the Na citrate as possible without drying out the tissue, using a pipette with a gel loading tip (Fig. 2a).
4. Quickly add 0.5 mL 37% paraformaldehyde fix solution. Leave tissue in fix for 15 min.
5. While viewing your tissue under a dissecting microscope, remove most of the fix, using a pipette with a standard pipette tip.
6. Replace with 50 uL 1× PBS as a quick wash.

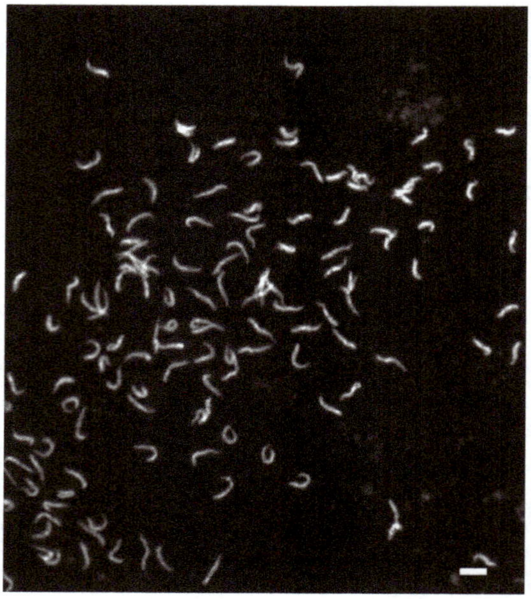

Fig. 1 DAPI-stained haploid sperm nuclei. Scale bar = 20 microns

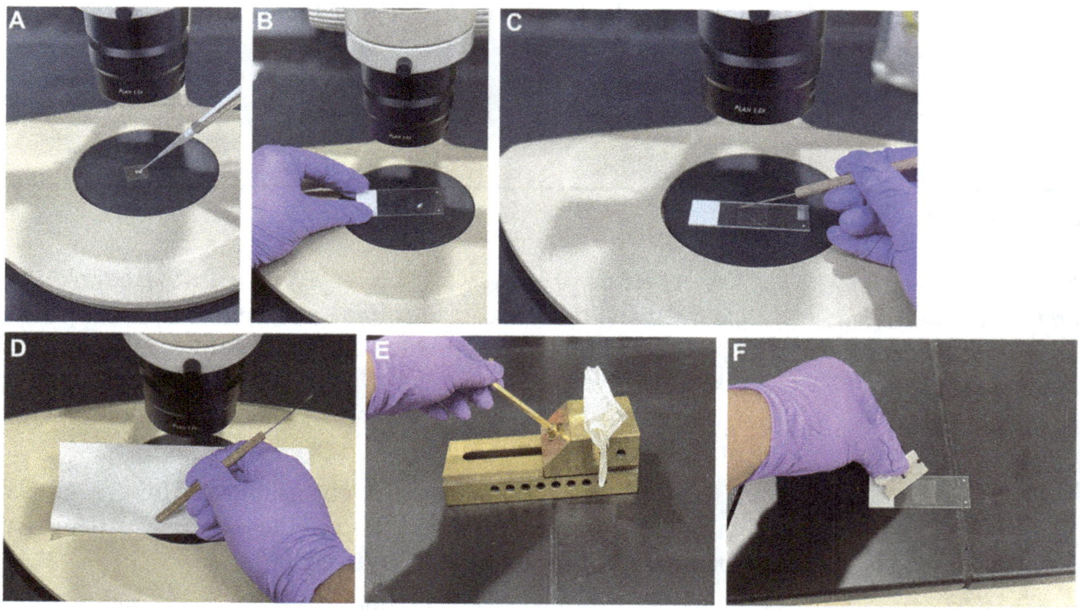

Fig. 2 The squash procedure. (**a**) Aspirating a bead of liquid on a siliconized coverslip with a gel loading tip. (**b**) Picking up the coverslip with a positively charged slide. (**c**) Tracing a grid over the coverslip with a metal probe. (**d**) Tracing a grid on the bottom of the slide (covered with a paper towel) with a probe handle. (**e**) Placing the slide in the vise. (**f**) Removing the coverslip with a razor blade

7. While viewing your tissue under a dissecting microscope, remove as much of the 1× PBS as possible using a pipette with a gel loading tip.
8. Replace with 7–10 uL of fresh 1× PBS.
9. Gently place a charged slide (charged slide facing down) over the coverslip with the drop of PBS. The coverslip should adhere to the slide, and you can then flip the slide/coverslip over (Fig. 2b).
10. With one hand, apply very gentle pressure to opposite corners of the coverslip using your index and middle fingers.
11. Use the needle of the probe to trace a grid-like pattern first horizontally and then vertically across the coverslip. Apply gentle pressure to the coverslip with the probe needle as you do this step (Fig. 2c).
12. Flip the slide over onto a paper towel. Place one finger at opposite ends of the slide and apply gentle pressure. Next, through the paper towel, take the handle end of the dissecting probe and again trace a grid-like pattern over the portion of the slide bottom that is opposite the coverslip on the other side. These **steps (10–13)** prevent slippage of the coverslip during squashing (Fig. 2d).
13. Fold the paper towel in half and place in the vise. The coverslip should be facing away from the handle side of the vise (Fig. 2e).
14. Gradually apply increased pressure to the slide. As a general rule, you should continue to increase pressure until it no longer becomes possible to easily tighten the vise handle. Leave the slide for 1–2 min in the vise.
15. Release the vise pressure and remove the slide.
16. Using tongs, submerge the slide in liquid nitrogen. After about 5 s, the bubbling of the nitrogen will stop.
17. Remove the slide and place coverslip up on a paper towel.
18. Quickly pick up the razor blade. While holding the slide down, pry one corner of the coverslip off with the razor blade using a flicking motion (flicking the blade away from you and your fingers (Fig. 2f).
19. Quickly transfer the slide to the Coplin jar containing −2 °C EtOH. The coverslip can be reused if still in one piece (remove all pieces of the coverslip if shattered during removal).
20. Slides can be stored in EtOH for 1 h to several days. Upon removal, allow the slide to air dry.
21. Rinse the tissue on the slide by pipetting 1 mL of 1× PBS and then quickly removing. Repeat 2–3 times.

22. Add 1 mL of 1× PBS + 5ug/mL DAPI for 5 min to the slide. Incubate in the dark for 15 min. To minimize the chance of the solution/sample from drying out, enclose the slide in a sealed Tupperware container with a moist paper towel.
23. Rinse the tissue on the slide by pipetting 1 mL of 1× PBS and then quickly removing. Repeat 2–3 times.
24. Apply a drop of mounting media to the tissue and place a 22 × 22 cm coverslip over the drop.
25. Seal the coverslip with nail polish.

3.1.2 Imaging and Ploidy Measurement

26. For each individual slide, image on an epifluorescence or confocal microscope. We typically use a 40× or 60× oil objective. Image a few candidate polyploid cells to determine an exposure setting that does not contain any saturated pixels. Set this exposure setting to be constant while imaging each slide. Remeasure again for each new slide.
27. Image several candidate polyploid cells, taking Z-stacks as needed to capture the entire nucleus. Due to the squash, your sample thickness should not be greater than 10–15 microns. We recommend a Z-stack step size of 1 micron.
28. Using the same exposure settings, image your internal ploidy control (e.g., elongated spermatid nuclei, Fig. 1) as well as nuclei in your tissue of interest.
29. Continue **steps 26–28** for all slides in an experiment.
30. Once imaging for an experiment is complete, open each Z-stack in ImageJ (ImageJ.nih.gov). Create a Z-projection for the DAPI channel using the "Sum-slices" option.
31. Within each projection, use the freehand ROI indicator to outline each nucleus (Fig. 3a). Only measure nuclei which are not on top/underneath a neighboring nucleus.
32. After drawing an ROI for an individual nucleus, use the "measure" function. Make sure that "Integrated Density" is being recorded.
33. Repeat **steps 31–32** for all nuclei that can be possibly measured in each Z-stack image. Measure both candidate polyploid cells and your known ploidy standard. Keep track of which nuclear measurements are for each cell type.
34. When you have completed imaging for a given slide, export the measurements into a database program (e.g., Excel). Transfer your notes about which measurements go with which cell type.
35. For the ploidy standard nuclei from each slide, calculate the median-integrated density value. Set the value for this internal standard to reflect the C value of the standard population (e.g., for haploid cells, set a value of 1.0, reflecting 1C). Divide all

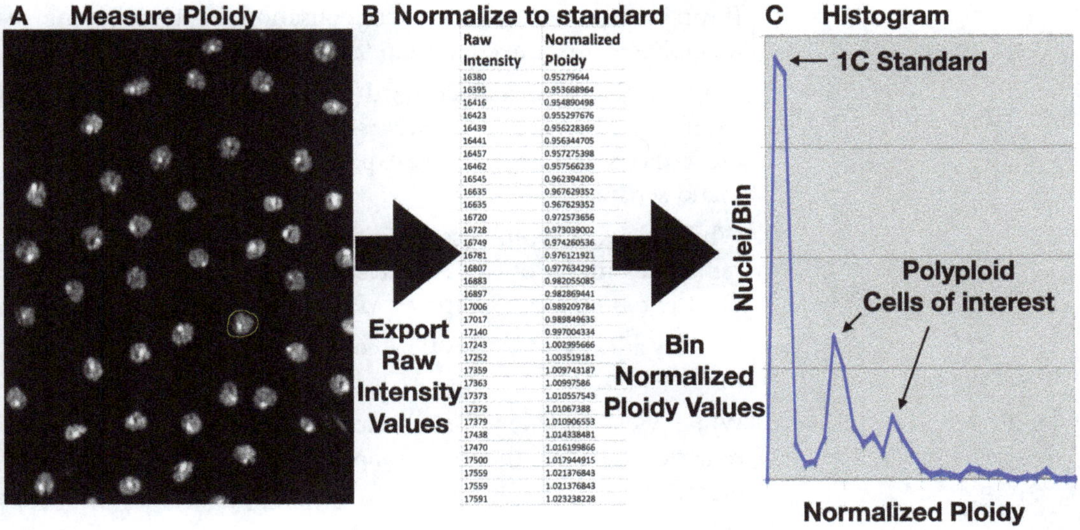

Fig. 3 Producing ploidy histogram data from samples prepared using Subheading 3.1. (**a**) ROI selection (yellow) of a DAPI-stained nucleus of interest in ImageJ. (**b**) Example range of raw integrated density and converted ploidy data (**c**), normalized to a haploid standard. (**c**) An example histogram plot that includes both a 1C standard and polyploid cells of interest

other nuclei measurements for the standard sample by the C value for the median value (Fig. 3b). This should give you a fairly narrow range of values. For a 1C cell population, we aim to have a range of values all between 0.5 and 1.5, though even narrower is better.

36. For the candidate polyploid nuclei, normalize all nuclei to the value of the median integrated density value of the ploidy standard.
37. Repeat **steps 34–36** for all slides in an experiment. This will now give you a set of values normalized to C, which enables you to combine data from different slides into your final dataset.
38. When your data analysis is complete, create a histogram from your data. Try several binning ranges to generate your desired plot (Fig. 3c).

3.2 Tissue Squash Protocol for Measuring Nuclear Ploidy from Condensed Chromosomes

3.2.1 Tissue Preparation

1. Dissect tissue as in Subheading 3.1, **step 1**.
2. If inducing premature chromosome condensation of interphase nuclei is desired, incubate tissue in Calyculin A for 30 min in a dissecting dish on a platform shaker for 30 min. If an enrichment of metaphase cells is desired, incubate tissue in colchicine for 30 min in a dissecting dish on a platform shaker for 30 min. Shake at a speed that gently circulates the solution.

If premature chromosome condensation or enrichment of metaphases is not needed, skip to the next step.

3. Transfer tissue to a siliconized coverslip with 50 uL of Na citrate for 10 min. This serves to swell the tissue. Longer incubations in Na citrate can experimentally induce sister chromatid separation.
4. While viewing your tissue under a dissecting microscope, remove as much of the Na citrate as possible without drying out the tissue, using a pipette with a gel loading tip (Fig. 2a).
5. Quickly add 11:11:2 MeOH:Acetic Acid:H_2O fix solution. Fix for 10–20 s.
6. While viewing your tissue under a dissecting microscope, remove most of the fix, using a pipette with a standard pipette tip.
7. While viewing your tissue under a dissecting microscope, remove as much of the fix as possible without drying out the tissue, using a pipette with a gel loading tip.
8. Add 7 microliters 45% acetic acid. Wait 1–2 min.
9. Repeat **steps 9–25** of Subheading 3.1 to squash the tissue (Fig. 2b–f).

3.2.2 Imaging and Chromosome Visualization

10. Image as described in Subheading 3.1, with the exception that Z-stacks may not be necessary, as one should focus on taking single images of each nucleus of interest where all chromosomes are in focus. The above protocol should enable visualization of condensed chromosomes. Ploidy analysis from such images is done simply by counting chromosome number.
11. Counting of individual chromosomes can be done from images to score for degree of polyploidy (Fig. 4) and/or aneuploidy. Each chromosome in a species typically has a distinct morphology [4]. N value can be determined by counting the number of chromosomes of each type, while C value can be determined by the number of distinct chromatids that are visible for each chromosome type. *Drosophila melanogaster* has a haploid number of 4 chromosomes and is a diploid organism. Due to both homologous chromosome pairing and the polytene nature of chromosomes produced by endocycles, it is possible for a cell of high ploidy to have a low N value. This is due to all chromosomes being part of a single giant polytene chromosome [14].

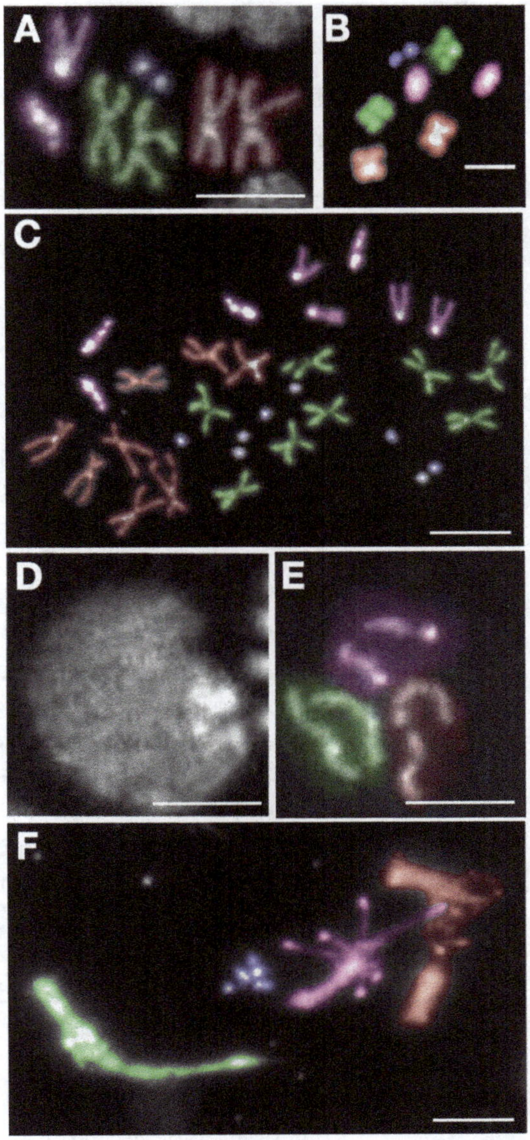

Fig. 4 Condensed chromosome imaging of *Drosophila melanogaster* cells, prepared using Subheading 3.2. (**a**) A metaphase spread from a 2 N/4C XY mitotic rectal epithelial cell. (**b**) A metaphase spread from a 2 N/4C XX mitotic midgut intestinal stem cell. (**c**) A metaphase spread from an 8 N/16C XY hindgut rectal papillar cell. (**d**) An interphase wing imaginal disc cell without Calyculin A. (**e**) An interphase XX wing imaginal disc cell treated with Calyculin A to reveal 2 N/2C ploidy (note the small fourth chromosomes are not present in this image). (**f**) An interphase hindgut rectal papillar cell of similar ploidy to the cell in (**c**), treated with Calyculin A. The centromeric regions on each chromosome become unpaired in this assay [8]. For all images where distinct chromosomes are seen, the chromosome type is labeled. Chromosome types are pseudo-colored as follows: X/Y, purple; second or third autosomes, green or red; autosome 4, blue. Scale bar = 25 microns

4 Notes

1. If you do not have a vise, you may apply strong pressure using your thumbs. However, in our experience, this results in more variability in tissue appearance between tissue preparations and between users.
2. It is generally more effective to use a smaller amount of tissue on many coverslips, rather than a large amount of tissue on a single coverslip. As a general guideline, we place 200 × 200 micron-sized tissue dissections from 2 to 3 *Drosophila melanogaster* per coverslip along with one adult testis for an internal ploidy control.
3. Cracking the slide, either during vise tightening or following submersion into liquid nitrogen, is a sign that you are tightening the vise too much.
4. Frequent reuse of siliconized coverslips (more than 3–4 times) usually leads to poor transfer of tissue to the slide and/or coverslip cracking. Make new siliconized coverslips as needed.
5. In our experience, it is easy to see the area where the 18 × 18 coverslip was after its removal. However, if this is not obvious to the experimenter, you may consider drawing the location of the coverslip on the bottom of the slide opposite the coverslip with a wax pencil.
6. Please be advised that any slide labeling should be done in pencil prior to removal from EtOH.
7. For Subheading 3.1, if the range of measured values for your ploidy standard exceeds +/− 50% of the median, this is a sign that your squash protocol did not work well and that the tissue is likely under-squashed. Try increasing the pressure on the vise.
8. For Subheading 3.1, in our experience, N values of 100 nuclei per tissue type are required to accurately reflect tissue ploidy distribution as visualized in a histogram.
9. For Subheading 3.1, in our experience, GFP fluorescence from a transgenic animal, expressed in a tissue of interest, generally survives the 3.7% fixation and squash protocol. This is not the case for Subheading 3.2, however.
10. For Subheading 3.2, we note that for *Drosophila* samples with prematurely condensed chromosomes, chromosome pairing may prevent resolution of N value, but will enable visualization of each chromosome type, including in cells with polytene chromosomes (Fig. 4d–f).

11. For Subheading 3.2, we caution against relying solely on absence of chromosomes from mitotic preparations as a sign of aneuploidy, as chromosomes can physically dissociate from nuclei during the squash procedure. Chromosome gains, in contrast, can be reliably scored as a readout of aneuploidy, and if the frequency of perceived chromosome losses is similar to that of chromosome gains, then the chromosome losses may then be reliably scored as a sign of aneuploidy.
12. Subheadings 3.1 and 3.2 can be adapted for antibody staining. The success of antibody staining in visualizing your protein of interest is highly dependent on the antibody. Many epitopes do not survive the freezing protocol or the MeOH fixation step in particular. We recommend trying antibody staining both before and after the squash procedure to identify optimal conditions for antibody signal. Optimal concentrations of primary antibody may differ from standard (non-squash) tissue preparations.
13. Subheadings 3.1 and 3.2 can be adapted for in situ hybridization. This may require different fixation conditions.

Acknowledgments

Study of polyploidy in the Fox laboratory is supported by a National Institutes of Health (NIH) grant R01GM118447. Delisa Clay is supported by a GRFP fellowship from the National Science Foundation (NSF). We thank Archan Chakraborty for comments on the manuscript.

References

1. Peterson NG, Fox DT (2021) Communal living: the role of polyploidy and syncytia in tissue biology. Chromosom Res. in press
2. Øvrebø JI, Edgar BA (2018) Polyploidy in tissue homeostasis and regeneration. Development 145:dev.156034
3. Fox DT, Soltis DE, Soltis PS et al (2020) Polyploidy: a biological force from cells to ecosystems. Trends Cell Biol 30:688–694
4. Gatti M, Bonaccorsi S, Pimpinelli S (1994) Looking at drosophila mitotic chromosomes. Methods Cell Biol 44:371–391
5. Kennison JA (2008) Dissection of larval salivary glands and polytene chromosomes preparation. Cold Spring Harb Protoc 3:pdb.prot4708
6. Johansen KM, Cai W, Deng H et al (2009) Polytene chromosome squash methods for studying transcription and epigenetic chromatin modification in Drosophila using antibodies. Methods 48:387–397
7. Gotoh E, Asakawa Y, Gotoh E, Kosaka H (1995) Inhibition of protein serine/threonine phosphatases directly induces premature chromosome condensation in mammalian somatic cells. Biomed Res 16:63–68
8. Stormo BM, Fox DT (2016) Distinct responses to reduplicated chromosomes require distinct Mad2 responses. elife 5:e15204
9. Fox DT, Gall JG, Spradling AC (2010) Error-prone polyploid mitosis during normal Drosophila development. Genes Dev 24:2294–2302
10. Schoenfelder KP, Montague RA, Paramore SV et al (2014) Indispensable pre-mitotic endocycles promote aneuploidy in the Drosophila rectum. Development 141:3551–3560

11. Stormo BM, Fox DT (2019) Interphase cohesin regulation ensures mitotic fidelity after genome reduplication. Mol Biol Cell 30:219–227
12. Cohen E, Allen SR, Sawyer JK, Fox DT (2018) Fizzy-related dictates a cell cycle switch during organ repair and tissue growth responses in the Drosophila hindgut. elife 7:e38327
13. (2007) 10X PBS. Cold Spring Harb Protoc. doi:https://doi.org/10.1101/pdb.rec10768
14. Stormo BM, Fox DT (2017) Polyteny: still a giant player in chromosome research. Chromosom Res 25:201–214

Chapter 22

Using Mosaic Cell Labeling to Visualize Polyploid Cells in the *Drosophila* Brain

Shyama Nandakumar and Laura Buttitta

Abstract

Traditional methods used to study endoreplication have limitations when used to identify rare events of polyploidization in complex, densely-packed tissues. Here, we describe a method to identify and visualize polyploid cells in situ using an existing mosaic, multicolor labeling technique named "CoinFLP" (Bosch et al., Development 142(3):597–606, 2015). CoinFLP allows easy visualization of polyploid cells *in situ* and can be combined with other techniques such as immunofluorescence for cell-type-specific labeling and flow cytometry to perform quantifications and can also be used for genetic manipulations. Further, by modifying the time of labeling, this technique can also be used to distinguish events of cell fusion from endocycle (Nandakumar et al., eLife 25:9, 2020)—allowing one to infer the method of polyploidization.

Key words Polyploidy, Endocycle, Endoreplication, Cell fusion, Mosaic labeling

1 Introduction

The study of endopolyploidy has progressed significantly in the last few decades. Early studies relied on histochemistry and radioactive isotope incorporation to observe and measure changes in levels of polyploidy. Advances in technology have allowed for more accurate and sensitive methods to be adopted such as high-resolution microscopy, flow cytometry, and the incorporation of thymidine analogues such as BrdU and EdU. While these methods continue to significantly advance polyploidy research, they have some limitations.

In densely packed, thick tissues with heterogeneous cell populations, traditional methods such as microscopy-based measurement of nuclear volume and DNA can become challenging. In tissues such as the adult *Drosophila melanogaster* brain, performing tissue squashes to measure DNA content of polyploid neurons is

Yves Van de Peer (ed.), *Polyploidy: Methods and Protocols*,
Methods in Molecular Biology, vol. 2545, https://doi.org/10.1007/978-1-0716-2561-3_22,

both technically challenging and cumbersome. Tissue squashes can destroy tissue structure and make it difficult to interpret location in complex organs, such as the brain. This becomes even more challenging when trying to detect rare populations of cells.

The use of thymidine analogs such as EdU or BrdU also have specific limitations. EdU and BrdU are known to be toxic to some cell types [1, 2]. Additionally, the timing of administration is critical—it is easy to miss slowly or transiently endocycling cells. Furthermore, it has been shown that only cells that have an active nucleotide salvage pathway are able to successfully incorporate these molecules into newly synthesized DNA [3].

While flow cytometry can be used to accurately measure even rare populations of polyploid cells, this technique has one major drawback. The preparation of tissue samples for flow cytometry requires a dissociation step and results in the loss of spatial information.

Here, we introduce the use of genetically encoded, binary lineage tracing as an approach to detect polyploid cells *in situ*. We use the transgenic mosaic labeling tool named CoinFLP which relies on two overlapping and exclusive flippase recombination target (FRT) sites [4] (Fig. 1). When the transgenic cassette

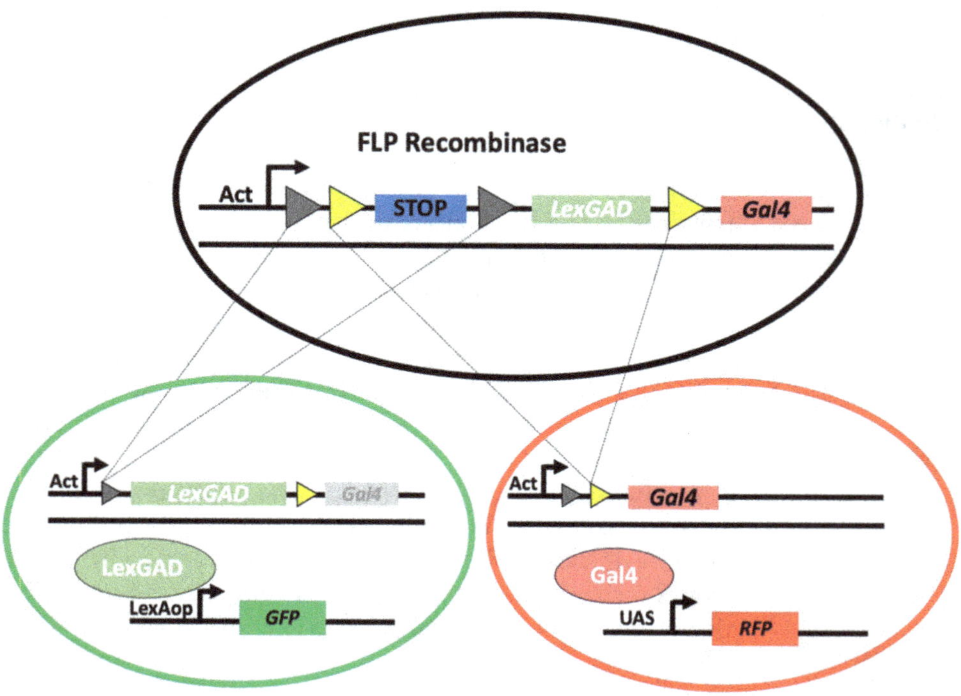

Fig. 1 CoinFLP scheme. CoinFLP uses overlapping and exclusive FRT sites (FRT in yellow and FRT3 in gray) that can be recombined to drive the expression of either LexGAD or a Gal4 driver in diploid cells. (Figure adapted from and based on the original publication describing CoinFLP [4])

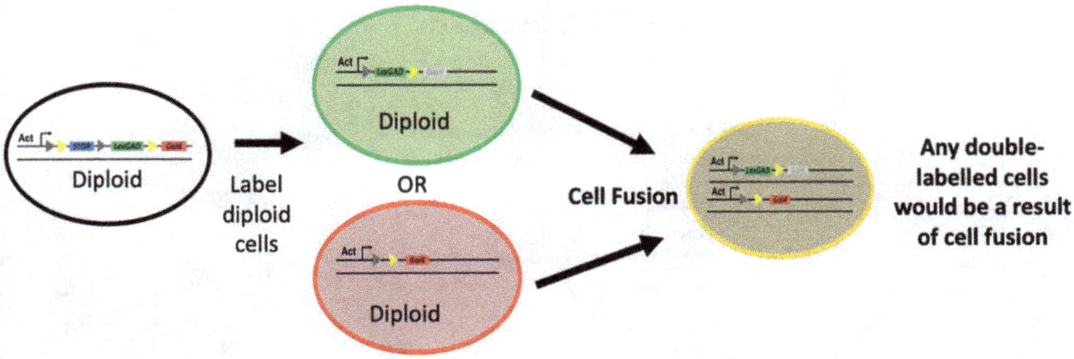

Fig. 2 "Early-FLP" paradigm to detect cell fusion in adult brains. Robust flipping is induced via heat-shock early in development when all cells are diploid to ensure nearly all cells are labeled exclusively with a red or a green fluorescent protein. Cell fusion events in the adult brain can be detected by the presence of cells that are red and green (yellow)—which result from the fusion of a red cell with a green cell

carrying the CoinFLP construct is in a heterozygous state in a diploid cell, it results in the cell being labeled only one of two colors. When cells in an otherwise diploid animal become polyploid, the multiple copies of the construct create the opportunity to label cells with multiple color states, allowing for the detection of a fraction of rare polyploid cells in situ. We describe below how to use the binary color system CoinFLP to detect polyploid cells in the adult *Drosophila* brain [5, 6]. We also demonstrate how this technique can be used to distinguish between events of cell fusion (Fig. 2) and cell cycle reentry (Fig. 3), by altering the timing of the somatic recombination to label cells.

Specific advantages to this approach for labeling polyploid cells in *Drosophila* are as follows:

- It works in any tissue or cell type that becomes polyploid by endomitosis, endoreplication, or cell fusion.
- One can infer the timing and method of polyploidization by altering the timing of the heat-shock and therefore induction of the labeling.
- It can detect rare events of polyploidy.
- Double-labeled polyploid cells co-express the transcriptional activators Gal4 and LexGAD, allowing for intersectional activation of transgenes.

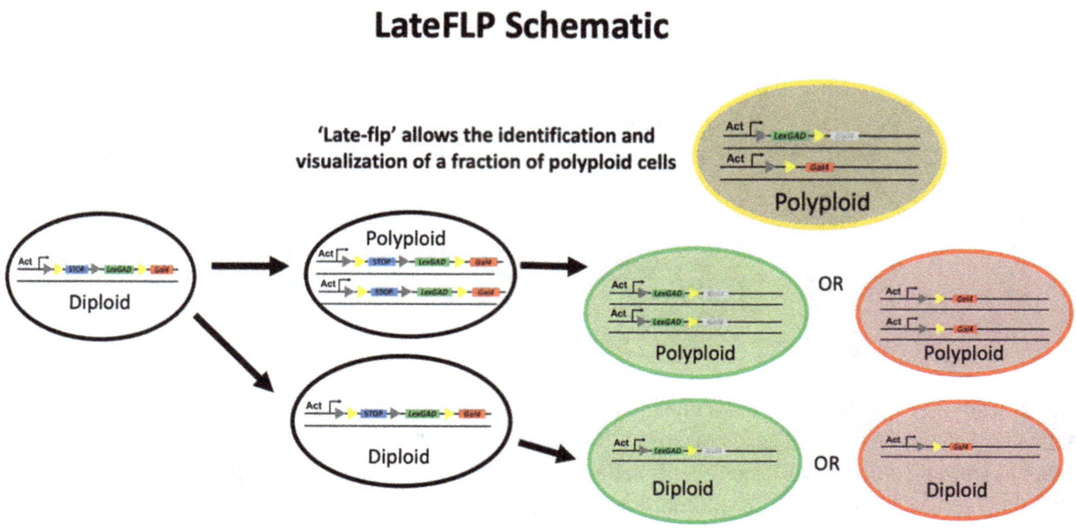

Fig. 3 "Late-FLP" paradigm to detect cell cycle reentry in adult brains. Unlabeled cells that become polyploid have multiple copies of the CoinFLP cassette, while diploid cells only have one copy. When labeling is induced via heat-shock after cells become polyploid, cells will have multiple copies of the transgenic construct. A polyploid cell can "flip" each copy of the transgene either "red" or "green." Thus, some polyploid cells can flip both red and green (yellow), while diploid cells can only flip red OR green. Since the adult fly brain exhibits very low levels of cell fusion, we can infer that most polyploid cells in the adult brain result from cell cycle reentry and endoreplication

2 Materials

1. Fly stocks used.

Genotype	Source
y,w,UAS-mCD8::RFP,LexAop-mCD8::GFP; CoinFLP-LexA::GAD.Gal4	BDSC 58754
y,w,hs-FLP;+;+	Buttitta Lab, derived from BDSC 1929
y,w,hs-FLP;LexAop-RFPnls/CyO-GFP; UAS-GFPnls/TM6B	Butitta Lab, derived from BDSC 29956 and BDSC 84278

2. *Drosophila* culturing vials with fly food, vial closures, and baker's yeast. The recipe used in this protocol is the Bloomington Cornmeal formulation without malt extract.
3. Fly anesthesia equipment: CO_2 source, fly pad.
4. Water bath for performing heat-shock.
5. 10X phosphate-buffered saline (PBS) pH 7.2 (1.4 M NaCl, 2.6 mM KCl,17.5 mM potassium phosphate, 120 mM sodium phosphate, pH ~7.2).

6. 1X PBS pH 7.2 made from 10X PBS stock with dH_2O.
7. 16% paraformaldehyde (PFA) (Electron Microscopy Sciences).
8. Deionized or milliQ water (dH2O).
9. PBS + Triton-X-100 0.1% (v/v) (PBS-T).
10. PBS + Triton-X-100 0.5% (v/v) (PBS-T).
11. PBS + Triton-X-100 0.1% (v/v) + 1% bovine serum albumin (BSA) (PAT).
12. PBS-T + 2% normal goat serum (NGS) (PBT-X).
13. Primary and secondary antibody for cell type-specific identification or labeling of nuclear lamina (e.g., anti-Repo to identify glia, anti-Elav to identify neurons, both from Developmental Studies Hybridoma Bank).
14. Hoechst 33258 or DAPI for fixed tissue nuclear counterstain.
15. Two pairs of sharp No. 5 forceps (Fine Science Tools).
16. Glass dissection dishes (small or large "embryo dish"—Electron Microscopy Sciences).
17. Kimwipes.
18. 10X Trypsin-EDTA (Sigma T4174) for tissue dissociation for analysis by flow cytometry.
19. Vybrant Dye-cycle violet™ for nuclear staining of live-dissociated nuclei (Life Technologies).
20. Standard 1.7 mL microcentrifuge tubes, pipettes and pipette tips, and transfer pipettes for performing washes for immunostaining.
21. Stereo dissecting microscope.
22. Glass microscope slides, coverslips, transparent binder protector rings (Avery), nail polish.
23. Vectashield H-1000 (Vector Labs) for mounting.
24. Epifluorescence microscope with deconvolution software or confocal microscope for high-resolution imaging.
25. Flow cytometer.

3 Methods

3.1 Fly Husbandry and Cell Labeling In Vivo

3.1.1 To Label Cells During Development "Early-FLP": This Will Allow Detection of Fusion Events in the Adult Brain

- Set up fly crosses as described in the schemes in Fig. 4. Crosses are set up in fly vials with 15 males and 15 females per vial. The parental generation is transferred to fresh food with baker's yeast every 2 days.
- 48 h after transferring F0 to a fresh vial, perform heat-shock labeling of the vial with the F1 larvae in a 37 °C water bath. F1 animals will be in the L2 larval stage.

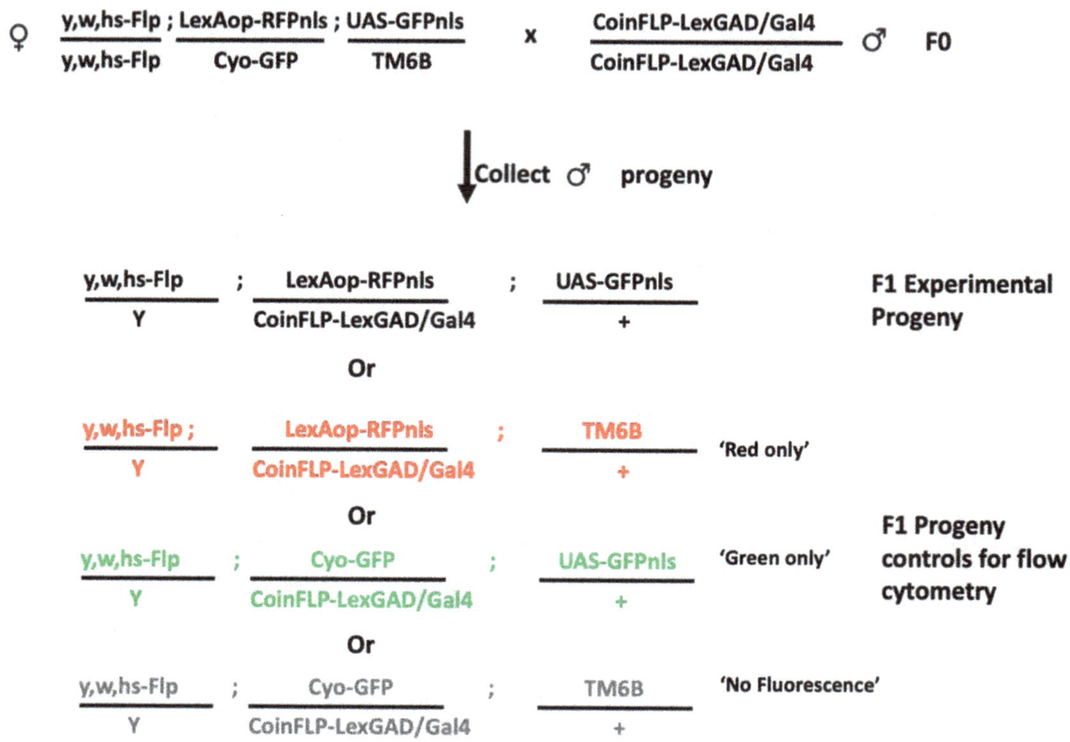

Fig. 4 Genetic crossing scheme. The genetic crosses to perform this protocol are shown. Since our method uses a nuclear GFP and a nuclear RFP which are on separate chromosomes, it is possible to isolate progeny which only express red or green fluorescent protein ("red only" or "green only"). These red and green only progeny can serve as ideal controls for flow cytometry compensation and gating

- Perform heat-shock for 30 min by pressing the vial closure into the vial, 1 cm away from the surface of the food. Submerge the vial in the water bath such that the sides of the vial are entirely in contact with the water. Weights can be used to ensure that the vials are submerged. This ensures that all animals in the vial are exposed to the heat-shock.
- After the heat-shock, pull the vial closure away from the surface of the food back to its normal position, and return the vial to room temperature.
- Allow adults to emerge. Collect emerging adult males daily in fresh vials with fly food without baker's yeast added. Transfer adults to fresh food every 5–7 days while aging.
- Dissect fly tissues at the desired age of adult F1 for further analysis by immunostaining or flow cytometry.

3.1.2 To Label Cells in the Adult "Late-FLP": This Will Allow Detection of Cell Cycle Reentry Events in the Adult Brain

1. Set up fly crosses as described in the schemes in Fig. 4. Crosses are set up in fly vials with 15 males and 15 females per vial. The parental generation is transferred to fresh food with baker's yeast every 2 days.

2. Crosses are maintained at room temperature (22–23 °C in the laboratory).
3. Emerging F1 adult males are collected daily and transferred into fresh vials without baker's yeast sprinkled on the surface.
4. Perform heat-shock at the desired age of the animal to label cells with CoinFLP.
5. Perform heat-shock for 15 min by pressing the vial closure into the vial, 1 cm away from the surface of the food. Submerge the vial in a 37 °C water bath, such that the sides of the vial are entirely in contact with the water. Weights can be used to ensure that the vials are submerged. This ensures that all animals in the vial are exposed to the heat-shock.
6. After the heat-shock, pull the vial closure away from the surface of the food back to its normal position, and return the vial to room temperature.
7. F1 adults are transferred to fresh food every 5–7 days.
8. Dissect fly tissues at the desired age of adult F1 for further analysis by immunostaining or flow cytometry.

3.2 Dissection, Fixation, Immunostaining (for Nuclear Lamina or Cell Type-Specific Markers), and Imaging

1. All incubations are with rotation on a nutator, unless being stored at 4 °C.
2. Dissect in 1X PBS.
3. Fix 30 min at room temp. in 4% paraformaldehyde freshly made from PFA ampules stored at 4 °C, 10X PBS, and dH2O to a final concentration of 4% PFA in 1X PBS.
4. Wash in 1X PBS 0.5% Triton X-100, 2 times 10 min each (can be left in 1X PBS 0.1% Triton X-100 overnight at 4 °C).
5. Incubate tissues for 10 min. in PAT.
6. Allow tissues to settle and remove PAT. Resuspend tissues in fresh PAT containing desired primary antibody diluted to the appropriate concentration. Incubate *either* overnight at 4 °C or 4 h at room temperature (RT).
7. Wash two times in 1X PBS 0.5% Triton X-100, 10 min.
8. Incubate in "PBT-X" with Normal goat serum added to a final conc. of 2% for 10 min.
9. Incubate tissues with the desired secondary antibody (diluted 1:2000 in "PBT-X" +2% goat serum) for 4 h at room temp (this can also be done O/N at 4 °C) the 2° antibodies are extremely light sensitive, so minimize stock solution exposure to light and wrap the incubating tubes in aluminum foil.
10. Wash two times in 1X PBS 0.1% Triton X-100 for 10 min.
11. Incubate with Hoechst 33258 to stain the DNA. Hoechst stock is diluted 1:1000 in PBS-Triton X. Incubate tissues with Hoechst solution for 10 min at RT.

12. Wash three times in PBS-Triton X 100 for 10 min.
13. Place tissues back into the dish w/PBS-Triton X 100. Remove trachea and any excess fat that may remain attached to the brain carefully and put them in a pile. On a slide, place two clear binder protector rings on top of each other creating a "well." Using a cut-off P20 tip, pipet tissues into the well. Using a non-cut-off pipette tip, aspirate excess solution, or use a Kimwipe to wick off excess tissues. Pipet 13 μL of Vectashield around or near tissues, allowing it to flow onto tissues. Tissues will turn translucent in the mounting medium.
14. Carefully cover the tissues with a coverslip and seal with clear fingernail polish.
15. Proceed to image slide on high-resolution microscope. Examples of images of polyploid tissues labeled by CoinFLP are shown in Fig. 5. Figure 6 shows how CoinFLP can be used to visualize polyploid neurons in the *Drosophila* brain.

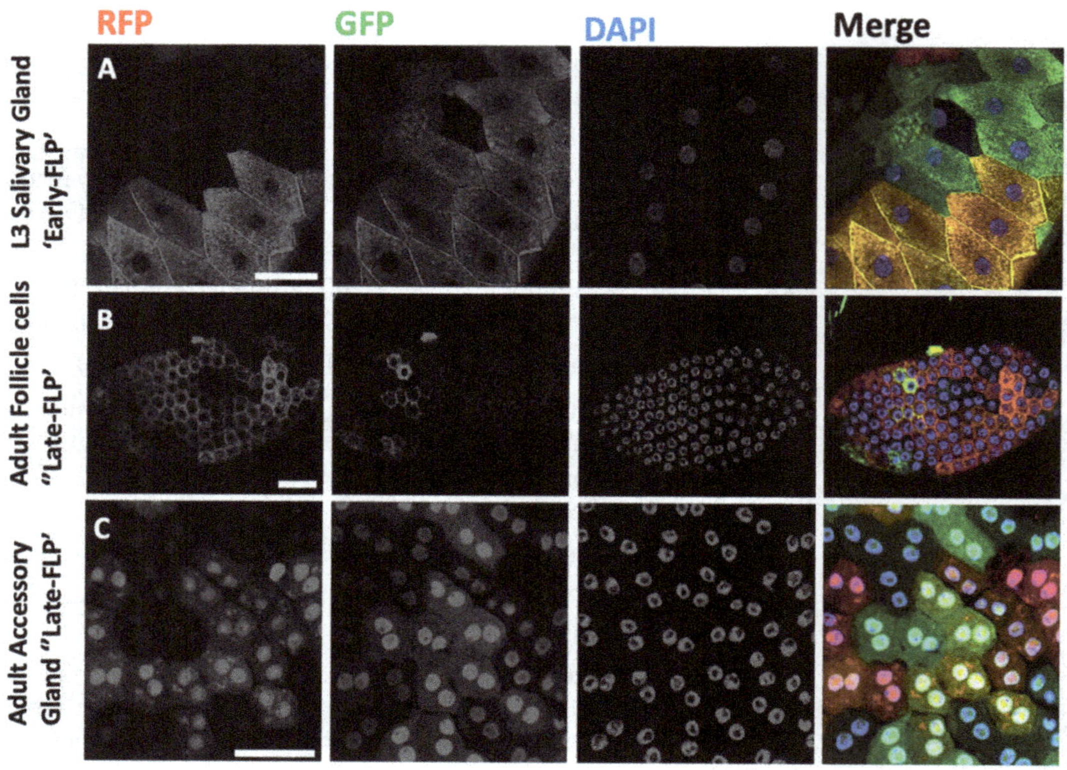

Fig. 5 CoinFLP can label known polyploid cells in larvae as well as adults. (**a**) Larval polytene salivary gland labeled with "early-FLP." Adult (**b**) endocycling follicle cells in the female and (**c**) binucleate accessory gland cells in the male labeled with "late-FLP." Scale bars = 100 μM

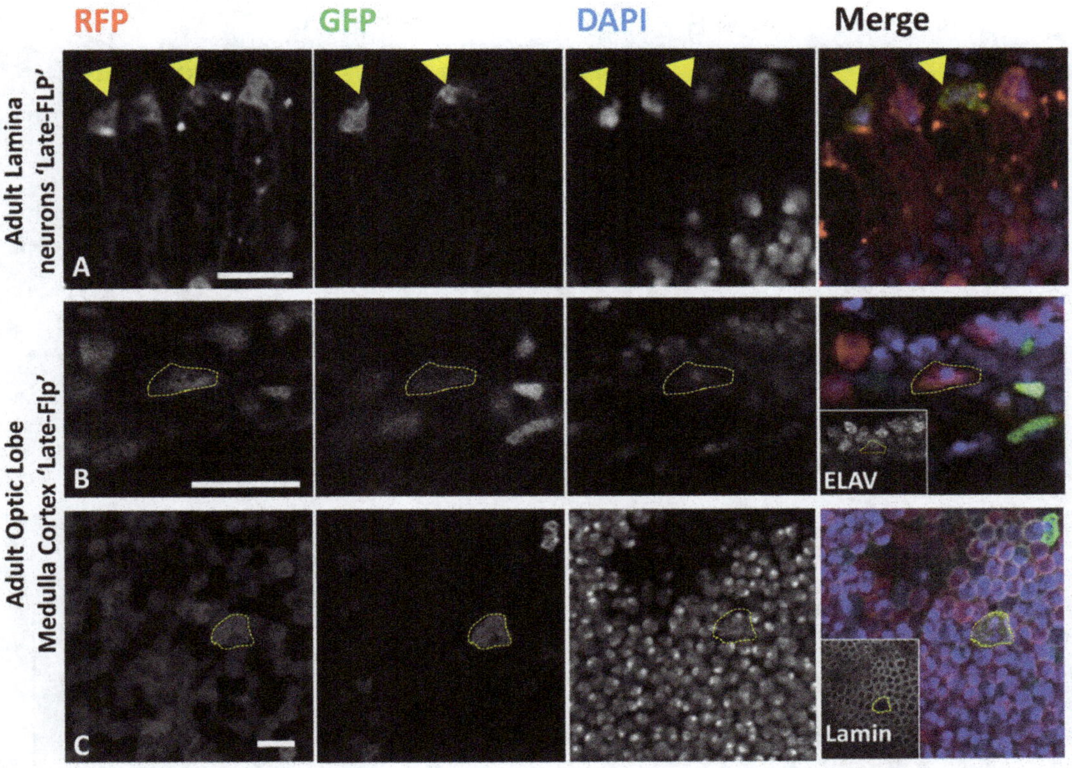

Fig. 6 CoinFLP can be used to identify rare and sparse polyploid cells in the adult brain. Results from a "Late-FLP" experiment to label polyploid cells in the adult brain. (**a**) Polyploid neurons in the fly lamina (yellow arrowheads). (**b**) Polyploid neurons (ELAV positive, inset) in the fly medulla cortex, yellow dotted line outlining cell body. (**c**) Large polyploid cell in the fly medulla cortex counterstained with nuclear lamin (inset) to observe nuclear morphology, yellow dotted line outlining cell body. Scale bars = 10 μM

3.3 Preparing Brain Tissues for Flow Cytometry Analysis

1. Prepare staining solution prior to dissection. 1 mL of staining solution should contain 900 μL of 10X trypsin (aliquoted stored in −20 °C to minimize freeze/thaw cycles), 100 μL of 10X PBS, 2 μL of Vybrant DyeCycle violet.
2. Anaesthetize the flies quickly with a pulse of CO_2 and place on a wet Kimwipe on ice.
3. Dissect adult brain in 1X PBS. Remove as much trachea and adipose tissue as possible. Dissections should be performed within 20 min to limit cell death.
4. Using forceps, carefully transfer the dissected brain to a microcentrifuge cap containing 100 μL of the staining solution. Incubate for 20 min.
5. Using a plastic P200 tip, pressed against the plastic cap, triturate to dissociate the brain.
6. Transfer the 100 μL from the cap with cells to the remaining 400 μL of staining solution in the microcentrifuge tube. This makes a 500 μL total.

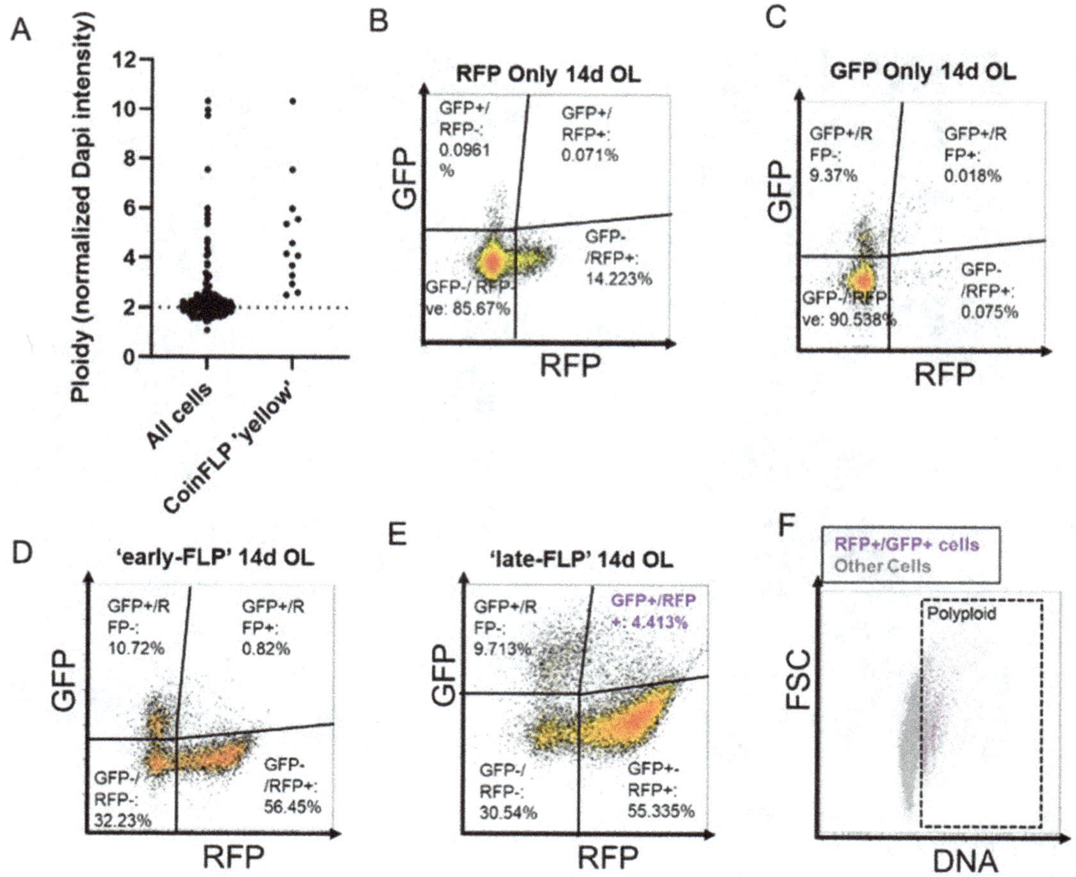

G

	Expected theoretical outcome (100% flipping)	Expected theoretical outcome (70% flipping)	Expected theoretical outcome (50% flipping)	Actual outcome from imaging & manual counting	Actual outcome measured by flow cytometry (~70% labelled)
Diploid cells flipping red	80%	56%	40%	ND	54.3%
Diploid cells flipping green	20%	14%	10%	ND	11.9%
Ratio of red:green	4:1	4:1	4:1	ND	4.6:1
% double labelled nuclei per OL	6.4%	4.48%	3.2%	~1%	4.4%
'Yellow' cells per OL	3424	2936	1712	~317	2354

Fig. 7 CoinFLP "yellow" cells are polyploid. (**a**) DAPI-integrated intensity was quantified for 132 nuclei from CoinFLP-labeled 14-day optic lobes using FIJI. Nuclear boundaries were guided by lamin staining for Dapi quantifications from at least 10 randomly chosen cells from 11 different representative images, from eight different brains. CoinFLP double-labeled cells are polyploid. (**b**, **c**) Single color controls used for gating. (**b**)

7. Incubate in the dark, (no mixing/no vortexing) for 45 min–1 h.
8. Prior to running, add 500 μL of 1X PBS, 0.1% BSA to the sample. This stops the trypsinization and dilutes the sample.
9. Do not pipet to mix (cells stick to the plastic pipet at this point). The addition of the PBS will mix the sample and dilute it to a concentration appropriate to maximize individual nuclei passing the laser.
10. Vortex gently before running on the flow cytometer for analysis. Use a flow cytometer with appropriate band-pass filters to allow detection of fluorophores in the 405 nm (DAPI), 488 nm (FIT-C or GFP or Alexa Fluor 488) and 555–580 nm (mCherry or RFP) emission wavelength range. Specifically, we used the Attune (violet and blue laser configuration) and the Attune NxT (violet, blue, red, and yellow lasers) flow cytometers.
11. On the flow cytometer, set up appropriate gates to discern stained events and remove doublets. Nonexperimental F1 siblings from the cross can be used as full minus one (FMO) controls to establish appropriate gates for GFP-positive and RFP-positive events. These can also be used to perform compensation if necessary. Sample flow cytometry plots are shown in Fig. 7.

Fig. 7 (continued) shows RFP only and (**c**) shows GFP only controls. (**d**) CoinFLP animals were heat-shocked at day of eclosion to induce labeling and brains were harvested at 14 days post eclosion. Inducing labeling at day one shows very few (0.82%) of cells in the OL that are double labeled. (**e**) CoinFLP labeling was performed 24 h before harvesting. 4.413% of the OL are double labeled. (**f**) DNA content dot plot showing all cells in the brain in gray and the double-labeled cells from (**e**) in purple. 97% of double-labeled (GFP+/RFP+) cells from "late-FLP" OL show polyploid DNA content. Single color controls in (**b**) and (**c**) are siblings from the cross *y,w, hsflp;lexAop-nlsRFP/Cyo;UAS-nlsGFP/TM6B* × *y,w; CoinFLP*. "RFP only" and "GFP only" are progeny that respectively did not inherit the *UAS-GFPnls* and *LexAop-RFPnls*. Cassettes and therefore contain balancer chromosomes. These controls were only used for gating purposes and not to estimate the efficiency of labeling. (**g**) Table showing the calculations of cell labeling. Assuming 100% of cells "flip" to label with CoinFLP, the flipping of LexA (red):Gal4 (green) ratio is biased 4:1 [4]. Thus, if flipping is complete, we would expect 80% of diploid cells to label red and ~20% to label green. If we assume that all the polyploid cells are tetraploid, and that 20% of the cells in the OL (20% of ~53,500 = 10,600 cells) are tetraploid [5], 4% of all polyploid cells will label green/green (probability of green is 0.2 therefore $0.2 \times 0.2 = 0.04 \times 100$), 64% red/red (probability of red is 0.8 thus, $0.8 \times 0.8 = 0.64 \times 100$) and 32% will label green/red or red/green and appear yellow or "double-labeled." The expectation would be then, that 3424 cells will label yellow per optic lobe under complete 100% flipping conditions. The amount of flipping is variable from animal to animal, and we estimate that in our samples with the lowest flipping we flip about one third of cells and with our strongest flipping we label about three fourth cells. If we label ~70% of cells, we expect about 2900 yellow cells per optic lobe, which is much closer to what we observe for CoinFLP by FACS. Calculations are based on *Drosophila* adult brain total cell counts from [7]. ((**a–f**) Figure and figure legend reproduced from [5])

4 Notes

1. If mounting non-brain *Drosophila* samples, one can make a "well" with nail polish and allow it to dry before mounting samples to elevate the coverslip and to prevent the sample from getting crushed. Alternatively, two strips of double-sided tape, one on top of the other on each side of the slide, can be used to elevate the coverslip. For larval imaginal disc or other flat tissues, no spacer is necessary, and 5 μL of mounting medium (or less) is sufficient.
2. When imaging thick and dense specimens such as the brain, it is recommended that a confocal microscope be used with the pinhole set to 1 Airy unit for the clearest images. We also recommend that the Z-stack step size be set to the optimal system recommendations (depends on the objective and the microscope).
3. For nuclear lamina labeling, we use mouse anti-Dm0 (ADL67.10) antibody diluted 1:100. For labeling neurons, we use rat anti-ELAV (7E8A10) 1:100, and for labeling glia, we use mouse anti-Repo (8D12) 1:100. All primary antibodies are from the Developmental Studies Hybridoma Bank. We recommend use of secondary antibodies conjugated to Alexa 633 (Invitrogen) in the far-red channel to avoid overlap with GFP and RFP from the CoinFLP labeling.
4. For dissociation to quantify labeled cells by flow cytometry, the amount of time the specimen is in the trypsin staining solution is critical. Note that less than 45 min will lead to DNA under staining and smearing artifacts in FACS, while incubations over 1 h lead to over digestion and excessive cell death.
5. CoinFLP and other similar multicolor labeling approaches have also been shown to work to detect polyploidy in other tissues [8–11].

Acknowledgments

This work in the Buttitta Lab was supported by the NIH (R01GM127367), the American Cancer Society (RSG-15-161-01-DDC), and a Barbour Scholar Award from the University of Michigan to S.N.

References

1. Lehner B, Sandner B, Marschallinger J, Lehner C, Furtner T, Couillard-Despres S et al (2011) The dark side of BrdU in neural stem cell biology: detrimental effects on cell cycle, differentiation and survival. Cell Tissue Res 345(3):313–328
2. Ross HH, Rahman M, Levkoff LH, Millette S, Martin-Carreras T, Dunbar EM et al (2011) Ethynyldeoxyuridine (EdU) suppresses in vitro population expansion and in vivo tumor progression of human glioblastoma cells. J Neurooncol 105(3):485–498
3. Vernis L, Piskur J, Diffley JFX (2003) Reconstitution of an efficient thymidine salvage pathway in Saccharomyces cerevisiae. Nucleic Acids Res 31(19):e120
4. Bosch JA, Tran NH, Hariharan IK (2015) CoinFLP: a system for efficient mosaic screening and for visualizing clonal boundaries in Drosophila. Development 142(3):597–606
5. Nandakumar S, Grushko O, Buttitta LA (2020) Polyploidy in the adult Drosophila brain. eLife 9:e54385
6. Nandakumar S (2020) Flow cytometry dissociation protocol for Individual Drosophila brains [Internet]. [cited 2021 Jun 28]. Available from: https://bio-protocol.org/prep687
7. Raji JI, Potter CJ (2021) The number of neurons in Drosophila and mosquito brains. PLoS One 16(5):e0250381
8. Peterson NG, Stormo B, Schoenfelder K, King J, Lee RR, Fox DT (2020) Cytoplasmic sharing through apical membrane remodeling. eLife 9:e58107
9. Sanders DW, Jumper CC, Ackerman PJ, Bracha D, Donlic A, Kim H et al (2021) SARS-CoV-2 requires cholesterol for viral entry and pathological syncytia formation. eLife 10:e65962
10. Matsumoto T, Wakefield L, Tarlow BD, Grompe M (2020) In vivo lineage tracing of polyploid hepatocytes reveals extensive proliferation during liver regeneration. Cell Stem Cell 26(1):34–47.e3
11. Rujano MA, Briand D, Đelić B, Spéder P (2021) Morphogenesis of a complex glial niche requires an interplay between cellular growth and fusion. BioRxiv. https://doi.org/10.1101/2021.05.09.443326

Part VI

Sequencing, Assembling, Editing and Engineering Polyploid Genomes

Chapter 23

Sequencing and Assembly of Polyploid Genomes

Yibin Wang, Jiaxin Yu, Mengwei Jiang, Wenlong Lei, Xingtan Zhang, and Haibao Tang

Abstract

Polyploidy has been observed throughout major eukaryotic clades and has played a vital role in the evolution of angiosperms. Recent polyploidizations often result in highly complex genome structures, posing challenges to genome assembly and phasing. Recent advances in sequencing technologies and genome assembly algorithms have enabled high-quality, near-complete chromosome-level assemblies of polyploid genomes. Advances in novel sequencing technologies include highly accurate single-molecule sequencing with HiFi reads, chromosome conformation capture with Hi-C technique, and linked reads sequencing. Additionally, new computational approaches have also significantly improved the precision and reliability of polyploid genome assembly and phasing, such as HiCanu, hifiasm, ALLHiC, and PolyGembler. Herein, we review recently published polyploid genomes and compare the various sequencing, assembly, and phasing approaches that are utilized in these genome studies. Finally, we anticipate that accurate and telomere-to-telomere chromosome-level assembly of polyploid genomes could ultimately become a routine procedure in the near future.

Key words Polyploidy, Sequencing, Genome assembly, Haplotype phasing

1 Introduction

Polyploid organisms consist of more than two homologous sets of chromosomes, resulting from either whole-genome duplication (WGD) events (autopolyploidy) or interspecific hybridization followed by chromosome doubling (allopolyploidy). Polyploidizations have been frequent in the evolution of plants and are considered to have occurred in the lineage of virtually all angiosperms [1–3]. Many well-known crop species are natural polyploids, which humans heavily depend on for food, feed, and fuel. Examples of polyploid crops include triploids of *Musa* spp. (banana); tetraploids of *Gossypium hirsutum* L. (cotton), *Solanum tuberosum*

Yibin Wang and Jiaxin Yu contributed equally with all other contributors.

Yves Van de Peer (ed.), *Polyploidy: Methods and Protocols*,
Methods in Molecular Biology, vol. 2545, https://doi.org/10.1007/978-1-0716-2561-3_23,

(potato), and *Arachis hypogaea* (peanut); hexaploids of *Triticum aestivum* (wheat) and *Avena sativa* (oat); and octoploids of *Fragaria ananassa* (strawberry) or *Saccharum spontaneum* (sugarcane), among many others. Although polyploidy is much more widespread in plants than in animals, there are numerous instances of both ancient and recent polyploidizations in invertebrates, insects, fishes, and amphibians [3–5].

Polyploid organisms often exhibit increased genetic and phenotype plasticity, with polyploidy being one of the most significant driving forces during plant evolution that has facilitated a rapid adaptation to dramatically changing environments [3, 6, 7]. To fully understand the functional and evolutionary impact of polyploidy, accurate assembly of a high-quality and chromosomal-scale reference genome plays a foundational role. However, considerable challenges exist when assembling high-quality reference sequences for polyploids, which often have more complex genomes than diploids because of large amounts of homologous sequences (both due to genome duplications and transposable element-mediated duplications) within the genome [3]. Failure to resolve the different copies of homologous sequences can often result in a suboptimal, fragmented assembly that blocks downstream applications.

Recently, third-generation sequencing technologies, such as single molecule real-time (SMRT) sequencing from PacBio or nanopore sequencing from Oxford Nanopore Technologies (ONT), have greatly improved the quality of reference genome assemblies at a fraction of the cost and time compared to just several years ago [8]. Combined with long-range scaffolding approaches, such as high-throughput chromatin conformation capture (Hi-C) [9] or 10X Genomics linked reads technology [10], these nascent yet powerful techniques promise to construct chromosome-level toward telomere-to-telomere assemblies [11]. Nevertheless, there have been disproportionately fewer polyploids with high-quality chromosome-level reference genome assemblies available compared to diploids. This is partly due to a lagging development of software tools—indeed, most genome assembly or phasing algorithms were initially designed for diploid genomes and are not yet fully compatible or effective in the case of complex polyploid species [12, 13].

Nonetheless, a few novel genome assembly algorithms have been designed specifically for polyploids, such as ALLHiC [14], PolyGembler [15], TRITEX [16], and have been applied to a few important polyploid crops. The adoption of these new approaches largely achieved the chromosome-level assemblies for these species, for example, for the genomes of the autooctoploid sugarcane [14] and the allohexaploid wheat [17]. Additionally, novel sequencing technologies have enabled the assembly of polyploid genomes with high repeat contents due to recent transposon proliferation [18]. For example, the HiFi (highly accurate long-read sequencing) reads from the PacBio platform typically achieve reads with average base accuracy >99.9% while still retaining a relatively long length of 15–20 Kbps to span most of the single repeats in plants. Consequently, high-quality chromosome-level assembly of polyploid genomes has rapidly become more accessible in recent years. Nonetheless, challenges remain for achieving the level of quality and contiguity of a polyploid genome assembly that is comparable to typical diploid genomes, which requires the continuous development of and investment in more cost-effective and sophisticated sequencing technologies and assembly algorithms.

This review aims to provide an overview of recently released polyploid genomes, the emerging sequencing technologies used, as well as the "state-of-the-art" computational approaches in conjunction with these new technologies, as adopted in recent polyploid genome studies. In closing, we also discuss limitations of the current approaches, open problems, and prospects of the field going forward.

2 Overview of Polyploid Genomes

2.1 Early History of Polyploid Genome Assembly

In the mid-1970s, using the dideoxy ribonucleotide chain-termination method [19], Sanger sequenced the first genome of phage phiX-174 [20] and ushered in the era of genome sequencing. The first plant genome of *Arabidopsis thaliana* was completed in 2000 [21], which was consisted of a "golden path" through the BAC library (bacterial artificial chromosome) completed at a considerable cost. While the initial set of assembled plant genomes were almost all diploid species, researchers explored various ways to attempt to decipher more complex polyploid genomes that represent a significant portion of our staple crops. However, the high level of repeats, high ploidy, and extreme genome sizes of polyploid genomes have continuously undercut assembly and phasing efforts.

The diploidized tetraploid of *Glycine max* was assembled through Sanger sequencing, also at a substantial cost [22]. Moreover, the natural allohexaploid of bread wheat (*Triticum aestivum*) was initially sequenced by Sanger through whole-genome shotgun (WGS) completed in 2014 [23]. Next-generation sequencing technologies, such as Illumina and Ion Proton, entered the arena circa 2005. While offering tremendous sequencing throughput at a significantly low cost, these technologies are generally less powerful than even the Sanger sequencing in resolving complex genomes, primarily due to their short reads that leave numerous sequencing gaps in a draft genome. As a result, very few high-quality polyploid genomes were released prior to 2015, and the studies of polyploid genomes were relatively stagnant during this period until the single molecule sequencing became mature enough for de novo genome assembly.

2.2 Rapid Development of Polyploid Genome Assembly in Recent Years

Single molecule sequencing and long-range scaffold technologies emerged in recent years and were quickly applied to plant genomes, resulting in a massive acceleration of de novo genome sequencing projects. More than 500 plant genomes have been published (as of December 2020; https://www.plabipd.de). Furthermore, studies of complex polyploid genomes entered a period of rapid development after 2018 because of the simultaneous development of new sequencing technologies and novel algorithms that are specially tailored to polyploid genomes. While there have already been excellent reviews before 2018 [24], we primarily focus on polyploid genome projects that were completed over the last 3 years that represent a brand new chapter in polyploid research. As a summary, we compiled a list of released assemblies of polyploid genomes since 2017 (Table 1).

Allopolyploid genome assembly can be resolved in a number of different ways and can utilize sequences from diploid progenitors to provide a structural basis for phasing, as was done in cotton [25] and tobacco [26]. The most widely cultivated crop of wheat is a hexaploid genome containing A, B, and D subgenomes with a large genome size and high level of heterozygosity [27, 28]. A hybrid strategy was employed to tackle the hexaploid wheat genome [28]. A draft assembly was initially constructed from Illumina short read through NRGene DenovoMagic2, with the draft contigs anchored onto 21 chromosomes that have been integrated with a high-density genetic map, sequencing from BAC libraries, as well as Hi-C data [28]. A high-quality and near-complete assembly was recently published along with a set of high-confidence annotations [28]. However, constructing a physical map or a high-density genetic map are both costly and time-consuming (due to the dependency on a physical library or a linkage mapping population, respectively). In comparison, single-molecule sequencing and long-range scaffolding technologies are generally much more cost-effective in practice. Two cotton cultivars, *Gossypium hirsutum* and *Gossypium barbadense*, are both allotetraploids derived from unknown diploid progenitors [29]. The newly completed genomes of the two cultivated allotetraploid cotton were assembled in higher quality by integrating PacBio SMRT, BioNano optical mapping, and Hi-C techniques [29].

While several allopolyploid genomes have been successfully sequenced and assembled using different hybrid strategies, challenges to the accurate assembly and complete phasing of autopolyploid exist. For example, the sweet potato (*Ipomoea batatas*) has a hexaploid genome consisting of six haplotypes generated from a hexaploidization event in which a diploid merged with a tetraploid in its lineage. In 2017, based on a novel single-nucleotide polymorphism (SNP)-based phasing approach, the sweet potato genome was separated into six distinct haplotypes [30]. However, this approach has limitations in finding major structural variations

Table 1
Statistics of polyploid genomes assembled in recent years from 2017 to 2021

Organism name	Ploidy level	Genome size (Mbps)	Status	Sequencing technology	GC%	Contig N50 (Kbps)	Release date	Citation
Xerophyta viscosa	$2n = 8\times = 48$	295.5	Scaffold	PacBio + Illumina	36.3	1,136	2017.03	[97]
Eleusine coracana	$2n = 4\times = 36$	1,196	Scaffold	Illumina + SOLiD	44.8	23.73	2017.06	[98]
Triticum dicoccoides	$2n = 4\times = 28$	10,079	Chromosome	Illumina + Hi-C	45.5	57	2017.07	[99]
Arachis monticola	$2n = 4\times = 40$	2,618.7	Chromosome	PacBio + BioNano + Hi-C + Illumina	37.4	106.66	2018.06	[100]
Triticum aestivum	$2n = 6\times = 42$	15,418.8	Chromosome	Hi-C + genetic mapping + Illumina	46.0	51.8	2018.08	[28]
Saccharum spontaneum	$2n = 4\times = 32$	3,133.3	Chromosome	PacBio + BAC + Hi-C + Illumina	44.9	45	2018.10	[31]
Panicum miliaceum	$2n = 4\times = 18$	848.3	Scaffold	PacBio + Hi-C + BioNano + Illumina	46.9	2,611	2019.01	[101]
Strawberry (*Fragaria × ananassa*)	$2n = 8\times = 56$	805.5	Chromosome	10X Genomics + PacBio + Hi-C + Illumina	39.2	17	2019.02	[102]
Gossypium hirsutum (TM1)	$2n = 4\times = 52$	2,296.6	Chromosome	PacBio + Illumina + Hi-C + BioNano + 10X Genomics	34.4	4,761	2019.04	[29]
Gossypium barbadense (ZM24)	$2n = 4\times = 52$	2,224.9	Chromosome	PacBio + Hi-C + BioNano + 10X Genomics + Illumina	34.9	1,983	2019.04	[29]
Arachis hypogaea	$2n = 4\times = 40$	2,529	Chromosome	PacBio + Illumina	36.4	696.6	2019.05	[103]
Spirodela polyrhiza	$2n = 4\times = 40$	138	Chromosome	PacBio + BAC	42.4	831	2019.08	[104]
Bonia amplexicaulis	$2n = 6\times = 72$	849	Scaffold	Illumina	43.2	24	2019.10	[105]
Guadua angustifolia	$2n = 4\times = 46$	1,614	Scaffold	Illumina	43.9	3.7	2019.10	[105]
Brassica napus (ZS11)	$2n = 4\times = 19$	960.8	Chromosome	PacBio + Hi-C + BioNano + Illumina	37.4	3,174	2020.01	[106]
Eragrostis tef	$2n = 4\times = 40$	577	Chromosome	PacBio + Hi-C + Illumina	45.5	1,536	2020.02	[107]
Medicago sativa L.	$2n = 4\times = 32$	2,738	Chromosome	PacBio HiFi + Hi-C + Illumina	32.7	459	2020.05	[15]
Gossypium mustelinum	$2n = 4\times = 52$	2,315	Chromosome	PacBio + Hi-C + Illumina	34.4	2,355	2020.05	[108]

(continued)

Table 1
(continued)

Organism name	Ploidy level	Genome size (Mbps)	Status	Sequencing technology	GC%	Contig N50 (Kbps)	Release date	Citation
Gossypium tomentosum	$2n = 4\times = 52$	2,194	Chromosome	PacBio + Hi-C + Illumina	34.2	10,240	2020.05	[108]
Gossypium darwinii	$2n = 4\times = 52$	2,183	Chromosome	PacBio + Hi-C + Illumina	34.1	9,318	2020.05	[108]
Echinochloa oryzicola	$2n = 4\times = 36$	950	Scaffold	PacBio + Illumina	42.5	1,915	2020.06	[109]
Echinochloa crus-galli	$2n = 6\times = 54$	1,340	Scaffold	PacBio + Illumina	45.91	1.57	2020.06	[109]
Fonio millet (*Digitaria exilis*)	$2n = 4\times = 36$	716	Chromosome	10X Genomics + BioNano + Hi-C + Illumina	45.5	78	2020.08	[110]
Cleistogenes songorica	$2n = 4\times = 40$	540.12	Chromosome	PacBio + Hi-C	/	21,790	2020.10	[111]
Cenchrus purpureus Schumach	$2n = 4\times = 40$	1,970	Chromosome	Nanopore + Illumina + Hi-C	48.8	1,873	2020.10	[112]
Carassius auratus	$2n = 4\times = 100$	1,730	Chromosome	PacBio RSII + BioNano + Hi-C	37.7	483	2020.11	[113]

among haplotypes and inadequate phasing accuracy in particular regions due to uncertainties in the read alignment prior to phasing [13, 14] and thus considered less accurate and hence less useful as a general assembly method for other polyploids.

In response to the set of challenges in polyploid genome assembly, an integrated approach, namely, ALLHiC (Fig. 2), combined whole-genome sequencing and Hi-C data and was utilized in the first haplotype-phased autopolyploid genome assembly of a wild relative of sugarcane *S. spontaneum* AP85-441 ($1n = 4\times = 32$) [31]. In this work, we initially generated a draft assembly from a BAC library and individual BAC pools. Additionally, PacBio long-read data were generated to reduce fragmentation in the draft contig assembly, which was then combined with Hi-C technology for scaffolding. ALLHiC was tailored to assemble polyploid genomes at a chromosomal level and can distinguish haplotypic contigs ("haplotigs") and successfully resolved complex polyploid genomes that contain very similar sequences between the homologous chromosomes [14]. In the *S. spontaneum* tetraploid genome, 89% of contigs were successfully anchored onto 32 chromosomes corresponding to four sets of monoploid each with eight chromosomes [31]. While undoubtedly useful in polyploid genome studies, the results of ALLHiC are still limited by the quality of the initial contig assembly that can be largely affected by collapsed contigs and chimeric contigs [14]. Chimeric contigs are often caused by false connections of different haplotypes, which would confuse the process of partitioning of contigs into distinct chromosomes because interchromosomal Hi-C signals would otherwise be detected between those chimeric contigs that contain pieces that belong to different chromosomes. Consequently, ALLHiC was later improved to incorporate a procedure to correct the chimeric contig identified in the initial contig set.

Cultivated alfalfa (*Medicago sativa* L.) is a self-incompatible, outcrossing autotetraploid ($2n = 4\times = 32$) plant with tetrasomic inheritance [32]. In recent work on the alfalfa genome, an allele-aware chromosome-level genome was assembled by integrating the HiFi and Hi-C technology using Canu and ALLHiC. First, 70 Gbp of PacBio CCS long-read data and 126 Gbp of Illumina short reads were generated. Then, Canu was used to assemble to contig-level, followed by ALLHiC algorithm that was used for chromosome-level assembly. A total of 32 superscaffolds were assembled, including eight homologous groups with each group containing a set of four alleles of chromosomes [32]. Although some misassemblies exist in the alfalfa genome, compared with the *S. spontaneum* assembly, the newly assembled alfalfa genome used a lower coverage of long-read data but showed a longer contig N50.

In another genome project of the autotetraploid potato, the researchers used HiFi long reads integrated with single-cell pollen genome sequencing and Hi-C technology [33]. The gamete sequencing was critical in the grouping of the hifiasm draft assembly into different haplotypes, and then each haplotype was subsequently scaffolded into chromosomes through Hi-C. This novel process resulted in a high-quality, haplotype-resolved, and chromosome-scale assembly of potato through genome sequencing with fewer switch errors than read-based haplotype phasing [33]. In summary, the combination of HiFi, Hi-C technologies, and/or other complexity-reduction techniques has shown substantial quality improvement for polyploid genome assembly.

2.3 Lack of High-Quality Reference Genomes of Polyploid Organisms

As the plummeting cost and advances in sequencing technologies and novel algorithms have alleviated the impact of genome complexity, several polyploid genomes have been sequenced and assembled, particularly in autopolyploid of *S. spontaneum*, alfalfa, and potato. However, vastly more polyploid taxa still lack chromosome-level assemblies, often with a limited degree of haplotype phasing. For instance, the leading sugarcane cultivars are derived from interspecific hybrids by crossing *S. officinarum* and *S. spontaneum,* so they have even higher genome complexity compared to *S. spontaneum.* Beyond the hybrid nature of cultivated sugarcane, they also have a prominent feature of the "Saccharum complex" and can often contain 10–12 allele copies [34]. Up until now, there is only a single reference genome of the wild sugarcane *S. spontaneum* AP85-441 [31]. Many more complex sugarcane genomes, including the higher ploidy leading cultivar varieties, are still not tackled. It seems highly likely that the single molecule sequencing integrated with long-range scaffolding technologies, with continuously improving algorithms for polyploid assembly and phasing, has the potential to ultimately overcome the challenges for the most extreme cases of polyploids.

3 Sequencing of Polyploid Genomes

3.1 Overview of the Sequencing Technologies

Sequencing technologies have constantly been evolving, starting from the initial dideoxy "Sanger" sequencing to now widespread short-read Illumina sequencing. Short-read sequencing is often called next-generation sequencing (NGS) technology, which has become a somewhat misplaced term nowadays as the term was initially coined with the intention to replace Sanger sequencing, a mission that has already been accomplished. Illumina sequencing has high per-base accuracy and very high throughput (i.e., high cost-efficiency), laying the foundation for various applications, including de novo genome assembly. Nonetheless, the relatively short length, with the mode at 2×150 bps, up to 2×250 bps, has apparent limitations for sequencing polyploids due to reads being too short to span the repeat-rich regions of the genome (such as centromeres, telomeres, MHC regions, GC-rich regions) and homologous or paralogous regions, which are often abundant and difficult to distinguish in polyploid genomes. Therefore, short reads are neither used alone to assemble polyploid plant genomes nor suited to compile a comprehensive catalog of structural variations (SV).

Nowadays, for de novo assemblies and identifications of SVs, sequencing reads longer than 10 Kbps can help to resolve the repetitive regions because they can span repeats that are shorter than the reads. Since most single LTR retrotransposons and centromeric repeats are often shorter than such long reads, assemblies utilizing long reads could readily walk through those problematic regions that were otherwise impossible to close with short reads (although, groups of nested transposons or huge tandem arrays may still be longer than such read lengths, and therefore not completely resolved). A commonly used metric contig N50—which measures the relative contiguity of draft assemblies—typically ranges between several Kbps to tens of Kbps for a short-read-based assembly. However, using long reads can often drastically increase assembly contiguity to a range between hundreds of Kbps to tens of Mbps.

3.2 Long-Read Sequencing

Single molecule sequencing nicely complements the shortcomings of the NGS to further allow a better and more accurate assembly of the polyploid genomes due to its ability to walk over repeats and much higher resolution toward discerning structural variations between the alleles within the genome. However, the current per-base accuracy of single-molecule sequencing is still lower than NGS sequencing technologies due to the intrinsic inaccuracies during signal measurements or the fact that they tend to use an error-prone polymerase (often to assist the signal emission). Single molecule sequencing technologies include two dominant players nowadays, including Pacific Biosciences (PacBio), whose technology relies on the optical signal, and Oxford Nanopore, which relies on the electronic signal when assaying the base composition of a sequencing read.

The widespread deployment of single-molecule sequencing is from PacBio. PacBio launched the Sequel I sequencing platform in 2011, which used SMRT technology to reach long reads with high efficiency. When a phosphor-linked nucleotide forms a cognate association with a DNA template, a fluorescence signal would be generated and then captured by a zero waveguide (ZMW) laser beam. The fluorescence signal linked to nucleotides disappears as each nucleotide gets incorporated into the DNA strand. By observing the changing of fluorescence in this process, we can identify the exact base sequence being incorporated during DNA synthesis while achieving the advertised goal of being "real time" [35]. Nevertheless, PacBio has a drawback of high per-base error rate, at 11–14%, depending on the version of the chemistry and polymerase. However, notably, most of the errors appear randomly (i.e., no biases toward one base or sets of bases) and can be resolved by sequencing multiple times or corrected by a hybrid polish with NGS. In contrast, errors in NGS are caused by PCR amplification errors which are multiplicative and biased in nature [36].

The Oxford Nanopore Technologies (ONT) was primarily introduced in 2014, which became more widely known as nanopore sequencing [8], which relies on a nanopore to generate signals from the nucleotide sequences without a synthesis step. Single DNA molecules are pulled through a nanopore during sequencing, which directly detects the different electrophysical characteristics of the nucleotide bases as they run through the pore. The sequencing length of ONT can easily hit tens of Kbps or hundreds of Kbps, with some ultra-long-read cases going to Mbps in lengths that appear only limited by the length of DNA molecules. While PacBio is limited by enzyme activity from its DNA synthesis, which can be terminated by the inadvertent damage of the enzyme from the ZMW laser beam, ONT reads have much longer reads but comparable or slightly higher error rates with also nonrandom errors compared to single-pass PacBio CLR reads.

Both the PacBio SMRT (CLR) and ONT produce long reads (>10 Kbps) but are characterized by low read accuracy (75–90%), which can achieve higher accuracy by combining with read-to-read error correction, but they are still vulnerable to misassembly from mis-mapping or mixing haplotypes especially in polyploid genomes [18, 35]. Moreover, a long-read correction requires high sequencing coverage or combined with high-accuracy sequencing such as Illumina sequencing, resulting in increased sequencing and computation costs [37, 38].

The upgraded, PacBio Sequel II sequencing platform has much higher throughput as well as the ability to produce highly accurate circular consensus long sequencing reads (HiFi or CCS reads). In the CCS model, the consensus base accuracy can reach 99%, with an average length of reads up to 20 Kbps, effectively trading accuracy over read length. The polymerase used in HiFi has a higher life span and better enzyme characteristics than that in the PacBio CLR. It is designed to pass the native dumbbell-shaped DNA templates multiple times, thereby achieving high-accuracy reads by building a consensus among the multiple passes ("subreads") for the same DNA molecule [18]. The high accuracy of the HiFi reads offers unique strengths to genome assembly and structural variation detection since we can often rely on faster, more accurate sequence alignment algorithms rather than fuzzy, sketch-based methods for string matching. In addition, HiFi reads significantly reduce the computation cost and can nicely separate repetitive regions and haplotype phasing [18], making it a desired approach to tackle polyploid genomes.

Moreover, increasing accuracy allows for assembling the genome while removing the need for read correction. Therefore, the introduction of HiFi sequencing technology has significantly accelerated the assembly of polyploid and complex genomes that used to take a substantial amount of time during the initial overlapping phase. A hexaploid genome, redwood, with a size of 27 Gbp, was sequencing by HiFi with a mere 22-fold genome coverage and 6 days to finish the high-quality assembly (https://www.pacb.com/blog/tackling-a-giant-genome), which would have otherwise been an impossible task to achieve.

3.3 Linked-Read Sequencing

Like long-read sequencing, the platform of Chromium using the GemCode technology introduced by 10X Genomics can also be used for haplotype phasing [39]. Linked reads generate conceptually "long" reads by labeling multiple linked reads that are derived from the same DNA molecules before making the sequencing library and associating the reads through common labels after sequencing. The Chromium system is optimized on the GemCode system, and both need to be combined with the Illumina sequencer. The core of the technology is partitioning numerous DNA fragments, gel beads, and enzyme mix into oil droplets inside GEM, which is a custom-designed microfluidic device so that each GEM could be considered as a mini-PCR system. Each GEM includes a part that binds to a sequencing library primer, one barcode, and then a random primer that binds to DNA fragments. After sequencing, we can obtain the linked reads that provide long-range information using barcodes that can associate short reads back to original long input molecules. Separating linked reads with different SNPs enables the genotyping of homologous chromosomes and detecting large segment structural variations between haplotypes in polyploids [39]. The 10X genomics nicely complements the shortcomings of Illumina sequencing, reducing the overall cost of obtaining long-distance information and presenting complete SNP differences that exist on separate DNA molecules to help better phase polyploid genomes.

3.4 Long-Range Technologies for Genome Scaffolding

Sometimes it may not be sufficient to assemble complete chromosome-level polyploid genomes using sequencing reads alone, and the final piece of the puzzle can be fulfilled by an integrated approach combined with other genome scaffolding technologies. For an overview, we summarize the overall strengths and limitations of various sequencing and genome scaffolding technologies in detail (Table 2).

Traditional technologies used to anchor contigs or scaffolds to chromosomes are based on high-density physical or genetic maps to provide long-distance linkage information among the contigs on the same chromosomes [40]. More advanced mapping technology

Table 2
Overview of sequencing and genome scaffolding technologies used in de novo genome sequencing

Technology	Reads	Highlights	Limitations
PacBio CLR	Single molecule long reads, average length ~ 20 kb	Long-read sequencing technology widely used to assemble genome in the past years	The high base error requires correction with high coverage or with short reads, also may introduce switch error into polyploid assembly
PacBio CCS (HiFi)	Single molecule long reads, circular sequencing, and consensus by multi passes, with average length ~ 15–20 kb	Generate long reads that have high base accuracy (>99%) and long length, reduced need for error correction, and lower sequencing coverage	Reads length only reach ~20 Kbps, therefore sacrificing length for accuracy
Oxford Nanopore	Single-molecule long reads, directly detected by base signal, with average length ~10 Kbps, ultra-long reads can reach 2.3 Mbps in the current record	Direct sequencing by measuring electric base signal without limited by enzyme activity	High base error and occurrence of homopolymer errors Requirement to prepare long-length DNA molecules
BioNano Genomics	Optical mapping of DNA fingerprint-labeled fluorescently in enzyme sites, with average length up to multiple hundred kilobases	Produce long-range linkage information for all chromosome	Limitations in sparse enzyme site regions and pool algorithms for integrating optical mapping and genome assembly
10X Genomics	Short reads from Illumina paired-end sequencing, but linked reads through GEMs, which provide linkage information with average length reach to 100 Kbps	Experimental methods to construct a microenvironment to sequencing a single molecule DNA fragment, capable of linked reads over long-range and ability to sequence using mainstream Illumina sequencers	Pool algorithms available for polyploid, and it is based on NGS, therefore inherits certain limitations in repetitive regions, especially for local tandem repeats where the labels are the same (i.e., unable to differentiate between tandem copies)
Hi-C	Derived short reads from Illumina paired-end sequencing can provide linkage information up to 100 Kbps for chromosomes based on proximity ligation	Methods can detect the chromosome interaction, can provide the whole-genome contact information	Requires (and sensitive to) a high-quality input haplotig set To achieve better results, chimeric contigs to be removed from the input

can provide additional long-range information, for instance, BioNano optical mapping technology and Hi-C. Lastly, scaffolding hints could also be obtained through interspecies collinearity with certain evolutionary assumptions [40], but often used as a last resort.

The optical mapping technologies represented by BioNano Genomics make fingerprints of large genome fragments available from measuring the physical distance of specific enzyme cutting or nicking sites along a DNA molecule. DNA molecules measured by BioNano have an average length of approximately 225 Kbps, which could provide long-range information of genomes for improving the continuity of genome assembly [41], with the ability to scaffold contigs to chromosomal sizes. Successful genome assemblies using the optical mapping approaches have been reviewed in detail in [42]. However, challenges exist in integrating sequencing and optical mapping in polyploid genome assembly because the optical mapping has a bias to break at closely spaced restriction site regions [43]. Moreover, since polyploid genomes contain highly similar allelic regions that show similar enzymatic site distributions, optical mapping technologies can be challenging to use in practice. The homologous chromosomes often could not phase very well since the optical map fingerprints have limited ability to resolve between haplotypes due to biases and lack of resolution in the restriction patterns. BioNano also has some practical limitations in achieving good results for certain types of plant DNA materials, likely due to certain polysaccharides inhibiting the optical mapping reactions.

A better tool for scaffolding plant genomes in practice is Hi-C. Hi-C was initially used to study the three-dimensional structure of chromatins, which was proposed in 2009 by combining chromosome conformation capture (3C technology) with high-throughput sequencing [9]. The core of Hi-C technology is based on proximity ligation of DNA fragments that cross-link adjacent sequences physically. First, the spatially adjacent chromatin segments are cut off from both sides, marked with a biotin label and connected. Then, the circular DNA becomes fragmented at random, which are then captured and sequenced so that proximal DNA segments can be obtained in pairs [9]. The Hi-C links of the contigs on the same chromatin are expected to have more contacts than those on different chromosomes. The frequency of Hi-C links decreases with the increase of genome distance, with an inverse power relationship ($k < -1$) between them [9]. By exploiting this inverse relationship, the contigs can be clustered into groups corresponding to individual chromosomes and then arranged in order according to the chromatin contact frequency (closer contig pairs share more chromatin contacts). Last, the orientation of each contig can be inferred by the relative positions of links between adjacent contigs, with the optimal orientation minimizing the overall distance between the links [44]. Hi-C technology was used successfully in the de novo assembly and haplotype phasing for the autotetraploid of *Saccharum officinarum* and allotetraploid of *Solanum tuberosum* [45].

4 Algorithmic Challenges During a Polyploid Genome Assembly

Genome assembly is a process of gradually assembling the sequenced reads into incrementally larger pieces to reconstruct a reference genome. The mainstream paradigm is that DNA fragments are assembled into contigs through overlap-based algorithms, or the "overlap-layout-consensus" algorithms (OLC), that work well with long reads [46, 47]. However, in most diploid genome projects [48, 49], two haplotypes are often co-assembled (or "collapsed") to form a monoploid reference and thereby have the potential to form mosaic or chimeric sequences. This simplistic monoploid reconstruction loses much of allelic information since it becomes difficult to recover the linkage between adjacent polymorphic sites due to this unfortunate "collapse," especially in highly heterozygous genomes.

Interspecific hybridization and whole-genome duplication (WGD) lead to many duplicate genes, or homeologs. These duplicate genes undergo sub-functionalization in a long period of evolution and participate in an elaborate regulatory mechanism, including dosage effect, gene silencing, *cis*- and *trans*-acting factors, and epistatic interactions [50–54]. Functional and evolutionary study of polyploids would require a full dissection of the different allele sequences. Therefore, a simple consensus that represented a monoploid reference, while conceptually simpler, is not informative for polyploid genomes, unlike the diploid genome where in many cases a single monoploid reference may be sufficient.

While the assembly of polyploid genomes is beset by the difficulty of highly similar regions between different haplotypes, we must describe $n - 1$ haplotypes before inferring all haplotypes theoretically, so the computational complexity that is required to reconstruct n haplotypes is much higher than that required for diploid assembly. In addition, homologous segments are very similar in polyploids, and there are even multiple identical haplotypes in some local regions, similar to autozygosity in a diploid genome, which cause switching errors, i.e., contigs that are misclassified into the wrong haplotypes. Such misassemblies are even more severe in autopolyploids since fused haplotypes become more likely due to the similarity between the homologous chromosomes [24, 55, 56].

A typical allopolyploid assembly approach can utilize sequences of the surviving diploid progenitors and then assign the polyploid genome's contigs to each subgenome with the help of parental genome information. However, the exact progenitors may be challenging to recover since the diploid progenitors may have accumulated genomic changes, like translocations, partial chromosomal recombinations, and gene losses in either the progenitor or the polyploid [57, 58]. For example, in allotetraploid *Brassica napus*, there were regions that were subject to widespread homologous exchange (HE) and gene loss between the A subgenome and the C subgenome [59]. This phenomenon brings difficulties to the assembly of polyploids through its diploid progenitors and may even receive adverse impact when a wrong progenitor genome has been assumed.

Additionally, polyploid genomes often experienced transposon amplification events in their lineages, especially in plant genomes. Active transposons in a recent lineage can rapidly result in many repetitive regions in the genome. The assembly of repetitive segments (including tandem repeat, segmental duplication, and centromeres [60–62]) may cause partial genome collapse, similar to the collapses that occurred between allelic sequences. When polymorphic regions appear in the assembly graphs, we often get shorter assembled contigs due to the ambiguity that appears as branching points between polymorphic and monomorphic regions [63]. If we visualize the case of two haplotypes in a diploid forming "bubbles" with the interleaving between homozygous and heterozygous regions, the case of *n* haplotypes would be combinatorically more complex with potential "bubbles" that can form between any two haplotypes (leading to bubbles nested inside other bubbles), depending on the topological ordering of the structural variation and the evolutionary distance between the haplotypes. To sum up, complete and accurate reconstruction of a polyploid genome from sequence data is still an open problem, due to these algorithmic challenges.

5 Computational Approaches of Assembly of Polyploid Genomes

Despite the algorithmic challenges, several polyploid genomes have been successfully resolved and studied by combining the improvement of sequencing technology and computational methods. We will discuss two popular classes of computational methods of polyploid genome assembly, including reference-based phasing and de novo assembly, that "heuristically" address the algorithmic difficulties of accurate polyploid assemblies. Furthermore, we compiled a list of polyploid genome assembly tools suitable for various sequencing data types (Table 3). The details of the reference-based phasing, local, and whole-genome assembly are visualized in Figs. 1 and 2.

Table 3
Overview of software used in polyploid genome assemblies

Software	Algorithm	Input data	Highlights	Limitations	URL	Citation
H-PoP & H-PoPG	The Polyploid Balanced Optimal Partition (PBOP) model	Short reads or long reads. When using H-PoPG it needs genotype information addition	Better performance compared with HapTree, HapCompass and SDhap	It is limited in identical handle haplotypes and repetitive regions. Difficult to be applied in high ploidy	https://github.com/MinzhuXie/H-PoPG	[65]
Ranbow	Graph-based phase algorithm, which makes use of multiple types of variation where Illumina reads overlaps	Short reads	It can generate accurate haplotype-resolved assembly	Relatively short, phased block N50 length	https://github.com/moeinzadeh/Ranbow	[70]
WhatsHap-polyphase	Partition read sets to multiple clusters which are reconstructed into haplotype per cluster	PacBio long reads	It can yield relatively precise phased haplotypes	It cannot produce a chromosome-scale haplotype. Moreover, it is a challenge to handle a genome that ploidy more than 5	https://github.com/whatshap/whatshap	[55]
nPhase	Based on SNPs, this method clusters long reads and generates consensus haplotigs iteratively	Long reads and short reads and reference genome	It doesn't require a priori knowledge of ploidy and enables resulting haplotigs with both accuracy and continuity	It can be affected by uneven depth distribution of reads from different haplotypes	https://github.com/OmarOakheart/nPhase	[71]
Hap10 & Hap++	Hap++ is a pre-process util in Hap10. It converts 10X Genomics data to fragment formation. Then either SDhaP or SDPNAL+ will be used.	10X Genomics	It yields relatively precise phased haplotypes	Similar to WhatsHap-polyphase in phasing performance. Furthermore, it requires more computational costs	https://github.com/smajidian/Hap10	[72]

POLYTE	Overlap-layout-consensus (OLC) de novo assembly	Short reads	Good accuracy in local haplotype assembly	Can't deal with complex assembly regions. When working on whole chromosomes, it needs to split reads into independent bins	https://github.com/HaploConduct/HaploConduct	[74]
HiCanu	Error correction for reads through base compression	PacBio HiFi reads	Capable of resolving diverse alleles and repeats, especially in segmental duplications	HiCanu only generates "pseudo haplotypes" and it needs additional linkage information to distribute contigs to the accurate haplotypes	https://github.com/marbl/canu	[75]
Hifiasm	Construct a string graph retaining all information by identifying haplotype in the reads overlap and correction process	PacBio HiFi data, trio information	Capable of resolving diverse alleles and repeats. Furthermore, it can generate alternative haplotypes with more continuity. Moreover, it simplifies the assembly process. If trio information is available, it yields chromosome-scale haplotype assembly in diploid	For polyploid, hifiasm assembly must be combined with Hi-C or strand-seq to generate chromosome-scale assembly	https://github.com/chhylp123/hifiasm	[77]
PolyGembler	HMM model. Scaffold partition to linkage group through graph-based cluster algorithm. Order and orientation are conducted by multidimensional scaling (MDS) algorithm and solving traveling salesman problems	Genome-wide genotyping data collected on F1 outbred mapping population & whole-genome sequence data on a reference sample or reference contig/scaffold	Allows the high-level missing data and genotyping error to generate accurate haplotype-resolve assembly on chromosome-scale	It struggles in handling high-ploidy genome. (>6). Rely on high-quality reference assembly	https://github.com/c-zhou/polyGembler	[15]
ALLHiC	Prune/optimize/Genetic Algorithm	Contig assembly, Hi-C	Wide application toward a spectrum of polyploid taxa and able to yield chromosome-scale assembly	Rely on high-quality local phased haplotype assembly and must know the ploidy in advance	https://github.com/tangerzhang/ALLHiC	[14]

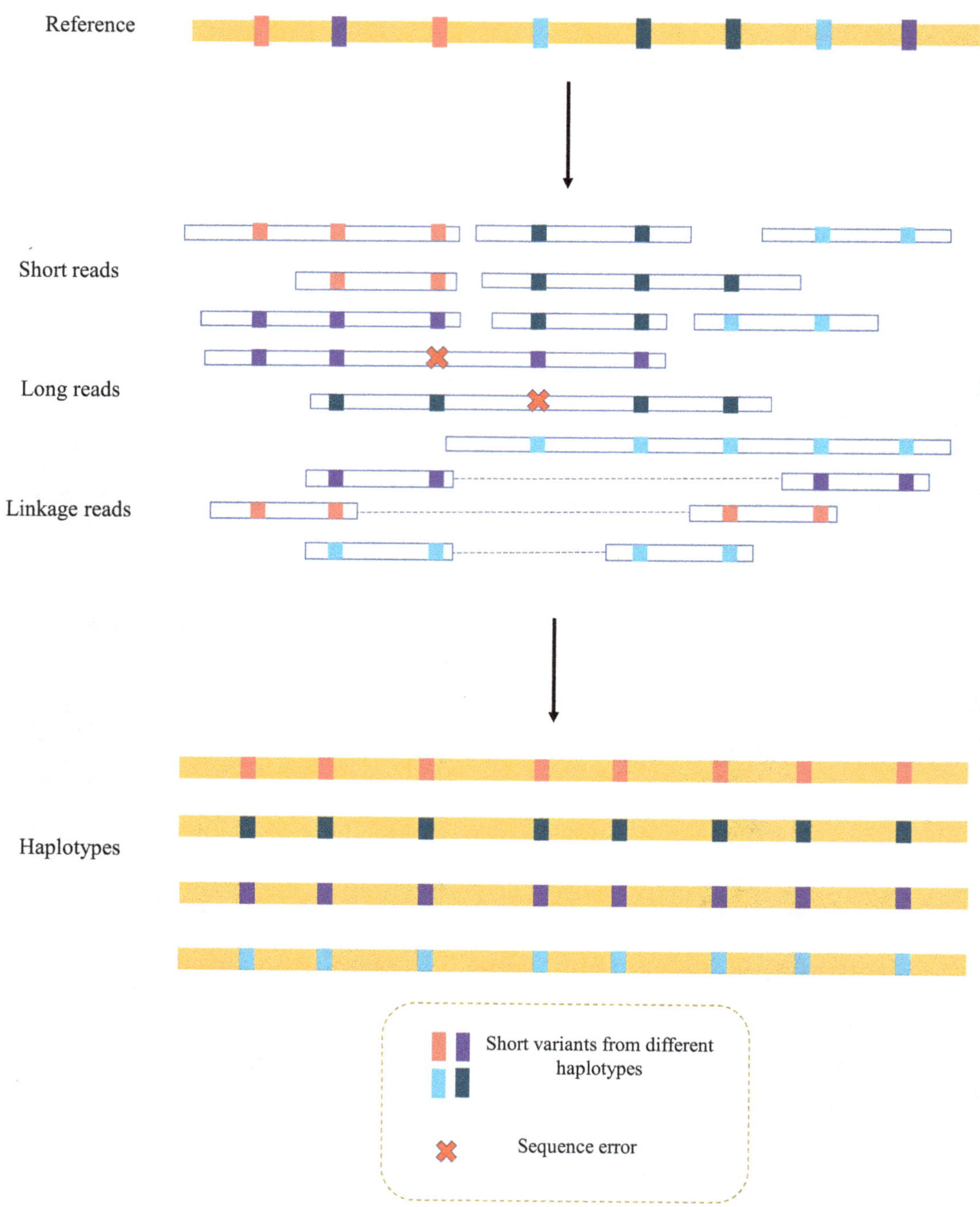

Fig. 1 Overview of reference-based phasing strategy. Distinct haplotypes are constructed from aligning various sequencing data against the reference genome assembly. Short reads and long reads and linked reads can generate dense and accurate local phase haplotypes

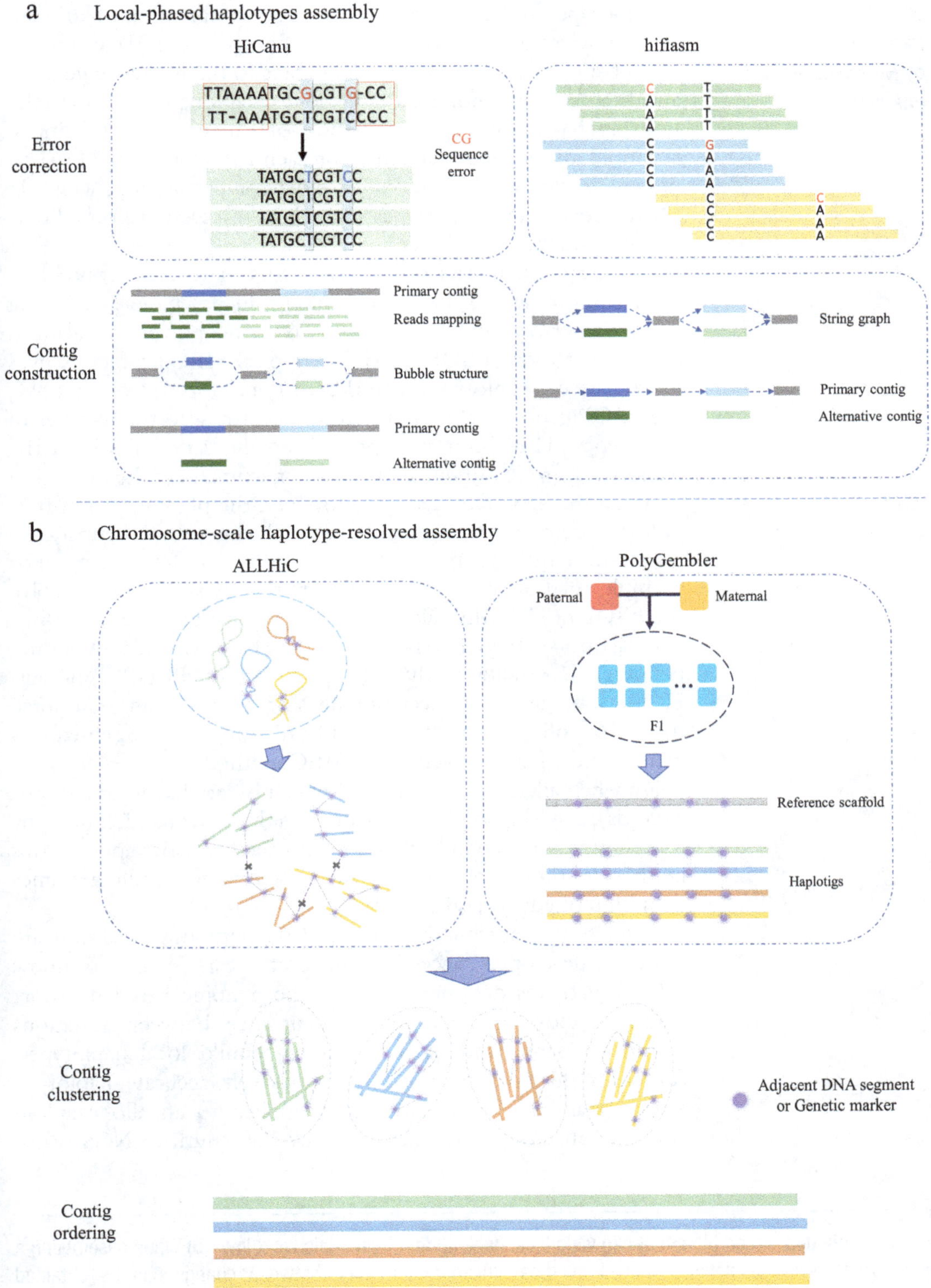

Fig. 2 Overview of de novo assembly strategy. (**a**) For local-phased haplotype assembly, error correction of sequencing reads and graph-based contig reconstruction are often used. HiCanu and hifiasm are two

5.1 Resolving Haplotype Assembly by Reference-Based Phasing

A haplotype can be regarded as a specific combination of adjacent single nucleotide variation (SNV) in a narrow sense. We can reveal variant sites by aligning sequencing reads to the reference genome and then reconstructing the polyploid's haplotype from the relationship based on the combination of variants after the initial assembly into a single reference for each haplotype (Fig. 1). This process is commonly referred to as haplotype phasing, which in essence treats the *n* haplotypes similarly as *n* accessions from different individuals.

In the past few years, several frameworks have been proposed to solve the problem of haplotype phasing in polyploids, including minimum error correction (MEC) [64] and polyploid balanced optimal partition (PBOP) [65]. The typical MEC problem strives to reconstruct haplotypes with the least amount of "corrections"; the PBOP problem clusters the reads according to the number of haplotypes. They both attempt to keep the variation within the haplotype group minimized while maximizing the variation between the haplotype groups. However, the phasing problem is, unfortunately, NP-hard, which can only be solved heuristically given the large number of SNPs involved in the phasing. For example, H-PoP uses a linear programming algorithm to solve the PBOP model, while also proposing a Polyploid Balanced Optimal Partition with Genotype constraint (PBOPG) model by adding genotype information. By solving PBOP or PBOPG models, H-PoP can get more accurate phasing results than traditional tools [66–68]. However, the PBOP model is essentially a generalized MEC problem. The MEC method is known to overphase when alleles are too similar (which is very likely for autopolyploids), resulting in phasing errors. On the other hand, limited by its computational complexity even with the heuristics applied, this method is still computationally too expensive to handle genomes with high ploidy levels [55, 69].

Due to the shortcomings of the MEC method, several software has been developed successively in recent years, such as Ranbow [70], which was designed to phase the genome based on short reads. Ranbow exploits the co-occurrence between variations within the same read or read pair to rebuild local haplotypes. Although this method can obtain relatively accurate haplotype assembly and has been successfully applied to an allohexaploid sweet potato [71], it is still limited by the length of NGS reads,

Fig. 2 (continued) assemblers that can yield local haplotig that follow this workflow. (**b**) When assembling a chromosome-scale haplotype-resolved genome, linkage information between contigs can be obtained through Hi-C technology or genetic mapping. Afterward, the set of haplotypes was constructed by partitioning contigs followed by ordering and orientation of contigs within each partition. ALLHiC and PolyGembler utilize Hi-C and genetic mapping to construct chromosome-scale haplotypes, respectively

which is either 2×150 bps or 2×250 bps. Ranbow cannot handle large-scale variations such as SVs well and variations covering longer physical distance, resulting in relatively small phasing blocks that are easily interrupted by long stretches of autozygous regions. WhatsHap-polyphase [55] partially addresses this problem, by clustering the reads based on similarity first to avoid the defects of the MEC framework, and then threads the haplotypes afterward.

In a recent study, nPhase [71] was designed to iterate the process of clustering similar long reads and generate consensus sequences while maintaining the identity of each group in each iteration. Long reads were eventually partitioned into groups that significantly differed from each other. The software claims better assembly accuracy and continuity than WhatsHap-polyphase [71]. For phasing with 10X genomics linked reads, the hap10 software has been developed and shows the high accuracy of phasing [72].

The efficacy of the reference-based phasing method mentioned above directly depends on the alignment between reads and the reference genome. Hence, reference quality, read length, sequencing error, and read depth variation all strongly impact the accuracy of phasing results. Moreover, repetitive regions will continue to confuse the alignment of reads, similar to their impact during the initial assembly of the reference genome. Therefore, it appears that we have not avoided the intrinsic difficulties assembling these regions but merely deferred in resolving these regions until alignment. While most phasing algorithms are designed based on small-scale genetic variation (SNP, Indel <50 bp) and work well toward small variations, they are relatively weak when dealing with polyploidy with complex variations such as SVs. In conclusion, applying this alignment-based strategy to phase polyploid genomes requires certain assumptions and experimentation to find the proper parameter settings during alignments, variant calling, and phasing.

5.2 De Novo Haplotype Assembly by Local Phasing

A more direct way is to de novo assemble contigs from different haplotypes with the hope of producing a non-collapsed polyploid assembly containing genetic information from each allele (Fig. 2a). This method tackles the central issue of repeats and haplotype collapse directly and can preserve genetic variations between different haplotypes to obtain a more accurate assembly.

To handle this problem, researchers have developed a series of more sophisticated de novo assemblers recently [24, 73]. POLYTE [74] uses Illumina short reads to perform de novo assembly through an overlap-layout-consensus (OLC) algorithm. This tool is yet incapable of directly yielding a whole-genome assembly for a large genome. However, it has a high degree of accuracy at the gene level and can be used for de novo assembly of specific genes or small genomes with relatively low-coverage short reads.

With the development of PacBio HiFi [18] technologies and ultra-long nanopore [37], de novo assembly of the polyploid genome has gradually become the mainstream, with a suite of assembly tools recently developed and potentially more in the pipeline. As discussed earlier, HiFi reads have both high accuracy (>99%) and long read length (15–25 Kbps), which offer the best combination for genome assembly, improving both the performance and accuracy for a typical assembly.

Assemblers optimized for the HiFi read have been developed and successfully applied to various taxa. HiCanu [75] is an optimized version of Canu [76], which builds the best overlap graph with reads as node and overlap as edge, in which the maximal path without branches form a unitig then a contig. HiCanu recognizes the bubble structure in the graph by identifying overlaps between contig and can separate allelic contigs. This idea is similar to FALCON_UNZIP [63]. HiFi reads may be subject to sequencing errors where there are multiple same consecutive bases, known as "homopolymers." HiCanu compresses the homopolymers in the raw reads and corrects errors based on compressed reads, which could significantly rectify such sequencing errors. HiCanu can also assemble repetitive sequences in the genome with a high degree of accuracy and perform well in regions of segmental duplications [75].

Another HiFi read assembler, hifiasm [77], tries to phase and correct reads at the stage of reads overlap and constructs an assembly string graph, striving to retain all duplications and variations. Therefore, the "alternative" assembly of hifiasm is generally less interrupted by the heterozygous runs. Additionally, hifiasm includes an innovative trio labeling algorithm that can accurately generate the haplotypes of diploid progenitor genomes when trio information is available.

5.3 Chromosome-Scale Haplotype-Resolved Assembly

So far, for most complex genomes, it is not sufficient to obtain a chromosomal assembly of polyploid genomes by most "state-of-the-art" sequencing technology alone. Specifically, even with long reads, contigs could still end at the border of large repetitive regions or regions close to centromeric repeats. Therefore, to determine the ordering and orientation of contigs of a draft assembly, we need to integrate long-range linkage information to assemble at the chromosome level (Fig. 2b). The most common strategies are a combined computational approach integrating with the genetic map, physical map, optical map, and Hi-C chromatin contact map.

First, the genetic map, or linkage map, describes the relative distance between genetic markers [78]. It can link certain contigs that contain those genetic markers onto chromosomes, based on their genetic positions [40]. Obtaining a genetic map with high-density genetic markers is the key that determines whether it can be valuable for scaffolding a genome assembly (denser or accurate maps tend to be much more useful) [40]. However, the application of traditional genetic mapping software in polyploids is quite limited. One reason is most linkage mapping methods are designed based on inbred populations, which are often tricky to construct for polyploids. They also require high accuracy of genotyping data, resulting in a significant decrease in the density of genetic markers that ultimately pass quality filtration [79–82]. The size of the mapping population determines the resolution of the map, but the current genetic mapping method may not be scalable due to the cost of developing and maintaining a medium to large mapping population.

PolyGembler [15] utilize genetic maps to assist the assembly of the polyploid genome. Firstly, the reference scaffolds were phased using genotyping data of the mapping population through a hidden Markov model (HMM). The occurrence of recombination events within and between haplotigs are then recorded to calculate their recombination fractions (RF). Based on the RF values, a graph partition algorithm and a heuristic algorithm that approximately solves the traveling salesman problem (TSP) are used to cluster and order haplotigs and finally generate a reliable polyploid chromosome-level assembly. PolyGembler is also capable of identifying potential misassemblies through the analysis of the RF values between adjacent markers. PolyGembler is sensitive to the input contigs, so that when the quality of the initial reference assembly is poor, the PolyGembler is naturally unable to produce an exemplary assembly.

Optical mapping and Hi-C contact maps assisted assembly do not require the construction of a mapping population, thereby dramatically reducing the required resources and cost for a genome assembly. In addition, as previously mentioned, optical mapping technology usually produces fragments that span multiple Mbps in size that go beyond the typical read lengths and hence more power in resolving large repeat arrays. Details for the computational methods employed in optical mapping have been reviewed elsewhere [83].

For Hi-C data, pairs of adjacent contigs or scaffolds could be linked together to form a chromosome-scale assembly. This assembly strategy led to the development of a number of assembly tools [44, 84, 85] and was successfully applied to the assembly of many diploid genomes [86, 87]. Due to the presence of multiple homeologs, the typical methods usually lead to misjoined contigs derived from disparate haplotypes in polyploids. ALLHiC is specifically designed to correct such misjoins, especially in an autopolyploid assembly, yet it is also suitable for haplotyping in highly heterozygous diploid and allopolyploid genomes [88, 89].

The ALLHiC pipeline contains several key steps, which can be briefly described in flow charts (Fig. 2b). First, ALLHiC prunes the adjacent signals between the homologous segments based on a heuristic method that classifies intra-haplotype vs. inter-haplotype contacts. Then, based on the strength of Hi-C contact, a single-linkage algorithm is used to partition contigs from the disparate haplotype into *k* groups where *k* is the number of chromosomes and a *rescue* algorithm that brings back the contigs that have inadequate interaction signal in the initial partition round. Subsequently, a genetic algorithm (GA) is used to optimize the orientation and order of contigs within each group. Since Hi-C data does not require a complex experimental setup with the only requirement of Hi-C data input, it is applicable to many species where such data is already available. Therefore, the ALLHiC algorithm has become widely used for polyploid genome assemblies. Like other Hi-C-based assembly methods, ALLHiC is dependent on the quality of the draft assembly and often requires correction and quality control steps.

Due to the complexity of a polyploid assembly, it is sometimes difficult to use a one-size-fits-all strategy and it is often vital to use a hybrid assembly pipeline to collect multiple modalities of data and combine the evidence using an integrated approach. TRITEX [16] is another good example of a hybrid computational approach. It uses Minia3 [90] software to increase the *k*-mer for iterative assemblies. TRITEX also combines various data, including 10X genomics, Hi-C, and genetic markers, to iteratively merge contig into the scaffold to generate chromosome-level assembly. The pipeline is mostly centered around utilizing short reads but has been successfully used in several genome projects of polyploid (bread wheat) or diploid with large genome size (rye, barley) [16, 91, 92].

6 Role of High-Quality Polyploid Genome Assembly and Prospects

Accurate phased assembly of the polyploidy is critical for the analysis of haplotype structure variants and allele-specific expression (ASE) for complex traits in polyploids. Allele-specific expression (ASE) or allelic imbalance (AI) has been speculated to play a major role in heterosis. Differential expression between the alleles likely resulted from the presence or absence of the sequence variants located in *cis*-regulatory elements upstream of the gene alleles [93]. Furthermore, the three-dimensional chromatin structure plays a vital role in biological processes in polyploid genomes, such as DNA replication, gene expression regulation, cell development, and chromatin recombination [94]. Taken together, the gene regulation in polyploids is far more complex than the diploid genome since the whole-genome duplication and subsequent gene loss often introduce gene dosage imbalances, both occurring before and after transcription.

A recent 3D genome study for the allotetraploid cotton and their diploid ancestors has revealed that the spatial proximity between homoeologous genes is possibly associated with coordinated gene expression [95]. The 3D genome structure of the artificial autopolyploid of *A. thaliana* has an increased level of inter-chromosome contacts with accompanying transcriptional changes [96]. There have been few studies thus far on the 3D genome of naturally occurring autopolyploid, partially due to the lack of high-quality and complete haplotype-phased assemblies until recently. With the advent of high-quality and complete haplotype phasing of the polyploid genome, it has become possible to investigate the key mechanism of allelic dominance in various polyploid crops.

Lately, many polyploid genomes have been assembled or upgraded to the chromosome level based on the latest sequencing and assembly methods designed for polyploid genomes. Owing to the HiFi reads that have an excellent balance of accuracy and length, the polyploid assembly has become cost-effective and much less computationally demanding. The various algorithms specifically tailored for polyploids, including ALLHiC, have successfully assembled the first autotetraploid sugarcane *S. spontaneum* AP85-441 and successfully used it in other polyploid genomes. We have now entered a golden age for chromosome-level or even telomere-to-telomere (T2T) assemblies and complete haplotype phasing. Nonetheless, there continue to be open challenges in polyploidy genome assembly that require more sophisticated sequencing technologies and algorithms. On the contig level, both HiCanu and hifiasm still have difficulties at fully resolving complex, nested, or highly tandem repeat regions. On the chromosomal level, Hi-C-based assembly algorithms, such as ALLHiC, are still heavily dependent on the initial assembly quality and are impacted by chimerism and repeat/allele collapse. Simultaneous improvement of both contig- and chromosomal-level assembly approaches would be desirable to overcome these challenges.

Finally, the assembly of a highly accurate telomere-to-telomere and haplotype-phased assembly for polyploid organisms is the ultimate dream come true for polyploid researchers. Better assemblies will undoubtedly provide a better understanding of the functional landscape and evolutionary dynamics in polyploids that have become widely successful across the plant tree of life. In agriculture, researchers need to understand the genome to allow more precision during breeding and better understand regulatory mechanisms for allelic dominance, subgenome dominance, and heterosis in polyploid crops. Driven both by technological advances and a wide interest in polyploids, we anticipate that more polyploid genomes will be deciphered and phased into high-quality haplotypes in the near future.

References

1. Doyle JJ, Flagel LE, Paterson AH et al (2008) Evolutionary genetics of genome merger and doubling in plants. Annu Rev Genet 42:443–461. https://doi.org/10.1146/annurev.genet.42.110807.091524
2. Soltis DE, Albert VA, Leebens-Mack J et al (2009) Polyploidy and angiosperm diversification. Am J Bot 96(1):336–348. https://doi.org/10.3732/ajb.0800079
3. Van de Peer Y, Mizrachi E, Marchal K (2017) The evolutionary significance of polyploidy. Nat Rev Genet 18(7):411–424. https://doi.org/10.1038/nrg.2017.26
4. Mable BK, Alexandrou MA, Taylor MI (2011) Genome duplication in amphibians and fish: an extended synthesis. J Zool 284(3):151–182. https://doi.org/10.1111/j.1469-7998.2011.00829.x
5. Otto SP, Whitton J (2000) Polyploid incidence and evolution. Annu Rev Genet 34(1):401–437
6. Sattler MC, Carvalho CR, Clarindo WR (2016) The polyploidy and its key role in plant breeding. Planta 243(2):281–296. https://doi.org/10.1007/s00425-015-2450-x
7. Adams KL, Wendel JF (2005) Polyploidy and genome evolution in plants. Curr Opin Plant Biol 8(2):135–141. https://doi.org/10.1016/j.pbi.2005.01.001
8. Jain M, Olsen HE, Paten B et al (2016) The Oxford Nanopore MinION: delivery of nanopore sequencing to the genomics community. Genome Biol 17(1):239. https://doi.org/10.1186/s13059-016-1103-0
9. Lieberman-Aiden E, van Berkum NL, Williams L et al (2009) Comprehensive mapping of long-range interactions reveals folding principles of the human genome. Science 326(5950):289–293. https://doi.org/10.1126/science.1181369
10. Zheng GX, Lau BT, Schnall-Levin M et al (2016) Haplotyping germline and cancer genomes with high-throughput linked-read sequencing. Nat Biotechnol 34(3):303–311. https://doi.org/10.1038/nbt.3432
11. Nurk S, Koren S, Rhie A et al (2021) The complete sequence of a human genome. bioRxiv:2021.2005.2026.445798. https://doi.org/10.1101/2021.05.26.445798
12. Zhang X, Wu R, Wang Y et al (2020) Unzipping haplotypes in diploid and polyploid genomes. Comput Struct Biotechnol J 18:66–72. https://doi.org/10.1016/j.csbj.2019.11.011
13. Tang H (2017) Disentangling a polyploid genome. Nat Plants 3(9):688–689. https://doi.org/10.1038/s41477-017-0001-0
14. Zhang X, Zhang S, Zhao Q et al (2019) Assembly of allele-aware, chromosomal-scale autopolyploid genomes based on Hi-C data. Nat Plants 5(8):833–845. https://doi.org/10.1038/s41477-019-0487-8
15. Zhou C, Olukolu B, Gemenet DC et al (2020) Assembly of whole-chromosome pseudomolecules for polyploid plant genomes using outbred mapping populations. Nat Genet 52(11):1256–1264. https://doi.org/10.1038/s41588-020-00717-7
16. Monat C, Padmarasu S, Lux T et al (2019) TRITEX: chromosome-scale sequence assembly of Triticeae genomes with open-source tools. Genome Biol 20(1):284. https://doi.org/10.1186/s13059-019-1899-5
17. Walkowiak S, Gao L, Monat C et al (2020) Multiple wheat genomes reveal global variation in modern breeding. Nature 588(7837):277–283. https://doi.org/10.1038/s41586-020-2961-x
18. Wenger AM, Peluso P, Rowell WJ et al (2019) Accurate circular consensus long-read sequencing improves variant detection and assembly of a human genome. Nat Biotechnol 37(10):1155–1162. https://doi.org/10.1038/s41587-019-0217-9
19. Sanger F, Coulson AR (1975) A rapid method for determining sequences in DNA by primed synthesis with DNA polymerase. J Mol Biol 94(3):441–448. https://doi.org/10.1016/0022-2836(75)90213-2
20. Sanger F, Air GM, Barrell BG et al (1977) Nucleotide sequence of bacteriophage phi X174 DNA. Nature 265(5596):687–695. https://doi.org/10.1038/265687a0
21. The Arabidopsis Genome Initiative (2000) Analysis of the genome sequence of the flowering plant *Arabidopsis thaliana*. Nature 408(6814):796–815. https://doi.org/10.1038/35048692
22. Schmutz J, Cannon SB, Schlueter J et al (2010) Genome sequence of the palaeopolyploid soybean. Nature 463(7278):178–183. https://doi.org/10.1038/nature08670
23. Mayer KFX, Rogers J, Dolezel J et al (2014) A chromosome-based draft sequence of the hexaploid bread wheat (*Triticum aestivum*) genome. Science 345(6194):1251788. https://doi.org/10.1126/science.1251788
24. Kyriakidou M, Tai HH, Anglin NL et al (2018) Current strategies of polyploid plant

genome sequence assembly. Front Plant Sci 9: 1660. https://doi.org/10.3389/fpls.2018.01660

25. Li F, Fan G, Lu C et al (2015) Genome sequence of cultivated Upland cotton (*Gossypium hirsutum* TM-1) provides insights into genome evolution. Nat Biotechnol 33(5): 524–530
26. Sierro N, Battey JN, Ouadi S et al (2014) The tobacco genome sequence and its comparison with those of tomato and potato. Nat Commun 5(1):1–9
27. International Wheat Genome Sequencing Consortium (IWGSC) (2014) A chromosome-based draft sequence of the hexaploid bread wheat (*Triticum aestivum*) genome. Science 345(6194):1251788. https://doi.org/10.1126/science.1251788
28. International Wheat Genome Sequencing Consortium (IWGSC), Investigators IRp, Appels R et al (2018) Shifting the limits in wheat research and breeding using a fully annotated reference genome. Science 361(6403):eaar7191. https://doi.org/10.1126/science.aar7191
29. Wang M, Tu L, Yuan D et al (2019) Reference genome sequences of two cultivated allotetraploid cottons, *Gossypium hirsutum* and *Gossypium barbadense.* Nat Genet 51(2):224–229. https://doi.org/10.1038/s41588-018-0282-x
30. Yang J, Moeinzadeh M-H, Kuhl H et al (2017) Haplotype-resolved sweet potato genome traces back its hexaploidization history. Nat Plants 3(9):696–703
31. Zhang J, Zhang X, Tang H et al (2018) Allele-defined genome of the autopolyploid sugarcane *Saccharum spontaneum* L. Nat Genet 50(11):1565–1573. https://doi.org/10.1038/s41588-018-0237-2
32. Chen H, Zeng Y, Yang Y et al (2020) Allele-aware chromosome-level genome assembly and efficient transgene-free genome editing for the autotetraploid cultivated alfalfa. Nat Commun 11(1):2494. https://doi.org/10.1038/s41467-020-16338-x
33. Sun H, Jiao W-B, Krause K et al (2021) Chromosome-scale and haplotype-resolved genome assembly of a tetraploid potato cultivar. bioRxiv:2021.2005.2015.444292. https://doi.org/10.1101/2021.05.15.444292
34. Souza GM, Berges H, Bocs S et al (2011) The sugarcane genome challenge: strategies for sequencing a highly complex genome. Trop Plant Biol 4(3–4):145–156. https://doi.org/10.1007/s12042-011-9079-0
35. Eid J, Fehr A, Gray J et al (2009) Real-time DNA sequencing from single polymerase molecules. Science 323(5910):133–138. https://doi.org/10.1126/science.1162986
36. D'Amore R, Johnson J, Haldenby S et al (2017) SMRT Gate: a method for validation of synthetic constructs on Pacific Biosciences sequencing platforms. BioTechniques 63(1): 13–20. https://doi.org/10.2144/000114565
37. Jain M, Koren S, Miga KH et al (2018) Nanopore sequencing and assembly of a human genome with ultra-long reads. Nat Biotechnol 36(4):338–345. https://doi.org/10.1038/nbt.4060
38. Chaisson MJ, Huddleston J, Dennis MY et al (2015) Resolving the complexity of the human genome using single-molecule sequencing. Nature 517(7536):608–611. https://doi.org/10.1038/nature13907
39. Ott A, Schnable JC, Yeh CT et al (2018) Linked read technology for assembling large complex and polyploid genomes. BMC Genomics 19(1):651. https://doi.org/10.1186/s12864-018-5040-z
40. Tang H, Zhang X, Miao C et al (2015) ALLMAPS: robust scaffold ordering based on multiple maps. Genome Biol 16(1):1–15
41. Lam ET, Hastie A, Lin C et al (2012) Genome mapping on nanochannel arrays for structural variation analysis and sequence assembly. Nat Biotechnol 30(8):771–776. https://doi.org/10.1038/nbt.2303
42. Tang H, Lyons E, Town CD (2015) Optical mapping in plant comparative genomics. Gigascience 4:3. https://doi.org/10.1186/s13742-015-0044-y
43. Jiao WB, Schneeberger K (2017) The impact of third generation genomic technologies on plant genome assembly. Curr Opin Plant Biol 36:64–70. https://doi.org/10.1016/j.pbi.2017.02.002
44. Burton JN, Adey A, Patwardhan RP et al (2013) Chromosome-scale scaffolding of *de novo* genome assemblies based on chromatin interactions. Nat Biotechnol 31(12): 1119–1125. https://doi.org/10.1038/nbt.2727
45. Zhou Q, Tang D, Huang W et al (2020) Haplotype-resolved genome analyses of a heterozygous diploid potato. Nat Genet 52(10): 1018–1023. https://doi.org/10.1038/s41588-020-0699-x
46. Whibley A, Kelley J, Narum S (2021) The changing face of genome assemblies: guidance on achieving high-quality reference genomes. Mol Ecol Resour 21(3):641–652

47. Sedlazeck FJ, Lee H, Darby CA et al (2018) Piercing the dark matter: bioinformatics of long-range sequencing and mapping. Nat Rev Genet 19(6):329–346
48. Zhang L, Chen F, Zhang X et al (2020) The water lily genome and the early evolution of flowering plants. Nature 577(7788):79–84
49. Warren WC, Harris RA, Haukness M et al (2020) Sequence diversity analyses of an improved rhesus macaque genome enhance its biomedical utility. Science 370(6523): eabc6617
50. Gallagher JP, Grover C, Hu G et al (2016) Insights into the ecology and evolution of polyploid plants through network analysis. Mol Ecol 25(11):2644–2660
51. Shi X, Zhang C, Kwan KD et al (2015) Genome-wide dosage-dependent and -independent regulation contributes to gene expression and evolutionary novelty in plant polyploids. Mol Biol Evol (9):2351–2366
52. Roulin A, Auer PL, Libault M et al (2013) The fate of duplicated genes in a polyploid plant genome. Plant J 73(1):143–153
53. Adams KL, Wendel JF (2005) Novel patterns of gene expression in polyploid plants. Trends Genet 21(10):539–543
54. Osborn TC, Pires JC, Birchler JA et al (2003) Understanding mechanisms of novel gene expression in polyploids. Trends Genet 19(3):141–147
55. Schrinner SD, Mari RS, Ebler J et al (2020) Haplotype threading: accurate polyploid phasing from long reads. Genome Biol 21(1):252. https://doi.org/10.1186/s13059-020-02158-1
56. Browning SR, Browning BL (2011) Haplotype phasing: existing methods and new developments. Nat Rev Genet 12(10): 703–714
57. Gaeta RT, Chris Pires J (2010) Homoeologous recombination in allopolyploids: the polyploid ratchet. New Phytol 186(1):18–28
58. Hufton AL, Panopoulou G (2009) Polyploidy and genome restructuring: a variety of outcomes. Curr Opin Genet Dev 19(6): 600–606
59. Chalhoub B, Denoeud F, Liu S et al (2014) Early allopolyploid evolution in the post-Neolithic *Brassica napus* oilseed genome. Science 345(6199):950–953
60. Tørresen OK, Star B, Mier P et al (2019) Tandem repeats lead to sequence assembly errors and impose multi-level challenges for genome and protein databases. Nucleic Acids Res 47(21):10994–11006
61. Hartley G, O'Neill RJ (2019) Centromere repeats: hidden gems of the genome. Genes (Basel) 10(3):223
62. Bailey JA, Yavor AM, Massa HF et al (2001) Segmental duplications: organization and impact within the current human genome project assembly. Genome Res 11(6): 1005–1017
63. Chin C-S, Peluso P, Sedlazeck FJ et al (2016) Phased diploid genome assembly with single-molecule real-time sequencing. Nat Methods 13(12):1050–1054
64. Lippert R, Schwartz R, Lancia G et al (2002) Algorithmic strategies for the single nucleotide polymorphism haplotype assembly problem. Brief Bioinform 3(1):23–31
65. Xie M, Wu Q, Wang J et al (2016) H-PoP and H-PoPG: heuristic partitioning algorithms for single individual haplotyping of polyploids. Bioinformatics 32(24):3735–3744
66. Das S, Vikalo H (2015) SDhaP: haplotype assembly for diploids and polyploids via semi-definite programming. BMC Genomics 16(1):1–16
67. Berger E, Yorukoglu D, Peng J et al (2014) HapTree: a novel Bayesian framework for single individual polyplotyping using NGS data. PLoS Comput Biol 10(3):e1003502
68. Aguiar D, Istrail S (2012) HapCompass: a fast cycle basis algorithm for accurate haplotype assembly of sequence data. J Comput Biol 19(6):577–590
69. Motazedi E, Finkers R, Maliepaard C et al (2018) Exploiting next-generation sequencing to solve the haplotyping puzzle in polyploids: a simulation study. Brief Bioinform 19(3):387–403. https://doi.org/10.1093/bib/bbw126
70. Moeinzadeh M-H, Yang J, Muzychenko E et al (2020) Ranbow: a fast and accurate method for polyploid haplotype reconstruction. PLoS Comput Biol 16(5):e1007843
71. Abou Saada O, Tsouris A, Eberlein C et al (2021) nPhase: an accurate and contiguous phasing method for polyploids. Genome Biol 22(1):1–27
72. Majidian S, Kahaei MH, de Ridder D (2020) Hap10: reconstructing accurate and long polyploid haplotypes using linked reads. BMC Bioinformatics 21(1):253. https://doi.org/10.1186/s12859-020-03584-5
73. Garg S (2021) Computational methods for chromosome-scale haplotype reconstruction. Genome Biol 22(1):1–24
74. Baaijens JA, Schönhuth A (2019) Overlap graph-based generation of haplotigs for

diploids and polyploids. Bioinformatics 35(21):4281–4289

75. Nurk S, Walenz BP, Rhie A et al (2020) HiCanu: accurate assembly of segmental duplications, satellites, and allelic variants from high-fidelity long reads. Genome Res 30(9):1291–1305
76. Koren S, Walenz BP, Berlin K et al (2017) Canu: scalable and accurate long-read assembly via adaptive *k*-mer weighting and repeat separation. Genome Res 27(5): 722–736
77. Cheng H, Concepcion GT, Feng X et al (2021) Haplotype-resolved *de novo* assembly using phased assembly graphs with hifiasm. Nat Methods 18(2):170–175
78. Collard BC, Jahufer M, Brouwer J et al (2005) An introduction to markers, quantitative trait loci (QTL) mapping and marker-assisted selection for crop improvement: the basic concepts. Euphytica 142(1):169–196
79. Bourke PM, van Geest G, Voorrips RE et al (2018) polymapR—linkage analysis and genetic map construction from F1 populations of outcrossing polyploids. Bioinformatics 34(20):3496–3502
80. Hackett CA, Boskamp B, Vogogias A et al (2017) TetraploidSNPMap: software for linkage analysis and QTL mapping in autotetraploid populations using SNP dosage data. J Hered 108(4):438–442
81. Grandke F, Ranganathan S, van Bers N et al (2017) PERGOLA: fast and deterministic linkage mapping of polyploids. BMC Bioinformatics 18(1):1–9
82. Broman KW, Wu H, Sen Ś et al (2003) R/qtl: QTL mapping in experimental crosses. Bioinformatics 19(7):889–890
83. Yuan Y, Chung CY-L, Chan T-F (2020) Advances in optical mapping for genomic research. Comput Struct Biotechnol J 18: 2051–2062
84. Ghurye J, Rhie A, Walenz BP et al (2019) Integrating Hi-C links with assembly graphs for chromosome-scale assembly. PLoS Comput Biol 15(8):e1007273
85. Dudchenko O, Batra SS, Omer AD et al (2017) *De novo* assembly of the Aedes aegypti genome using Hi-C yields chromosome-length scaffolds. Science 356(6333):92–95
86. Wang K, Wang J, Zhu C et al (2021) African lungfish genome sheds light on the vertebrate water-to-land transition. Cell 184(5): 1362–1376.e1318
87. Matthews BJ, Dudchenko O, Kingan SB et al (2018) Improved reference genome of *Aedes aegypti* informs arbovirus vector control. Nature 563(7732):501–507
88. Wang P, Yu J, Jin S et al (2021) Genetic basis of high aroma and stress tolerance in the oolong tea cultivar genome. Hortic Res 8(1):1–15
89. Hu W, Ji C, Shi H et al (2021) Allele-defined genome reveals biallelic differentiation during cassava evolution. Mol Plant 14(6):851–854
90. Chikhi R, Limasset A, Medvedev P (2016) Compacting de Bruijn graphs from sequencing data quickly and in low memory. Bioinformatics 32(12):i201–i208
91. Rabanus-Wallace MT, Hackauf B, Mascher M et al (2021) Chromosome-scale genome assembly provides insights into rye biology, evolution and agronomic potential. Nat Genet 53(4):564–573. https://doi.org/10.1038/s41588-021-00807-0
92. Jayakodi M, Padmarasu S, Haberer G et al (2020) The barley pan-genome reveals the hidden legacy of mutation breeding. Nature 588(7837):284–289. https://doi.org/10.1038/s41586-020-2947-8
93. Shao L, Xing F, Xu C et al (2019) Patterns of genome-wide allele-specific expression in hybrid rice and the implications on the genetic basis of heterosis. Proc Natl Acad Sci U S A 116(12):5653–5658. https://doi.org/10.1073/pnas.1820513116
94. Eagen KP (2018) Principles of chromosome architecture revealed by Hi-C. Trends Biochem Sci 43(6):469–478. https://doi.org/10.1016/j.tibs.2018.03.006
95. Wang M, Wang P, Lin M et al (2018) Evolutionary dynamics of 3D genome architecture following polyploidization in cotton. Nat Plants 4(2):90–97. https://doi.org/10.1038/s41477-017-0096-3
96. Zhang H, Zheng R, Wang Y et al (2019) The effects of Arabidopsis genome duplication on the chromatin organization and transcriptional regulation. Nucleic Acids Res 47(15): 7857–7869. https://doi.org/10.1093/nar/gkz511
97. Costa MD, Artur MA, Maia J et al (2017) A footprint of desiccation tolerance in the genome of *Xerophyta viscosa*. Nat Plants 3: 17038. https://doi.org/10.1038/nplants.2017.38
98. Hittalmani S, Mahesh HB, Shirke MD et al (2017) Genome and Transcriptome sequence of Finger millet (Eleusine coracana (L.) Gaertn.) provides insights into drought tolerance and nutraceutical properties. BMC Genomics 18(1):465. https://doi.org/10.1186/s12864-017-3850-z

99. Akpinar BA, Biyiklioglu S, Alptekin B et al (2018) Chromosome-based survey sequencing reveals the genome organization of wild wheat progenitor *Triticum dicoccoides*. Plant Biotechnol J 16(12):2077–2087. https://doi.org/10.1111/pbi.12940
100. Yin D, Ji C, Ma X et al (2018) Genome of an allotetraploid wild peanut *Arachis monticola*: a *de novo* assembly. Gigascience 7(6):giy066. https://doi.org/10.1093/gigascience/giy066
101. Shi J, Ma X, Zhang J et al (2019) Chromosome conformation capture resolved near complete genome assembly of broomcorn millet. Nat Commun 10(1):464. https://doi.org/10.1038/s41467-018-07876-6
102. Edger PP, Poorten TJ, VanBuren R et al (2019) Origin and evolution of the octoploid strawberry genome. Nat Genet 51(3): 541–547. https://doi.org/10.1038/s41588-019-0356-4
103. Bertioli DJ, Jenkins J, Clevenger J et al (2019) The genome sequence of segmental allotetraploid peanut Arachis hypogaea. Nat Genet 51(5):877–884. https://doi.org/10.1038/s41588-019-0405-z
104. An D, Zhou Y, Li C et al (2019) Plant evolution and environmental adaptation unveiled by long-read whole-genome sequencing of Spirodela. Proc Natl Acad Sci U S A 116(38):18893–18899. https://doi.org/10.1073/pnas.1910401116
105. Guo ZH, Ma PF, Yang GQ et al (2019) Genome sequences provide insights into the reticulate origin and unique traits of Woody Bamboos. Mol Plant 12(10):1353–1365. https://doi.org/10.1016/j.molp.2019.05.009
106. Song JM, Guan Z, Hu J et al (2020) Eight high-quality genomes reveal pan-genome architecture and ecotype differentiation of *Brassica napus*. Nat Plants 6(1):34–45. https://doi.org/10.1038/s41477-019-0577-7
107. VanBuren R, Man Wai C, Wang X et al (2020) Exceptional subgenome stability and functional divergence in the allotetraploid Ethiopian cereal teff. Nat Commun 11(1):884. https://doi.org/10.1038/s41467-020-14724-z
108. Chen ZJ, Sreedasyam A, Ando A et al (2020) Genomic diversifications of five Gossypium allopolyploid species and their impact on cotton improvement. Nat Genet 52(5): 525–533. https://doi.org/10.1038/s41588-020-0614-5
109. Ye CY, Wu D, Mao L et al (2020) The genomes of the allohexaploid Echinochloa crus-galli and its progenitors provide insights into polyploidization-driven adaptation. Mol Plant 13(9):1298–1310. https://doi.org/10.1016/j.molp.2020.07.001
110. Abrouk M, Ahmed HI, Cubry P et al (2020) Fonio millet genome unlocks African orphan crop diversity for agriculture in a changing climate. Nat Commun 11(1):4488. https://doi.org/10.1038/s41467-020-18329-4
111. Zhang J, Wu F, Yan Q et al (2021) The genome of *Cleistogenes songorica* provides a blueprint for functional dissection of dimorphic flower differentiation and drought adaptability. Plant Biotechnol J 19(3): 532–547. https://doi.org/10.1111/pbi.13483
112. Yan Q, Wu F, Xu P et al (2021) The elephant grass (*Cenchrus purpureus*) genome provides insights into anthocyanidin accumulation and fast growth. Mol Ecol Resour 21(2): 526–542. https://doi.org/10.1111/1755-0998.13271
113. Chen D, Zhang Q, Tang W et al (2020) The evolutionary origin and domestication history of goldfish (*Carassius auratus*). Proc Natl Acad Sci U S A 117(47):29775–29785. https://doi.org/10.1073/pnas.2005545117

Chapter 24

Genome Editing by CRISPR/Cas9 in Polyploids

Carlos Sánchez-Gómez, David Posé, and Carmen Martín-Pizarro

Abstract

CRISPR/Cas system has been widely used for genome editing in the past few years. Even though it has been performed in many polyploid species to date, its efficient accomplishment in these organisms is still a challenge. The presence of multiple homoeologous genes as targets for their editing requires more rigorous work and specific needs to assess successful genome editing. Here, we describe a general stepwise protocol to select target sites, design sgRNAs, indicate vector requirements, and screen CRISPR/Cas9-mediated genome editing in polyploid species.

Key words Genome editing, CRISPR/Cas9, sgRNA, Off-targets, Polyploids, Homoeologous genes

1 Introduction

Since the beginning of the twentieth century, when the first non-directed genome editing was performed in barley (*Hordeum vulgare* L.) with X-rays and radium mutagens, biotechnology-based tools research has developed systems to perform DNA editing in a more efficient and targeted way [1]. This achievement has been possible thanks to an increase in the number of sequenced genomes and the discovering of site-specific nucleases (SSNs) and their applications. The use of SSNs is based on their ability to cleave DNA in a specific sequence in the genome and generate a DNA double-strand break (DSB). DSBs activate two different mechanisms to repair DNA damage. In the absence of a homologous repair template, the error-prone nonhomologous end joining (NHEJ) pathway generates insertions/deletions (InDel mutations) or substitutions at the DSB site, generally resulting in the disruption of the gene function [2]. However, when an exogenous donor template is available, the homology-directed repair (HDR) pathway is activated, repairing the DSB by homologous recombination and, therefore, allowing to perform precise modification of the locus of interest.

Yves Van de Peer (ed.), *Polyploidy: Methods and Protocols*,
Methods in Molecular Biology, vol. 2545, https://doi.org/10.1007/978-1-0716-2561-3_24,

SSN-based tools to create targeted mutations are mainly divided into four classes according to their chronological order of discovery: (i) meganucleases (MNs) or customized homing nucleases, (ii) zinc-finger nucleases (ZFNs), (iii) transcription activator-like effector nucleases (TALENs), and (iv) clustered regularly interspaced short palindromic repeats (CRISPR) with CRISPR-associated system (Cas) proteins (CRISPR/Cas). MN, ZNF, and TALEN systems are based on DSBs generated by site-specific nucleases, and therefore they require complex customization of the DNA-binding protein to achieve the desired sequence-specific DNA-binding module linked to the DNA cleavage domain [3]. However, CRISPR/Cas system is based on an easily engineered single guide RNA (sgRNA) that specifies a targeted nucleic acid sequence and the Cas nucleases, which generate the DSB. Therefore, CRISPR/Cas constitutes a much easier and efficient system to design, simply requiring to select a single RNA of around 20 nucleotides that are complementary to a DNA target sequence which must be adjacent to the PAM (protospacer adjacent motif) sequence motif, that follows the consensus NGG for the most commonly used Cas protein, Cas9 [3]. Specifically, *Streptococcus pyogenes* Cas9 (SpCas9) is the most widely used as it was the first CRISPR/Cas system used for genome editing in eukaryotic cells [4].

Originally discovered as part of the adaptive immunity against invading genetic elements in archaea [5] and in bacteria [6], the CRISPR/Cas system was adapted as a biotechnological tool to precisely edit DNA firstly in prokaryotic cells [7] and later in eukaryotic cells [4] and plants [8–10]. In recent years, CRISPR/Cas-based genome editing has been performed in a great variety of species, but polyploids remain difficult targets for this technique. Traditionally, polyploidy has been defined as the heritable condition of possessing more than two complete sets of chromosomes [11], although a lot of prokaryotes with more than one copy of a circular chromosome has been discovered over the past decade. Therefore, this definition may change as they have been also considered polyploids [12]. Even though examples of polyploids can be found within the three domains in the tree of life and within the five kingdoms of eukaryotes [12–14], they have been deeper studied in plants, as many cultivated species are polyploids [15]. The inherent characteristics of polyploid species, such as the presence of homoeologous genes, high heterozygosity, repetitive DNA, and genome irregularity, make polyploids genome editing a challenging goal [16], requiring a refined experimental design with more parameters to take into account than in diploid species. Nevertheless, CRISPR/Cas-mediated genome editing was successfully performed in a polyploid plant species, the hexaploid bread wheat (*Triticum aestivum*), in 2013 for the first time [17]. This work was followed by many other reports of genome editing in polyploid plant species, such as the triploid species banana (*Musa acuminata*) [18], the tetraploids soybean (*Glycine max*) [19], tobacco (*Nicotiana tabacum*) [20], potato (*Solanum tuberosum*) [21], cotton (*Gossypium hirsutum*) [22], rapeseed (*Brassica napus*) [23], switchgrass (*Panicum virgatum*) [24], and tragopogon (*Tragopogon mirus*) [25], the hexaploid camellia (*Camelina sativa*) [26], the octoploid strawberry (*Fragaria* × *ananassa*) [27], and the deca-/tridecaploid sugarcane (*Saccharum* spp. hybrid) [28].

In this chapter, a general CRISPR/Cas9 editing protocol to generate knockout stable lines will be explained, using *TM6* MADS-box gene from the octoploid species *Fragaria* × *ananassa* as an example [27]. As the transformation protocols and constructs design are very specific depending on the species, vectors selected, and nature of the experiment, we will focus on sgRNA design and posterior gene editing assessing. Although we will mainly focus on tools and resources for genome editing in plants, the general requirements and technical aspects included in this chapter are applicable to other polyploid organisms.

2 Materials

2.1 Sequenced Genome Databases

Ideally, the genome of the species of interest or at least a closely related species should be sequenced and annotated. Although the NCBI database can be used, other databases that unify sequenced genomes of plant species, like PlantGDB or Phytozome, or that centralize genome sequencing data from specific plant families that include polyploid species, such as SolGenomics Network (https://solgenomics.net/) for *Solanaceae* or Genome Database for *Rosaceae* (https://www.rosaceae.org/), are favored. Most importantly, we recommend the database with the most recently published genome available for the species of interest.

2.2 Tools for sgRNA Selection

There are plenty of bioinformatics tools to predict CRISPR target sites and to select the sgRNAs. While choosing a tool, it is recommendable to take into account some features that not all of them offer, like a calculation of the on-target efficiency, the possible off-targets, what algorithm it is used, or the possibility of uploading a reference genome of interest when this is not available in the database. There is a great variety of algorithms that forecast the specificity and efficiency of the predicted guide sequences. The specificity of SpCas9-mediated DNA cleavage is sequence- and locus-dependent and governed by the quantity, position, and identity of mismatching bases [29]. However, the prediction of the efficiency parameter is more complicated as it is a multifactorial value that depends on GC content, folding energy, microstructures, or even preferences of specific nucleotides in positions proximal to the PAM sequence [30, 31].

Out of the several online tools for target sites prediction and sgRNA design, there are some such as CRISPR-P2.0 (http://crispr.hzau.edu.cn/CRISPR2/) [32] that still have limitations in polyploids [16]. We suggest using some other tools, such as CRISPOR (http://crispor.tefor.net/) [33] and CHOPCHOP (https://chopchop.cbu.uib.no/) [34], since they include the previously mentioned recommended features. Another suitable tool for polyploids is CRISPR MultiTargeter (https://multicrispr.net/) [35] as it allows to use multiple DNA sequences or similar genes as the input data. Thus, this tool allows predicting both common and unique sgRNAs among them. Finally, Cas-OFFinder (http://www.rgenome.net/cas-offinder/) [36] offers a more precise calculation of possible off-targets within the given genome, so it can be used as a second filter while choosing target sequences (*see* **Note 1**).

2.3 CRISPR/Cas9 Vector Requirements

The mutation efficiency of the CRISPR/Cas9 system not only depends on the sgRNA sequence but also on the expressions of Cas9 and sgRNAs [37]. Thus, it is important to select a CRISPR/Cas9 plasmid that includes the proper promoters for the experiment we want to perform. Typically, RNA polymerase III promoters are used to guide sgRNA expression, as they present a high and stable small nuclear RNA transcriptional activity [38]. In particular, the U6 promoter has been widely used for sgRNA expression, since it was already known to be efficient in other RNA-mediated tools like RNA interference (RNAi) [39, 40]. Despite many available CRISPR/Cas9 plasmids include U6 promoters from Arabidopsis (AtU6) to drive the sgRNA, these promoters are commonly used for genome editing of other plant species. Nevertheless, although the efficiency and mutation frequency was high in some of those cases [22], a great increase in these parameters has been achieved when endogenous U6 promoter sequences were used [41–43]. Therefore, the identification of highly and constitutively expressed RNA polymerase III endogenous promoters is strongly recommended [37, 38, 43].

The selection of the promoter to drive Cas9 expression is also critical for achieving high-efficiency gene targeting [44]. In most cases, the Cauliflower Mosaic Virus *35S* (*CaMV35S*) and ubiquitin promoters are used to drive the constitutive expression of Cas9, but the efficiency they achieve largely depends on the species they are used. For instance, ubiquitin promoter works better than the *CaMV35S* promoter in rice [37] and alfalfa [45], while none of them presents a high efficiency in Arabidopsis or citrus, in which *YAO* promoter is more recommended [46, 47]. Therefore, some bibliographic research to investigate the best option for the use of promoters in the target species or related genera is highly recommended (*see* **Note 2**).

2.4 Tools and Approaches for CRISPR/Cas9-Mediated Genome Editing Screening

When more than one sgRNA have been used to generate the edited plant (*see* **Note 1**), a screening of the possible InDels generated should be performed to analyze the success of the CRISPR/Cas9 genome editing. For that purpose, it is necessary to design oligonucleotides spanning the target sites to perform a PCR and an electrophoresis.

The Cas9 nuclease introduces the blunt cut in the DNA predominantly 3 bp upstream from the PAM. Therefore, if there is a restriction enzyme site overlapping these 3 bp 5′ of the PAM, it will be disrupted during DNA repair of the DSB. Thus, a restriction analysis of a PCR product flanking the target site allows to easily screen if the editing has been successful in that particular site.

There are different ways to analyze the CRISPR/Cas9-induced mutations at the targeted sequences. One of the most common approaches involves amplifying the targeted region by PCR and sequencing the amplicon by Sanger. Then, the sequences are analyzed with tools such as DSDecodeM (http://skl.scau.edu.cn/dsdecode/) [48], which automatically decodes the Sanger sequencing chromatograms with biallelic, heterozygous, and homozygous mutations into allelic sequences. However, this tool cannot be easily applied to decode genome editing in polyploid genomes. The other common approach requires to clone the PCR products into a vector and sequence several clones by Sanger, allowing to identify single alleles in each clone. However, this latter approach might lead to bias in the amplification and cloning of the different edited alleles coming from the several subgenomes in the polyploid species. Thus, there is a risk of an overrepresentation of a particular homoeologous target while missing information of others. To minimize this risk, it is important to sequence a high number of clones aiming to increase the chances that all the different homoeologous-edited loci are represented in the analysis.

The main alternative to the cloning and Sanger sequencing approaches is performing an amplicon sequencing of the targeted loci by next-generation amplicon sequencing (NGS). Although more expensive than PCR cloning and Sanger sequencing approaches, PCR-seq-based screening will deeply cover all the different edited loci in the sample. However, to the best of our knowledge, all the user-friendly tools designed to analyze CRISPR/Cas9 mutations using raw NGS data as an input, such as CRISPR-GA (http://crispr-ga.net/) [49] or Cas-analyzer (http://www.rgenome.net/cas-analyzer/#!) [50], are design for diploid species and do not allow to consider several sequences as possible targets, so the discrimination between homoeologs is not feasible. Therefore, an NGS-based approach requires designing custom scripts for sequences identification and quantification.

As an alternative to all the previously mentioned strategies, it would be recommendable to design primers that might specifically amplify each homoeolog when possible, allowing easier analyses of the genome editing in each subgenome by any of the Sanger sequencing-based approaches. For this purpose, we recommend PolyOligo online tool (http://ec2-52-52-41-39.us-west-1.compute.amazonaws.com/) [51], which allows a very easy and fast design of standard PCR or Sanger sequencing primers on polyploid and complex genomes. Using the genomic coordinates as input data for the locus of interest, this tool provides specific primers for the different subgenomes as output. Furthermore, it is possible to add reference genomes.

Finally, to analyze and align the sequencing results, we recommend the software MEGA [52], as it is a user-friendly tool to generate DNA sequence alignments.

3 Methods

3.1 Target Gene Sequences Retrieving

Download the genomic and coding sequences (CDS) of all homoeologs of the selected target gene from the chosen database.

3.2 sgRNA Prediction

We next will include the steps to predict and select sgRNAs using two different suitable tools: CRISPOR (http://crispor.tefor.net/) and MultiTargeter (https://multicrispr.net/).

3.2.1 CRISPOR

1. Enter a single genomic sequence with a maximum length of 2.3 kb in the box located in the **step 1** section. The sequence of one of the homoeologous genes must be selected. When the size of the gene of interest is longer than 2.3 kb, the whole process should be repeated as many times as the total gene length requires in order to get all the possible sgRNAs.
2. Select a genome in the **step 2** section. If the genome of interest is not available in the database of the web tool, it is possible to request CRISPOR technical support to upload it.
3. Select a protospacer adjacent motif (PAM) and Cas protein in the **step 3** section. Select "20bp-NGG – SpCas9, SpCas9-HF1, eSpCas9 1.1" for standard CRISPR/Cas9 knockout experiment and submit.
4. Predicted guide sequences for PAMs shall be listed by default in decreasing order of their specificity score, in particular, the MIT score [29] (Fig. 1) (*see* **Note 3**). The sgRNAs are displayed in green, yellow, and red colors depending on whether they are predicted to have a high, medium, or low specificity score, respectively. It also shows the position of the PAMs within the input sequence and whether it is in the forward or reverse strand. Furthermore, it offers information about the restriction enzymes that might be used to screen mutations induced by the particular sgRNA, as previously mentioned in the Methods section. Other information displayed is the

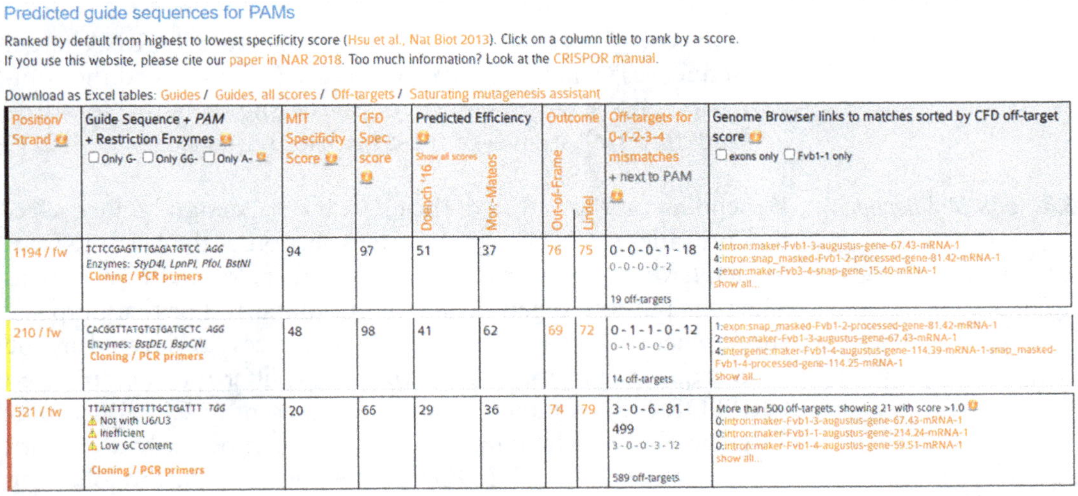

Fig. 1 Screenshot of the CRISPOR-predicted guide sequences output. The image has been modified to show only one example of sgRNAs with different specificity scores, i.e., high (green), medium (yellow), and low (red) score. Position of the PAM site in the input sequence and strand position of the sgRNAs is firstly displayed. The predicted guide sequences are listed by default in decreasing order of their MIT specificity score. Enzymes shown here might be used for the edit screening since their sites overlap the mainl cleavage site 3 bp upstream to the PAM. Information about the CFD specificity score, predicted efficiencies, and probabilities to obtain out-of-frame and InDel edits is also included. Finally, the output includes the number of off-targets with 0 to 4 mismatches and links to the location of the predicted off-targets in the Genome Browser

predicted efficiency of the sgRNAs, using two different scores, being the Doench '16 [53] for guides expressed under the U6 promoter (*see* **Note 4**). Finally, it offers information about the off-targets, indicating the number of them and the position of the mismatches located from 0 to 4 bp upstream of the PAM site. The last column provides links to the Genome Browser location of the predicted off-targets, allowing to locate those and predict the biological relevance of the putative predicted off-targets.

5. Download as Excel tables: "Guides, all scores" and "Off-targets."

3.2.2 CRISPR MultiTargeter

1. Select "Search for common and unique CRISPR guide RNA targets in multiple similar sequences" tool.
2. Select NN as 5′ dinucleotide sequence, 20-nucleotide target length, 3' PAM sequence orientation. To obtain common sgRNAs for all the homoeologous sequences, click on "No" in the mismatch allowance in the first 8 nucleotides section. Select NGG as PAM sequence, enter the input sequences, i.e., all homoeologous genomic sequences with a maximum length of 50 kb as FASTA format, and submit.
3. Predicted common guide sequences will be listed separately depending on the strand they are, and in order of their position

within the consensus sequence. As no mismatches were allowed, "Target sequences" and "Computed sgRNA target sequences" will be the same (*see* **Note 5**). Download the common sgRNA target sequences by clicking in both "forward" and "reverse" strands of the consensus sequence.

3.3 sgRNA Selection

Depending on the nature of the experimental design, guides selection priorities could change. For the selection of the best sgRNAs, we first have to identify those sgRNAs that are located within exons when the purpose is getting knock-out mutants. For that purpose, check whether the selected sgRNA sequences are located in the CDS sequence previously retrieved. Next, filter out the predicted CRISPOR sgRNA targets that are not present in CRISPR Multi-Targeter output. The intersection of data can be easily done with Microsoft Excel using VLOOKUP function. Then, the following parameters should be taken into account to prioritize desired targets:

1. The highest predicted specificity and efficiency score possible. Avoid sgRNAs with a specificity score below 50 in CRISPOR (marked in yellow and red). By doing so, sgRNAs with the lowest number of off-targets and/or sgRNAs with the higher number of mismatches with the putative off-targets are prioritized.
2. Location of off-targets displayed in the genome browser. Ideally, predicted off-targets should be located in intergenic regions to avoid disrupting other non-targeted CDSs.
3. Presence of commonly used lab restriction enzymes in the candidate sgRNA sequence for further mutation screening.
4. If using U6 promoters during sgRNA expression, guides must start with a G. Therefore, select those that naturally starts with G or alternatively change the first nt for a G in the primer designing when there is no optimal G-starting sgRNA.

To generate the final vector, a forward oligonucleotide including the 20 nts sequence of the selected sgRNA (without the PAM sequence) and its complementary reverse oligonucleotide will be designed in order to generate a double-stranded DNA fragment that will be inserted in the desired vector. As previously mentioned, neither the detailed construct designing nor the transformation of the species of interest will be discussed in this chapter since it depends on the researcher's preferences or goals.

3.4 Electrophoresis-Based Detection of Mutations

Once the plants transformed with the final sgRNAs-CRISPR/Cas9 vector have been obtained, isolate genomic DNA from control and several CRISPR lines. In order to determine whether the genome editing was successful, we propose two complementary approaches based on PCR and PCR followed by a restriction analysis:

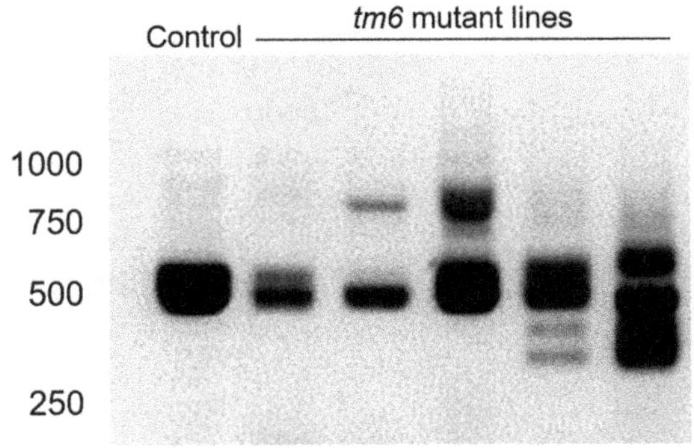

Fig. 2 PCR amplification of the CRISPR/Cas9 target site of *FaTM6* in control and CRISPR/Cas9 mutant lines of *Fragaria* x *ananassa*. Oligonucleotides flanking the target sites in a conserved region among all the homoeologous subgenomes were used in order to identify InDels in all of them. Numbers denote the sizes of the bands

3.4.1 Large InDels Detection by PCR

Large InDels might be generated when the experimental setup includes more than one sgRNA. In this case, design oligonucleotides spanning the target sites and perform a PCR in the control and the transformed lines (*see* **Note 6**). InDels size corresponding to the distance between two sgRNAs are expected to be found in the electrophoresis when these two sgRNAs have induced simultaneous DSBs (Fig. 2).

3.4.2 Detection of Genome Editing by Restriction Analysis of the PCR Product

As previously mentioned in the Subheading 2.4 section of Materials and in the Subheading 3.3 section of Methods, it is recommended to select sgRNAs that contain restriction enzyme sites near the PAM sequence. If this is the case, design oligonucleotides spanning the target site, amplify by PCR, and digest with the indicated restriction enzyme. Uncleaved bands indicate that the CRISPR/Cas9-mediated edition was successful.

3.5 Homoeolog-Specific Primer Design

In polyploids, it is important to take into account that the analysis by sequencing or restriction of the edited loci in the different homoeologous targets is hindered when they contain conserved sequences. Thus, to avoid this issue, it is essential to design specific primers to amplify and further analyze each homoeologous target locus separately. For this purpose, we recommend PolyOligo online tool, which allows an easy designing of oligonucleotides for PCR and sequencing on polyploid and complex genomes.

1. Select "PCR/Sanger – Standard PCR or Sanger sequencing primers" method
2. Indicate the target regions to amplify, providing the coordinates of the region of each homoeologous chromosome. Depending on the species genome, the annotation of the coordinates must be differently displayed

3. Select the reference genome and at least 20 as the maximum number of primer pairs to return. Regarding search options, we recommend *Deep—SLOW* depth in search options, 55–65 °C for the Tm, 20–25 nts for primer sizes, and 20–80% of GC-content. The only parameter to adjust off-target allowance is the size of 3′-end nucleotides seed used by the algorithm to search for possible off-targets. We recommend this parameter to be as long as the primer size (*see* **Note 7**). View results or download them as .txt or .bed format. Results will be displayed with target and designed primers information (*see* **Note 8**)

3.6 Analysis of Genome Editing by Sequencing

To study the CRISPR/Cas9-mediated sequences edits, the easiest and cheapest way is to amplify the targeted region by PCR, purify the bands, clone them into pGEM®-T Easy vector (Promega), and sequence by Sanger.

1. Perform a PCR with a high-fidelity DNA polymerase and homoeolog specific primers flanking the sgRNAs.
2. Purify the band from the agarose gel. If more than one band is obtained, purify each of them separately (*see* **Note 9**).
3. Since proofreading DNA polymerases generally generates blunt-ended PCR products, A-tailing reaction must be performed for cloning into pGEM®-T Easy following the manufacturer's instructions.
4. Clone the A-tailed products into pGEM®-T Easy following the manufacturer's instructions.
5. Select several positive clones for each homoeolog target for further Sanger sequencing. If it is not possible to design specific primers for each subgenome, select a high number of clones to try to cover at least five times each possible homoeolog target. Therefore, the number of clones to select depends on the ploidy of the species of interest.

3.7 Sequences Alignment by MEGA

In order to analyze the edits generated, an alignment of the homoeologous sequences of the CRISPR/Cas9 lines with the reference genome must be done. For that purpose, we recommend using MEGA software performing the following steps:

1. Download and install MEGA (https://www.megasoftware.net/).
2. Run MEGA.
3. Select Align function > Edit/Build Alignment > Create a new alignment.
4. Paste reference and transgenic lines DNA sequences in FASTA format.

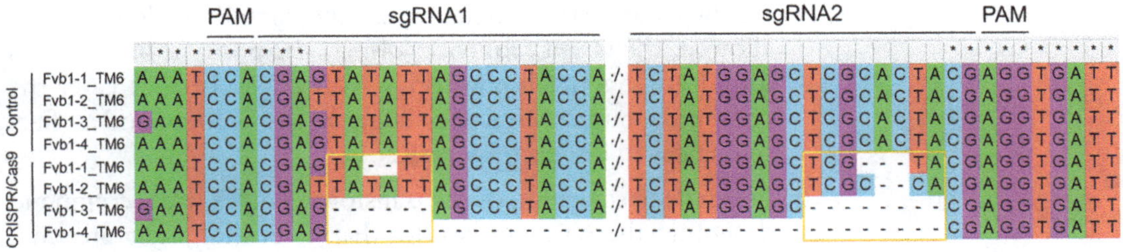

Fig. 3 DNA sequences alignment of the homoeologous genes of *FaTM6* in a control and a CRISPR/Cas9 mutant line. The homoeolog of *TM6* in the chromosome Fvb1–2 contains an SNP (marked with an arrow) 4 nts upstream of the PAM sequence in the target sequence of the sgRNA1, avoiding the DNA strand break by the CRISPR/Cas9 machinery. The rest of homoeologs were correctly edited for both sgRNAs in the positions next to PAM (orange squares), including a big deletion in the target of the Fvb1–4 chromosome

5. Select Alignment > Align by ClustalW. Figure 3 shows the alignment using as an example data from the genome editing of *FaTM6*, a gene encoding for a B class of MADS-box transcription factor, which plays a key role in another development in the octoploid strawberry *Fragaria* × *ananassa* [27].

3.8 CRISPR/Cas9 Edited Line Selection

Finally, once the mutation screenings and analyses have been successfully performed, selection of the most suitable generated lines shall be done according to the experiment goals. Since this protocol is focused on obtaining CRISPR/Cas9-mediated knockout lines, the selection of those should be done prioritizing premature stop codon mutations, InDels that are not multiples of three, and/or big deletions between two sgRNA target sequences.

4 Notes

1. Select and clone in the final vector more than one sgRNAs targeting the loci of interest. This strategy will improve the efficiency of the mutagenesis and will allow generating independent editions and/or large deletions [27, 54].
2. It is recommendable that both sgRNAs and Cas9 expression cassettes are included in the same binary vector, since the efficiency of coexpression from a single vector is higher than if both elements are separated into different vectors.
3. CRISPOR also display an alternative specificity score, the cutting frequency determination score (CFD) [53], which correlates better than MIT score with the total off-target cleavage fraction of a guide [55], so we recommend considering the former for the selection of the sgRNAs.
4. CRISPOR manual mentions that efficiency parameters are not very accurate, so it is preferable to select the sgRNAs based on the specificity and desired position.

5. In the case that mismatches in the first 8 nucleotides are allowed, the target sequences display the position of the SNPs found among the homoeologs with an X. In that case, the computed target sequences suggest the recommended final sequences to design those sgRNAs.
6. For general screening of the transformed lines, we recommend designing the oligonucleotides that flank the target sites in a conserved region among all the homoeologous subgenomes in order to identify InDels in all of them.
7. Nucleotides ($n = 12$ by default) from the 3′-end of the primer are aligned to the full genome and all matches are considered potential locations where the primer may bind. If primers bind in both directions relatively close to one another, then off-target amplification may occur. By that definition, an increment on the 3′-end seed size will have fewer potential binding sites on the genome, and therefore, fewer potential off-targets.
8. During the primer selection, check that the specific primers do not bind to the rest of the homoeolog targets. As the algorithm only penalizes off-targets based on the 3′-end sequence of primers, unspecific primer binding with mismatches at the 5′-end may occur if the input sequence does not present many possible primer combinations. This is the main limitation of the tool, as the algorithm does not reward those primers with mismatches near the 3′-end in the rest of homoeologs.
9. When specific primers for each homoeologous subgenome have been used, a single band is expected if there are not chimerism in the transgenic organism. Moreover, if there is no possibility to design specific primers for each homoelog, several amplicons are expected in the PCR. Therefore, each of them must be purified separately and characterize. An alternative to the gel purification step is to purify the PCR reaction and clone it directly into pGEM®-T Easy and later analyze the different edits by restriction and size analysis and/or Sanger sequencing.

Acknowledgments

CSG was supported by a grant from the Spanish Ministries of Science and Innovation (MICINN, RTI2018-09309-A-I00). CMP was supported by a grant from the European Research Council (ERC-2014-StG 638134).

References

1. Songstad DD, Petolino JF, Voytas DF, Reichert NA (2017) Genome editing of plants. CRC Crit Rev Plant Sci 36:1–23. https://doi.org/10.1080/07352689.2017.1281663
2. Zhu H, Li C, Gao C (2020) Applications of CRISPR–Cas in agriculture and plant biotechnology. Nat Rev Mol Cell Biol 21:661–677. https://doi.org/10.1038/s41580-020-00288-9
3. Gaj T, Gersbach CA, Barbas CF III (2013) ZFN, TALEN and CRISPR/Cas-based methods for genome engineering. Trends Biotechnol 31:397–405. https://doi.org/10.1016/j.tibtech.2013.04.004.ZFN
4. Cong L, Ran FA, Cox D et al (2013) Multiplex genome engineering using CRISPR/Cas systems. Science 339:819–823. https://doi.org/10.1126/science.1231143.Multiplex
5. Mojica FJM, Díez-Villaseñor C, García-Martínez J, Soria E (2005) Intervening sequences of regularly spaced prokaryotic repeats derive from foreign genetic elements. J Mol Evol 60:174–182. https://doi.org/10.1007/s00239-004-0046-3
6. Sorek R, Kunin V, Hugenholtz P (2008) CRISPR – A widespread system that provides acquired resistance against phages in bacteria and archaea. Nat Rev Microbiol 6:181–186. https://doi.org/10.1038/nrmicro1793
7. Jinek M, Chylinski K, Fonfara I et al (2012) A programmable dual-RNA-guided DNA endonuclease in adaptive bacterial immunity. Science (80-) 337:816–821. https://doi.org/10.1126/science.1225829
8. Feng Z, Zhang B, Ding W et al (2013) Efficient genome editing in plants using a CRISPR/Cas system. Cell Res 23:1229–1232. https://doi.org/10.1038/cr.2013.114
9. Mao Y, Zhang H, Xu N et al (2013) Application of the CRISPR-Cas system for efficient genome engineering in plants. Mol Plant 6: 2008–2011. https://doi.org/10.1093/mp/sst121
10. Jiang W, Zhou H, Bi H et al (2013) Demonstration of CRISPR/Cas9/sgRNA-mediated targeted gene modification in Arabidopsis, tobacco, sorghum and rice. Nucleic Acids Res 41:1–12. https://doi.org/10.1093/nar/gkt780
11. Woodhouse M, Burkart-Waco D, Comai L (2009) Polyploidy Nat Educ 2:1
12. Soppa J (2014) Polyploidy in archaea and bacteria: about desiccation resistance, giant cell size, long-term survival, enforcement by a eukaryotic host and additional aspects. J Mol Microbiol Biotechnol 24:409–419. https://doi.org/10.1159/000368855
13. Baatout S (1999) Molecular basis to understand polyploidy. Hematol Cell Ther 41:169–170. https://doi.org/10.1007/s00282-999-0169-5
14. Yin F, Liu W, Chai J et al (2018) CRISPR/Cas9 application for gene copy fate survey of polyploid vertebrates. Front Genet 9:1–7. https://doi.org/10.3389/fgene.2018.00260
15. Fang Z, Morrell PL (2016) Polyploidy boosts domestication. Nat Plants 2:1–2. https://doi.org/10.1038/NPLANTS.2016.116
16. Zaman QU, Li C, Cheng H, Hu Q (2019) Genome editing opens a new era of genetic improvement in polyploid crops. Crop J 7: 141–150. https://doi.org/10.1016/j.cj.2018.07.004
17. Shan Q, Wang Y, Li J et al (2013) Targeted genome modification of crop plants using a CRISPR-Cas system. Nat Biotechnol 31:684–686. https://doi.org/10.1038/nbt.2652
18. Kaur N, Alok A, Shivani et al (2020) CRISPR/Cas9 directed editing of lycopene epsilon-cyclase modulates metabolic flux for β-carotene biosynthesis in banana fruit. Metab Eng 59:76–86. https://doi.org/10.1016/j.ymben.2020.01.008
19. Sun X, Hu Z, Chen R et al (2015) Targeted mutagenesis in soybean using the CRISPR-Cas9 system. Sci Rep 5:1–10. https://doi.org/10.1038/srep10342
20. Gao J, Wang G, Ma S et al (2015) CRISPR/Cas9-mediated targeted mutagenesis in Nicotiana tabacum. Plant Mol Biol 87:99–110. https://doi.org/10.1007/s11103-014-0263-0
21. Andersson M, Turesson H, Nicolia A et al (2017) Efficient targeted multiallelic mutagenesis in tetraploid potato (Solanum tuberosum) by transient CRISPR-Cas9 expression in protoplasts. Plant Cell Rep 36:117–128. https://doi.org/10.1007/s00299-016-2062-3
22. Li C, Unver T, Zhang B (2017) A high-efficiency CRISPR/Cas9 system for targeted mutagenesis in Cotton (Gossypium hirsutum L.). Sci Rep 7:1–10. https://doi.org/10.1038/srep43902
23. Braatz J, Harloff HJ, Mascher M et al (2017) CRISPR-Cas9 targeted mutagenesis leads to simultaneous modification of different homoeologous gene copies in polyploid oilseed rape (Brassica napus). Plant Physiol 174:935–942. https://doi.org/10.1104/pp.17.00426

24. Park JJ, Yoo CG, Flanagan A et al (2017) Defined tetra-allelic gene disruption of the 4-coumarate:coenzyme A ligase 1 (Pv4CL1) gene by CRISPR/Cas9 in switchgrass results in lignin reduction and improved sugar release Mike Himmel. Biotechnol Biofuels 10:1–11. https://doi.org/10.1186/s13068-017-0972-0
25. Shan S, Mavrodiev EV, Li R et al (2018) Application of CRISPR/Cas9 to Tragopogon (Asteraceae), an evolutionary model for the study of polyploidy. Mol Ecol Resour 18: 1427–1443. https://doi.org/10.1111/1755-0998.12935
26. Morineau C, Bellec Y, Tellier F et al (2017) Selective gene dosage by CRISPR-Cas9 genome editing in hexaploid Camelina sativa. Plant Biotechnol J 15:729–739. https://doi.org/10.1111/pbi.12671
27. Martín-Pizarro C, Triviño JC, Posé D (2019) Functional analysis of the TM6 MADS-box gene in the octoploid strawberry by CRISPR/Cas9-directed mutagenesis. J Exp Bot 70:949–961. https://doi.org/10.1093/jxb/ery400
28. Eid A, Mohan C, Sanchez S et al (2021) Multiallelic, targeted mutagenesis of magnesium chelatase with CRISPR/Cas9 provides a rapidly scorable phenotype in highly polyploid sugarcane. Front Genome Ed 3. https://doi.org/10.3389/fgeed.2021.654996
29. Hsu PD, Scott DA, Weinstein JA et al (2013) DNA targeting specificity of RNA-guided Cas9 nucleases. Nat Biotechnol 31:827–832. https://doi.org/10.1038/nbt.2647
30. Moreno-Mateos MA, Vejnar CE, Beaudoin JD et al (2015) CRISPRscan: designing highly efficient sgRNAs for CRISPR-Cas9 targeting in vivo. Nat Methods 12:982–988. https://doi.org/10.1038/nmeth.3543
31. Xu H, Xiao T, Chen CH et al (2015) Sequence determinants of improved CRISPR sgRNA design. Genome Res 25:1147–1157. https://doi.org/10.1101/gr.191452.115
32. Liu H, Ding Y, Zhou Y et al (2017) CRISPR-P 2.0: an improved CRISPR-Cas9 tool for genome editing in plants. Mol Plant 10:530–532. https://doi.org/10.1016/j.molp.2017.01.003
33. Concordet JP, Haeussler M (2018) CRISPOR: intuitive guide selection for CRISPR/Cas9 genome editing experiments and screens. Nucleic Acids Res 46:W242–W245. https://doi.org/10.1093/nar/gky354
34. Labun K, Montague TG, Krause M et al (2019) CHOPCHOP v3: expanding the CRISPR web toolbox beyond genome editing. Nucleic Acids Res 47:W171–W174. https://doi.org/10.1093/nar/gkz365
35. Prykhozhij SV, Rajan V, Gaston D, Berman JN (2015) CRISPR multitargeter: a web tool to find common and unique CRISPR single guide RNA targets in a set of similar sequences. PLoS One 10:1–18. https://doi.org/10.1371/journal.pone.0119372
36. Bae S, Park J, Kim JS (2014) Cas-OFFinder: a fast and versatile algorithm that searches for potential off-target sites of Cas9 RNA-guided endonucleases. Bioinformatics 30:1473–1475. https://doi.org/10.1093/bioinformatics/btu048
37. Feng C, Su H, Bai H et al (2018) High-efficiency genome editing using a dmc1 promoter-controlled CRISPR/Cas9 system in maize. Plant Biotechnol J 16:1848–1857. https://doi.org/10.1111/pbi.12920
38. Qi X, Dong L, Liu C et al (2018) Systematic identification of endogenous RNA polymerase III promoters for efficient RNA guide-based genome editing technologies in maize. Crop J 6:314–320. https://doi.org/10.1016/j.cj.2018.02.005
39. Miyagishi M, Taira K (2002) U6 promoter-driven siRNAs with four uridine 3′ overhangs efficiently suppress targeted gene expression in mammalian cells. Nat Biotechnol 20:497–500. https://doi.org/10.1038/nbt0502-497
40. Xia XG, Zhou H, Ding H et al (2003) An enhanced U6 promoter for synthesis of short hairpin RNA. Nucleic Acids Res 31:e100. https://doi.org/10.1093/nar/gng098
41. Zhou J, Wang G, Liu Z (2018) Efficient genome editing of wild strawberry genes, vector development and validation. Plant Biotechnol J 16:1868–1877. https://doi.org/10.1111/pbi.12922
42. Long L, Guo DD, Gao W et al (2018) Optimization of CRISPR/Cas9 genome editing in cotton by improved sgRNA expression. Plant Methods 14:1–9. https://doi.org/10.1186/s13007-018-0353-0
43. Johansen IE, Liu Y, Jørgensen B et al (2019) High efficacy full allelic CRISPR/Cas9 gene editing in tetraploid potato. Sci Rep 9:1–7. https://doi.org/10.1038/s41598-019-54126-w
44. Gil-Humanes J, Wang Y, Liang Z et al (2017) High-efficiency gene targeting in hexaploid wheat using DNA replicons and CRISPR/Cas9. Plant J 89:1251–1262. https://doi.org/10.1111/tpj.13446
45. Wolabu TW, Cong L, Park J et al (2020) Development of a highly efficient multiplex genome editing system in outcrossing tetraploid alfalfa (*Medicago sativa*). Front Plant Sci 11:1–9. https://doi.org/10.3389/fpls.2020.01063

46. Yan L, Wei S, Wu Y et al (2015) High-efficiency genome editing in arabidopsis using YAO promoter-driven CRISPR/Cas9 system. Mol Plant 8:1820–1823. https://doi.org/10.1016/j.molp.2015.10.004
47. Zhang F, LeBlanc C, Irish VF, Jacob Y (2017) Rapid and efficient CRISPR/Cas9 gene editing in Citrus using the YAO promoter. Plant Cell Rep 36:1883–1887. https://doi.org/10.1007/s00299-017-2202-4
48. Liu W, Xie X, Ma X et al (2015) DSDecode: a web-based tool for decoding of sequencing chromatograms for genotyping of targeted mutations. Mol Plant 8:1431–1433. https://doi.org/10.1016/j.molp.2015.05.009
49. Güell M, Yang L, Church GM (2014) Genome editing assessment using CRISPR Genome Analyzer (CRISPR-GA). Bioinformatics 30: 2968–2970. https://doi.org/10.1093/bioinformatics/btu427
50. Park J, Lim K, Kim JS, Bae S (2017) Cas-analyzer: an online tool for assessing genome editing results using NGS data. Bioinformatics 33:286–288. https://doi.org/10.1093/bioinformatics/btw561
51. Ledda M, Cobo N, Lorant A, et al (2019) PolyOligo: a bioinformatic platform for identifying target DNA sequences for the development of sub-genome specific DNA markers in polyploid/complex genomes
52. Tamura K, Stecher G, Kumar S (2021) MEGA11: molecular evolutionary genetics analysis version 11. Mol Biol Evol 1–13. https://doi.org/10.1093/molbev/msab120
53. Doench JG, Fusi N, Sullender M et al (2016) Optimized sgRNA design to maximize activity and minimize off-target effects of CRISPR-Cas9. Nat Biotechnol 34:184–191. https://doi.org/10.1038/nbt.3437
54. Belhaj K, Chaparro-Garcia A, Kamoun S, Nekrasov V (2013) Plant genome editing made easy: targeted mutagenesis in model and crop plants using the CRISPR/Cas system. Plant Methods 9:1–10. https://doi.org/10.1186/1746-4811-9-39
55. Tycko J, Wainberg M, Marinov GK et al (2019) Mitigation of off-target toxicity in CRISPR-Cas9 screens for essential non-coding elements. Nat Commun 10:1–14. https://doi.org/10.1038/s41467-019-11955-7

Chapter 25

Developing a CRISPR System in Nongenetic Model Polyploids

Shengchen Shan, Bing Yang, Bernard A. Hauser, Pamela S. Soltis, and Douglas E. Soltis

Abstract

The genetic consequences following polyploidy (i.e., whole-genome duplication; WGD) vary greatly across organisms and through time since polyploidization. At the gene level in allopolyploids, changes include loss/retention of both parental gene copies, function/expression divergence between the two parental copies, and silencing of one parental copy. Functional studies of genes with different retention patterns contribute to a better understanding of the genetic factors underlying the success of polyploids. Most research on gene functions to date focuses on a few well-established genetic models or crops. However, many species that best exemplify the polyploidy process are nongenetic models; the lack of an efficient genome editing system hinders functional studies in these systems. In this chapter, we discuss the considerations of developing CRISPR, a robust and efficient genome editing system, in polyploid plants that are not genetic models. We use diploid and polyploid *Tragopogon* (Asteraceae) as examples of a well-studied evolutionary model system for which abundant genetic and genomic resources are lacking. Using this system, we provide our protocols for sgRNA design, plasmid construction, a useful protoplast transient assay, and a plant transformation method we developed for this system. We also provide suggestions for possible modifications to these protocols to help promote successful application to other non-models. With the rapid applications of CRISPR in plant sciences, the broad adaptation of CRISPR in studies of the evolutionary significance of WGD holds enormous potential. We hope our studies and methods developed for polyploid *Tragopogon* will provide a guideline for establishing a CRISPR system in other nongenetic model polyploids of evolutionary or other interest.

Key words CRISPR, Genome editing, Plasmid construction, Polyploidy, Nongenetic model, Transformation, Transient assay

---

Supplementary Information The online version contains supplementary material available at [https://doi.org/10.1007/978-1-0716-2561-3_25].

Yves Van de Peer (ed.), *Polyploidy: Methods and Protocols*,
Methods in Molecular Biology, vol. 2545, https://doi.org/10.1007/978-1-0716-2561-3_25,

Abbreviations

BAP	6-benzylaminopurine
BSA	Bovine serum albumin
DSB	Double-stranded DNA break
dsOligo	Double-stranded oligonucleotides
HDR	Homology-directed repair
IBA	Indole-3-butyric acid
NAA	Naphthaleneacetic acid
NHEJ	Non-homologous end joining
PAM	Protospacer adjacent motif
RT	Room temperature
sgRNA	Single guide RNA
spCas9	Cas9 from *Streptococcus pyogenes*
WGD	Whole-genome duplication

1 Introduction

Polyploidy (i.e., whole-genome duplication; WGD) is a major evolutionary force in all eukaryotes, especially plants [1, 2]. All living angiosperms (i.e., flowering plants) share one common WGD event, and many lineages have experienced multiple rounds of polyploidization [3, 4]. The genetic consequences following WGD may lead to physiological and phenotypic novelty and contribute to the evolutionary success of polyploids [5–7]. Gene fate following WGD varies greatly across different polyploid lineages as well as through time since polyploidization. The various genetic modifications following WGD include the loss or retention of both parental gene copies, expression/function divergence of the two parental copies, and silencing of one copy. Over time, the duplicated genome of a polyploid undergoes genome fractionation and shifts in gene expression, both of which are components of diploidization [1, 8, 9]. Genomic studies across multiple plant species indicate that genome fractionation is not a random process, and the function of a gene may determine its fate following WGD: highly connected genes that are part of the same pathways and transcription factors are often preferentially retained (i.e., the gene balance hypothesis) [10–12]; many other genes are quickly returned to single copy following WGD and are conserved singletons across diverse plant genomes [13, 14]. In addition, the rate of gene loss may differ between the two subgenomes in an allopolyploid, with one parental genome containing more genes than the other (i.e., genome dominance) [15, 16].

For all these reasons, functional studies of genes showing various patterns of gene retention and loss following WGD will help to

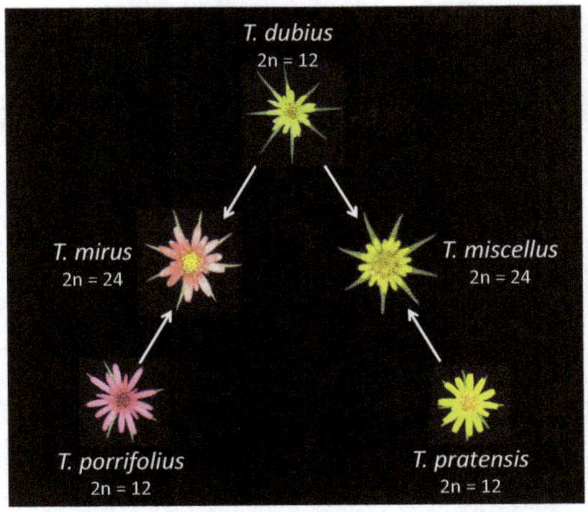

Fig. 1 The *Tragopogon* triangle. The diploid parents of allotetraploid *T. mirus* are diploids *T. dubius* and *T. porrifolius*; those of allotetraploids T. miscellus are T. dubius and *T. pratensis*. Reproduced from 2009 with permission from John Wiley and Sons

reveal the underlying genetic factors that contribute to the success of polyploids. Effectively and efficiently conducting these much-needed functional studies requires the development of an efficient genome editing system. To date, such functional studies have largely been confined to crops and well-developed model plant systems [17–20]. However, rich genomic resources and genome editing systems are either lacking or severely limited for many naturally occurring species that best exemplify various aspects of the polyploidy process. One such evolutionary model system for the study of genome doubling is found in *Tragopogon* (Asteraceae) [21–24]. *Tragopogon mirus* and *T. miscellus* are naturally formed allotetraploids, the parents of which are diploid *T. dubius* and *T. porrifolius* and diploid *T. dubius* and *T. pratensis*, respectively [22] (Fig. 1). Both allopolyploids are only 90–100 years old and therefore provide a unique opportunity to investigate the immediate consequences following WGD [21, 23].

CRISPR can be engineered to provide efficient and robust genome editing systems [19, 20, 25–29]. The CRISPR toolbox comprises various Cas proteins, including Cas9, Cas12a, Cas13, and their engineered variants [30]. Beyond genome editing, CRISPR technology has been applied in regulating gene expression and epigenome editing [30]. Compared to other CRISPR systems, CRISPR/Cas9 engineered from *Streptococcus pyogenes* (i.e., CRISPR/SpCas9) is the most popular and robust [30]. Therefore, this chapter focuses on the application of CRISPR/SpCas9, shortened as CRISPR/Cas9, in genome editing. The CRISPR/Cas9 system is a ribonucleoprotein complex consisting of Cas9 endonuclease and engineered CRISPR single guide RNA (sgRNA)

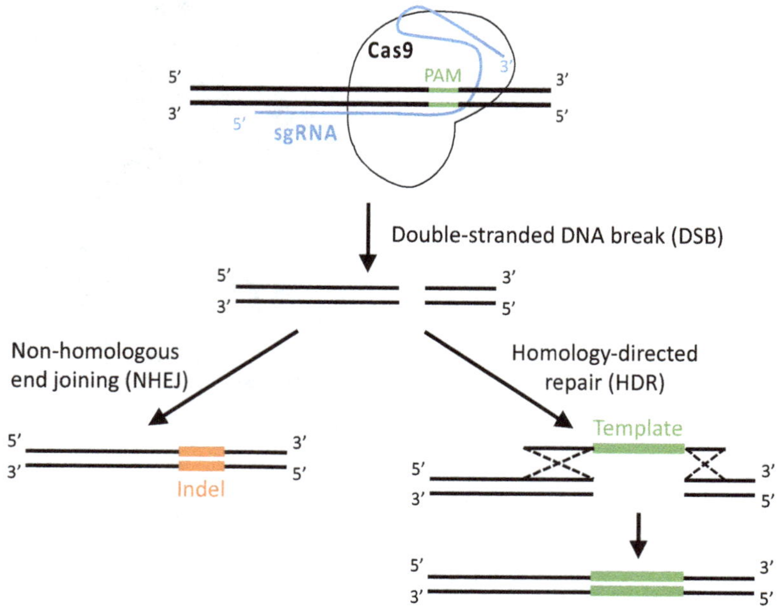

Fig. 2 Schematic description of the mechanism of CRISPR/Cas9-mediated genome editing. The Cas9-sgRNA complex recognizes and binds to any genome region with a PAM sequence (shown in green). If the spacer sequence (the first 20 nt at the 5' end) of the sgRNA matches the genomic sequence immediately upstream of PAM, a DSB will be introduced by the Cas9 endonuclease. If the DSB is repaired by the error-prone NHEJ pathway, an indel is introduced at the DSB site. Alternatively, with the presence of a template DNA, which is flanked by sequences that are identical to the genomic regions adjacent to the DSB (indicated by the dashed lines), the DSB is repaired by the HDR pathway

(Fig. 2). The Cas9-sgRNA complex recognizes and binds to any genomic regions with a protospacer adjacent motif (PAM) sequence (NGG for SpCas9; N represents any nucleotide). If the first 20 nucleotides (nt) at the 5′ end (i.e., the spacer sequence) of the sgRNA match the genomic sequence immediately upstream of PAM, Cas9 endonuclease cleaves both strands, leaving a double-stranded DNA break (DSB) [31]. The DSB will be intrinsically repaired by either a non-homologous end joining (NHEJ) pathway or a homology-directed repair (HDR) pathway [32]. The efficient but error-prone NHEJ pathway introduces small insertions or deletions at the DSB site, and this process has been widely adopted for gene knock-out [33–35]. In the presence of a template DNA, the less efficient but accurate HDR pathway has been used for gene replacement or targeted DNA insertion [36–38].

In this chapter, we use *Tragopogon* as an example [35] and provide detailed protocols of (1) sgRNA design, (2) construction of expressing plasmids for sgRNA and Cas9, (3) development and implementation of a transient assay to evaluate effectiveness (e.g., utilizing the *GFP* reporter system [20]), and (4) development of an

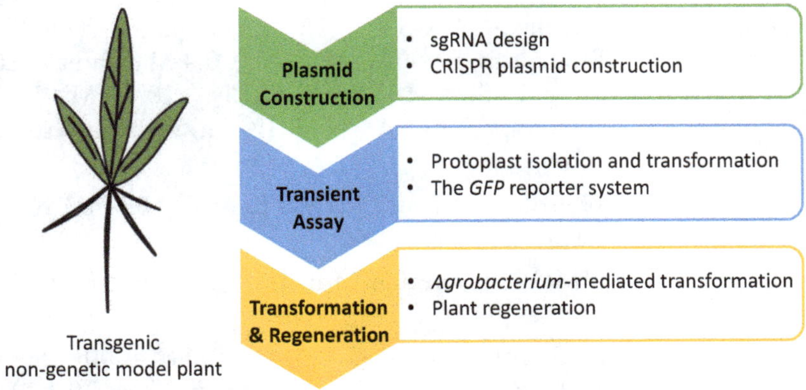

Fig. 3 The process of applying CRISPR in a non-genetic model plant species

effective plant transformation system—all of which are required for applying CRISPR (Fig. 3). We realize that each new non-model system will present new methodological challenges that will have to be solved by the investigator. Therefore, in the Notes section (*see* **Notes 1–11**), various suggestions and considerations that may be helpful in developing CRISPR more broadly across diverse plant systems (including polyploids) are discussed.

2 Materials

2.1 Molecular Cloning Reagents

1. Entry vector pENTR4-AtU6.1-AtU6.29 (Supplementary Fig. S1).
2. Destination vector pCAMBIA1300-Cas9-GFP (Supplementary Fig. S2).
3. Restriction enzymes: *Bsa*I, *BsmB*I, *Btg*ZI, *Xba*I, and *Xho*I (New England Biolabs, Ipswich, MA, USA).
4. Plasmid extraction kit (Promega, Madison, WI, USA).
5. 1 kb DNA ladder.
6. *E. coli* XL1-Blue chemically competent cells.
7. Electrocompetent *Agrobacterium tumefaciens* strain EHA105.
8. Luria-Bertani (LB) broth medium.
9. T4 DNA ligase.
10. Electroporator.
11. LR recombinant clonase.
12. 30 and 37 °C incubators.
13. 30 and 37 °C shaker incubators.
14. Gel electrophoresis apparatus.

2.2 Protoplast Transient Assay

1. Enzyme solution: 1.5% (m/v) cellulase RS (Desert Biologicals, Phoenix, AZ, USA), 0.75% (m/v) macerozyme R-10 (Desert Biologicals, Phoenix, AZ, USA), 0.4 M mannitol, 10 mM MES at pH 5.7, 10 mM $CaCl_2$, 20 mM KCl, 5 mM 2-mercaptoethanol, and 0.1% (m/v) bovine serum albumin (BSA).
2. W5 solution: 2 mM MES at pH 5.7, 154 mM NaCl, 125 mM $CaCl_2$, and 5 mM KCl.
3. MMG solution: 0.4 M mannitol, 15 mM $MgCl_2$, and 4 mM MES at pH 5.7.
4. PEG solution: 40% (m/v) PEG 4000 (Sigma-Aldrich, St. Louis, MO, USA), 0.4 M mannitol, and 0.1 M $CaCl_2$.
5. Clear plastic ice cream cups (5–8 oz) and lids (Solo, Lake Forest, IL, USA).
6. 1/2 MS medium: 2.22 g/L MS basal medium with vitamins (PhytoTech Labs, Lenexa, KS, USA), 30 g/L sucrose, 5 g/L Agargellan (PhytoTech Labs, Lenexa, KS, USA), pH 5.8.
7. Three plasmids used in the *GFP* reporter system [20]: (1) a plasmid expressing the *Cas9* gene; (2) a plasmid expressing the sgRNA; and (3) a plasmid expressing the mutant *GFP* gene.
8. Incubator (14 h light/10 h dark, 25 °C).
9. 15-mL round bottom falcon tube.
10. Bleach.
11. Tween-20.
12. Filter paper.
13. Blade.
14. Vacuum.
15. Nylon mesh (40 μm).
16. Centrifuge.
17. Bovine serum albumin (BSA).
18. Cell culture plate (12 wells, flat bottom).

2.3 Agrobacterium-Mediated Transformation

1. LB solid medium: 40 g/L LB agar; if adding antibiotics: 50 mg/L kanamycin and 50 mg/L rifampicin.
2. Callus induction medium: 4.44 g/L MS basal medium with vitamins, 30 g/L sucrose, 0.5 mg/L naphthaleneacetic acid (NAA), 1 mg/L 6-benzylaminopurine (BAP), 5 g/L Gelrite (PhytoTech Labs, Lenexa, KS, USA), pH 5.8. For selective callus induction medium, add 0.3 g/L timentin, and 15 or 25 mg/L hygromycin.
3. Co-cultivation medium: 4.44 g/L MS basal medium with vitamins, 30 g/L sucrose, 0.5 mg/L NAA, 1 mg/L BAP, 5 g/L Gelrite, 0.1 mM acetosyringone (filter-sterilized), pH 5.2.

4. COCO-1 medium: 2.41 g/L Woody plant basal medium with vitamins (PhytoTech Labs, Lenexa, KS, USA), 30 g/L sucrose, 2.5 mg/L BAP, 8.5 g/L Agargellan, 5% (v/v) coconut water (filter-sterilized), pH 5.7. For selective COCO-1 medium, add 0.3 g/L timentin, and 15 or 25 mg/L hygromycin.
5. WPM medium: 4.44 g/L MS basal medium with vitamins, 30 g/L sucrose, 5 g/L Gelrite, 0.4 g/L timentin, 0.2 g/L cefotaxime, 15 or 25 mg/L hygromycin, pH 5.8.
6. TY medium: 5 g/L tryptone, 3 g/L yeast extract, 50 mg/L rifampicin, 50 mg/L kanamycin, 0.1 mM acetosyringone, pH 5.5.
7. AAM infection medium: 2.95 g/L KCl, 0.25 g/L $MgSO_4 \cdot 7H_2O$, 0.15 g/L $CaCl_2 \cdot 2H_2O$, 0.15 g/L $NaH_2PO_4 \cdot H_2O$, 10 mg/L $MnSO_4 \cdot 4H_2O$, 3 mg/L H_3BO_3, 2 mg/L $ZnSO_4 \cdot 7H_2O$, 0.75 mg/L KI, 0.25 mg/L $Na_2MoO_4 \cdot 2H_2O$, 0.025 mg/L $CuSO_4 \cdot 5H_2O$, 0.025 mg/L $CoCl_2 \cdot 6H_2O$, 0.876 g/L L-glutamine, 0.266 g/L L-aspartic acid, 0.174 g/L L-arginine, 7.52 mg/L glycine, 0.1 g/L myo-inositol, 0.5 mg/L nicotinic acid, 0.5 mg/L pyridoxine hydrochloride, 0.1 mg/L thiamine hydrochloride, 27.8 mg/L $FeSO_4 \cdot 7H_2O$, 37.26 mg/L Na_2EDTA, 0.5 g/L casamino acids, 68.5 g/L sucrose, 36 g/L glucose, 0.1 mM acetosyringone, pH 5.2.
8. COCO-R medium: 2.41 g/L Woody plant basal medium with vitamins, 30 g/L sucrose, 1.5 mg/L indole-3-butyric acid (IBA), 8.5 g/L Agargellan, pH 5.7.
9. WPM medium: 2.41 g/L Woody plant basal medium with vitamins, 30 g/L sucrose, 8.5 g/L Agargellan, 0.3 g/L timentin, 15 or 25 mg/L hygromycin, pH 5.7.
10. Incubator and shaker incubator for bacterial culture.
11. Forceps.
12. Razor blade and blade holder.
13. Spectrophotometer.
14. Incubator (14 h light/10 h dark, 25 °C).
15. Growth chamber (14 h light/10 h dark, 25 °C).

3 Methods

3.1 sgRNA Design and CRISPR Plasmid Construction

1. Predict the intron-exon structure of the target gene. *Geneious* (version 8.1.8; [39]) can be used to map the cDNA sequence to the genomic sequence (maximum gaps per read: 95%; maximum gap size: 1000).

2. CRISPR/Cas9 target site is manually identified and is usually constrained to the first few exons for gene knock-out. Locate a PAM sequence (NGG for SpCas9); the 20 nt directly upstream of PAM is a potential CRISPR/Cas9 target site. The spacer sequence (the first 20 nt at the 5′ end) of the sgRNA is identical to the sequence of the CRISPR/Cas9 target site. A pair of complementary oligonucleotides is synthesized for each CRISPR/Cas9 target site.
3. Anneal one pair of complementary oligonucleotides using the ramping down program in a thermocycler (Supplementary Table S1). Based on the sequence at the multiple cloning sites of plasmid pENTR4-AtU6.1-AtU6.29, the overhang of the double-stranded oligonucleotides (dsOligo) is 5′-GATT-3′ on the sense strand and 5′-AAAC-3′ on the antisense strand.

Component	20 μL reaction
Oligonucleotide 1 (100 μM)	1 μL
Oligonucleotide 2 (100 μM)	1 μL
H_2O	18 μL

4. Digest 1 μg of pENTR4-AtU6.1-AtU6.29 vector with *Btg*ZI at 60 °C for 3 h. Inactivate the enzyme at 80 °C for 20 min.

Component	50 μL reaction
pENTR4-AtU6.1-AtU6.29	x μL (1 μg)
*Btg*ZI	2 μL (5 units/μL)
10× CutSmart buffer	5 μL
H_2O	50 μL

5. Ligate *Btg*ZI-digested and purified pENTR4-AtU6.1-AtU6.29 vector with the first dsOligo (e.g., gTraPDS1). Incubate the ligation reaction at room temperature (RT) for 3 h or 4 °C overnight. Use half of the ligation product (5 μL) to transform *E. coli* XL1-Blue competent cells.

Component	10 μL reaction
*Btg*ZI-digested pENTR4-AtU6.1-AtU6.29	x μL (50 ng)
dsOligo 1 (e.g., gTraPDS1)	5 μL
T4 ligase	2 μL
10× T4 ligase buffer	1 μL
H_2O	10 μL

6. Colony-PCR-based (Supplementary Table S2) confirmation of the positive clones. Use forward primer AtU6-F3 (located at the AtU6–1 promoter region; the sequences of the primers used in this chapter are listed in Supplementary Table S3) and the reverse primer at the insert region (here using gTraPDS-R1). The expected size of the PCR product is 883 bp. Extract the plasmid from the cultures· of positive colonies (i.e., pENTR4-AtU6.1-AtU6.29-gTraPDS1).
7. pENTR4-AtU6.1-AtU6.29-gTraPDS1 can be double digested by *Xho*I and *Xba*I. The sizes of the two expected bands are 471 bp and 2759 bp.

Component	50 μL reaction
pENTR4-AtU6.1-AtU6.29-gTraPDS1	x μL (1 μg)
*Xho*I	2 μL (5 units/μL)
*Xba*I	2 μL (5 units/μL)
10× CutSmart buffer	5 μL
H_2O	50 μL

8. Sequence the positive candidate clone by using primer AtU6-F2 (located at the AtU6.29 promoter).
9. Insert the second dsOligo. Digest 1 μg pENTR4-AtU6.1-AtU6.29-gTraPDS1 by using *Bsa*I at 37 °C for 4 h. Inactivate the enzyme at 60 °C for 20 min.

Component	50 μL reaction
pENTR4-AtU6.1-AtU6.29-gTraPDS1	x μL (1 μg)
*Bsa*I	2 μL (5 units/μL)
10× CutSmart buffer	5 μL
H_2O	50 μL

10. Ligate the second dsOligo, i.e., gTraPDS2, with *Bsa*I-digested and purified pENTR4-AtU6.1-AtU6.29-gTraPDS1.

Component	10 μL reaction
*Bsa*I-digested pENTR4-AtU6.1-AtU6.29-gTraPDS1	x μL (50 ng)
dsOligo 2 (e.g., gTraPDS2)	5 μL
T4 ligase	2 μL
10× T4 ligase buffer	1 μL
H_2O	10 μL

11. Colony-PCR-based confirmation of the positive clones. Forward primer, gTraPDS-F2, and reverse primer, gTraPDS-R1, are used. The expected size of the PCR product is 524 bp. Culture the positive colonies and extract plasmid DNA.
12. *Bsm*BI restriction enzyme cuts pENTR4-AtU6.1-gTraPDS2-AtU6.29-gTraPDS1 into four segments: 236 bp, 528 bp, 773 bp, and 1667 bp. Incubate the digestion reaction at 55 °C for 4 h.

Component	50 μL reaction
pENTR4-AtU6.1-gTraPDS2-AtU6.29-gTraPDS1	x μL (1 μg)
*Bsm*BI	2 μL (5 units/μL)
10× NEBuffer 3.1	5 μL
H_2O	50 μL

13. Sequence the positive candidate clones by using primer AtU6-R2. The final plasmid is named pENTR4-AtU6.1-gTraPDS2-AtU6.29-gTraPDS1, which will be used as the Gateway entry vector.
14. Perform Gateway LR reaction. The destination vector, pCAMBIA1300-Cas9-GFP, contains a Gateway recombination site: *attR*1-ccdB-*attR*2. On the entry vector (pENTR4-AtU6.1-gTraPDS2-AtU6.29-gTraPDS1), the sgRNA cassettes are flanked by the Gateway recombination sites: *attL*1 and *attL*2. Transfer the Gateway recombination reaction to *E. coli* strain XL1-Blue and identify the correct plasmid.

Component	5 μL reaction
pENTR4-AtU6.1-gTraPDS2-AtU6.29-gTraPDS1	1 μL
pCAMBIA1300-Cas9-GFP	1 μL
LR clonase	1 μL
H_2O	2 μL

15. Colony-PCR-based confirmation of the positive clones. Use forward primer gTraPDS-F1 (located at the sgRNA site) and the reverse primer GmUbi-R (located at the promoter region of the *GFP* gene). The expected size of the PCR product is 1585 bp. In addition, primers gTraPDS-F2 and AtU6-R2 can also be used for PCR confirmation. The expected size of the PCR product is 632 bp. Extract the plasmid from the positive colonies (i.e., pCAMBIA1300-Cas9-GFP-AtU6.1-gTraPDS2-AtU6.29-gTraPDS1).

16. pCAMBIA1300-Cas9-GFP-AtU6.1-gTraPDS2-AtU6.29-gTraPDS1 can be double digested by *Bam*HI and *Xho*I. The sizes of expected bands are: 472 bp, 1051 bp, 1094 bp, 6.9 kb, and 10 kb.
17. Transform *Agrobacterium* with the positive plasmid by using an electroporator (the protocol can be found at https://github.com/GatorShan/CRISPR_nongenetic_model). Extract plasmid from the positive *Agrobacterium* colonies and transform *E. coli* cells. Extract plasmids from the positive *E. coli* colonies (e.g., 4–5 colonies) and confirm the plasmids using PCR and restriction enzyme digestion as shown in **steps 15** and **16**.

3.2 Protoplast Isolation and Transformation

3.2.1 Prepare Plant Material

1. Remove the pericarp (fruit wall) of *Tragopogon* achenes to expose the seed; sterilize the seeds in 20% (v/v) bleach (containing a few drops of Tween-20) for 20 min (shake slowly). Rinse the seeds 4 times with autoclaved deionized water to remove excess bleach on the seed surface. Blot the seeds on the autoclaved filter paper.
2. Grow the seeds on 1/2 MS medium in a sterile ice cream box in an incubator (14 h light/10 h dark, 25 °C).
3. Cotyledons and true leaves are harvested 3 weeks after seed germination.

3.2.2 Protoplast Isolation

1. Cut cotyledons and true leaves into approximately 0.5-mm strips using a razor blade. Do not press the razor blade vertically to avoid any additional damage to the cells.
2. Transfer the leaf pieces to 15 mL 0.6 M mannitol (for plasmolysis), shake at 30–50 rpm, RT for 15 min.
3. Remove the 0.6 M mannitol solution and add 15 mL enzyme solution.
4. Vacuum for 45 min in the dark.
5. Enzyme digestion (shake at 30–40 rpm; 25–26 °C; in the dark for 5 hours).
6. Filter the enzyme solution (containing protoplasts) through 40-μm nylon mesh. Shake the flask slowly to avoid the protoplasts sinking to the bottom.
7. Wash the protoplasts using 21% (m/v) sucrose solution.
   (a) Prepare four 15-mL round bottom falcon tubes. Add 5 mL 21% sucrose to each tube; add ~5 mL filtered enzyme solution (containing protoplasts) on top of the 21% sucrose.
   (b) Centrifuge at 250 g for 3 min at RT.

8. Wash the protoplasts again using W5 solution.
   (a) Prepare four new 15-mL round bottom falcon tubes; add 3 mL W5 to each tube; transfer the ~2 mL middle layer (containing the protoplasts) from each tube (from **step 7b**) to the new tubes each containing 3 mL W5; mix gently and slowly; add another 2 mL W5 solution to each tube, mix gently and slowly; centrifuge at 250 g for 3 min at RT (protoplasts clump at the bottom), and remove the supernatant.
9. Wash the protoplasts again using W5 solution.
   (a) Add 2 mL W5 solution to each tube; mix gently and slowly. Add another 2 mL W5 solution; mix gently and slowly. Centrifuge at 100 g for 2 min at RT.
10. Remove the supernatant and add 1 mL W5 solution to each tube; mix the W5 solution from two tubes into one tube. Because four tubes are used at **step 7a**, there will be two tubes after this mixing step. Put the two tubes on ice for 30 min (in the dark).
11. Centrifuge at 100 g for 2 min at RT; remove the supernatant, and add 400 μL MMG solution to each of the two tubes that remain at this point; mix gently and slowly.

3.2.3 Protoplast Transformation

1. For each protoplast transformation experiment, in the 2-mL tube, add (in this order):
   (a) 30 μL plasmid (containing 10 μg plasmid; if using multiple plasmids in one transformation reaction, add 10 μg each). Transform the protoplasts with three plasmids if using the *GFP* reporter system.
   (b) 200 μL MMG (containing protoplasts) and mix.
   (c) 230 μL PEG solution (equal volume: 30 μL + 200 μL), mix gently and slowly.
2. Put the tube in the dark at RT for 15 min.
3. Wash by using W5 solution.
   (a) Add 460 μL (1× volume) W5 solution; mix gently and slowly.
   (b) Add 920 μL (2× volume) W5 solution; mix gently and slowly.
4. Centrifuge at 250 g for 3 min; remove the supernatant and add 400 μL W5 solution to the 2-mL tube; mix gently and slowly.
5. Coat the cell culture plate with 5% BSA.
   (a) Add ~600 μL 5% BSA to the well and pipette up and down.
6. Transfer 400 μL W5 solution (containing protoplasts) to the well; cover the cell culture plate with foil.

3.3 Tragopogon Transformation

3.3.1 Agrobacterium Culture

1. Streak the *Agrobacterium tumefaciens* strain EHA 105 (from **step 17** of Subheading 3.1) onto an LB plate containing antibiotics (e.g., 50 mg/L kanamycin and 50 mg/L rifampicin). Incubate the plate at 28 °C for 2 days.
2. Pick *Agrobacterium* colonies and inoculate into 20 mL liquid TY medium (containing antibiotics and acetosyringone). Shake at 100 rpm, 28 °C for 24 h.

3.3.2 Transformation

1. Prepare leaf explants: use leaf tissue from **step 3** of Subheading 3.2.1 and chop them into 1-cm leaf segments. Gently press the blade on the leaf surface to make wounded explants (this promotes transformation efficiency). Transfer the leaf explants to callus induction medium (without antibiotics) for pre-cultivation and place the plates in the dark for 1–2 days.
2. Prepare *Agrobacterium* suspension: centrifuge TY medium from **step 2** of Subheading 3.3.1 at 2000 rpm for 10 min; remove the TY medium and re-suspend the *Agrobacterium* pellet with AAM infection medium. For each transformation experiment, prepare 20 mL AAM infection medium with $OD_{600} = 0.1$.
3. Co-culture of explants and *Agrobacterium*: immerse explants in 20 mL AAM infection medium for 30 min; explants are thoroughly blotted dry on autoclaved filter paper and transferred to co-cultivation medium. Place the plates in the dark for 3 days.
4. Transfer the explants to selective callus induction medium and place the plates (covered by foil) inside the incubator for 3–5 weeks. Move explants to fresh medium biweekly. Callus will emerge from the cut edge of the explants. Successful transformants will survive on the selection medium and emit green fluorescence under blue light.
5. Transfer the callus to selective COCO-1 medium and place the plates inside the same incubator for 2 weeks.
6. Move the explants (with callus) to selective MS medium for shooting. Shoots will regenerate from the callus within 2 weeks.
7. Excise shoots from the explants/callus and transfer the shoots to COCO-R medium for rooting.
8. Transfer rooted shoots to soil inside a tray with a lid. Put the tray in the incubator for 2 weeks. Repot the seedlings and put the pots inside the growth chamber.

4 Notes

1. Bacterial strains and cloning plasmids are freely available for research purposes upon request from either the Yang or Soltis laboratory under a Material Transfer Agreement (MTA). A request can be submitted at https://github.com/GatorShan/CRISPR_nongenetic_model.
2. Choice of species. If a range of ploidal levels exists in the species/genus of interest, start with a diploid representative, and have all methods well-developed before applying CRISPR in the polyploids [35].
3. Target gene selection. Start with a single-copy gene within the genome; a marker gene for which a mutant has an obvious phenotype (such as an albino phenotype) is an excellent candidate for the first gene knock-out using CRISPR [20].
4. sgRNA design. In an allopolyploid with multiple homeologs, confine the sgRNA target region within the conserved sequence between the homeologs. By doing so, one sgRNA could knock out all homeologs. Detailed guidelines of sgRNA design can be found at [20].
5. Subheading 3.1, **step 14**: before setting up the Gateway LR recombination, linearize the entry vector with a restriction enzyme (e.g., *Aat*II, *Eco*NI) to prevent its replication in *E. coli* and increase the recombination efficiency [40].
6. Using transient assay to identify the most efficient CRISPR system [20]. For example, the *GFP* reporter system can be used to test the expression level of the CRISPR component. This method could circumvent the ploidal level of the host organism as the system utilizes CRISPR to knock out an exogenous gene.
7. The efficiencies of three CRISPR/Cas9 systems, tailored for rice, wheat, and *Arabidopsis*, can be compared in a nongenetic model species from protoplast transient assays (utilizing the *GFP* reporter system) [20]. These CRISPR/Cas9 systems differ in their species-specific codon-optimized *Cas9* gene and promoters driving *Cas9* and sgRNA [20]. The results of these comparisons using the protoplast transient assays will then enable the investigator to select the most efficient system for future use.
8. Subheading 3.2.2, **step 5**: the enzyme digestion time varies depending on the species and physiological characters of the leaf. Thick leaves (e.g., leaves with thick cuticle and recalcitrant cell walls) may require longer digestion time than soft leaves. If working on protoplast isolation for the first time, check the

status of the leaf strips under a microscope every 1.5 hours to ensure that cells are not over-digested (i.e., broken cells and extensive cell debris).

9. Plant transformation. Adapt from a phylogenetically closely related species if possible; if not available, adapt from the methods described here.

10. Subheading 3.3.2, **step 4**: the duration of placing explants on selective callus induction medium varies among *Tragopogon* species: 3 weeks for diploid *T. porrifolius*; 5 weeks for tetraploid *T. mirus*.

11. Evaluation of genome editing results. Sub-cloning is recommended as there are four possible alleles (two per diploid parent) for each locus in an allotetraploid. It is challenging to interpret a chromatogram [20].

References

1. Van de Peer Y, Mizrachi E, Marchal K (2017) The evolutionary significance of polyploidy. Nat Rev Genet 18:411–424. https://doi.org/10.1038/nrg.2017.26
2. Soltis DE, Visger CJ, Soltis PS (2014) The polyploidy revolution then...and now: Stebbins revisited. Am J Bot 101:1057–1078. https://doi.org/10.3732/ajb.1400178
3. Jiao Y, Wickett NJ, Ayyampalayam S et al (2011) Ancestral polyploidy in seed plants and angiosperms. Nature 473:97–100. https://doi.org/10.1038/nature09916
4. Leebens-Mack JH, Barker MS, Carpenter EJ et al (2019) One thousand plant transcriptomes and the phylogenomics of green plants. Nature 574:679–685. https://doi.org/10.1038/s41586-019-1693-2
5. Van de Peer Y, Ashman TL, Soltis PS, Soltis DE (2021) Polyploidy: an evolutionary and ecological force in stressful times. Plant Cell 33:11–26. https://doi.org/10.1093/plcell/koaa015
6. Soltis PS, Soltis DE (2016) Ancient WGD events as drivers of key innovations in angiosperms. Curr Opin Plant Biol 30:159–165. https://doi.org/10.1016/j.pbi.2016.03.015
7. Soltis PS, Liu X, Marchant DB et al (2014) Polyploidy and novelty: Gottlieb's legacy. Evolution 30:818–830. https://doi.org/10.2307/2407821
8. Wendel JF, Lisch D, Hu G, Mason AS (2018) The long and short of doubling down: polyploidy, epigenetics, and the temporal dynamics of genome fractionation. Curr Opin Genet Dev 49:1–7. https://doi.org/10.1016/j.gde.2018.01.004
9. Soltis PS, Marchant DB, Van de Peer Y, Soltis DE (2015) Polyploidy and genome evolution in plants. Curr Opin Genet Dev 35:119–125. https://doi.org/10.1016/j.gde.2015.11.003
10. Birchler JA, Veitia RA (2012) Gene balance hypothesis: connecting issues of dosage sensitivity across biological disciplines. Proc Natl Acad Sci 109:14746–14753. https://doi.org/10.1073/pnas.1207726109
11. Birchler JA, Veitia RA (2010) The gene balance hypothesis: implications for gene regulation, quantitative traits and evolution. New Phytol 186:54–62
12. Birchler JA, Veitia RA (2007) The gene balance hypothesis: from classical genetics to modern genomics. Plant Cell 19:395–402
13. Li Z, Defoort J, Tasdighian S et al (2016) Gene duplicability of core genes is highly consistent across all angiosperms. Plant Cell 28:326–344. https://doi.org/10.1105/tpc.15.0087
14. Tasdighian S, Van Bel M, Li Z et al (2017) Reciprocally retained genes in the angiosperm lineage show the hallmarks of dosage balance sensitivity. Plant Cell 29:2766–2785. https://doi.org/10.1105/tpc.17.00313
15. Woodhouse MR, Cheng F, Pires JC et al (2014) Origin, inheritance, and gene regulatory consequences of genome dominance in polyploids. Proc Natl Acad Sci 111:5283–5288. https://doi.org/10.1073/pnas.1402475111
16. Schnable JC, Springer NM, Freeling M (2011) Differentiation of the maize subgenomes by genome dominance and both ancient and ongoing gene loss. Proc Natl Acad Sci 108:

4069–4074. https://doi.org/10.1073/pnas.1101368108

17. Shan Q et al (2013) Targeted genome modification of crop plants using a CRISPR-Cas system. Nat Biotechnol 31:8–10. https://doi.org/10.1038/nbt.2652
18. Scheben A, Wolter F, Batley J et al (2017) Towards CRISPR/Cas crops – bringing together genomics and genome editing. New Phytol:682–698. https://doi.org/10.1111/nph.14702
19. Jaganathan D, Ramasamy K, Sellamuthu G et al (2018) CRISPR for crop improvement: an update review. Front Plant Sci 9:1–17. https://doi.org/10.3389/fpls.2018.00985
20. Shan S, Soltis PS, Soltis DE, Yang B (2020) Considerations in adapting CRISPR/Cas9 in nongenetic model plant systems. Appl Plant Sci 8:e11314. https://doi.org/10.1002/aps3.11314
21. Novak SJ, Soltis DE, Soltis PS (1991) Ownbey's *Tragopogons*: 40 years later. Am J Bot 78: 1586–1600
22. Ownbey M (1950) Natural hybridization and amphiploidy in the genus *Tragopogon*. Am J Bot 37:487–499
23. Tate JA, Symonds VV, Doust AN et al (2009) Synthetic polyploids of *Tragopogon miscellus* and *T. mirus* (Asteraceae): 60 years after Ownbey's discovery. Am J Bot 96:979–988. https://doi.org/10.3732/ajb.0800299
24. Soltis DE, Soltis PS, Pires JC et al (2004) Recent and recurrent polyploidy in *Tragopogon* (Asteraceae): cytogenetic, genomic and genetic comparisons. Biol J Linn Soc 82:485–501
25. Hsu PD, Lander ES, Zhang F (2014) Development and applications of CRISPR-Cas9 for genome engineering. Cell 157:1262–1278. https://doi.org/10.1016/j.cell.2014.05.010
26. Zhang F (2019) Development of CRISPR-Cas systems for genome editing and beyond. Q Rev Biophys 52. https://doi.org/10.1017/s0033583519000052
27. Zhu H, Li C, Gao C (2020) Applications of CRISPR-Cas in agriculture and plant biotechnology. Nat Rev Mol Cell Biol 21:661–677. https://doi.org/10.1038/s41580-020-00288-9
28. Chen K, Wang Y, Zhang R et al (2019) CRISPR/Cas genome editing and precision plant breeding in agriculture. Annu Rev Plant Biol 70:667–697. https://doi.org/10.1146/annurev-arplant-050718-100049
29. Gao C (2018) The future of CRISPR technologies in agriculture. Nat Rev Mol Cell Biol 19: 275–276. https://doi.org/10.1038/nrm.2018.2
30. Zhang Y, Malzahn AA, Sretenovic S, Qi Y (2019) The emerging and uncultivated potential of CRISPR technology in plant science. Nat Plants 5:778–794. https://doi.org/10.1038/s41477-019-0461-5
31. Jinek M, Chylinski K, Fonfara I et al (2012) A programmable dual-RNA-guided DNA endonuclease in adaptive bacterial immunity. Science 337:816–821
32. Symington LS, Gautier J (2011) Double-strand break end resection and repair pathway choice. Annu Rev Genet 45:247–271. https://doi.org/10.1146/annurev-genet-110410-132435
33. Morineau C, Bellec Y, Tellier F et al (2017) Selective gene dosage by CRISPR-Cas9 genome editing in hexaploid *Camelina sativa*. Plant Biotechnol J 15:729–739. https://doi.org/10.1111/pbi.12671
34. Alagoz Y, Gurkok T, Zhang B, Unver T (2016) Manipulating the biosynthesis of bioactive compound alkaloids for next-generation metabolic engineering in opium poppy using CRISPR-Cas9 genome editing technology. Sci Rep 6:1–9. https://doi.org/10.1038/srep30910
35. Shan S, Mavrodiev EV, Li R et al (2018) Application of CRISPR/Cas9 to *Tragopogon* (Asteraceae), an evolutionary model for the study of polyploidy. Mol Ecol Resour 18:1427–1443. https://doi.org/10.1111/1755-0998.12935
36. Yu QH, Wang B, Li N et al (2017) CRISPR/Cas9-induced targeted mutagenesis and gene replacement to generate long-shelf life tomato lines. Sci Rep 7:1–9. https://doi.org/10.1038/s41598-017-12262-1
37. Vu TV, Sivankalyani V, Kim EJ et al (2020) Highly efficient homology-directed repair using CRISPR/Cpf1-geminiviral replicon in tomato. Plant Biotechnol J 18:2133–2143. https://doi.org/10.1111/pbi.13373
38. Hahn F, Eisenhut M, Mantegazza O, Weber APM (2018) Homology-directed repair of a defective glabrous gene in *Arabidopsis* with cas9-based gene targeting. Front Plant Sci 9: 424. https://doi.org/10.3389/fpls.2018.00424
39. Kearse M, Moir R, Wilson A et al (2012) Geneious basic: an integrated and extendable desktop software platform for the organization and analysis of sequence data. Bioinformatics 28: 1647–1649. https://doi.org/10.1093/BIOINFORMATICS/BTS199
40. Char SN, Li R, Yang B (2019) CRISPR/Cas9 for mutagenesis in rice. In: Kumar S, Barone P, Smith M (eds) Transgenic plants. Methods in molecular biology 1864:279–293

Chapter 26

Efficiently Editing Multiple Duplicated Homeologs and Alleles for Recurrent Polyploids

Rui-Hai Gan, Li Zhou, and Jian-Fang Gui

Abstract

Research on the evolutionary fate of duplicated genes in recurrent polyploids is scarce due to the difficulties in disentangling the different homeologs and alleles of duplicated genes. This chapter describes the detailed procedures to identify different homeologs and alleles of duplicated genes, to analyze their molecular characteristics, and to reveal their functional divergence by gene editing with CRISPR/Cas9 (clustered regularly interspaced short palindromic repeats/CRISPR-associated system 9). Using the gene editing approach, we efficiently constructed multiple knockout mutant lines with single or simultaneously disrupted different homeologs or alleles in a recurrent polyploid fish, demonstrating its usability for targeting and mutating multiple divergent homeologs and alleles in recurrent duplicated genomes.

Key words Recurrent polyploid, Duplicated genes, Homeolog, Allele, CRISPR/Cas9

1 Introduction

Polyploids are organisms with three or more chromosome sets (genomes). All plants and most species of extant vertebrates seem to have evolved from polyploid ancestors. Therefore, it is believed that whole genome duplication (WGD) or polyploidy paves the way for evolutionary innovations and species speciation [1–3]. Besides the two rounds (2R) of WGDs that occurred during early vertebrate evolution, a third (3R) WGD shared by teleost was proposed to happen 226–316 million years ago [4, 5]. Additionally, WGDs occurred repeatedly in some taxonomic fish orders, such as Acipenseriformes, Cypriniformes, Salmoniformes, etc. [6–10]. A fourth autopolyploidy or allopolyploidy event has happened during the evolution of salmonids [11] or cyprinids [9, 10, 12, 13], respectively. In comparison with tetraploid goldfish (*Carassius auratus*), two extra rounds of polyploidy, an early allopolyploidy and a later autopolyploidy, were supposed to have taken place during hexaploid gibel carp (*C. gibelio*) evolution [14–16]. Duplicated

Yves Van de Peer (ed.), *Polyploidy: Methods and Protocols*,
Methods in Molecular Biology, vol. 2545, https://doi.org/10.1007/978-1-0716-2561-3_26,

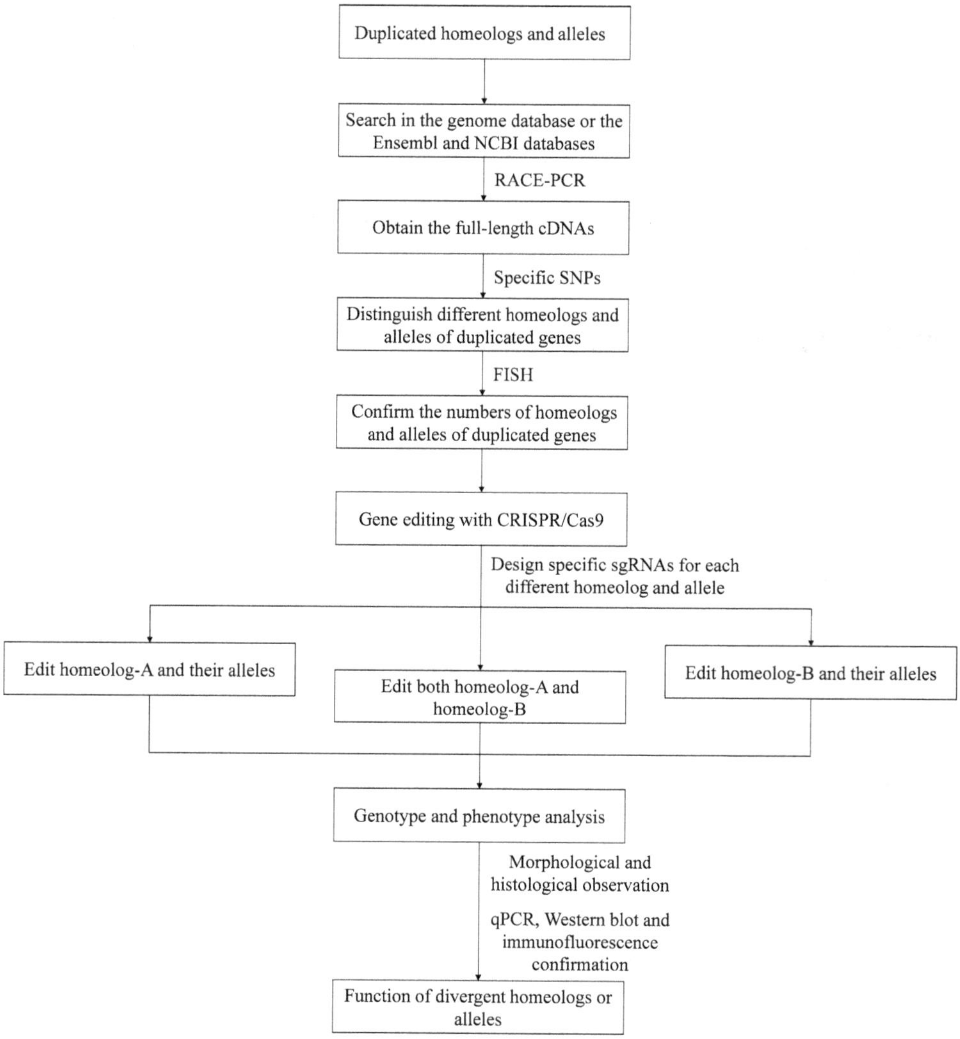

Fig. 1 Technical framework to edit multiple duplicated homeologs or alleles and to investigate the functional divergence in recurrent polyploids

homeologs or alleles can be produced by allopolyploidy or autopolyploidy events. Under the relaxed purifying selection, they might be pseudogenized, lost, and subfunctionalized or evolve a new function (neofunctionalization) [2, 17, 18]. In comparison with paleopolyploids with diploidized genomes, it is an important challenge to examine evolutionary fates of the duplicated homeologs or alleles in recurrent polyploids with higher ploidy levels because it is much more difficult to bioinformatically and experimentally disentangle the differential targets of duplicated copies. In this chapter, we outline a technical approach (Fig. 1) in a hexaploid fish and describe the detailed procedures to study the functional divergence of multiple duplicated homeologs and alleles [19].

The process begins with the identification of different copies of duplicated genes. Searching in assembled genome data or the Ensembl and NCBI databases, we can obtain the homologous genome sequences of the candidate gene. Specific or degenerate primers are designed to amplify the full-length cDNA transcripts. Rapid amplification of cDNA ends (RACE) is the most popular technology. It is necessary to use Taq DNA Polymerase High Fidelity and carefully disentangle different copies of the target gene. Fluorescence in situ hybridization (FISH) can be used to confirm the copy numbers and to directly visualize the physical location of duplicated genes on the homoeologous or homologous chromosomes. Taken the results of above analyses, homeolog/allele diversification and biased retention/loss of duplicated genes will be discussed.

Clustered regularly interspaced short palindromic repeats/CRISPR-associated system 9 (CRISPR/Cas9) gene editing [20–22] is a key technology for genetic manipulation. Using this genome editing technology, we efficiently constructed multiple knockout mutant lines with single or simultaneously disrupted different homeologs or alleles, demonstrating the possibility to target and mutate multiple divergent homeologs and alleles in recurrent duplicated genomes [19]. Quantitative PCR (qPCR), Western blot, or immunofluorescence can be used to conform the knockout effects. To reveal the functional divergence of different homeologs and alleles, detailed comparisons, such as morphological and histological observation, transcriptome profiling, and dynamic expression changes of marker genes, between wild-type (WT) and different mutant lines need to be carried out.

2 Materials

2.1 RNA Extraction

1. Microcentrifuge.
2. FastPrep®-24 (MP Biomedicals).
3. Tissue or embryo of interest.
4. SV total RNA isolation system (Promega).
5. TRIzol® reagent (Invitrogen).
6. Molecular biology-grade chloroform.

2.2 PCR Cloning

1. Thermocycler.
2. Microcentrifuge.
3. SMARTer® RACE 5′/3′ Kit (Clontech).
4. TransStart® FastPfu Fly DNA Polymerase (Transgen).
5. RevertAid First Strand cDNA Synthesis kit (Thermo Fisher Scientific).

6. Gel extraction kit (Omega).
7. Zero Background ZT4-Blunt Fast Cloning Kit (Zomanbio).
8. Trans5α Chemically Competent Cell (Transgen).

2.3 FISH

1. Upright fluorescence microscope (Carl Zeiss).
2. DIG-Nick Translation Mix (Roche).
3. Biotin-Nick Translation Mix (Roche).
4. Giemsa stain solution (Solarbio).
5. RNase solution: 100 μg/mL RNase A in 2× SSC.
6. Fixative: methanol:glacial acetic acid = 3:1.
7. DAPI (4′,6-Diamidine-2′-phenylindole): Keep a 1 mg/mL of DAPI in ddH_2O and store at −20 °C. Dilute in 1× PBS for a final concentration of 1 μg/mL.
8. 20× SSC stock: 300 mM sodium citrate and 3 M NaCl, adjusted to pH 7.0 with HCl and store at room temperature. This stock solution is diluted to 2× SSC and 0.1× SSC with ddH_2O.
9. Phosphate-buffered saline (PBS) (10× stock): 1.37 M NaCl, 27 mM KCl, 18 mM KH_2PO_4, 100 mM Na_2HPO_4, adjust to pH 7.4 with HCl and store at room temperature. This stock solution is diluted to 1× PBS with ddH_2O.

2.4 Gene Editing with CRISPR/Cas9

1. Thermocycler.
2. PLI-100 (Harvard).
3. Microcentrifuge.
4. TranscriptAid T7 High Yield Transcription Kit (Thermo Fisher Scientific).
5. Mmessage mMACHINE™ T7 (Thermo Fisher Scientific).
6. TransStart® FastPfu Fly DNA Polymerase (Transgen).
7. Zebrafish-Codon-Optimized Cas9 plasmid [23].
8. pUC19-gRNA vector [24].
9. XbaI digestion enzyme kit (NEB).
10. 70% and 100% nuclease-free ethanol.
11. NaOH solution: 50 mM NaOH in ddH_2O.

2.5 Histologic Section

1. Embedding machine.
2. Pathology slicer.
3. Frozen platform.
4. Dehydrator.
5. Hematoxylin and Eosin (H & E) Staining Kit (Servicebio).

6. Alcohol benzene.
7. 4% paraformaldehyde (PFA).
8. 75%, 85%, 95%, and 100% ethanol.

2.6 qPCR

1. CFX96TM Real-Time PCR System (Bio-Rad).
2. Hard-Shell Low-Profile 96-Well Skirted PCR Plates (Bio-Rad).
3. iTaqTM Universal SYBR® Green Supermix (Bio-Rad).

2.7 Western Blot

1. MiniProGel™ Protein Gel-Cast and Electrophoresis System (Beyotime).
2. MiniBlot™ Electrophorectic Transfer Cell (Beyotime).
3. ImageQuant LAS 4000mini (General Electric).
4. TBST: 0.1% Tween-20 in 1× TBS.
5. Blocking buffer: 5% skim milk powder in TBST.
6. Tris-buffered saline (TBS) (10× stock): 1.5 M NaCl, 250 mM Tris-base, adjust to pH 7.4 with HCl and store at room temperature. This stock solution is diluted to 1× TBS with ddH_2O.
7. 1× SDS-PAGE running buffer: 25 mM Tris-base, 250 mM glycine, 0.1% SDS.
8. 1× transfer buffer: 25 mM Tris-base, 192 mM glycine, 20% methanol.
9. Phosphate-Buffered Saline (PBS) (10× stock): 1.37 M NaCl, 27 mM KCl, 18 mM KH_2PO_4, 100 mM Na_2HPO_4, adjust to pH 7.4 with HCl and store at room temperature. This stock solution is diluted to 1× PBS with ddH_2O.
10. RIPA buffer: 50 mM Tris-HCl, PH7.4, 150 mM NaCl, 1% NP-40, 1 mM EDTA, 0.25% deoxycholic acid, 1 mM NaF, 1× proteinase inhibitor cocktail (Roche).

2.8 Immunofluorescence Staining

1. Upright fluorescence microscope (Carl Zeiss).
2. DAPI (4′,6-Diamidine-2′-phenylindole): Keep a 1 mg/mL of DAPI in ddH_2O and store at −20 °C. Dilute in 1× PBS for a final concentration of 1 μg/mL.
3. Phosphate-Buffered Saline (PBS) (10× stock): 1.37 M NaCl, 27 mM KCl, 18 mM KH_2PO_4, 100 mM Na_2HPO_4, adjust to pH 7.4 with HCl and store at room temperature. This stock solution is diluted to 1× PBS with ddH_2O.
4. PBST: 0.1% Tween-20 in 1× PBS.
5. 75%, 85%, and 100% ethanol.

6. Antibody dilution buffer (ADB): 0.05% Triton X-100, 10% normal donkey serum (or the serum corresponding to your secondary antibody), and 3% BSA in 1× PBS. Store in 1 mL aliquots at −20 °C.
7. Blocking solution: 10% ADB in 1× PBS.

3 Methods

3.1 RNA Extraction

The total RNA is extracted using a combination of TRIzol® reagent (Invitrogen) and SV total RNA isolation system (Promega) and following the recommendations of manufacturer's protocols.

1. Put 30–100 mg of tissue or embryos in a 1.5-mL nuclease-free microcentrifuge tube with 1 mL of TRIzol® reagent, and add one 3-mm and six 1-mm mill beads. The samples are homogenized in a FastPrep®-24 tissue homogenizer at a vertical velocity of 4 m/s for 20 s. Put the homogenate on ice for 5 min, and homogenize it again, 3 times in total. Finally, incubate at room temperature for 3 min.
2. Add 0.2 mL of chloroform to the mixture and vortex vigorously for 15 s. Then incubate at room temperature for 3 min. Centrifuge the microtube at 16,000 × g for 20 min at 4 °C.
3. Carefully transfer 400 μL of the colorless upper aqueous phase to a new nuclease-free microtube, add an equal volume of absolute ethanol, and mix thoroughly by pipetting up and down.
4. Transfer the mixture to a spin column (Promega) and place it in a collection tube (Promega). Centrifuge at 12,000 × g for 1 min at 4 °C, and discard the liquid from the collection tube.
5. Add 600 μL of RNA wash solution (Promega) to the spin column, centrifuge at 12,000 × g for 1 min at 4 °C, and discard the liquid from the collection tube.
6. Add 50 μL of freshly prepared DNase incubation solution (40 μL of Yellow Core Buffer, 5 μL of 0.09 M $MnCl_2$ and 5 μL of DNase I) to the spin column, and incubate at 37 °C for 25 min.
7. Add 200 μL of DNase stop solution to the spin column and centrifuge at 12,000 × g for 1 min at 4 °C, and discard the liquid from the collection tube.
8. Add 600 μL of RNA wash solution to the spin column, centrifuge at 12,000 × g for 1 min at 4 °C, and discard the liquid from the collection tube.
9. Add 250 μL of RNA wash solution to the spin column, centrifuge at 16,000 × g for 2 min at 4 °C.

10. Transfer the spin column to a new elution tube, and add 50 μL of nuclease-free water onto the membrane of the spin column. Centrifuge at 12,000 × g for 1 min at 4 °C and collect the RNA solution. The concentration of the purified RNA is measured with Nanodrop 2000c spectrophotometer and stored at −80 °C.

3.2 cDNA Synthesis

The first strand cDNAs are synthesized from total RNA using RevertAid First Strand cDNA Synthesis kit (Thermo Fisher Scientific) according to the manufacturer's protocol.

1. Add the following reagents into a nuclease-free microtube on ice: 0.1 ng to 5 μg of total RNA, 1 μL of Oligo $(dT)_{18}$ primer, 1 μL of Random Hexamer primer, and nuclease-free water to get a final volume of 12 μL. Mix gently, centrifuge briefly and incubate at 65 °C for 5 min, and chill on ice.
2. Add the following components in the indicated order: 4 μL of 5× Reaction Buffer, 1 μL of RiboLock RNase Inhibitor (20 U/μL), 2 μL of 10 mM dNTP Mix, 1 μL of RevertAid M-MuLV RT (200 U/μL). The final volume of the reaction is 20 μL. Mix gently and centrifuge briefly.
3. Place the tube in a thermocycler and run the reaction with the following conditions: primer annealing at 25 °C, for 5 min; extension at 42 °C, for 1 h; terminate the reaction at 70 °C for 5 min.
4. The product of the first strand cDNA synthesis can be directly used for PCR or qPCR or stored at −20 °C.

3.3 PCR Cloning and Sequence Analysis

1. A BLAST search is first performed against assembled genome data or against the Ensembl and NCBI databases (genomic information unavailable) using the homologous sequence of the human, mouse or zebrafish candidate gene as the query. According to the genome sequence or the conserved sequences among multiple species, the specific or degenerate primers are designed to amplify the cDNA fragment.
2. PCR amplification with high-fidelity polymerase is performed in a total volume of 25 μL containing 15.5 μL of ddH_2O, 5 μL of 5× TransStart® FastPfu Fly Buffer, 2 μL of 2.5 mM dNTPs, 0.5 μL of forward primer (at 10 μM), 0.5 μL of reverse primer (at 10 μM), 0.5 μL of TransStart® FastPfu Fly DNA Polymerase, 1 μL of cDNA. Amplification is performed in a PCR thermocycler with the following program: initial denaturation at 95 °C for 2 min, 35 cycles of (denaturation at 95 °C for 20 s, primer annealing at 56 °C for 20 s, extension at 72 °C for 1 min) and a final extension of 5 min at 72 °C.

3. Analyze 2 μL of the PCR product on a 1.2% agarose gel with a suitable DNA staining dye to make sure a DNA fragment of the correct size is observed. If not, adjust PCR conditions or redesign the primers.
4. The unpurified PCR product can be used to directly perform Sanger sequencing to confirm the sequence of the target gene.
5. The PCR products are purified using the Gel Extraction Kit (Omega) and ligated with the Zero Background ZT4-Blunt Fast Cloning Kit (Zomanbio) following the manufacturer's instruction.
6. The ligation products are transfected into Trans5α chemically competent cells (Transgen) according to the manufacturer's protocol.
7. Single positive colonies are confirmed by PCR using ZT4-Blunt primers according to the manufacturer's instruction of the Zero Background ZT4-Blunt Fast Cloning Kit and Sanger sequencing is performed (*see* **Note 1**).
8. Multiple alignment of the cDNA fragment sequences of the target gene is performed using DNAMAN version 7.0 to manually design gene-specific primers (GSPs) for 5′ and 3′ RACE-PCR.
9. Total RNA is used to construct a SMART cDNA library with the SMARTer® RACE 5′/3′ Kit (Clontech) according to the manufacturer's protocol.
10. The 5′ and 3′ RACE-PCRs are performed on the SMART cDNA library using GSPs and adaptor primers (UPM). The PCR amplification with a high-fidelity polymerase is performed and the PCR products are purified, cloned, and Sanger sequenced as described above (*see* **Note 1**).
11. To obtain the full-length cDNA and carefully disentangle different copies of duplicated genes, other GSPs are designed based on the sequences of the 5′ and 3′ cDNA ends of the target gene and used to perform 5′ end to 3′ end PCR with a high-fidelity polymerase from the cDNA. Then the PCR products are purified, cloned, and Sanger sequenced (*see* **Note 1**).
12. To classify homeologs and alleles of duplicated genes by specific single nucleotide polymorphisms (SNPs), multiple nucleotide alignment is performed using DNAMAN version 7.0 (*see* **Note 2**).

3.4 FISH

3.4.1 Preparation of Chromosome Metaphases

Using the Phytohemagglutinin (PHA) in vivo injection method (polyploid fish as an example):

1. Inject 2 μL of 10 μg/μL PHA (diluted with 0.75% saline) for each gram of fish at the base of the pectoral fin, once every 12 h, for a total of 2 times.

2. After another 12 h, 1 μL of 10 μg/μL colchicine (diluted with 0.75% saline) is injected into each gram of fish at the base of the pectoral fin.
3. After 4 h, the head kidney tissue is taken out and washed with 0.75% saline to prepare tissue slurry. Before sampling, the fish is anesthetized with MS-222 (Sigma).
4. Transfer the cell suspension to a 10-mL centrifuge tube, centrifuge at 350 × g for 8 min, and discard the supernatant.
5. The cells are resuspended with 8 mL of 0.075 M KCl (preheated to 37 °C in advance) and incubated at 37 °C for 50–60 min, shake gently every 20 min.
6. Add 2 mL of freshly prepared fixative at room temperature, and fix for 5 min. Centrifuge at 350 × g for 8 min at 4 °C, and discard the supernatant.
7. Resuspend the cells with 2 mL of pre-cooled fresh fixative and fix for 15 min at 4 °C. Centrifuge at 350 × g for 8 min at 4 °C, and discard the supernatant.
8. Repeat **step** 7 three more times.
9. Add 100–200 μL of pre-cooled fresh fixative and resuspend the cells by gentle pipetting.
10. Suck approximately 10 μL of fixed cell suspension with a pipette and drop it onto a pre-cooled slide. The slides are dried at room temperature for 6–8 h. Good slides for FISH should have 10–20 metaphase spreads per 1–2 mm^2.
11. Randomly take 1–2 slides, add 500 μL Giemsa stain solution on the slide, incubate them at room temperature for 20 min, wash them with water, air dry them, and examine the slides with a microscope to analyze the number and structure of chromosomes. Store at −20 °C until the use for FISH.

3.4.2 Probe Labeling

1. Probe DNA is labeled by Biotin-Nick Translation Mix (Roche) or DIG-Nick Translation Mix (Roche). Add the following reagents into a microcentrifuge tube on ice: 1 μg of probe DNA, 4 μL of Biotin-Nick Translation Mix or 4 μL of DIG-Nick Translation Mix, and ddH_2O to get a final volume of 20 μL. Mix gently, centrifuge briefly and incubate overnight (12–16 h) at 16 °C (*see* **Note 3**).
2. Store labeled probes at −20 °C or immediately prepare the hybridization mixture.

3.4.3 Hybridization

1. Take the slides with a chromosome metaphase and bake them at 65 °C for 1 h.
2. Evenly spread 200 μL of RNase solution on the slide, cover with Parafilm and incubate for 1 h at 37 °C in a wet box containing a small amount of water.

3. Remove the Parafilm and wash the slides in 2× SSC for 5 min on a shaking platform.
4. Successively place the slides in a Coplin jar containing 80%, 95% and 100% ethanol to dehydrate for 5 min each, and air dry at room temperature.
5. Denature the slides with a chromosome metaphase in 2× SSC with 70% formamide (pre-heated) for 3.5 min at 62 °C.
6. Immediately immerse the slides in ice-cold 70%, 90%, and 100% ethanol to dehydrate for 5 min each, and air dry at room temperature.
7. Prepare the probe hybridization mixture. The components of the mixture are as follows: 30 μL of 100% formamide, 6 μL of 20× SSC, 3 μL of 10% SDS, 3 μL of salmon sperm DNA, 8 μL of 50% dextran sulfate, labeled probe (200–300 ng) and ddH_2O to get a final volume of 60 μL per microcentrifuge tube (*see* **Note 4**).
8. Mix gently, centrifuge briefly and denature the probe mixture for 10 min at 98 °C, and immediately incubate on ice for at least 30 min.
9. Add 60 μL of the denaturation probe mixture to a completely dry slide with denatured metaphase chromosomes, cover it with Parafilm without bubbles, and hybridize it for 16–24 h at 37 °C in a wet box containing a small amount of water.

3.4.4 Post-Hybridization Washing and Signal Detection

1. Remove the Parafilm and wash the slides in 2× SSC for 5 min at room temperature on a shaking platform.
2. Wash the slides in 2× SSC with 20% formamide for 5 min at room temperature on a shaking platform.
3. Wash the slides in 2× SSC with 1% Tween-20 for 5 min at room temperature on a shaking platform.
4. Wash the slides in 0.1× SSC with 1% Tween-20 for 5 min at room temperature on a shaking platform.
5. Wash the slides in 1× PBS with 1% Tween-20 for 5 min at room temperature on a shaking platform.
6. Wash the slides in 1× PBS for 5 min at room temperature on a shaking platform.
7. Shake off any excess liquid, incubate the slides with 5 ng/μL of FITC-conjugated anti-digoxigenin antibody from sheep (Roche) or 5 ng/μL of ExtrAvidin-Cy3 antibody (Sigma) in 1× PBS containing 1% BSA, cover with Parafilm and incubate at 37 °C for 2 h in dark conditions (*see* **Note 5**).
8. Wash the slides in 1× PBS with 1% Tween-20 for 10 min at room temperature on a shaking platform.

9. Wash the slides 3 times with 1× PBS on a shaking platform for 5 min each time. Then incubate with 1 μg/mL of DAPI in 1× PBS containing 1% BSA for 10 min at room temperature in dark conditions.
10. Wash the slides in 1× PBS with 1% Tween-20 for 10 min at room temperature on a shaking platform.
11. Wash the slides 3 times with 1× PBS on a shaking platform for 5 min each time.
12. Remove the buffer. After the slides are slightly dried, seal the slides with anti-fade mounting medium. Capture images under an upright fluorescence microscope.

3.5 Gene Editing with CRISPR/Cas9

3.5.1 Target sgRNA Sequence Selection and Design

Single guide RNAs (sgRNAs) can be designed using the online service website http://zifit.partners.org/ZiFiT/ or https://www.benchling.com/crispr/ by entering a target sequence (*see* **Note 6**). Alternatively, sgRNAs can be designed manually by targeting 5′-N(20)-NGG-3′ (targeting template strand) or 5′-CCN-N(20)-3′ (targeting non-template strand), using following criteria:

1. For the coding genes, find the coding sequence (CDS) region of the target gene, analyze the corresponding genome structure, and identify the exon part of CDS.
2. The target site should be selected as much as possible in the first 2/3 region of the CDS and after the ATG, but not in the last exon. It is best to destroy important domains and/or all isoform/variant.
3. At the same time, consider avoiding the existence of additional in-frame start codons downstream of the first start codon. The target site can also be selected at the junction of intron and exon to disrupt gene splicing. In principle, do not choose the 3′- or 5′-untranslated regions.
4. Verify the target sites′ uniqueness: the 23-nt sequences including the 20-nt target sgRNA sequence and the -NGG PAM sequence are blasted against the reference genome sequence to detect the specificity of the target sites in the genome and avoid off-target genome editing events.

3.5.2 Synthesis of Target sgRNA Primers

A target sgRNA primer pair is synthesized as follows:

sgRNA-Forward: 5′-GTAATACGACTCACTATA+20-nt sgRNA target site+GTTTTAGAGCTAGAAATAGC-3′.

sgRNA-Reverse: 5′-AAAAGCACCGACTCGGTGCC-3′.

The 20-nt sgRNA target site sequence is the 20-nt sequence directly upstream of the PAM (5′-NGG-3′) but does not contain the -NGG PAM sequence.

3.5.3 Preparation of sgRNA

1. Prepare template for the sgRNA by in vitro transcription. A target sgRNA primer pair (sgRNA-Forward and sgRNA-Reverse) and pUC19-gRNA vector are used for PCR amplification with a high-fidelity polymerase. The PCR amplification is performed in a total volume of 50 μL containing 32 μL of ddH_2O, 10 μL of 5× TransStart® FastPfu Fly Buffer, 4 μL of 2.5 mM dNTPs, 1 μL of sgRNA-Forward (at 10 μM), 1 μL of sgRNA-Reverse (at 10 μM), 1 μL of TransStart® FastPfu Fly DNA Polymerase, 1 μL of pUC19-gRNA vector (20 ng/μL). Amplification is performed in a PCR thermocycler as follows: initial denaturation at 95 °C for 2 min, 35 cycles of (denaturation at 95 °C for 20 s, primer annealing at 58 °C for 20 s, extension at 72 °C for 10 s) and a final extension of 5 min at 72 °C.
2. Confirm the presence of a single PCR band of about 100 bp by analyzing 2 μL of the PCR product on a 1.2% agarose gel with a suitable DNA staining dye. The PCR products are purified using the Gel Extraction Kit (Omega) and finally eluted with 20 μL nuclease-free water.
3. Perform in vitro transcription of the sgRNA using the TranscriptAid T7 High Yield Transcription Kit (Thermo Fisher Scientific). The reaction is as follows: 4 μL 5× TranscriptAid Reaction Buffer, 2 μL of ATP, 2 μL of CTP, 2 μL of GTP, 2 μL of UTP, 2 μL of TranscriptAid Enzyme Mix, 6 μL of sgRNA fragment (<1 μg). After mixing, incubate at 37 °C for 2–8 h.
4. Add 1 μL DNase I, mix well and incubate at 37 °C for 15 min to digest the remaining DNA template.
5. Purify RNA by 4 M lithium chloride precipitation with the following reaction system: 30 μL of nuclease-free water, 2.5 μL of 4 M LiCl, 100 μL of nuclease-free pure ethanol. After mixing, incubate at −20 °C for at least 1 h or overnight to precipitate.
6. Centrifuge at 16,000 × g for 15 min at 4 °C, and discard the supernatant.
7. Wash once with 1 mL 70% ethanol, centrifuge at 12,000 × g for 5 min at 4 °C, and discard the supernatant.
8. Air dry at room temperature for about 10 min and resuspend in 40 μL of nuclease-free water.
9. After dissolution, measure the concentration with a Nanodrop 2000c spectrophotometer, and store the aliquots at −80 °C.

3.5.4 Preparation of Cas9 mRNA

1. Linearize the zebrafish-codon-optimized Cas9 plasmid. The reaction system is as follows: 1 μL of XbaI, 4 μL of CutSmart® buffer, 2 μL of zebrafish-codon-optimized Cas9 plasmid (1 μg/μL), 33 μL of ddH_2O. After mixing, incubate at 37 °C for 4 h.

2. Purify the linearized plasmid using the Gel Extraction Kit (Omega), and finally elute with 20 μL nuclease-free water.
3. Perform in vitro transcription of the Cas9 mRNA using Mmessage mMACHINE™ T7 (Thermo Fisher Scientific). The reaction is as follows: 10 μL of 2× T7 NTP/CAP, 2 μL of 10× T7 Reaction Buffer, 2 μL of T7 Enzyme Mix, 6 μL of linearized zebrafish-codon-optimized Cas9 plasmid (< 1 μg). After mixing, incubate at 37 °C for 6–8 h.
4. Add 1 μL TURBO DNase™, mix well and incubate at 37 °C for 15 min to digest the remaining DNA template.
5. Purify RNA by anhydrous lithium chloride precipitation with the following reaction system: 30 μL of nuclease-free water and 30 μL of lithium chloride precipitation solution. After mixing, incubate at −20 °C for at least 1 h or overnight to precipitate.
6. Centrifuge at 16,000 × g for 15 min at 4 °C, and discard the supernatant.
7. Wash once with 1 mL 70% ethanol, centrifuge at 12,000 × g for 5 min at 4 °C, and discard the supernatant.
8. Air dry at room temperature for about 10 min and resuspend in 20 μL of nuclease-free water.
9. After dissolution, measure the concentration with a Nanodrop 2000c spectrophotometer, and store the aliquots at −80 °C.

3.5.5 Microinjection

1. Prepare 5 μL of fresh injection mixture sufficient to inject at least 500–2000 embryos. 1 μL sgRNA (250–1500 ng/μL), 3 μL nuclease-free water (replace with sgRNA when multiple target sites are injected simultaneously) and 1 μL Cas9 mRNA (1000–1500 ng/μL) are mixed. Keep the mixture on ice until it is transferred into the injection needle for microinjection. The injection concentration of sgRNA can range from 50 to 300 ng/μL (*see* **Note** 7).
2. Inject the one-cell stage animal pole of embryo with 2 nL using a quantitative microinjection system (Harvard PLI-100) pressurized by nitrogen.

3.5.6 Generation and Identification of Mutant Lines

1. Analyze the modified target gene by PCR screening and sequencing. Extract genomic DNA separately from each injected embryo and WT embryo by heating the embryos at 95 °C for 20 min in 60 μL of 50 mM NaOH. Terminate the reaction by adding 1/10 volume of 1 M Tris-HCl (pH 8.0). Enough genomic DNA is obtained from a single embryo after blastocyst stage.
2. Amplify the supernatant by PCR with specific primers. Design the specific primers to be at least 100 bp from both sides of the sgRNA target site; the PCR product should not exceed 500 bp and be a single band.

3. Directly Sanger sequence the unpurified PCR product followed by sequence trace decomposition. Analyze Sanger sequencing trace files to identify mutant efficiency by comparing the WT reference trace and the CRISPR/Cas9-edited sample trace. Sanger sequencing traces accurately match the upstream region of the sgRNA target site between the WT samples and CRISPR/cas9-edited samples. However, the CRISPR/Cas9-edited sample trace shows mixed base calls downstream of this target site.
4. After confirming that the target site knockout is effective, clone and Sanger sequence the purified PCR products (*see* **Note 8**).
5. After verifying the presence of effective knockout genotypes, raise the remaining embryos to adulthood. Some polyploid teleost fishes can reproduce by unisexual gynogenesis, such as auto-allo-hexaploid gibel carp (*Carassius gibelio*) [10, 25] (Fig. 2). The F0 individuals carrying mutations with a premature stop codon are chosen to proliferate by gynogenesis to obtain F1 offspring with different or all mutated homeologs and alleles.
6. According to the specific SNPs in homeologs or alleles, the F1 gynogenetic offspring can be classified into different mutated genotypes.
7. If necessary, the F1 gynogenetic offspring with different mutated genotypes can be chosen to reproduce by gynogenesis to obtain F2 mutated clones, in which the F2 gynogenetic offspring possess the same mutated genotypes.

3.6 Histological Comparison Among WT and Mutant Lines

3.6.1 Paraffin Section Making

1. Fix the fresh tissue of WT and mutant lines in 4% PFA for more than 24 h.
2. Take the tissue out of the fixation solution, trim the target tissue in the fume hood with a scalpel and put the trimmed tissue and corresponding label into the dewatering tank.
3. Put the dehydration box into the dehydrator and dehydrate with gradient alcohol: 75% alcohol 4 h, 85% alcohol 2 h, 90% alcohol 2 h, 95% alcohol 1 h, absolute ethanol I 30 min, absolute ethanol II 30 min, alcohol benzene 10 min, xylene I 10 min, xylene II 10 min, 65 °C melting paraffin I 1 h, 65 °C melting paraffin II 1 h, and 65 °C melting paraffin III 1 h.
4. Embed the wax-soaked tissue in the embedding machine. First, put the melted wax into the embedding frame. Before the wax solidifies, remove the tissue from the dehydration box. According to the requirements of the embedding surface, put it into the embedding frame and affix the corresponding label. Cool on the −20 °C freezing table, remove the wax block from the embedded frame and trim the wax block after the wax solidifies.

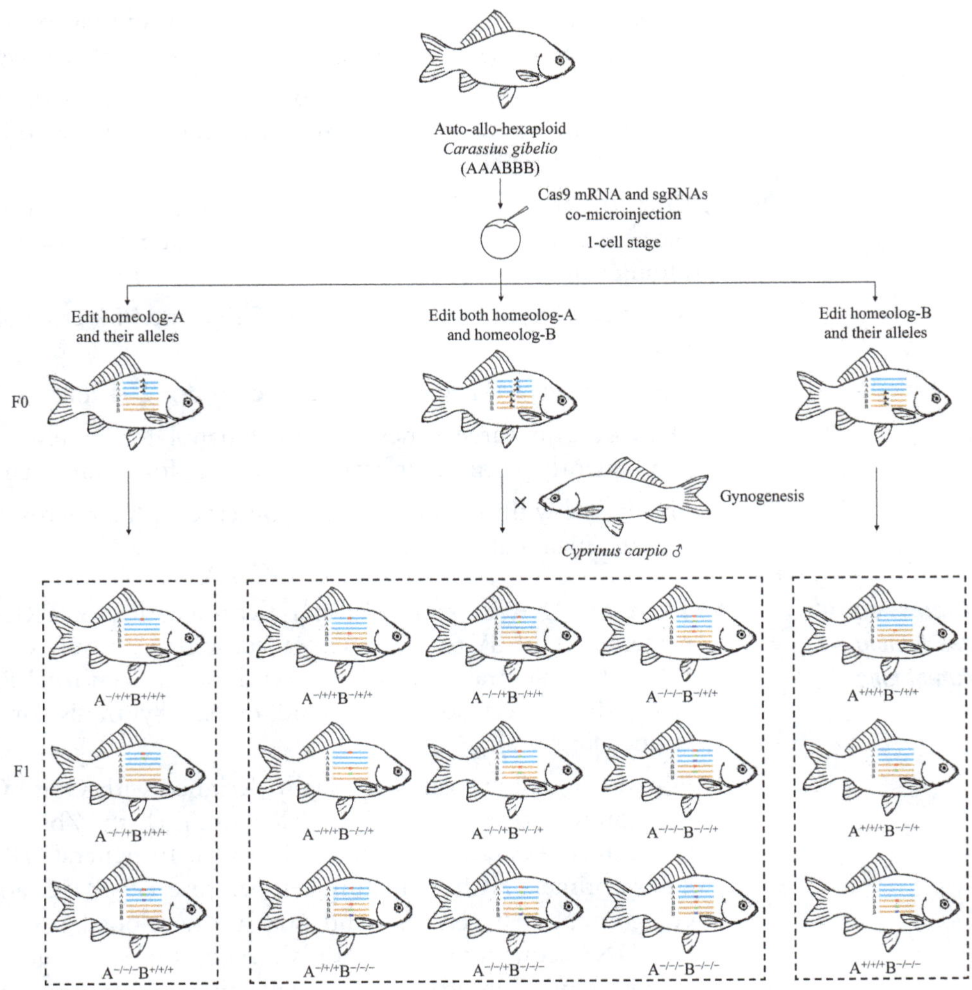

Fig. 2 Schematic diagram of multiple knockout mutant lines with single or simultaneously disrupted different homeologs or alleles by CRISPR/Cas9 in gynogenetic auto-allo-hexaploid gibel carp (*C. gibelio*). The mutant genotypes modified at the target sites of different homeologs and alleles are shown by different colors

5. Put the trimmed wax block into the −20 °C freezing table for cooling, and put the cooled wax block into the pathology slicer machine for slicing, with a thickness of 4 μm. Float the slices in warm water at 40 °C in the spreading machine to flatten the tissue, pick up the tissue with the glass slides and bake in the oven at 60 °C. After the water-baked dried wax is melted, take it out and store it at room temperature.

3.6.2 Hematoxylin and Eosin (H & E) Staining

1. Deparaffinize and rehydrate: culture the sections in two changes of xylene for 20 min each time. Dehydrate in two changes of pure ethanol for 5 min, subsequently dehydrate in 75% ethanol for 5 min. Wash in distilled water.

2. Stain sections with hematoxylin solution (Servicebio) for 2–4 min, and soak in running tap water to rinse excess dye.
3. Treat the sections with hematoxylin differentiation solution (Servicebio) for 2–5 s, and soak in running tap water to rinse excess differentiation solution.
4. Treat the section with hematoxylin Scott tap bluing (Servicebio) for 2–5 s, and soak in running tap water to rinse excess bluing solution.
5. Soak the sections in 85% ethanol for 5 min and 95% ethanol for 5 min.
6. Stain sections with eosin dye (Servicebio) for 2–4 min.
7. Dehydrate in three times of pure ethanol for 5 min each, subsequently permeate in xylene two times for 5 min each.
8. Finally seal with neutral gum, and observe with a microscope, acquire images and analyze.

3.7 Expression Changes Among WT and Mutant Lines by qPCR

1. The total RNA is extracted using a combination of TRIzol® reagent and the SV total RNA isolation system (*see* Subheading 3.1). The first strand cDNAs are synthesized from total RNA using the RevertAid First Strand cDNA Synthesis kit (*see* Subheading 3.2).
2. For qPCR, primer pairs are usually designed with Oligo Calc (Oligonucleotide Properties Calculator) (http://biotools.nubic.northwestern.edu/OligoCalc.html). In general, 18- to 24-bp primers with a melting temperature of 58 °C and a GC-content between 40% and 60% are developed based on the cDNA sequence of the candidate genes. Primers are located on different exons to avoid amplification of genomic DNA contaminations. The length of the amplified product should be between 100 and 300 bp. It is best to test the specificity of all primers on a few cDNA samples using RT-PCR prior to the qPCR and confirmed by Sanger sequencing (*see* **Note 9**).
3. qPCR is performed in a total volume of 20-μL reactions containing 10 μL of iTaq™ Universal SYBR® Green Supermix (Bio-Rad), 0.5 μL of each primer (10 μM), 1 μL of cDNA, and 8 μL of ddH_2O on a CFX96™ Real-Time PCR System (Bio-Rad). Run all samples in triplicate. Samples are run with the following program parameters: 95 °C for 30 s, followed by 40 cycles of 95 °C for 5 s and 60 °C for 30 s, with data acquisition at the end of every cycle, ending with the melt curve from 65 to 95 °C (reading every 0.5 °C, hold 5 s). Negative control (no-template DNA reaction) is always included.
4. qPCR data is normalized to the mRNA levels of optimal reference genes. Several housekeeping genes, including *actin, beta 1*

(*actb1*), *glyceraldehyde-3-phosphate dehydrogenase* (*gapdh*), *eukaryotic translation elongation factor 1 alpha 1, like 1* (*eef1a1l1*), etc., are selected to check their expression stability in the samples analyzed. Rank ordering of expression stability is performed using geNorm analysis [26]. The mRNA levels are calculated with the $2^{-\Delta\Delta CT}$ method [27]. For statistical analysis, Tukey's test is calculated with SPSS software (SPSS Inc.). A probability (p) of <0.05 is considered statistically significant.

3.8 Expression Changes Among WT and Mutant Lines by Western Blot

3.8.1 Extraction of Total Protein from Tissue

1. Wash 30–100 mg of tissue from WT and mutant lines with ice-cold 1× PBS for 2–3 times to remove blood stains, cut into small pieces and place in a microcentrifuge tube.
2. Add two 3-mm mill beads and 10 times of volume ice-cold RIPA buffer to homogenize the sample in a FastPrep®-24 tissue homogenizer at a vertical velocity of 4 m/s for 20 s.
3. Centrifuge the microtube at 12,000 × g for 20 min at 4 °C.
4. Carefully transfer the supernatant to a new microcentrifuge tube.
5. Add 5× SDS-PAGE protein loading buffer to the protein solution at a ratio of 4:1, denature it at 98 °C for 10 min, and either directly use for Western blot analysis or store at −20 °C.

3.8.2 Sodium Dodecyl Sulfate Polyacrylamide Gel Electrophoresis (SDS-PAGE)

1. Assemble the laminating and gel making frames according to the MiniProGel™ Protein Gel-Cast and Electrophoresis System (Beyotime) instructions.
2. Prepare 10 mL of 10% separation gel: 4 mL of ddH_2O, 3.3 mL of 30% acrylamide, 2.5 mL of 1.5 M Tris-HCl (pH 8.8), 0.1 mL of 10% SDS, 0.1 mL of 10% APS, and 4 μL of TEMED. After mixing, quickly add it to the gel making plate. Make sure that the liquid level is lower than the edge of the glass plate by 1–2 cm. Then, gently add water or absolute ethanol to the upper layer of the separation gel until it is full. Let it stand at room temperature for 30 min.
3. When the separating gel has solidified, discard the upper water phase or ethanol and absorb the remaining liquid with filter paper.
4. Prepare 3 mL of 5% stacking gel: 2.1 mL of ddH_2O, 0.5 mL of 30% acrylamide, 0.38 mL of 1 M Tris-HCl (pH 6.8), 30 μL of 10% SDS, 30 μL of 10% APS, and 3 μL of TEMED. After mixing, gently add to the solidified separating gel, carefully insert the 10- or 15-well gel comb to avoid bubbles. Let it stand at room temperature for 30 min.
5. When the stacking gel is solidified, it is ready to start electrophoresis.

6. Place the solidified SDS-PAGE gel plate in the electrophoresis tank, carefully remove the comb, add enough 1× SDS-PAGE running buffer, and clean the sample loading well. Load your samples and 10 μL protein marker as a size standard to the sample loading well.
7. Electrophorese at 80 V until the protein marker has entered the separation gel and then continue at 120 V till the blue dye bromophenol blue has reached the bottom of the gel.

3.8.3 Transferring the Protein Gel to a PVDF Membrane

1. Following electrophoresis, remove the SDS-PAGE gel from the gel plate and immerse it in 1× transfer buffer. According to the size of the protein marker, cut the desired size of the target protein gel block and remove other unnecessary parts.
2. Cut a piece of the PVDF membrane to the same size of the gel, which is activated in 100% methanol for at least 30 s, and then transfer to 1× transfer buffer.
3. Cut 6 pieces of filter paper with the same size as the PVDF membrane and soak the filter paper and foam pad (Beyotime) in 1× transfer buffer.
4. Assemble the transfer system according to the MiniBlot™ Electrophorectic Transfer Cell (Beyotime) instructions.
5. Transfer the gel for 50 min at 90 V (200–300 mA).

3.8.4 Immunodetection of Western Blot

1. Disassemble the transfer membrane splint, carefully take out the transferred membrane with clean tweezers, and place it into the incubation tank with TBST, followed by washing 3 times with TBST on a shaking platform for 10 min each time.
2. Incubate the transferred membrane with blocking buffer on a shaking platform for 2 h at room temperature.
3. Remove the blocking buffer, and incubate the membrane with appropriate dilutions of primary antibody (diluted with blocking buffer) for 2 h at room temperature or overnight at 4 °C on a shaking platform (*see* **Note 10**).
4. Recycle the primary antibody, and wash the membranes 3 times with TBST at room temperature with gentle agitation for 10 min each time.
5. Incubate the membrane with appropriate dilutions of HRP-conjugated secondary antibody (diluted with blocking buffer) for 1 h at room temperature on a shaking platform.
6. Wash the membranes 3 times with TBST at room temperature with gentle agitation for 10 min each time.
7. Add HRP substrate peroxide solution and HRP substrate luminol reagent in a ratio of 1:1 in the centrifuge tube to prepare a proper amount of chromogen solution. Place the membrane

on the tray of the ImageQuant LAS 4000 mini system (General Electric), and add the prepared chromogen solution into the system for 1 min. Obtain the image with the chemiluminescence imager.

3.9 Expression Changes Between WT and Mutant Lines by Immunofluorescence Staining of Paraffin Sections

1. Deparaffinize and rehydrate: the sections (*see* Subheading 3.6.1) are cultured in two changes of xylene for 15 min each time. Dehydrate in two changes of pure ethanol for 5 min, subsequently dehydrate in 85% and 75% gradient ethanol for 5 min each time. Wash in distilled water.
2. Soak the slides in citrate-EDTA antigen retrieval solution (Beyotime) and heat at 95–100 °C for about 20 min. After cooling at room temperature, wash 3 times with 1× PBS on a shaking platform for 5 min each time.
3. Shake off any excess liquid and mark the objective tissue with a liquid blocker pen. Incubate the slides with blocking solution for 10 min at room temperature.
4. Throw away the blocking solution. Incubate the slides with primary antibody (diluted with ADB) for 4 h at room temperature or overnight at 4 °C in a wet box containing a small amount of water (*see* **Note 11**).
5. Take out the slides and wash them 5 times with PBST on a shaking platform for 10 min each time, followed by washing them 2 times with 1× PBS for 5 min each time.
6. Add simultaneously different secondary antibodies (diluted with ADB) and incubate at 37 °C for 1 h in dark condition.
7. Take out the slides and wash them 3 times with 1× PBS on a shaking platform for 5 min each time. Then incubate with DAPI solution for 10 min at room temperature in dark conditions.
8. Take out the slides and wash them 5 times with 1× PBS on a shaking platform for 10 min each time.
9. Remove the buffer. After the slides are slightly dried, seal the slides with anti-fade mounting medium. Observe the slides under an upright fluorescence microscope and capture images.

4 Notes

1. For polyploids, there may be multiple alleles of the same target gene, and more clones (about 20–30 clones) are needed to be sequenced to obtain different copies of duplicated genes.
2. If one of the homeologs/alleles could not be found in the assembled genome or the transcripts of the polyploid animal, it may have been pseudogenized or lost during evolution.

Pseudogenization or gene loss may be confirmed by aligning homologous sequences of the target gene from other species against the assembled genome using GeneWise [28] and BLASTN, respectively.

3. PCR-amplified DNA fragments or bacterial artificial chromosome (BAC) DNA containing the target gene can be used as probe DNA. For allopolyploids, the DNA containing different homeologs is labeled by the Biotin-Nick Translation Mix (Roche) or DIG-Nick Translation Mix (Roche).
4. To identify the homoeologous chromosomes in allopolyploids, biotin- or dig-labeled probe DNA containing different homeologs are co-hybridized.
5. If biotin- and dig-labeled probes are co-hybridized, the antibodies used for detection are simultaneously incubated on the slides.
6. For polyploids, sgRNA should be selected at a conserved region of all homeologs and alleles to simultaneously disrupt the function of duplicated genes. On the other hand, sgRNA should be selected at a region specific to each different homeolog and allele to simply disrupt the function of each homeolog and allele. We usually design at least two different sgRNAs from different regions in the target gene to ensure more effective mutations.
7. There are multiple homeologs and alleles of duplicated genes in polyploids. In order to obtain more effective mutations and different mutant alleles, multiple different sgRNAs of target gene are simultaneously injected with about 200 ng/μL of each sgRNA. To reveal the function divergence of different homeologs and alleles in polyploids, it is necessary to construct multiple knockout mutant lines with single or simultaneously disrupted different homeologs and alleles.
8. There are multiple homeologs and alleles of duplicated genes in polyploids. Sanger sequencing is performed on at least 30 clones to analyze the genotypes of different mutant homeologs and alleles.
9. It is important to find the significant sequence differences to design primers specific for each different homeolog and allele. If not, the qPCR products amplified by a high-fidelity polymerase can be purified and cloned into Trans5α chemically competent cells. About 20–30 clones of each sample are sequenced and classified according to the specific SNPs in the alleles. The average ratios of sequencing clone numbers are used to show the expression levels.

10. It is also important to find the significant sequence differences to prepare antibodies specific for each different homeolog and allele.
11. It is also important to find the significant sequence differences to prepare antibodies specific for each different homeolog and allele. For co-localization analyses, the antigens specific for different homeologs and alleles should be obtained by immunization in different animals, such as rabbit, mouse, or goat.

Acknowledgments

This work was supported by the National Key Research and Development Program of China (2018YFD0900204), the Strategic Priority Research Program of Chinese Academy of Sciences (XDB31000000), the Key Program of Frontier Sciences of the Chinese Academy of Sciences (QYZDY-SSW-SMC025), and China Agriculture Research System of MOF and MARA.

References

1. Sessa EB (2019) Polyploidy as a mechanism for surviving global change. New Phytol 221:5–6
2. Cheng F, Wu J, Cai X et al (2018) Gene retention, fractionation and subgenome differences in polyploid plants. Nat Plants 4:258–268
3. Van de Peer Y, Mizrachi E, Marchal K (2017) The evolutionary significance of polyploidy. Nat Rev Genet 18:411–424
4. Meyer A, Van de Peer Y (2005) From 2R to 3R: evidence for a fish-specific genome duplication (FSGD). BioEssays 27:937–945
5. Amores A, Force A, Yan YL et al (1998) Zebrafish *hox* clusters and vertebrate genome evolution. Science 282:1711–1714
6. Taylor JS, Braasch I, Frickey T et al (2003) Genome duplication, a trait shared by 22000 species of ray-finned fish. Genome Res 13: 382–390
7. Leggatt RA, Iwama GK (2003) Occurrence of polyploidy in the fishes. Rev Fish Biol Fisher 13:237–246
8. Van de Peer Y, Taylor JS, Meyer A (2003) Are all fishes ancient polyploids? J Struct Funct Genom 3:65–73
9. Zhou L, Gui JF (2017) Natural and artificial polyploids in aquaculture. Aquac Fish 2: 103–111
10. Gui JF, Zhou L (2010) Genetic basis and breeding application of clonal diversity and dual reproduction modes in polyploid *Carassius auratus gibelio*. Sci China Life Sci 53:409–415
11. Lien S, Koop BF, Sandve SR et al (2016) The Atlantic salmon genome provides insights into rediploidization. Nature 533:200–205
12. Luo J, Chai J, Wen Y et al (2020) From asymmetrical to balanced genomic diversification during rediploidization: subgenomic evolution in allotetraploid fish. Sci Adv 6:eaaz7677
13. Chen D, Zhang Q, Tang W et al (2020) The evolutionary origin and domestication history of goldfish (*Carassius auratus*). Proc Natl Acad Sci 117:29775–29785
14. Li XY, Zhang XJ, Li Z et al (2014) Evolutionary history of two divergent *Dmrt1* genes reveals two rounds of polyploidy origins in gibel carp. Mol Phylogenet Evol 78:96–104
15. Zhu HP, Ma DM, Gui JF (2006) Triploid origin of the gibel carp as revealed by 5S rDNA localization and chromosome painting. Chromosom Res 14:767–776
16. Zhou L, Gui JF (2002) Karyotypic diversity in polyploid gibel carp, *Carassius auratus gibelio* Bloch. Genetica 115:223–232
17. Soltis PS, Soltis DE (2016) Ancient WGD events as drivers of key innovations in angiosperms. Curr Opin Plant Biol 30:159–165
18. Liu S, Li Z, Gui JF (2009) Fish-specific duplicated *dmrt2b* contributes to a divergent function through Hedgehog pathway and

maintains left-right asymmetry establishment function. PLoS One 4:e7261

19. Gan RH, Wang Y, Li Z et al (2021) Functional divergence of multiple duplicated *Foxl2* homeologs and alleles in a recurrent polyploid fish. Mol Biol Evol 38:1995–2013
20. Ren F, Lin Q, Gong G et al (2020) Igf2bp3 maintains maternal RNA stability and ensures early embryo development in zebrafish. Commun Biol 3:94
21. Jiang D, Chen J, Fan Z et al (2017) CRISPR/Cas9-induced disruption of *wt1a* and *wt1b* reveals their different roles in kidney and gonad development in Nile tilapia. Dev Biol 428:63–73
22. Wang H, Yang H, Shivalila CS et al (2013) One-step generation of mice carrying mutations in multiple genes by CRISPR/Cas-mediated genome engineering. Cell 153: 910–918
23. Liu D, Wang Z, Xiao A et al (2014) Efficient gene targeting in zebrafish mediated by a zebrafish-codon-optimized cas9 and evaluation of off-targeting effect. J Genet Genomics 41: 43–46
24. Chang N, Sun C, Gao L et al (2013) Genome editing with RNA-guided Cas9 nuclease in zebrafish embryos. Cell Res 23:465–472
25. Zhang J, Sun M, Zhou L et al (2015) Meiosis completion and various sperm responses lead to unisexual and sexual reproduction modes in one clone of polyploid *Carassius gibelio*. Sci Rep 5:10898
26. Vandesompele J, De Preter K, Pattyn F et al (2002) Accurate normalization of real-time quantitative RT-PCR data by geometric averaging of multiple internal control genes. Genome Biol 3:RESEARCH0034
27. Livak KJ, Schmittgen TD (2001) Analysis of relative gene expression data using real-time quantitative PCR and the 2(-Delta Delta C (T)) method. Methods 25:402–408
28. Birney E, Clamp M, Durbin R (2004) GeneWise and Genomewise. Genome Res 14: 988–995

INDEX

Yves Van de Peer (ed.), *Polyploidy: Methods and Protocols*,
Methods in Molecular Biology, vol. 2545, https://doi.org/10.1007/978-1-0716-2561-3,

CPSIA information can be obtained
at www.ICGtesting.com
Printed in the USA
LVHW011441190423
744480LV00029B/862

9 781071 625606